125.00

D0915488

PHYSICAL METHODS OF CHEMISTRY
Second Edition

Volume V

DETERMINATION OF STRUCTURAL FEATURES OF
CRYSTALLINE AND AMORPHOUS SOLIDS

PHYSICAL METHODS OF CHEMISTRY

Second Edition

Editors: Bryant W. Rossiter
John F. Hamilton

PHYSICAL METHODS OF CHEMISTRY

Second Edition

Edited by

BRYANT W. ROSSITER
JOHN F. HAMILTON

Research Laboratories
Eastman Kodak Company
Rochester, New York

Volume V
DETERMINATION OF STRUCTURAL FEATURES OF CRYSTALLINE AND AMORPHOUS SOLIDS

A WILEY-INTERSCIENCE PUBLICATION
John Wiley & Sons, Inc.
NEW YORK / CHICHESTER / BRISBANE / TORONTO / SINGAPORE

Library of Congress Cataloging in Publication Data:

Physical methods of chemistry.

 "A Wiley-Interscience publication"
 Includes bibliographies and indexes.
 Contents: v. 1. Components of scientific instruments
and applications of computers to chemical research—
[etc.]—v. 3 Determination of chemical composition
and molecular structure—
v. 5. Determination of structural features of crystalline
and amorphous solids.
 1. Chemistry—Manipulation—Collected works.
I. Rossiter, Bryant W., 1931– II. Hamilton,
John F.
QD61.P47 1986   542   85–6386

ISBN 0-471-52509-X (v.5)

Printed in the United States of America

10 9 8 7 6 5 4 3 2 1

CONTRIBUTORS

FRANK J. BERRY, Department of Chemistry, The University of Birmingham, Birmingham, England

ROBERT COMÈS, Laboratoire de Physique des Solides, Université Paris-Sud, Orsay-Cedex, France

DAVID E. COX, Brookhaven National Laboratory, Upton, New York

STEVE M. HEALD, Brookhaven National Laboratory, Upton, New York

P. MARK HENRICHS, Research Laboratories, Eastman Kodak Company, Rochester, New York†

JAMES M. HEWITT, Analytical Technology Division, Eastman Kodak Company, Rochester, New York

ROBERT A. JACOBSON, Department of Chemistry, Iowa State University, Ames, Iowa

MÖGENS S. LEHMANN, Institute Laue-Langevin, Grenoble Cedex, France

WILLIAM N. LIPSCOMB, Department of Chemistry, Harvard University, Cambridge, Massachusetts

CHUCK F. MAJKRZAK, The National Institute of Standards and Technology, Gaithersburg, Maryland

RAYMOND P. SCARINGE, Research Laboratories, Eastman Kodak Company, Rochester, New York

J.-MARTIN SPAETH, Universität Gesamthochschule Paderborn, Paderborn, Federal Republic of Germany

JOHN M. TRANQUADA, Brookhaven National Laboratory, Brookhaven National Laboratory, Upton, New York

†Present affiliation: Exxon Chemical Company, Baytown, Texas.

PREFACE TO PHYSICAL METHODS OF CHEMISTRY

This is a continuation of a series of books started by Dr. Arnold Weissberger in 1945 entitled *Physical Methods of Organic Chemistry*. These books were part of a broader series, *Techniques of Organic Chemistry*, and were designated Volume I of that series. In 1970, *Techniques of Chemistry* became the successor to and the continuation of the *Techniques of Organic Chemistry* series and its companion, *Techniques of Inorganic Chemistry*, reflecting the fact that many of the methods are employed in all branches of chemical sciences and the division into organic and inorganic chemistry had become increasingly artificial. Accordingly, the fourth edition of the series, entitled *Physical Methods of Organic Chemistry*, became *Physical Methods of Chemistry*, Volume I in the new *Techniques* series. The last edition of *Physical Methods of Chemistry* has had wide acceptance, and it is found in most major technical libraries throughout the world. This new edition of *Physical Methods of Chemistry* will consist of eight or more volumes and is being published as a self-standing series to reflect its growing importance to chemists worldwide. This series will be designated as the second edition (the first edition, Weissberger and Rossiter, 1970) and will no longer be subsumed within *Techniques of Chemistry*.

This edition heralds profound changes in both the perception and practice of chemistry. The discernible distinctions between chemistry and other related disciplines have continued to shift and blur. Thus, for example, we see changes in response to the needs for chemical understanding in the life sciences. On the other hand, there are areas in which a decade or so ago only a handful of physicists struggled to gain a modicum of understanding but which now are standard tools of chemical research. The advice of many respected colleagues has been invaluable in adjusting the contents of the series to accommodate such changes.

Another significant change is attributable to the explosive rise of computers, integrated electronics, and other "smart" instrumentation. The result is the widespread commercial automation of many chemical methods previously learned with care and practiced laboriously. Faced with this situation, the task of a scientist writing about an experimental method is not straightforward.

Those contributing to *Physical Methods of Chemistry* were urged to adopt as their principal audience intelligent scientists, technically trained but perhaps inexperienced in the topic to be discussed. Such readers would like an introduction to the field together with sufficient information to give a clear understanding of the basic theory and apparatus involved and the appreciation for the value, potential, and limitations of the respective technique.

Frequently, this information is best conveyed by examples of application, and many appear in the series. Except for illustration, however, no attempt is made to offer comprehensive results. Authors have been encouraged to provide ample bibliographies for those who need a more extensive catalog of *applications*, as well as for those whose goal is to become more expert in a *method*. This philosophy has also governed the balance of subjects treated with emphasis on the *method*, not on the results.

Given the space limitations of a series such as this, these guidelines have inevitably resulted in some variance of the detail with which the individual techniques are treated. Indeed, it should be so, depending on the maturity of a technique, its possible variants, the degree to which it has been automated, the complexity of the interpretation, and other such considerations. The contributors, themselves expert in their fields, have exercised their judgment in this regard.

Certain basic principles and techniques have obvious commonality to many specialties. To avoid undue repetition, these have been collected in Volume I. They are useful on their own and serve as reference material for other chapters.

We are deeply sorrowed by the death of our friend and associate, Dr. Arnold Weissberger, whose enduring support and rich inspiration motivated this worthy endeavor through four decades and several editions of publication.

BRYANT W. ROSSITER
JOHN F. HAMILTON

Research Laboratories
Eastman Kodak Company
Rochester, New York
March 1986

PREFACE

In Volume V the series *Physical Methods of Chemistry* turns its attention from techniques that are used mainly to determine elemental composition or molecular structure and properties to methods that are used primarily to determine structure in solids, mostly crystalline solids. The techniques used range from those such as X-ray diffraction, which is applicable to the long-range structure of the perfect crystal; to those probing the local order in less perfect crystals; to the magnetic spectroscopies, which we use mostly to study the immediate environment of impurities or defects.

Notable by their absence from this volume are the techniques of electron microscopy and diffraction, which have been used increasingly during the past decade or so to investigate the structure and defects of the crystal lattice. These electron microscopy and diffraction methods have been grouped along with certain other microscopy subjects in the soon-to-be-released Volume IV. Techniques that are used to study the optical, electrical, and dielectric properties of solids will be covered in Volume VIII, and techniques used for surface studies, in Volumes IXA and IXB.

Chapters have been written by world-class authors who are widely recognized in their fields. Authors have directed their writings to the competent, professional scientist who is interested in obtaining information provided by the technique, but who is perhaps not an expert in the use of the method. In each case, authors of chapters have supplied, either in the text or through liberal reference to monographs and other scientific literature, sufficient information for the investigator to apply the techniques successfully in the laboratory.

We acknowledge our deep gratitude to the contributors who have spent long hours working over manuscripts. We greet previous contributors, Professors Robert A. Jacobson and William N. Lipscomb, and welcome several new contributors to Volume V: Dr. Frank J. Berry, Dr. Robert Comés, Dr. David E. Cox, Dr. Steve M. Heald, Dr. P. Mark Henrichs, Mr. James M. Hewitt, Dr. Mögens S. Lehmann, Dr. Chuck F. Majkrzak, Dr. Raymond P. Scaringe, Professor Dr. J.-Martin Spaeth, and Dr. John M. Tranquada.

We are also extremely grateful to the many colleagues from whom we have sought counsel on the choice of subject matter and contributors. We express our gratitude to Mrs. Ann Nasella for her enthusiastic and skillful editorial assistance. In addition, we heartily thank the specialists whose critical readings of the manuscripts have frequently resulted in the improvements accrued from collective wisdom. For Volume V they are Dr. G. R. Apai, Professor C. R. Dybowski, Dr. R. S. Eachus, Professor C. H. W. Jones, Dr. J. D. Jorgensen, Professor T. J. Kistenmacher, and Dr. D. L. Smith.

BRYANT W. ROSSITER
JOHN F. HAMILTON

Rochester, New York
May 1990

CONTENTS

PHYSICAL METHODS OF CHEMISTRY
Second Edition

Volume V

DETERMINATION OF STRUCTURAL FEATURES OF
CRYSTALLINE AND AMORPHOUS SOLIDS

Chapter 1

X-RAY CRYSTAL STRUCTURE ANALYSIS

William N. Lipscomb and Robert A. Jacobson

Section 1 introduces the general principles of a complete structure determination; two specific methods of structure determination, symbolic addition and Patterson superposition, illustrate some of the methods that practicing crystallographers use to solve complex structures. Many other methods are available, and a few are also described briefly. Equations for the computation of amplitudes of scattering, the electron density, and the vector map are derived in Section 2. Mathematics beyond elementary calculus is not assumed, and the necessary properties of vectors, complex numbers, and Fourier series are presented in detail. Section 3 is a geometrical description of symmetry in crystals with emphasis on the symmetries most frequently encountered in complex crystals. This section owes a great deal to Professor J. H. Sturdivant, who kindly gave us permission to use his notes on symmetry, which formed the basis from which Section 3 was written. In Section 4 the relation between the experimental methods and the preceding sections are outlined, and one complete structure determination is described in detail.

Each section is independent of the others. Section 1 is designed primarily for the reader who wishes to obtain a general idea of how complex structures are determined. The later sections elucidate different aspects of these methods of structure investigation, and they were used during the past several years for notes in courses on X-ray crystallography.

1   INTRODUCTION AND ILLUSTRATION OF SOME TYPICAL METHODS FOR THE DETERMINATION OF COMPLEX STRUCTURES

1.1   Introduction

Some of the finest recent achievements of X-ray crystallography are undoubtedly the determinations of the structures of myoglobin [1], hemoglobin [2], lysozyme [3], carboxypeptidase A [4], ribonuclease [5], aspartate carbamoyltransferase [6], the southern bean mosaic virus [7], and other large biological molecules [8]. These structures are among the most complex elucidated at present and represent, of course, many years of research effort. Currently, the routine determination of almost any complex structure containing say 50–100 or more atoms can be expected from many crystallographic laboratories. This chapter introduces the general principles of X-ray crystallography and illustrates a few of the frequently used methods for the solution of such structures.

The complexity of structures solved by X-ray diffraction techniques increased greatly since about 1950. This increase paralleled the development of the digital computer, which plays a vital role in modern-day crystallography. Developments and extensions of the techniques by which the first approximately correct atomic arrangement (trial structure) in the unit cell of a previously unsolved structure is obtained coupled with ease of data collection and refinement have made this technique attractive to even the relative novice. Indeed, in many favorable cases it is possible to expect a complete structure determination within a week's time or less.

A structure that is established by careful X-ray diffraction techniques is very likely to be correct. (It should be noted that a few cases of homometric structures have been found for simple structures that have exactly the same array of three-dimensional vectors—and therefore X-ray intensities—from different arrangements. An example is the mineral bixbyite [9].) The crystallographer is in a very favorable position relative to other "more sporting" techniques for structure determination in that the ratio of observations to unknowns is quite large, ranging from 5 to 1, or even 10 to 1 when complete data are obtained. For example, in the structure determination of $[Rh(t-BuDiNC)_2]BPh_4 \cdot 1.5CH_3CN$ (DiNC = chelating isonitrile ligand) [10], some 5912 independent X-ray diffraction maxima were observed. The unknowns in this case are the x, y, and z coordinates of the rhodium, oxygen, carbon, nitrogen, and boron atoms of which there are a total of 86 atoms, a scale parameter, six anisotropic temperature parameters for each atom, giving 775 parameters in all. The equations relating these 775 unknowns to the observations are not simple, but they are readily tractable now with modern computers.

The final test of correctness of a structure is the agreement between the observed and the calculated intensities. The exact agreement obtained varies according to the nature of the structure and the methods used to acquire the data, among other things; but it should be comparable with that obtained in similar structural studies. The trial structure may be obtained by mathematical techniques involving few or no chemical assumptions or by other techniques involving intuition, guesses, and the background available in the *scientific method*; but the test of obtaining good agreement with large numbers of observations with relatively few parameters is so severe that the results can usually be considered equally as good with any method.

In this introduction we describe the direct determination of X-ray phases from the intensities and also from the analysis of the Patterson function, a three-dimensional map of all the vectors that can be drawn between pairs of atoms in the structure and then plotted on a common origin. This map can be computed directly from the observed data without chemical assumptions. Although these methods are not the only ones available for trial structure determination, they have proved to be the ones of most generality; hence, they are the most widely used.

The beginning stages of a structure determination are similar in all methods of attack; hence, we start with a brief description of what the chemist can learn from the preliminary stages. Briefly, these stages are (1) the identification from either powder photographs or a determination of the unit cell and molecular weight, (2) the unit cell symmetry, and (3) the complete structure determination.

For more detailed discussion of the topics of this section, consult [11].

1.2   Identification by Powder Photography

X rays are very short wavelength electromagnetic radiations produced when L or M electrons fall into the K shell of elements of moderately high or high atomic number, when M or N electrons fall into the L shell of elements of high atomic number, or when electrons are decelerated rapidly. These types of radiation are called *characteristic K, characteristic L,* or *white radiation,* respectively. The vacancies necessary for the transitions to occur are produced when electrons are accelerated through a high potential, about 50,000 V, toward a water-cooled target of a heavy metal such as copper (1.5418-Å radiation) or molybdenum (0.71069-Å radiation). X rays are not appreciably deviated by an ordinary lens system; hence, pinholes are used to produce a collimated beam, or a self-focusing arrangement is used in which the source, a large area of the sample, and the radiation detector (e.g., photographic film or photon counter) all lie on the circumference of a circle, commonly called the *Rowland circle.* If a flat sample is used in this way, absorption problems are usually negligible. However, if a cylindrical specimen of powder or a single crystal is used, the X rays must pass through part of the sample, and excessive absorption of X rays can occur if the sample is too large. Given a sample of a particular size, the amount of intensity reduction is proportional to the intensity incident upon the sample and to the thickness of the sample; that is, $dI = -\mu I\, dx$, where μ is termed the *linear absorption coefficient.* It can be calculated from $\mu = \rho[f_A \mu_A^{(m)} + f_B \mu_B^{(m)} + f_C \mu_C^{(m)} + \cdots]$, where ρ is the density of the sample and f_A is the weight fraction of element A having mass absorption coefficient $\mu_A^{(m)}$. Tables of $\mu^{(m)}$ are found in the *International Tables for X-ray Crystallography* (The Kynoch Press, Birmingham, England, 1952). Integration of this expression gives $I = I_0 e^{-\mu x}$, where I is the resultant intensity for a sample of thickness x on which an initial intensity of I_0 falls. The intensity of diffracted radiation depends on I and on the total size of the specimen, which is proportional to x^2 for a cylindrical specimen. Hence, $I_{\text{diff}} = KIx^2$. The maximum value of I_{diff} is obtained by setting $dI_{\text{diff}}/dx = KI_0 d(x^2 e^{-\mu x})/dx = 0$, from which the optimum size sample is found to be $x = 2/\mu$. Typical values of $1/\mu$ are 0.5 mm for Cu K_α radiation and 4 mm for Mo K_α radiation for diamond, and 0.006 mm for Cu K_α and 0.05 mm for Mo K_α radiation for I_2. To avoid appreciable corrections for absorption it is necessary to keep the sample thickness to about one-quarter or less of $1/\mu$. However, more often one does not wish to sacrifice intensities of weak reflections; hence, absorption corrections are made.

   When K radiation is produced from a target, the K_α doublet ($L \rightarrow K$, resolved only at high angles) is the desired line. However, an additional line (K_β, $M \rightarrow K$) is produced that can be reduced effectively by the use of special absorption filters or by a crystal monochromator. For example, a 10-μm ($1\,\mu$m $= 10^{-4}$ cm) nickel filter reduces the Cu K_α line to about $\frac{1}{2}$ its original intensity, while at the same time reducing the Cu K_β line to $\frac{1}{85}$ of its original intensity or about $\frac{1}{300}$ of the intensity of the reduced Cu K_α intensity. For molybdenum radiation a zirconium filter of 80–100 μm is employed to obtain similar results. A crystal monochromator is a crystal that strongly diffracts X rays according to Bragg's law:

$$n\lambda = 2d'_{hkl} \sin \Theta$$

Bragg's law (discussed in greater detail below) arises from the condition that X rays diffracted by equivalent planes spaced d'_{hkl} apart only coherently reinforce one another at certain angles of deviation (2Θ) from the direct beam of X rays. In this relation n is an integer (the order of the reflection) and λ is the wavelength. By appropriate selection of the scattering angle and the d'_{hkl}, the K_α radiation can be selected over the K_β and at the same time the background radiation produced from the white radiation significantly reduced. Typical monochromating crystals are LiF, Si, and graphite. Although graphite is ordered only in the $00l$ direction, its efficiency of diffraction is very high, and thus it makes a good monochromator for many applications.

A given diffraction maximum from a small crystal is usually sharply defined. The reflection profile is intrinsically narrow because of the many unit cells the crystal contains, for reasons discussed below; for practical purposes, most of the profile width comes from experimental conditions associated with producing the diffraction maxima, such as the X-ray beam divergence, and from crystal imperfections. If a finely powdered crystalline material is used as a sample in a diffraction experiment, all orientations of crystals can occur; thus a given plane diffracting in a given order, say $n = 1$, gives a diffraction maxima whenever it is oriented so that Bragg's law is satisfied for specular reflection. All reflections from a given plane in a powdered sample will lie on a cone making an angle of 2Θ with respect to its axis (Figure 1.1). Because 2Θ can have values from $0°$ to $180°$, a cylindrical strip of film is usually more convenient to use than a flat plate (Figure 1.2) on a film instrument.

A powder sample that is not sufficiently randomly oriented or that is too coarse will give discrete spots instead of continuous powder lines. Hence it is usual to grind the sample until it passes through about a 300-mesh screen, which gives a crystallite size of about 40 μm or less. In addition, in some experimental arrangements, the sample is rotated during data collection to increase the

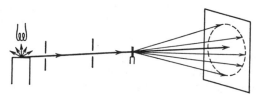

Figure 1.1   Experimental arrangement showing a single powder line at an angle of deviation 2Θ from the X-ray beam.

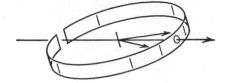

Figure 1.2   Use of a cylindrical strip of film to record a powder pattern.

randomness of the orientation. The several acceptable techniques for mounting the sample include introduction into a 0.1–1.0-mm Lindemann or special glass tube, similar to a melting point tube, or adhering the material to such a glass tube or rod. These special glasses give a particularly low absorption and background.

A satisfactory film exposure or countertrace can usually be obtained in one-half to 10 h, and position-sensitive multidetectors are becoming available that yield data of counterquality in significantly reduced times. A pattern of discrete sharp lines indicates that the material is crystalline; broadened lines indicate small particle size; and diffuse halos indicate amorphous material, glassy material, or liquid crystals.

The method of X-ray powder photography is at its best when the identity of two samples is being tested. If the patterns are identical and if this identity is supported by other physical and chemical techniques, this method is one of the most powerful. Computer programs now exist to match an unknown pattern with those previously recorded and on file in the computer memory. (The Joint Committee on Powder Diffraction Standards is very active in this area.) For the usual crystalline compound both the positions and intensities of all lines must be identical; hence, the test of identity is a severe one. If, on the other hand, the two samples give obviously different diffraction patterns, no positive statements can be made. Probably the two compounds are different, but there always remains the possibility that the same compound has merely crystallized in two different crystal structures (*polymorphism*). Samples as small as 10–100 μg give satisfactory patterns, but the quality of the diffraction pattern is usually a bit better if approximately 1 mg of material is available. The sample is nearly always left unchanged; therefore, recovery of the material is usually complete.

Powder diffraction techniques, coupled with computer programs to assign indices to the reflections, afford an excellent way to obtain precise unit cell parameters, especially when single crystals cannot be readily obtained.

For the simpler inorganic compounds, some investigators have been remarkably successful in the solution of structures from powder data alone. Pauling's early work on the structures of minerals and Zachariasen's studies on compounds of the thoride and actinide series are among the most extensive and impressive studies of complete structure determinations from powder photographs. This area has experienced a resurgence of interest with the use of profile refinement and other related techniques for the refinement of trial structures to obtain a best fit to accurate counterbased data [12]. For complex structures, though, the chemist must look to studies of single crystals to obtain more than a method of characterization or a comparison that tests the identity of two samples.

1.3   The Unit Cell and Molecular Weights

Within a few hours to a few days, the dimensions of the unit cell of a previously unknown structure can be determined from a single crystal of dimensions 0.05–0.8 mm. The unit cell is the smallest unit of volume that shows the full symmetry of the crystal system and from which the complete three-dimensional arrangement of molecules is generated by translations along the axes. Excluding cases of translational disorder, which are very rare among complex crystals, the unit cell necessarily contains an integral number (Z) of molecules of molecular weight M. The volume V can be obtained, as described below, from the unit cell dimensions and angles. An accurate determination of the density ρ of the crystal then yields ZM, an integer times the molecular weight; that is, $\rho = ZM/(NV)$, where N is Avogadro's number. If a rough measurement of the molecular weight is known, say from colligative property measurements, then the value of Z can be found, and an accurate molecular weight can be determined from the equation above. X-ray measurements of the unit cell parameters can be made so precisely that the final accuracy of the molecular weight is limited only by the accuracy of the density determination, which in favorable cases is a few tenths of a percent. In fact, a few of the present atomic weights were determined by this method.

The internal symmetry of the unit cell frequently gives a lower limit to Z, often 2 or 4; and the known molecular volumes or densities of similar compounds can be used to give either the value of Z or an upper limit to Z. Frequently, these are the same; then the molecular weight can be determined without the aid of the cryoscopic method.

1.4   The Symmetry of the Unit Cell

The determination of the unit cell size and its symmetry is the first stage of a complete structure determination. However, such a study often yields valuable information in its own right; hence, it is frequently carried out even when a complete structural solution is not contemplated. The study usually takes less than a day, and it yields (1) the molecular weight as described previously, (2) a method of characterizing the compound that is more compact and more nearly unique than X-ray powder patterns, and (3) frequently some information about molecular symmetry. Compilations such as *Crystal Data* (United States Department of Commerce; National Institute of Standards and Technology; and the Joint Committee on Powder Diffraction Standards, USA) contain information on unit cell size, and symmetry when known, for most materials on which structure investigation of any kind has been reported; and such compilations are updated periodically. The information is given in terms of crystal system and cell axes ratios. Therefore, it affords an easy and valuable check to determine whether the compound under study was investigated earlier.

Section 3 gives a detailed discussion of the general topic of symmetry of crystals; hence, only a few general conclusions are stated here. Molecular symmetry information can sometimes be inferred from the symmetry of the unit cell; this is most often true when the molecule has a center of symmetry since this molecular center is frequently also used as a center of symmetry in the crystal structure and can often be established uniquely from unit cell and space group information only. For example, many centrosymmetric molecules crystallize in the ubiquitous space group $P2_1/c$, described below, with two molecules in the unit cell. Establishment of this crystallographic information, in less than a day from a single 0.2-mm crystal, then proves that the molecule has a center of symmetry, providing no disorder is present. (Orientational disorder is rare; and when it occurs, it is often apparent from the abnormally rapid decline in intensities with increasing angle of scattering and from the presence of abnormal streaks of radiation associated with only partial ordering of neighboring molecules.) When there are four molecules present in a unit cell of symmetry $P2_1/c$, no conclusion can be drawn about the molecular symmetry. However, if $4n + 2$ ($n = 0, 1, \cdots$) molecules are present, at least two, and hence presumably all, have centrosymmetric molecular symmetry. One example of this is biphenylene [13], for which a possible structure involving one six-membered and two five-membered rings was eliminated immediately by these very simple symmetry arguments. This study was also carried to the final stages, and it proved the correctness of the presently accepted structure with benzene rings fused to a four-membered ring in the centrosymmetric arrangement.

Unfortunately, molecules do not often make use of other symmetry elements, such as a plane or an axis of symmetry, even if present in its crystal structure. Indeed, centrosymmetric molecules do not always make use of the centers of symmetry in the crystal structure. Therefore, proof that a center of symmetry is absent in the unit cell tells us nothing about the molecular symmetry. However, noncentrosymmetric molecules frequently crystallize in the orthorhombic symmetry group $P2_12_12_1$ with four molecules in the cell, or the monoclinic symmetry group $P2_1$ with two molecules in the unit cell, particularly if they are optically active organic molecules. These symbols are identified in Section 3.3.

1.5   General Principles of a Complete Structure Determination

To determine a complex molecular structure by the X-ray diffraction method, the complete three-dimensional arrangement of all atoms in the unit cell must be found. Normally this type of study is carried out under the direction of an experienced crystallographer, with a considerable amount of computational aid. If the heavy-atom method (following) is used, then a single crystal of dimensions 0.05–0.8 mm is required of a single heavy-atom derivative, such as a chloride, a bromide, a sodium or potassium salt, or a molecular crystal containing another molecule with heavy atoms. It usually makes little difference to the crystallographer whether the heavy atom in the unit cell is bound to the molecule, and it is often helpful to make several different types of heavy-atom derivatives. A few days' survey may show that one of these derivatives can be solved with far less work than the others. Even though these derivatives may seem little different to the chemist, they may present very different problems to the crystallographer.

The accumulation of a few thousand intensities by countermethods usually takes a few days. Relevant crystallographic computations (as follows) and the complete solution of the structure require a few days to a few weeks or more, depending on the complexity of the structure, with the aid of high-speed computers. The improvements in computers and the advances being made in the automatic recording of intensities, including the use of position-sensitive multidetectors, promise further reductions in this total time; already the amount of effort required is often much less than that expended by, for instance, the organic chemist using classical methods on a new natural product.

1.6   Scattering by Electrons

The periodically varying field of an electromagnetic wave is capable of accelerating an electron. But an electron so accelerated emits an electromagnetic wave, which has a frequency essentially identical to that of the incident wave and a definite phase relation that we shall take as identical to that of the incident wave. Thus the maximum of the incident wave corresponds to the maximum of the scattered wave. The energy of X rays is usually large in comparison with all the binding energies, except for the K electrons of the elements of high atomic number and the K and L electrons of the very heavy elements. Hence to the approximation usually employed all electrons are said to scatter X rays independently of all other electrons, with the same amplitude and phase. However, some interference can occur in an atom between scattered waves from different electrons because of their different positions in space.

The scattered waves from K and L electrons in heavy elements can exhibit a shift in phase, and this effect is increasingly important as a basis for solving certain complex structures as the accurate measurement of X-ray intensity increases, and as both a wide variety of target materials for generating X rays of different wavelengths and synchrotron sources of X rays become available.

1.7   Scattering by an Atom

Consider an atom at the origin of a coordinate system. If all the electrons of this atom were at the origin, the amplitudes of scattering from them would then be additive. Then the atomic form factor, the ratio of the amplitude of scattering of the atom to that of a single electron at the origin, would be Z, the atomic number.

The acceleration produced by the electric vector of the electromagnetic wave has a strong inverse dependence on the mass of the particle. Hence, the nucleus produces a negligible scattering of X rays despite its larger charge.

However, the electron density of an atom is spread out in space to an extent consistent with the coulombic binding energies. Except for the relatively small effects of chemical binding and the usually more important effects of anisotropy

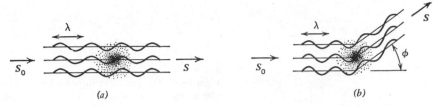

Figure 1.3 (*a*) Scattering from an atom at zero angle, showing no path difference; (*b*) scattering at angle ϕ, showing path difference.

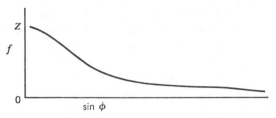

Figure 1.4 Decrease of amplitude of scattering from an atom of atomic number Z as a function of angle of scattering ϕ.

of thermal vibrations, atoms may be regarded as spherically symmetric. They have an approximately exponential decrease of electron density from the center and an effective radius of about 1–3 Å. Hence, interference can take place between different parts of the electron cloud about the atom. As illustrated in Figure 1.3*a*, at zero angle of scattering there is no path difference between the incident and the scattered waves; hence, the atomic form factor is just Z, the total number of electrons. As the scattering angle increases (Figure 1.3*b*) some interference occurs, and the form factor is less than Z. When one takes into account the manner in which the density decreases as distance from the nucleus increases, a resulting form factor is obtained that decreases as the angle of scattering increases (see Figure 1.4). In addition, temperature vibrations and other disorders further increase the diffuseness of the electron distribution. These effects together with the form factor account for the general decline of intensity as the angle of scattering increases, a feature that is characteristic of all X-ray diffraction intensity distributions. The observed reflections are obtained as small, very localized diffraction maxima because of the periodic nature of the crystal, as will be discussed next.

1.8 Scattering by Two or More Atoms

Waves scattered from an atom at $x = 0$ and an atom at $x = a$ reinforce each other when the path difference is $h\lambda$, where h is an integer and λ is the wavelength of the X rays (Figure 1.5). If we now add a third atom at $x = 2a$, the maximum for these atoms still occurs at the same angle ϕ, but it is sharper because slight

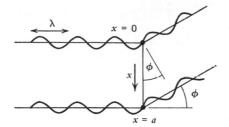

Figure 1.5 The condition for a maximum when two atoms scatter X rays is that the path difference is $h\lambda$. In this example $h = 1$. In general, the condition for a maximum is that $a \sin \phi = h\lambda$.

deviations from the correct value of ϕ produce some interference between atoms at $x = 0$ and $x = 2a$. As more and more atoms are added to the row at intervals a distance a apart, the maximum becomes sharper and sharper. This result is proved in Section 2, where it is generalized to directions of incidence at angles other than 90° to the row of atoms. Further generalization to three dimensions, also discussed in Section 2, then shows that when the crystal structure is resolved into planes (called *equivalent planes*) that have identical atomic environments, the Bragg law $n\lambda = 2d'_{hkl} \sin \Theta$ gives the condition for a maximum (Figure 1.6).

Many identical atoms spaced a distance a apart as in Figure 1.5 simply show successive maxima for increasing values of h. These maxima occur at increasing values of Θ and show the normal decline in intensity because of atomic shape and thermal vibrations. Suppose, now, that we add another atom of the same or different kind at a point X, a certain distance between 0 and a, and similarly at $X + a$, in accord with the repeat unit a of the one-dimensional row (Figure 1.7). If, for example, the two larger atoms were at $X = a/2$ and $3a/2$, they would send out waves exactly out of phase with respect to the waves scattered with a path difference of λ from the atoms at 0 and a. Or, if the path difference between any two atoms at 0 and a is 2λ, then the waves from the atoms at $a/2$ and $3a/2$ will be in phase with respect to the atoms at 0 and a. Thus for odd orders interference

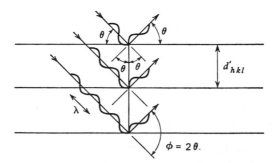

Figure 1.6 The Bragg law for the case $n = 2$. The condition for a maximum is $n\lambda = 2d'_{hkl} \sin \theta$, where $2d'_{hkl} \sin \theta$ is the path difference associated with equivalent planes, which are spaced by a distance d'_{hkl}. The letters *hkl* identify the planes.

occurs, and for even orders reinforcement occurs. When the larger atoms are not greatly different from the smaller ones (Figure 1.7), an alternation of intensities occurs as one examines successive orders of reflection, with the odd orders weak and the even orders strong. If all the atoms were identical, the odd orders would be missing completely for this choice of atomic positions.

Another simple example occurs when the atoms are identical and the atoms designated by the larger circles are not quite at $X = a/2$ and $3a/2$. Then interference is not quite complete, and the values of the intensities are used to locate the positions of these atoms relative to those at $X = 0$ and a. The quantitative relation for the amplitude of scattering is derived in Section 2.4; but for this simple example it is $F_h = 2f_1 + 2f_2 e^{-2\pi i h x}$, where f_1 and f_2 are the amplitudes of scattering from the two kinds of atoms, including the effect of thermal motion; h is the order of the reflection; $x = X/a$ is the atomic position expressed as a fraction of the unit-cell dimension; and $i = \sqrt{-1}$. An explanation for the use of the complex exponential notation instead of the trigonometric function follows, but for the simple case in which all atoms are identical $(f_1 = f_2 = f)$ and a new origin is chosen between the two types of atoms, F_h reduces to the trigonometric function $4f \cos(2\pi h x/2)$, where now x refers to the new origin. It is reasonable to expect the amplitude of scattering to be proportional to the amplitude f of the individual atoms and to be a trigonometric function of both the position and the order of the reflection, since path differences vary with both x and h.

The generalization of the preceding argument to three dimensions and any number N of atoms in the unit cell leads to the expression

$$F_{hkl} = \sum_{j=1}^{N} f_j \exp[-2\pi i(hx_j + ky_j + lz_j)] \tag{1a}$$

For a crystal structure with a center of symmetry taken as the origin of the coordinate system, (1a) reduces to

$$F_{hkl} = \sum_{j=1}^{N/2} 2f_j \cos 2\pi(hx_j + ky_j + lz_j) \tag{1b}$$

where this sum extends over all atoms not related by the center of symmetry. The intensity of scattering, the quantity observed after routine correction for

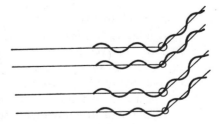

Figure 1.7 The values of x coordinate for these lines are 0, X, a, and $X + a$, reading from the bottom. The larger atoms at X and $X + a$ send out waves that interfere with those from the atoms at 0 and a.

polarization and time-of-reflection factors, is simply proportional to either the square of F_{hkl}, when F_{hkl} is real, or $|F_{hkl}|^2$, if F_{hkl} is complex. Both correction factors and the method of their computation are described in Section 2.6. For our present purpose we assert that the amplitudes or intensities of scattering can be computed from a set of atoms at known positions x_j, y_j, and z_j in the unit cell. Comparison with the corresponding observed quantities constitutes a test of the correctness of these atomic positions.

1.9   Fourier Series for the Electron Density

In an X-ray diffraction study the investigator really wishes to start with the intensities and to derive the atomic positions x_j, y_j, and z_j. Unfortunately this procedure can be quite difficult, and it is not merely a computational problem. The problem is most clearly illustrated by the crystal structure with a center of symmetry chosen as the origin. Because the crystal structure is a periodic function, it is possible and convenient to express the electron density as a three-dimensional Fourier series. For the centrosymmetric case, the expression is

$$\rho(x, y, z) = \frac{1}{V} \sum F_{hkl} \cos 2\pi(hx + ky + lz) \tag{2}$$

where the sum extends over all planes and orders of scattering, that is, all observed reflections, and V is the volume of the unit cell. The proof that the coefficients in this expansion are F_{hkl}/V and the derivation of the general expression for the noncentrosymmetric case are given in Section 2.7. The maxima in the function $\rho(x, y, z)$ are at or very near the atomic positions.

If the amplitudes of scattering F_{hkl} are known, the complete electron-density map of the unit cell can be computed from (2). But one observes F_{hkl}^2 in this centrosymmetric example. Although the magnitude of F_{hkl} is easily found by taking its square root, the sign of F_{hkl} is unknown. A random selection of signs or phases followed by an electron-density map calculation is impractical because of the many possible combinations (even in the centrosymmetric case, 2^N combinations are possible for N reflections; $2^{10} \approx 10^3$). Direct methods for selection of the appropriate signs have had considerable, but not universal, success; and we discuss one such method in Section 1.14, as well as a few other methods that are somewhat more indirect. If the signs of most of the large terms in (2) are determined and attached to the observed F_{hkl}, the series "refines"; that is, the peaks in $\rho(x, y, z)$ occur at positions somewhat shifted from those atomic positions used to calculate the signs from (1b), and new peaks can occur corresponding to atoms not yet included. If the majority of signs are correct and because the observed F_{hkl} are used with these computed signs, the atomic shifts are ultimately toward the correct final atomic positions. Unfortunately, one must have nearly all the atoms within a few tenths of an angstrom of their correct positions before this refinement process converges to the correct structure; that is, a structure that produces a set of F_{hkl}^c from (1b), the magnitudes of which agree with the observed F^o_{hkl}. We give the term *trial structure* to a structure that is sufficiently close to the correct one. The following sections offer a few comments about structure refinement and describe a few methods used to obtain a correct trial structure for complex molecules.

1.10   Least-Squares Refinement Methods

Unless the structure is prohibitively large, such as a protein, the final stages of a structure determination usually involve a least-squares refinement of the positions and associated thermal parameters of the atoms to get the best fit to the magnitude of the structure factor expression given by (1). With the use of anisotropic thermal parameters for the atoms, it is common for structures whose data are measured by film techniques to refine to a crystallographic residual index, $R = \Sigma\, ||F^o| - |F^c||/\Sigma\, |F^o|$, of 15% or less and with good quality counter-data for a structure to refine to a residual index of approximately 5% or better.

1.11   The Heavy-Atom Method

The heavy-atom method is used for both centrosymmetric and noncen-trosymmetric crystals. Because the description of the method is simpler for centrosymmetric crystals, it is discussed first.

Equation (1) shows that the total amplitude of scattering depends on numbers f_j, proportional to the number of electrons associated with each atom j, as well as on their relative positions. If one atom is of high atomic number compared with the rest, for example, an organometallic complex containing iron with carbon, nitrogen, or oxygen, the iron largely determines both the signs and magnitudes of the amplitudes of scattering F_{hkl}. Suppose that by trial-and-error methods the positions of the few heavy atoms in the unit cell are found. These heavy atoms will then give correct signs to many of the larger F_{hkl} values; and when these signs are used with the corresponding observed magnitudes of the F_{hkl} values in the computation of the electron-density map (2), the positions of the light atoms as well as slightly better positions of the heavy atoms can be found. If this computation is repeated with revised heavy-atom positions as well as the positions of such light atoms as are clearly defined, a better approximation to the true electron density is obtained, and after a few cycles the positions of all atoms are known.

When the structure is not centrosymmetric, the principles are the same, but the statements are not as simple because the mathematical expressions are more complex. These expressions are derived in detail in Section 2.

The limitation of the heavy-atom method as described occurs when the number of light atoms becomes so great that the heavy atoms do not determine enough signs for the electron-density map to give a recognizable molecular structure. It is unnecessary that the number of electrons in the heavy atom be greater than the total in all the light atoms because these light atoms are distributed throughout the unit cell. Thus their contributions to the total amplitude of scattering are smaller, on the average, because of interference effects than is the contribution of an equal number of electrons in a few heavy atoms. From statistical arguments we conclude that the foregoing heavy-atom method should succeed when, roughly,

$$2Z^2 \geqslant \sum_j z_j^2$$

where Z is the atomic number of the heavy atom and z_j is the atomic number of the jth light atom. The vitamin B_{12} structure [14], determined partly by the use of the heavy-atom method, is sufficiently complex that here the use of this method is marginal, and the structural solution represents a major scientific accomplishment.

1.12   The Patterson Map

The location of heavy atoms and the solution of complex structures is greatly facilitated by the computation of the vector map or Patterson map, developed by Patterson [15].

$$P(u, v, w) = \frac{1}{V} \sum |F_{hkl}|^2 \cos 2\pi(hu + kv + lw) \tag{3}$$

This function can be computed directly from the observed $|F_{hkl}|$ without any knowledge or assumptions about the signs or phases of the amplitudes F_{hkl}. The Patterson map (3) is obtained from the product $\rho(x, y, z)\rho(x + u, y + v, z + w)$ of the electron densities at x, y, z and $x + u, y + v, z + w$. When the expression in (2) is used for the electron density and all contributions at different x, y, z are added, (3) results. These mathematical details are described in Section 2.7.

Physically, the function $P(u, v, w)$ is large whenever both $\rho(x, y, z)$ and $\rho(x + u, y + v, z + w)$ are large. Hence, $P(u, v, w)$ has maxima whenever the difference u, v, w between the coordinates at these two points corresponds to a vector between a pair of atoms. Moreover, the maxima in $P(u, v, w)$ are proportional to the product of the scattering amplitudes (approximately the atomic numbers) of the pair of atoms. Hence, $P(u, v, w)$ is a map of all interatomic vectors redrawn from a single origin. Its great advantage is that it can be computed directly from observed data. Its disadvantage is that it contains many peaks that are often poorly resolved, particularly for complex structures. If the unit cell contains N atoms, the vector map contains $N(N - 1)$ peaks, plus the origin peak. Sharpening procedures discussed in more detail in Section 2.7.2 can help improve peak resolution, but only to a limited extent.

Consider Ni(L)Br, where L is the tridentate ligand, 6,6'-dimethyl-di(2-pyridylmethyl)amine, as an example. There are four formula weights in the unit cell. If we ignore hydrogen atoms, there are four Ni, eight Br, and $4 \times 17 = 68$ light atoms in the unit cell of dimensions $a = 11.23$, $b = 16.19$, and $c = 8.75$ Å; $\beta = 92.67°$; and $\alpha = \gamma = 90°$. Thus there are $80 \times 79 = 6320$ peaks in $P(u, v, w)$; but $12 \times 11 = 132$ are prominent Br—Br, Ni—Ni, or Br—Ni vectors; and the others are sufficiently small and randomly distributed for the most part, so that to a first approximation they can be ignored. Hence, a reasonable procedure for solving this structure is to analyze the 132 prominent peaks to determine the Ni and Br atom positions. These positions and the known amplitudes of scattering can then be substituted into (1a) to yield the amplitudes of scattering for the structure containing only Br and Ni atoms. The magnitudes of these amplitudes will not agree well with the observed amplitudes because the contributions of all the light atoms are omitted, but most of the phases will be correct, and these phases along with the observed amplitudes can then be used to compute an electron-density map that should show a number of the light atoms in approximately the correct positions along with better positions for the Br and Ni atoms.

Thus once the heavier atoms are located, the positions of the lighter atoms can be readily determined. However, with 132 peaks in the Patterson function, the determination of the heavier atom positions is not trivial. These positions can be found, though, by a systematic method of Patterson analysis called the *Patterson superposition technique*. We consider this method in the next section, along with examples of Patterson maps for both an artificial and a real structure.

1.13   Patterson Superposition Method

1.13.1   Illustration of the Principle

Many complex structures have been solved by this method. As a very simple example we consider a hypothetical two-dimensional structure in Figure 1.8*a*. If the vectors between pairs of atoms in this structure are drawn and then plotted from a common origin, the resulting vector or Patterson map is that shown in Figure 1.8*b*. For a given pair of atoms A and B one must remember that there is a vector from A to B and also one in the opposite direction from B to A. Hence,

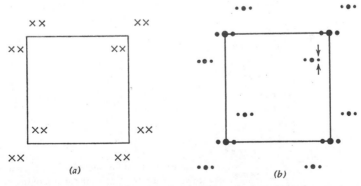

Figure 1.8   (*a*) Unit cell containing four atoms, related in pairs by center of symmetry at origin. (*b*) Vector map of the structure of (*a*). The single interaction to be used for superposition is indicated by arrows. Next largest peaks are double interactions, and the largest peak is the origin peak that indicates interaction within atoms.

the Patterson map always has a center of symmetry at the origin whether the original structure is centrosymmetric or not. In addition, unless specially modified, a Patterson map always has a very large peak at the origin that represents the interactions between different parts of electron density within each atom. Suppose in this example that we take atomic scattering amplitudes proportional to the number of electrons and assume that each atom has five electrons. Aside from the factor V in (3), the origin peak has the weight $4 \times 5^2 = 100$; the double interactions have weight $2 \times 5^2 = 50$; and the single interactions have weight $5^2 = 25$. For this centrosymmetric structure there are four single interactions plus four double interactions in the unit cell of the vector map, exclusive of the origin peak. In general there are $n(n - 1)$ vectors if there are n atoms in the unit cell, since vectors can be drawn from each of these n atoms in turn at the origin. If the structure is centrosymmetric, there are n single interactions, and the remaining interactions are double, giving rise to $n(n - 2)/2$ double peaks. From the absence of certain reflections (see Section 3.4) it is often apparent that either the three-dimensional structure or one of its projections onto two dimensions has a center of symmetry. If so, the preceding analysis can be helpful in starting to decipher the Patterson map. In particular, from the knowledge of which atoms are present, a single interaction can often be located that will form a good starting point for the deconvolution of the Patterson map.

In the Patterson superposition or vector convergence method one image of the structure can result if a single interaction in the Patterson map is located and used and the structure is centrosymmetric; otherwise, additional images are also present. To perform a superposition, a second copy of the Patterson map is overlayed with its origin on this single interaction, as illustrated in Figure 1.9. When the two maps are combined, usually using a minimum function, the highest peaks indicated by the arrows give the structure [16]. In the following sections we give an illustration of this method for a real crystal, the foregoing Ni—Br complex.

Figure 1.9   Result of slightly displaced superposition of vector map on single interaction of the same vector map. Heaviest peaks of heights 125 to 100 are indicated by arrows; next highest peaks, 50 or lower, are ignored. The heavy peaks represent the structure (Figure 1.8).

1.13.2   Dibromo-6,6'-dimethyl-di(2-pyridylmethyl)amine Nickel(II)

The Patterson superposition method is not restricted to a centrosymmetric crystal. The structure of $Ni(L)Br_2$ is a striking example of the power of the method in a noncentrosymmetric application.

A brief algebraic analysis may be helpful in describing this method. Assume that N atoms are arranged in a unit cell at positions $A_1, A_2, \ldots, A_N$. Then the ideal Patterson function is described by the vector set $(A_j - A_i)$, where both i and j take on any values from 1 to N. Assume that we have selected a well-resolved single peak that represents the vector $A_\beta - A_\alpha$. Now shift the Patterson function by this vector, and select the resultant vectors that are common to the unshifted vector set. The origin is taken as that for the unshifted map. Thus for $(A_j - A_i) + (A_\beta - A_\alpha)$, the subsets $(A_j - A_\alpha)$, $j = 1, \ldots, N$ and $(A_\beta - A_i)$, $i = 1, \ldots, N$ are in common as well as points resulting from any other cases where vector triangles are formed using $(A_\beta - A_\alpha)$ as one side. If point atoms and a single interaction are assumed, there are essentially no points of the latter type. The two subsets represent, respectively, an image of the atomic arrangement in the unit cell with atom α at the origin and an inverted image of the atom arrangement with atom β at the origin. It can be shown readily that these subsets are related by a center of symmetry halfway along the vector $A_\beta - A_\alpha$. Thus if $A_\beta - A_\alpha$ is a single vector in a centrosymmetric structure (β and α being related through the center of symmetry), the two images coincide, and one superposition is sufficient to determine the atomic arrangement in the unit cell.

For the noncentrosymmetric structure, a second peak in either subset is selected and the superposition process is repeated. If, for example, $A_\gamma - A_\alpha$ is selected, only the subset $(A_j - A_\alpha)$ remains.

However, in practice, we are not dealing with point atoms, and Patterson peaks have appreciable width. Therefore even if sharpened Pattersons, described in Section 2.7.2, are employed, many extraneous peaks usually remain; and great care must be taken to avoid selecting one of these peaks after the first superposition. Of course, more than two superpositions can be carried out as long as vectors belonging to the same subset are selected consistently. Recognizable structural features or use of space-group symmetry are valuable aids in ensuring that vectors in the same subset are chosen.

As an illustration, consider again the $Ni(L)Br_2$ complex, the unit cell of which is monoclinic, acentric, and contains four molecules [17]. In addition, the absence of certain reflections ($h0l$, $h + l$ odd) indicates the presence of a symmetry element called an n glide, which means that for every atom at x, y, z in the unit cell, an equivalent atom must be located at $\frac{1}{2} + x$, $-y$, $\frac{1}{2} + z$. Such symmetry elements are considered in more detail in Section 3.3; for now we need recognize only that the coordinates of the atoms of two of the four molecules must be specified.

A projection of the sharpened Patterson along the c axis for this Ni—Br complex is shown in Figure 1.10. Since there must be 132 heavy-atom vectors in

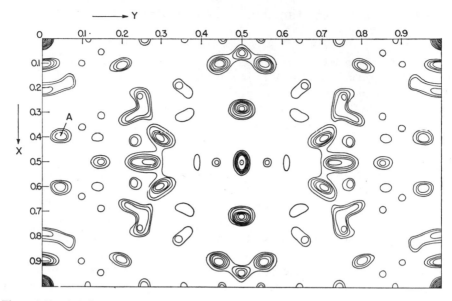

Figure 1.10 A sharpened Patterson from the observed data for Ni(L)Br$_2$, where L is the tridentate ligand 6,6'-dimethyldi(2-pyridylmethyl)amine, and projected along the c axis. The peak selected for the first superposition is indicated by an A.

this map, many peaks must be overlapping or partially overlapping in this projection. If a multiple peak of multiplicity m is chosen for the first superposition, $2m$ images result. Therefore it is best to select a well-resolved peak of as low a multiplicity as possible. The peak marked with an A was selected since this peak is small and well resolved. The result of the superposition with this peak is shown in Figure 1.11. For this computer calculation, the minimum value of the Patterson and the shifted Patterson was chosen at each grid point. There must be other atoms that are produced by n-glide symmetry operating on the head and tail atoms of the shift vector, namely, at positions displaced by one-half in the x direction. The y coordinate is determined by the position of the glide plane relative to the superposition map origin (which in general is not the same as the origin in the electron-density space); when this position is selected, all atoms must conform to this symmetry operation.

The two peaks marked B and C were selected as those related to the head and the tail of the first shift vector, and a further superposition was carried out using the peak marked B. This result is shown in Figure 1.12a, which illustrates that the Ni and Br atom positions can, for the most part, be determined readily since these atoms should correspond to peaks with at least two contours. However, it should also be noted that there are two extraneous peaks in this map that are related also by the n-glide operation. Such an effect is common, and often more than two superpositions are required to eliminate enough accidental coincidences to obtain a good trial structure. Such accidental coincidences are less

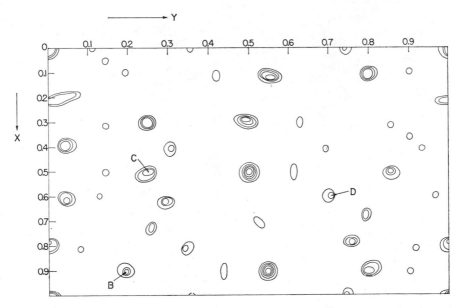

Figure 1.11   Result of the first superposition using peak A of Figure 1.10. A minimum function was used to compute this map. Peaks marked B and C were selected as those related to the origin and to A by the n-glide operation.

frequent in three dimensions than in two. Other peaks can be selected for the second superposition, which will give fewer extraneous peaks; for example, if peak D (Figure 1.11) is used, the map shown in Figure 1.12b results. Figure 1.13 shows a projection of the final electron-density map using the Ni and Br atom positions so determined. Light-atom positions can be determined from a superposition carried out in three dimensions where overlap will not so hinder the search for this type of vector.

The Patterson superposition method has been applied successfully to many different categories of structures; structures such as glycyl-L-trypotophan dihydrate [18], B_8Cl_8 [19], cellobiose [20], α-L-sorbose [21], L-aspartic acid [22], $(NPCl_2)_5$ [23], $(CH_5)_2Fe_2(CO)_3CS$ [24], and many others [25] were determined by this technique.

1.13.3   Extensions of the Patterson Superposition Method

The Patterson superposition method was extended by the complete use of all the space-group symmetry of the unit cell. To illustrate, in the foregoing example, for every atom at x, y, z there must be a related atom at $\frac{1}{2} + x$, $\bar{y}$, $\frac{1}{2} + z$; and the peaks on the $(\frac{1}{2}, v, \frac{1}{2})$ line in the Patterson give evidence for the possible presence or at least for the definite absence of atoms at the corresponding positions in electron-density space. If more than one symmetry element is present, further restrictions can be placed on the probability of appreciable electron density

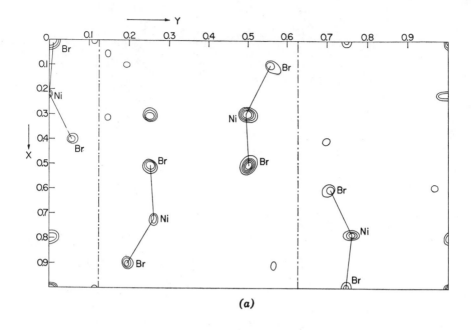

(a)

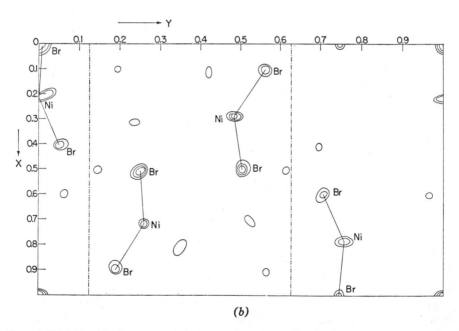

(b)

Figure 1.12 (a) Result of a second superposition using the peak marked B of Figure 1.11. Note that of those peaks higher than one contour, one extraneous peak remains per asymmetric unit. (b) Result of a second superposition using the peak marked D in Figure 1.11. The image of the structure remains with no extraneous peaks of any reasonable height.

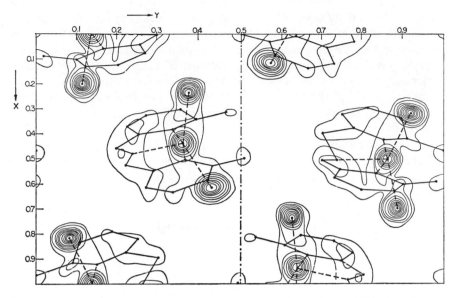

Figure 1.13 A *c*-axis projection of the electron-density map of the unit cell of Ni(L)Br$_2$. The origin of the unit cell has been shifted to the *n*-glide position. The ligand atom positions were determined from a three-dimensional electron-density calculation.

being located at any particular point in the unit cell. From these considerations a map called a *symmetry map* or *symmetry minimum function* can be produced and can be used with the Patterson in a superposition procedure [26]. An alternate use of symmetry is (1) carry out one or more superpositions without regard to symmetry, and (2) computer analyze the resultant map testing appropriate peaks as possible symmetry partners using the space group symmetry [27].

It was also shown that significant improvement is obtained in the resultant superposition map if, when the chosen shift vector is between atoms of different atomic numbers, one Patterson is weighted differently from the other [28], according to the ratio of the atomic numbers involved. Such a procedure results in fewer ambiguities in the resultant map, and partially destroys the pseudocenter of symmetry introduced by the superposition technique.

Known geometry can also be used, and this approach is discussed briefly under the topic of search methods in Section 1.16.

1.14   Direct Methods

1.14.1   The Symbolic Addition Method—Centrosymmetric Case

As we have shown, the Patterson superposition method can be used to attempt to obtain a trial structure for either a centrosymmetric or a noncentrosymmetric structure. Another method has become the prime method of use in many laboratories and can be used with great success to determine structures containing essentially equal atoms, especially for centrosymmetric structures, and we discuss this method and its application. This is the *symbolic addition method,* one of a closely related group of direct methods, which, as the name suggests, involves the direct use of intensities rather than their indirect (Patterson) use. Since it is possible to deconvolute the Patterson function to obtain an image of the structure, it should be possible to mathematically manipulate the magnitudes of the structure amplitudes directly to obtain the same result.

The symbolic addition method started from an equation suggested by Sayre [29] and was developed into an addition procedure and extended by Karle and Hauptman; see, for example, [30]. Consider the structure factor written in the form

$$F_{\mathbf{k}} = \sum_j f_j \exp(-2\pi i \mathbf{k} \cdot \mathbf{r}_j)$$

where f_j contains the temperature factor and, anticipating a later discussion, $\mathbf{k} \cdot \mathbf{r}_j = hx_j + ky_j + lz_j$. Now define

$$|E_{\mathbf{k}}|^2 = \frac{|F_{\mathbf{k}}|^2}{\varepsilon \sum f_j^2} \tag{4}$$

where ε is a factor to account for degeneracy, but which for most reflections can be assumed to be unity [30]. For N equal atoms,

$$E_{\mathbf{k}} = \frac{1}{N^{1/2}} \sum_{j=1}^{N} \exp(-2\pi i \mathbf{k} \cdot \mathbf{r}_j) \tag{5}$$

Then

$$E_{\mathbf{k}} E_{\mathbf{h}-\mathbf{k}} = \frac{1}{N} \left[\sum_j \exp(-2\pi i \mathbf{h} \cdot \mathbf{r}_j) + \sum_i \sum_{\substack{j \\ i \neq j}} \exp\{-2\pi i [\mathbf{k} \cdot \mathbf{r}_j + (\mathbf{h} - \mathbf{k}) \cdot \mathbf{r}_i]\} \right]$$

where the vector quantities $\mathbf{h}$ and $\mathbf{k}$ refer in general to different sets of indices. Depending on the value of $\mathbf{k}$, the second term can assume either positive or negative values at random, and therefore averaging

$$E_{\mathbf{h}} = N^{1/2} \langle E_{\mathbf{k}} E_{\mathbf{h}-\mathbf{k}} \rangle_{\mathbf{k}} \tag{6}$$

In the unequal atom case it can be shown that

$$E_{\mathbf{h}} \approx \sigma_2^{3/2} \sigma_3^{-1} \langle E_{\mathbf{k}} E_{\mathbf{h}-\mathbf{k}} \rangle_{\mathbf{k}} \tag{7}$$

where

$$\sigma_n = \sum_{j=1}^{N} Z_j^n$$

However, it is not necessary to average over all values of $\mathbf{k}$ to determine the sign of $|E_{\mathbf{h}}|$. In fact, for $|E|$ values sufficiently large, $S_{\mathbf{h}} = S_{\mathbf{k}}S_{\mathbf{h}-\mathbf{k}}$, where $S_{\mathbf{h}}$ is the sign of the corresponding $E_{\mathbf{h}}$. This can be seen by considering the restriction placed on the individual terms making up $E_{\mathbf{k}}$, if $|E_{\mathbf{k}}|$ is to be very large:

$$E_{\mathbf{k}} = \frac{1}{N^{1/2}} [\exp(-2\pi i \mathbf{k} \cdot \mathbf{r}_1) + \exp(-2\pi i \mathbf{k} \cdot \mathbf{r}_2) + \cdots + \exp(-2\pi i \mathbf{k} \cdot \mathbf{r}_N)]$$

For $|E_{\mathbf{k}}|$ to approach its maximum value, all terms in the summation must be nearly equal so that a common phase factor can be factored out; that is,

$$E_{\mathbf{k}} \approx N^{1/2} \exp(-2\pi i \mathbf{k} \cdot \mathbf{r})$$

The exponential term represents the phase of $|E_{\mathbf{k}}|$. (Remember, $|e^{-i\theta}|^2 = 1$.) Thus if $|E_{\mathbf{h}-\mathbf{k}}|$, $|E_{\mathbf{k}}|$, and $|E_{\mathbf{h}}|$ are very large,

$$E_{\mathbf{k}} E_{\mathbf{h}-\mathbf{k}} \approx [N^{1/2} \exp(-2\pi i \mathbf{k} \cdot \mathbf{r})]\{N^{1/2} \exp[-2\pi i(\mathbf{h} - \mathbf{k}) \cdot \mathbf{r}]\}$$

$$\approx N \exp(-2\pi i \mathbf{h} \cdot \mathbf{r})$$

$$\approx N^{1/2} E_{\mathbf{h}}$$

In other words, the larger the $|E|$ values, the greater is the probability that the relation $S_{\mathbf{h}} = S_{\mathbf{k}}S_{\mathbf{h}-\mathbf{k}}$ will hold. Cochran and Woolfson [31] showed that this is indeed the case and derived the following probability expression for positive $|E_{\mathbf{h}}|$

$$P_{+}(\mathbf{h}) \approx \frac{1}{2} + \frac{1}{2} \tanh \sigma_3 \sigma_2^{-3/2} |E_{\mathbf{h}}| \sum_{\mathbf{k}} E_{\mathbf{k}} E_{\mathbf{h}-\mathbf{k}} \tag{8}$$

To illustrate the usefulness of this method and to describe it in somewhat more detail, we consider its application to the structure determination of a small molecule, tetrahydrofurandione. This compound crystallizes [32] in the centrosymmetric orthorhombic space group *Pnma* or in the noncentrosymmetric space group *Pn2$_1$a* with four molecules per unit cell. Ordinarily, when such a space group ambiguity occurs and when the expected molecular symmetry is compatible with the centrosymmetric space group, one proceeds to use the more symmetrical space group. Only if this fails is the noncentrosymmetric space group used. Statistical tests on the intensities also indicated the centrosymmetric space group. Here we begin by assuming that the molecular plane of symmetry coincides with the plane of symmetry in the space group *Pnma*. The $|E_{\mathbf{h}}|$ values are listed in Table 1.1. Listings of the $|E_k||E_{h-k}|$ were made for each E value (only the larger E values were considered); and after referring to this listing, the signs of the following three reflections were chosen to fix an origin: $S_{211} = -$, $S_{441} = +$, and $S_{174} = -$. In addition, the signs of three reflections were represented by the algebraic symbols $a-c$. Taking only those probabilities greater than 97%, the signs of as many as possible of the largest E values were determined from $|E_k||E_{h-k}|$ products. (Of course one must keep in mind that the signs of all symmetry-related reflections can be referred to those selected in the

Table 1.1   List of Reflections with $|E| \geqslant 2.30$

h	k	l	$|E|$	Sign
2	1	1	3.95	-1
4	0	2	3.23	-1
8	4	4	2.87	a
0	8	2	2.87	c
5	0	3	2.86	c
9	0	4	2.86	ab
4	4	1	2.82	1
4	1	6	2.80	$-a$
1	5	8	2.72	$-ac$
2	0	5	2.68	a
2	4	4	2.62	a
6	8	5	2.61	$-ab$
2	7	3	2.60	b
2	4	2	2.53	$-a$
4	10	5	2.52	-1
8	12	1	2.51	$-b$
5	13	4	2.50	$-b$
3	14	3	2.48	$-b$
0	5	5	2.46	$-a$
1	7	4	2.45	-1
1	12	6	2.44	
3	1	7	2.43	ac
4	13	5	2.42	ab
5	4	3	2.39	ab
8	0	6	2.34	
10	8	1	2.34	$-ab$
6	0	5	2.33	a
5	0	5	2.32	$-c$
1	0	7	2.32	c
3	0	2	2.31	$-ac$
0	0	2	2.31	-1
2	4	6	2.30	$-a$
6	0	3	2.30	$-a$

set.) Using this approach, we obtained the signs indicated in the last column of Table 1.1. Relations are also usually obtained between some or all of the algebraic symbols, and thus these symbols can be reduced in number or sometimes they can be determined completely. In this particular example it was possible to determine signs for two of the three symbols used, and the signs for 171 of the largest E values were thus obtained. A typical listing for the 244 reflection is shown in Table 1.2. From this table it is apparent that the sign corresponding to the letter b is probably $-$ and that the internal consistency of the sign indications for this 244 reflection is excellent. If p unknown symbols were left, then $2p$ electron-density maps could be computed and investigated for structural features characteristic of the compound under investigation. Other tests involving internal sign consistencies can be employed that also help reduce the number of possibilities. In this example the sign corresponding to the letter a was still in doubt, but the calculation of electron-density maps clearly indicated that the negative sign was the most appropriate choice.

Figure 1.14 is a composite Fourier map computed from the large E values and projected down the c axis. The four tetrahydrofuran molecules can be readily seen in this map. Of course this example is quite simple. In more complicated structures expect to find many more extraneous peaks in an E map and to experience more difficulty in locating the correct atomic positions. In general, structures with symmetry planes or other features that tend to give rise to larger E values are more susceptible to solution by the symbolic addition method.

1.14.2 Noncentrosymmetric Algebraic Methods and the Tangent Formula

These methods were extended to produce powerful algebraic-statistical methods for noncentrosymmetric structures that are similar to the methods just described for centrosymmetric structures [29, 33, 34]. The problem is the determination of the phase angle $\phi_{\mathbf{h}}$ for each reflection $\mathbf{h}$, where $\mathbf{h}$ is some hkl and the phase angle

Table 1.2 Some Products Determining S_{244}

$\mathbf{k}$	$\mathbf{h} - \mathbf{k}$	$S_{\mathbf{k}}$	$S_{\mathbf{h}-\mathbf{k}}$	$S_{\mathbf{h}} = S_{\mathbf{k}}S_{\mathbf{h}-\mathbf{k}}$	$\beta|E_{\mathbf{h}}|\,|E_{\mathbf{h}-\mathbf{k}}|\,|E_{\mathbf{k}}|^a$
$44\bar{1}$	$\bar{2}05$	-1	$-a$	a	3.79
$\bar{4}\bar{4}\bar{1}$	605	$+1$	$+a$	a	3.30
$44\bar{1}$	$\overline{6}85$	$+1$	$-ab$	$-ab$	3.69
211	033	-1	ab	$-ab$	4.10
$40\bar{2}$	$\bar{2}46$	-1	$-a$	a	3.74
082	$2\bar{4}2$	b	$-a$	$-ab$	3.64
$08\bar{2}$	$24\bar{6}$	b	$-a$	$-ab$	3.31

$^a\beta = \sigma_3\sigma_2^{3/2}.$

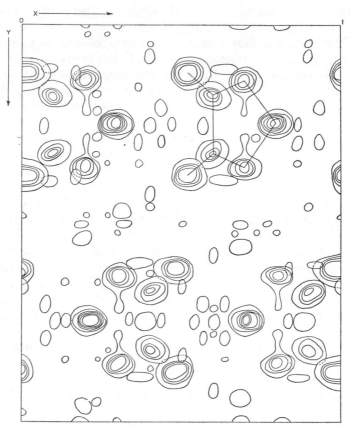

Figure 1.14 A Fourier map computed using the large E values for tetrahydro-3,4-furandione. Signs for these E values were determined by the symbolic addition method.

is in the range from 0 to 2π rad (or $-\pi$ to π). The main formulas and procedure for the most widely used such method are summarized by Karle and Karle [30, 35] as follows:

$$\phi_{\mathbf{h}} \approx \langle \phi_{\mathbf{k}} + \phi_{\mathbf{h}-\mathbf{k}} \rangle_{\mathbf{k}_r} \tag{9a}$$

$$\phi_{\mathbf{h}} \approx \frac{\Sigma_{\mathbf{k}_r} |E_{\mathbf{k}} E_{\mathbf{h}-\mathbf{k}}| (\phi_{\mathbf{k}} + \phi_{\mathbf{h}-\mathbf{k}})}{\Sigma_{\mathbf{k}_r} |E_{\mathbf{k}} E_{\mathbf{h}-\mathbf{k}}|} \tag{9b}$$

$$\tan \phi_{\mathbf{h}} \approx \frac{\Sigma_{\mathbf{k}} |E_{\mathbf{k}} E_{\mathbf{h}-\mathbf{k}}| \sin(\phi_{\mathbf{k}} + \phi_{\mathbf{h}-\mathbf{k}})}{\Sigma_{\mathbf{k}} |E_{\mathbf{k}} E_{\mathbf{h}-\mathbf{k}}| \cos(\phi_{\mathbf{k}} + \phi_{\mathbf{h}-\mathbf{k}})} \tag{10}$$

where $\mathbf{k}$ refers to hkl other than $\mathbf{h}$, and $\mathbf{k}_r$ refers to those $\mathbf{k}$ for which the unitary structure factors $|E_{\mathbf{k}}|$ are large. Equation (10) is known as the *tangent formula*. A simple physical basis for sign and phase relations was discussed by several authors, for example, by Kainuma and Lipscomb [36].

To initiate the procedure, phases are assigned arbitrarily to up to three reflections (the number depending on the symmetry), thus fixing the origin. Equation (8) is then applied with the introduction of symbols for unknown phases where necessary, as in the centrosymmetric symbolic addition method. When sufficient starting phases are obtained, (10) is applied reiteratively by machine. One arbitrary phase can be restricted to choose the enantiomorph.

In the applications by the Karles (see [37]) the method produced solutions for molecular structures having 50 or more first-row atoms with no heavier atoms in the asymmetric unit. Some examples are L-arginine dihydrate [38], panamine [35], and reserpine [39]. For a given complexity of structure this method is less dependent on chemical information than any other; therefore, it is particularly useful where chemical evidence is minimal or misleading. One example is veratrobasine [40], in which the chemical information previously led to an incorrect ring system and therefore to failure of Patterson, vector superposition, rigid body search, and molecular packing methods. This molecular structure, having the formula $C_{27}H_{41}O_3N$, was solved very easily by use of the tangent formula. There are four molecules in the unit cell, and the crystal symmetry is $P2_12_12_1$. Phases were assigned to three reflections to fix the origin. Signs of six structure invariants were obtained from the Σ_1 formula [41, 33] (e.g., for $h0l$ reflections in this centric projection)

$$SE_{2h,0,2l} = S \sum_k (-1)^{k+l}(E_{hkl}^2 - 1) \qquad (11)$$

and two phases were assigned symbols of unknown values, one of which determined the enantiomorph. In (11) S denotes the "sign of" what follows. Fifty-two phases were then established by the symbolic addition procedure, with the use of (8); and then these phases were used as input to the tangent formula, which gave phases for 636 reflections, each with a value of $|E_h| > 1.1$. This procedure converged quickly on a computer and led to the full structure based on 3105 unique reflections.

1.14.3   Automatic Direct Method Procedures

In recent years there were considerable efforts directed toward the development of a completely automated direct method procedure [42]. Probably the best known of these is the MULTAN procedure. The "best" $|E|$ values in terms of those giving high probability of predicting phases and also entering into many combinations are automatically selected as the starting set, including reflections to fix the origin. By allowing various signs or phases ($\pm\pi/4$, $\pm 3\pi/4$) to be given to starting reflections, it is possible to calculate and then to refine several phase sets (usually ≥ 32) and to use the sets giving the "best figures of merit" to calculate electron-density maps. Often the structure is recognized readily from these maps, and the coordinates obtained are immediately used in a least-squares refinement procedure.

1.15   Protein Structure Determination

Following the crystallization of urease by Summer in 1926, Bernal and Hodgkin (née Crowfoot) [43] showed in 1934 that high-quality diffraction maxima can be obtained from a crystal of pepsin maintained in a small amount of its mother liquor. Perutz began a systematic study of hemoglobin crystals in 1938 and was joined shortly thereafter by Kendrew who began a parallel investigation of myoglobin. Heavy-atom techniques became powerful enough for scientists to attempt to determine protein structure only when Bokhoven, Schoone, and Bijvoet [44] proposed the *multiple isomorphous replacement method*, which is described later. An initial important achievement by Perutz and his co-workers [45] was the attachment of heavy atoms to hemoglobin and the demonstration that the derivatives were isomorphous with the protein. The structure of myoglobin (MW 17,000) reached 6-Å resolution [46] in 1958 and atomic resolution (2.0 Å) [47] in 1960. Hemoglobin [2] (MW 67,000) was solved at 5.5 Å in 1960 and at atomic resolution in 1968. The first enzyme brought to atomic resolution was lysozyme [4] (MW 14,600) at 2.0 Å in 1965. Carboxypeptidase A [48] (MW 34,600) was at atomic resolution (2.8 Å) in 1966, and probable mechanisms of action were proposed [49]. The structures at atomic resolution of ribonuclease [5], ribonuclease-S [50], chymotrypsin [51], papain [52], and subtilisin [53] (with molecular weights of 13,700; 11,500; 25,000; 23,000; and 27,600, respectively) were published as of mid-1969; and since then the number and complexity of structures so determined continues to increase dramatically [8]. Generally, at a resolution of about 3 Å or less, it is possible to obtain atomic coordinates with the combination of the chemical sequence and the electron-density map obtained by isomorphous replacement methods. Thus even though individual atoms are not resolved, the solution of the structure can usually be obtained in the range from 3.5 to 2.0-Å resolution. We now consider some of the simpler aspects of the isomorphous replacement method.

Equation (36) will illustrate that the electron density $\rho(x, y, z)$ at every point x, y, and z can be computed from (1) the indices h, k, and l, which are known from the geometry of the experiment; (2) the amplitudes of X-ray scattering F_{hkl}, which are directly obtainable from the measured diffraction maxima; and (3) the phase angle α_{hkl} for each maximum. Thus the determination of α_{hkl} completes the solution to the problem.

Suppose the structure is known, and we also have located the heavy atoms in two isomorphous derivatives. Then the structure factor (a complex number) is

$$F_P = \sum f_j \exp[-2\pi i(hx_j + ky_j + lz_j)]$$

$$F_{H_1} = \sum f_j \exp[-2\pi i(hx_j + ky_j + lz_j)] + \sum f_{j'} \exp[-2\pi i(hx_{j'} + ky_{j'} + lz_{j'})]$$

and

$$F_{H_2} = \sum f_j \exp[-2\pi i(hx_j + ky_j + lz_j)] + \sum f_{j''} \exp[-2\pi i(hx_{j''} + ky_{j''} + lz_{j''})]$$

for the protein and heavy-atom derivatives 1 and 2, respectively. We rewrite the last two equations as

$$F_P = -f_{H_1} + F_{H_1}, \qquad F_P = -f_{H_2} + F_{H_2}$$

where $f_{H_1} = \Sigma f_{j'} \exp[-2\pi i(hx_{j'} + ky_{j'} + lz_{j'})]$ can be computed for each hkl if we locate the positions of heavy atoms in the first derivative. A similar statement applies to the second heavy-atom derivative. This situation is illustrated in Figure 1.15a.

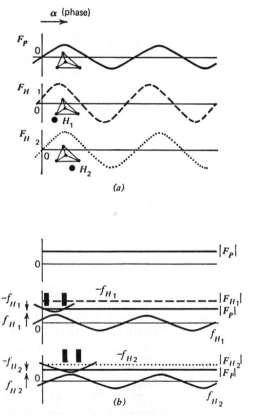

Figure 1.15 (a) The wave scattered by the four small atoms, representing the protein, has a maximum coinciding with most of the scattering matter. If a heavy atom H_1 (or H_2) is added, the maximum shifts in a direction that reflects its contribution. (b) The X-ray experiment yields knowledge only of the magnitudes of the scattering amplitude (F_P) for the protein (solid lines), (F_{H_1}) for the first derivative (dashed lines), and (F_{H_2}) for the second derivative (dotted lines). Location of the heavy atoms yields both their amplitudes and phases (sine curves) with maxima as indicated from their average contribution at the phase position indicated by large dots above. If these sine waves are redrawn as $-f_{H_1}$ and $-f_{H_2}$ and measured from the lines (F_{H_1}) and (F_{H_2}) as indicated, then their intersections with the (F_P) line satisfy the equations $(F_P) = -f_{H_1} + (F_{H_1})$ and $(F_P) = -f_{H_2} + (F_{H_2})$ in the approximation that F_{H_1}, F_{H_2} and F_P have the same phase angle α.

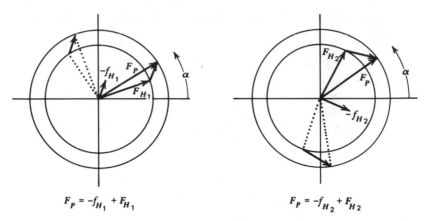

$$F_P = -f_{H_1} + F_{H_1}$$ $$F_P = -f_{H_2} + F_{H_2}$$

Figure 1.16   The information in Figure 1.15b plotted as phase circles. Only the magnitudes of F_P, F_{H_1}, and F_{H_2} are known from the X-ray experiment. Location of the heavy atoms yields both the magnitude and the phase of f_{H_1} and f_{H_2}, each of which can be placed only in the two indicated positions to satisfy the vector equations $F_P = -f_{H_1} + F_{H_1}$ and $F_P = -f_{H_2} + F_{H_2}$. The phase α that is common to both is correct.

Now let us assume the actual situation: We have located the heavy atoms in each of these two derivatives, but we know only the magnitudes $|F_P|$, $|F_{H_1}|$, and $|F_{H_2}|$ from the measured intensities of maxima from the protein and the two heavy-atom derivatives. Thus the situation as it exists in practice is illustrated in Figure 1.15b. However, when the known complex numbers $-f_{H_1}$ and $-f_{H_2}$ are added to $|F_{H_1}|$ and $|F_{H_2}|$, respectively, they must yield $|F_P|$, but they can only do so at certain phase angles α, as shown in Figure 1.15b. Usually there are two solutions for each heavy-atom derivative, but only one solution in common between two heavy-atom derivatives. The common solution is the correct one for each hkl. Inasmuch as the phase repeats after 2π, the information in Figure 1.15 can be plotted as shown in Figure 1.16. Because the phases of both $|F_P|$ and $|F_{H_1}|$ are initially unknown, but the magnitudes $|F_P|$ and $|F_{H_1}|$ are known, a complete circle having a line width corresponding to the error in $|F_P|$ (or $|F_{H_1}|$) represents our initial knowledge. But remember, both magnitude and phase of f_{H_1} are known from the heavy-atom positions, so we must find where to place $-f_{H_1}$ (in either Figure 1.15 or 1.16) to solve the equation $F_P = -f_{H_1} + F_{H_1}$. As before, there are usually two solutions for the other derivative, but only one solution for the phase angle α in common for these two derivatives for each reflection hkl. Finally, we show in Figure 1.17 the more usual method of replotting the phase circles and of obtaining a graphical solution. In practice, careful treatment of errors and least-squares procedures are required to find the best, properly weighted, α_{hkl} values for a given set of heavy-atom derivatives. A computer program for this procedure written at Harvard University by H. M. Muirhead is in general use.

Heavy-atom derivatives can be prepared by immersing single crystals into a solution of a heavy-atom salt, but it is frequently desirable to use a dialysis

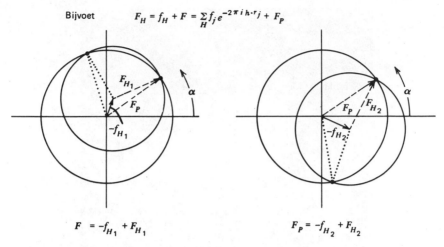

Bijvoet $F_H = f_H + F = \sum\limits_{H} f_j e^{-2\pi i h \cdot r_j} + F_P$

$F = -f_{H_1} + F_{H_1}$ $F_P = -f_{H_2} + F_{H_2}$

Figure 1.17 Phase circles in the complex plane representing the structure factor (scattering amplitude) with its real component to the right and imaginary component vertical. Location of the heavy-atom positions and occupancies gives the vectors f_{H_1} for derivative 1 and f_{H_2} for derivative 2. The completely unknown phase, but known magnitude leads to the circles for F_P for the protein and F_{H_1} and F_{H_2} for the total scattering of each reflection for the derivatives. The intersections of circles are the only points that satisfy the vector equation $F_{H_1} = F_P + f_{H_1}$ and $F_{H_2} = F_P + f_{H_2}$. The intersection common to both derivatives is the correct phase angle α.

procedure (Figure 1.18) to make gradual changes in concentration of salt or pH. For a fairly large protein molecule, such as carboxypeptidase A, the intensity changes produced by the heavy atoms are not always large, but are frequently quite observable (Figure 1.19). Some 20,600 independent diffraction maxima were obtained for the protein, but only the 6000 largest reflections of the protein were measured in the 2.0–3.0-Å range (14,000 reflections possible) on the three best heavy-atom derivatives. A region of the electron-density map at 2.0-Å

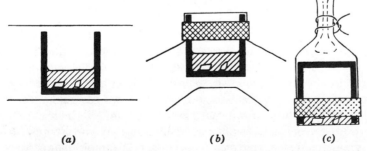

(a)   *(b)*   *(c)*

Figure 1.18 Dialysis cell, 0.8–1.0 cm in diameter by 1.5–2.0 cm, (a) placed in a length of dialysis tubing, which, in (b), is pulled tightly over the open end of the cell, and then secured by means of a polyethylene ring (cross-hatched). In (c) the cell is inverted, and the dialysis tubing tied. [W. N. Lipscomb, J. C. Coppola, J. A. Hartsuck, M. L. Ludwig, H. Muirhead, J. Searl, and T. A. Steitz, *J. Mol. Biol.*, **19**, 423 (1966).]

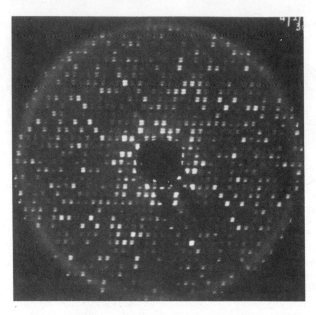

Figure 1.19   The *h0l* X-ray reflections of carboxypeptidase A are shown as the lower one of each pair of spots, while the corresponding reflections from a derivative having two lead atoms per molecule are shown as the upper one of each pair.

resolution is shown in Figure 1.20. Even the waterlike structure can be seen in the region between four neighboring molecules (Figure 1.21). The large amount of water in protein crystals is important in that poor substrates or inhibitors can be diffused into the crystals, and the binding can be deduced from X-ray diffraction data on the complex with the use of phases α_{hkl} of the parent protein. One example of the excess density associated with binding of glycyl-L-tyrosine to carboxypeptidase A is given in Figure 1.22, and the mechanistic deductions from this study are summarized in Figure 1.23. Finally, a drawing of the polypeptide backbone is shown in Figure 1.24.

1.16   Search Methods

Other types of methods that have received considerable attention over the past few years can generally be classified as *search methods*. Such methods work best when the approximate molecular structure, or at least a reasonable fragment of the molecular structure, can be assumed. Either of two general approaches can be followed. One is to use known space group and unit cell information and to try to devise *reasonable* packing arrangements that are consistent with such a unit-cell symmetry. Reasonable means either that intramolecular contacts must be larger than some prescribed minimum [54], or that the configurations found lead to a minimum in the intermolecular energy [55].

A second approach employs the fact that a Patterson function contains all interatomic vectors and therefore contains the vectors between atoms in the

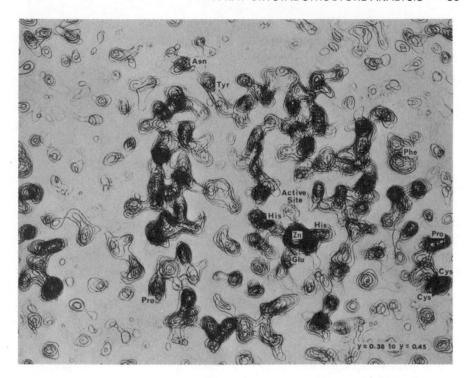

Figure 1.20 Composite of sections $y = 0.38$ to 0.45 of the electron density at 2.0-Å resolution of carboxypeptidase A. A helix extends along the left-hand side of the map from Pro 288 up to the C-terminal Asn 307, which is hydrogen bonded to residue Tyr 265. The enzyme contains zinc, which has three ligands to the protein (His 69, Glu 72, and His 196), and one more ligand to H_2O, which is displaced by peptide substrates and by some inhibitors. An extended chain, part of an extensive β-pleated sheet structure, is parallel to the helix, and at the lower right a disulfide bond can be seen.

fragment in question. By appropriate methods the resultant vectors can be fitted to the Patterson; thus, the orientation of the fragment can be determined [56–58].

1.17 Interpretive Problems

Structure determination is not always a completely routine procedure. Frequently the heavy atoms occur at or near special positions in the unit cell of a crystal in such a way that they do not contribute greatly to a reasonably large fraction of the intensities. Usually the resulting electron-density maps, computed either with the contributions of the heavy atoms only or by use of another method described earlier, are difficult to interpret. Sometimes the problem is due to two superimposed molecules with different orientations. Sometimes it is difficult to recognize the light-atom part of the molecule when it is very badly resolved. Great care must be exercised in identifying possible atoms to be

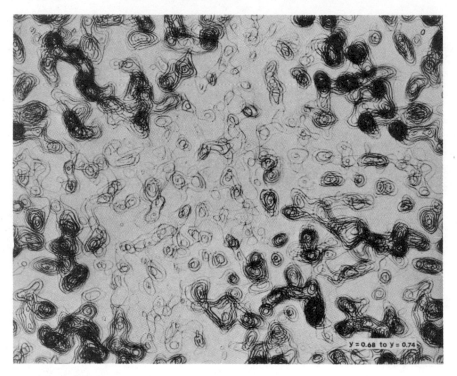

Figure 1.21   The four molecules are shown only in part as the denser regions in the four corners. The major central region is the solvent (mostly 0.2 M LiCl solution) in which the protein is present in the crystalline phase.

included in the structure; a false structure results unless high demands are placed on: (1) the detailed agreement of observed and calculated intensities, (2) equally good resolution of atoms in the map, (3) equally normal behavior of all coordinate and temperature parameters, and (4) good chemical sense.

We illustrate only one of the interpretative problems here. The solution of the iresin p-bromobenzoate diester structure was accomplished by X-ray diffraction methods [59] and involved the measurement of 1432 diffraction maxima. The chemical structure is shown in Figure 1.25. The final electron-density map is shown in Figure 1.26, in which each atom is taken from the particular section of electron density corresponding to its z coordinate, and a composite projection is then made from the various sections. The three stages immediately preceding the final stage of Figure 1.26 are shown as F-2, F-3, and F-4 in Figure 1.27.

Successive refinements of the structure lead to the penultimate stage F-4. The model from which F-4 was computed included the contributions of all atoms except O(5) and C(12). Phases from this model were combined with the observed data, which of course contain contributions from all atoms, to give map F-4. This map shows very clearly that atoms not yet included tend to appear with only about one-third of their proper electron density.

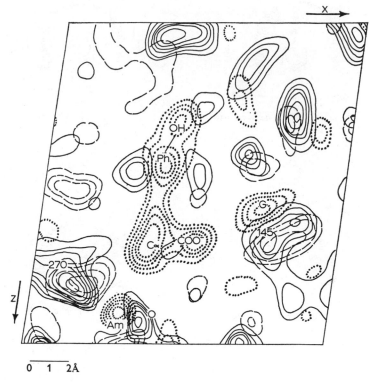

Figure 1.22 The solid contours are those of carboxypeptidase A at 2.0-Å resolution. The dotted contours indicate the excess electron density when glycyl -L-tyrosine is added to the active site, and negative contours are shown as dashed lines. In this section at $y = 0.50$ the guanidinium group of Arg 145 moves about 2Å from the negative contours to the dotted contours. Also the carboxyl group of Glu 270 moves out of this section by about 2 Å, as indicated by the negative contours. Features of the substrate Gly-Tyr are part of the phenyl ring (Ph), the OH group, the CO_2^- group, the C_α position, and the probable position of the amino group (Am).

At earlier stages, therefore, atomic positions can easily be assigned to false detail. This happened in F-2, which was computed at an earlier stage of the investigation when the positions of the two bromine atoms and ten carbon or oxygen atoms were known, none of which was in this five-membered ring area. A five-membered ring, shown in F-3, was then *forced* onto the set of peaks of F-2, and the electron-density map was recomputed with the result shown in F-3. Only the lack of resolution of atoms O(5) and C(12), and especially the nonspherical nature of C(12), prevented an erroneous interpretation. Note that inclusion of the contribution of oxygen at the position of O(5) in F-3, in a totally erroneous position, actually produces a peak more than two-thirds of its proper height. Thus in the computation of the next stage, F-4, the atoms labeled O(5) and C(12) were omitted.

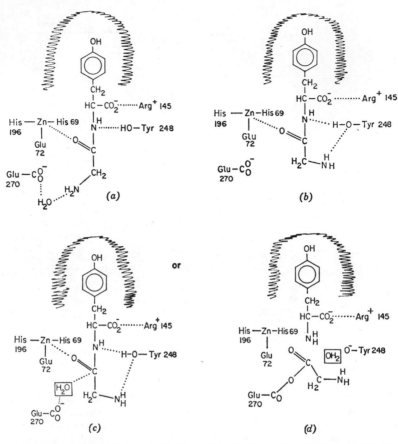

Figure 1.23 (*a*) Details of binding of Gly-Tyr. (*b*) Probable mode of binding of Gly-Tyr when carboxypeptidase A cleaves the peptide bond. (*c*) General base mode of attack by H_2O, promoted by Glu 270. (*d*) Nucleophilic attack by Glu 270 on the carbon atom of the carbonyl group of the substrate, followed by attack of H_2O. It is not certain whether the amino terminus of Gly-Tyr is NH_2 or NH_3^+.

Therefore it is a common, but not general, experience that in the latter stages of an investigation atoms entered into false positions appear at about two-thirds of their proper height, and atoms that have not yet been entered appear at only about one-third of their proper height especially for noncentrosymmetric structures. If the background of the maps is not sufficiently smooth for the deviations to be recognized, the interpretation of the maps is more difficult. Even at best the procedure is dangerous. Fortunately, there are other guideposts, as mentioned at the beginning of this section, and the method is powerful to the point of being very nearly certain in the hands of an experienced and careful investigator.

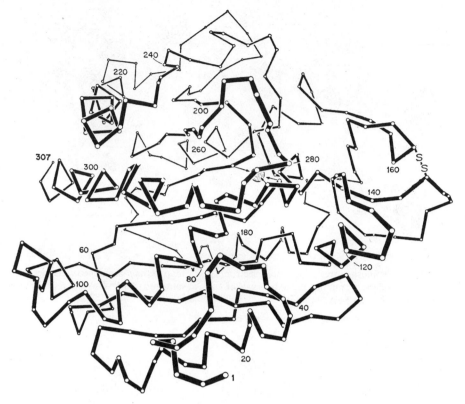

Figure 1.24   Backbone model, showing each peptide unit as a rod, and each C_α as a dot. The zinc atom is a shaded circle near the center, and arrows indicate the three ligands from the protein to zinc.

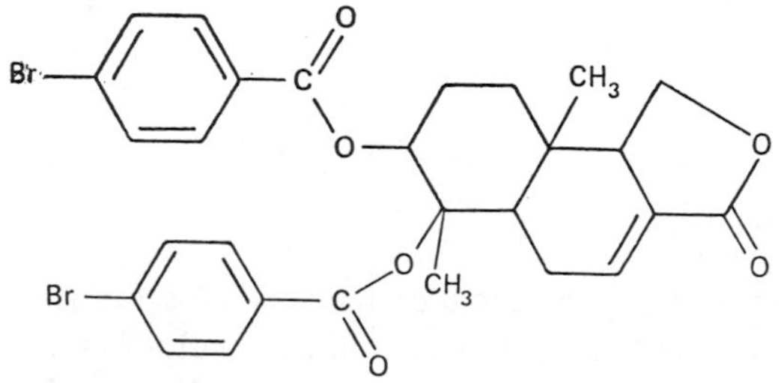

Figure 1.25   The *p*-bromobenzoate diester of the sesquiterpenoid iresin.

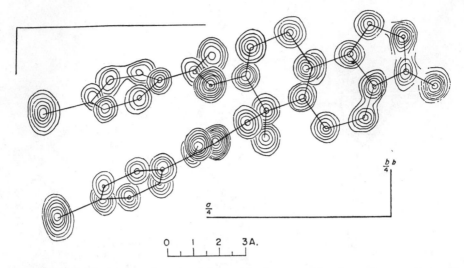

Figure 1.26   Composite three-dimensional electron density map of the *p*-bromobenzoate diester of iresin. Electron-density contours are at intervals of $1\,e\,\text{Å}^{-3}$ starting at $2\,e\,\text{Å}^{-3}$, except for bromine, for which intervals are $5\,e\,\text{Å}^{-3}$.

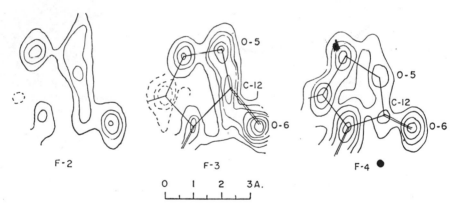

Figure 1.27   Successive stages in the refinement of the region of the five-membered ring. Solid contours are in the section $14/20\,c$, nearly the plane of the ring, while dashed contours are in the nearby section $13/20\,c$ along the z axis.

2   KINEMATIC THEORY OF X-RAY SCATTERING FROM CRYSTALS

In this section we outline the simpler aspects of scattering of X rays from crystals. When mathematics beyond ordinary calculus is required we outline it in some detail, and we often appeal to physical grounds in preference to mathematical rigor. The reader will find that this approach suffices to give a basic understanding of the process, since the assumptions of the kinematic theory—that X rays suffer no change in wavelength, that boundary effects and refractive index effects are negligible, and that multiple scattering effects are negligible—have generally proved to be quite adequate in the determination of crystal and molecular structures. When difficulties arise their effects usually are easy to circumvent experimentally or to take into account, within the accuracy that is usually desired. For a more detailed discussion of the topics in this section, see [60].

2.1   Mathematical Introduction

2.1.1   Vectors

A physical quantity that has both magnitude and direction can be represented by a vector. The length is taken as proportional to the magnitude. Addition of vectors **A** and **B** to give a resultant **C** is carried out by placing the tail of **B** on the head of **A**, or vice versa:

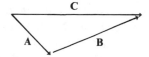

Multiplication of **A** by a scalar g gives a new vector g**A**, which has the same direction as **A** and has length $g|$**A**$|$ or g**A**, where $|$**A**$|$ or A represents the length of **A**. A unit vector can be defined by dividing **A** by its magnitude A. It is possible to define three mutually orthogonal unit vectors: **i** along x, **j** along y, and **k** along z. Let the components of **A** be A_x, A_y, and A_z; then $\mathbf{A} = \mathbf{i}A_x + \mathbf{j}A_y + \mathbf{k}A_z$. From the Pythagorean theorem, $|\mathbf{A}| = (A_x^2 + A_y^2 + A_z^2)^{1/2}$ is the length of **A**.

The addition of $\mathbf{A} + \mathbf{B} = \mathbf{C}$ then becomes $A_x + B_x = C_x$, $A_y + B_y = C_y$, and $A_z + B_z = C_z$ in components. Subtraction can be considered a special case of addition, if one defines $-\mathbf{A}$ as a vector with the same magnitude as **A** but with opposite direction. The associative law, $(\mathbf{A} + \mathbf{B}) + \mathbf{D} = \mathbf{A} + (\mathbf{B} + \mathbf{D})$; the commutative law, $\mathbf{A} + \mathbf{B} = \mathbf{B} + \mathbf{A}$; and the distributive law for scalar multiplication, $g(\mathbf{A} + \mathbf{B}) = g\mathbf{A} + g\mathbf{B}$, are proved easily.

The angle between **A** and **B** is defined as positive if measured from **A** to **B** when **A** and **B** are drawn from the same point. If Θ is the angle **A**, **B**, $-\Theta$ is the angle **B**, **A**.

2.1.2   The Scalar (Dot) Product

The scalar, or dot, product is represented by $\mathbf{A} \cdot \mathbf{B}$, and by definition it is equal to $AB \cos(\mathbf{A}, \mathbf{B})$. Geometrically this product is A times the projection of **B** on **A**. The reverse product $\mathbf{B} \cdot \mathbf{A} = BA \cos(\mathbf{B}, \mathbf{A})$; but the angle $(\mathbf{B}, \mathbf{A})$ is $-(\mathbf{A}, \mathbf{B})$ and $\cos(-\Theta) = \cos(\Theta)$, so that $\mathbf{A} \cdot \mathbf{B} = \mathbf{B} \cdot \mathbf{A}$. The reader can show geometrically that $\mathbf{A} \cdot (\mathbf{B} + \mathbf{C}) = \mathbf{A} \cdot \mathbf{B} + \mathbf{A} \cdot \mathbf{C}$, where **A**, **B**, and **C** are any three vectors. This is the distributive law.

Note that for our $\mathbf{i}, \mathbf{j}, \mathbf{k}$ set of orthogonal vectors $\mathbf{i} \cdot \mathbf{i} = \mathbf{j} \cdot \mathbf{j} = \mathbf{k} \cdot \mathbf{k} = 1$, and $\mathbf{i} \cdot \mathbf{j} = \mathbf{j} \cdot \mathbf{k} = \mathbf{k} \cdot \mathbf{i} = 0$. The distributive law then gives $\mathbf{A} \cdot \mathbf{B}$ in terms of components as $\mathbf{A} \cdot \mathbf{B} = (\mathbf{i}A_x + \mathbf{j}A_y + \mathbf{k}A_z) \cdot (\mathbf{i}B_x + \mathbf{j}B_y + \mathbf{k}B_z) = A_x B_x + A_y B_y + A_z B_z$. A special case is $A^2 = \mathbf{A} \cdot \mathbf{A} = A_x^2 + A_y^2 + A_z^2$, from which the length of $\mathbf{A}$ may be obtained if the components are known. A relation useful for calculating bond angles in an orthogonal system is

$$\cos(\mathbf{A}, \mathbf{B}) = \frac{\mathbf{A} \cdot \mathbf{B}}{AB} = \frac{A_x B_x + A_y B_y + A_z B_z}{(A_x^2 + A_y^2 + A_z^2)^{1/2}(B_x^2 + B_y^2 + B_z^2)^{1/2}}$$

The equation $\cos(\mathbf{A}, \mathbf{B}) = (\mathbf{A} \cdot \mathbf{B})/AB$ may also be expanded in nonorthogonal systems.

2.1.3   The Vector (Cross) Product

The vector, or cross, product is designated by $\mathbf{A} \times \mathbf{B}$ and is defined as $\mathbf{p}AB \sin(\mathbf{A}, \mathbf{B})$, where $\mathbf{p}$ is a unit vector perpendicular to the plane of $\mathbf{A}$ and $\mathbf{B}$, and $\mathbf{A}, \mathbf{B},$ and $\mathbf{p}$ form a right-handed system. First note that $AB \sin(\mathbf{A}, \mathbf{B})$ is the area of the parallelogram defined by $\mathbf{A}$ and $\mathbf{B}$. Note also since $\sin(\mathbf{A}, \mathbf{B}) = -\sin(\mathbf{B}, \mathbf{A})$, the cross product is not commutative; that is, $\mathbf{A} \times \mathbf{B} = -\mathbf{B} \times \mathbf{A}$. However, the cross product is distributive: $\mathbf{A} \times (\mathbf{B} + \mathbf{C}) = \mathbf{A} \times \mathbf{B} + \mathbf{A} \times \mathbf{C}$. An interesting way to restate these equations is in terms of a Cartesian system of unit vectors $\mathbf{i}, \mathbf{j},$ and $\mathbf{k}$. The definition then gives $\mathbf{i} \times \mathbf{i} = \mathbf{j} \times \mathbf{j} = \mathbf{k} \times \mathbf{k} = 0$, and $\mathbf{i} \times \mathbf{j} = \mathbf{k} = -\mathbf{j} \times \mathbf{i}$, $\mathbf{k} \times \mathbf{i} = \mathbf{j} = -\mathbf{i} \times \mathbf{k}$, $\mathbf{j} \times \mathbf{k} = \mathbf{i} = -\mathbf{k} \times \mathbf{j}$. The product $\mathbf{A} \times \mathbf{B}$ then becomes, for example,

$$\mathbf{A} \times \mathbf{B} = (\mathbf{i}A_x + \mathbf{j}A_y + \mathbf{k}A_z) \times (\mathbf{i}B_x + \mathbf{j}B_y + \mathbf{k}B_z)$$

$$= \mathbf{i}(A_y B_z - A_z B_y) + \mathbf{j}(A_z B_x - A_x B_z) + \mathbf{k}(A_x B_y - A_y B_x)$$

$$= \begin{bmatrix} \mathbf{i} & \mathbf{j} & \mathbf{k} \\ A_x & A_y & A_z \\ B_x & B_y & B_z \end{bmatrix}$$

The use of the unit vector $\mathbf{p}$ becomes obvious in later discussions. One important application is to three noncoplanar vectors, where the scalar

$$\mathbf{A} \cdot \mathbf{B} \times \mathbf{C} = \begin{bmatrix} A_x & A_y & A_z \\ B_x & B_y & B_z \\ C_x & C_y & C_z \end{bmatrix}$$

triple product is the volume of the parallelepiped defined by $\mathbf{A}, \mathbf{B},$ and $\mathbf{C}$ (Figure 1.28). This follows because $|\mathbf{B} \times \mathbf{C}|$ is the area of the base, and $\mathbf{A} \cdot \mathbf{p}$ is the altitude measured along a line perpendicular to the base. An expression for the volume of the unit cell in terms of the measured values of $a, b, c, \alpha, \beta,$ and γ can now readily

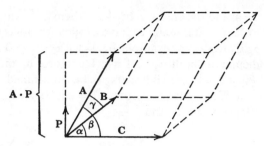

Figure 1.28   The volume of the parallelepiped defined by **A**, **B**, and **C** is $\mathbf{A} \cdot \mathbf{B} \times \mathbf{C}$.

be found as follows:

$$V^2 = (\mathbf{a} \cdot \mathbf{b} \times \mathbf{c})^2 = \begin{bmatrix} a_x & a_y & a_z \\ b_x & b_y & b_z \\ c_x & c_y & c_z \end{bmatrix} \begin{bmatrix} a_x & b_x & c_x \\ a_y & b_y & c_y \\ a_z & b_z & c_z \end{bmatrix}$$

$$= \begin{bmatrix} a_x^2 + a_y^2 + a_z^2 & a_x b_x + a_y b_y + a_z b_z & a_x c_x + a_y c_y + a_z c_z \\ a_x b_x + a_y b_y + a_z b_z & b_x^2 + b_y^2 + b_z^2 & b_x c_x + b_y c_y + b_z c_z \\ a_x c_x + a_y c_y + a_z c_z & b_x c_x + b_y c_y + b_z c_z & c_x^2 + c_y^2 + c_z^2 \end{bmatrix}$$

$$= \begin{bmatrix} \mathbf{a} \cdot \mathbf{a} & \mathbf{a} \cdot \mathbf{b} & \mathbf{a} \cdot \mathbf{c} \\ \mathbf{b} \cdot \mathbf{a} & \mathbf{b} \cdot \mathbf{b} & \mathbf{b} \cdot \mathbf{c} \\ \mathbf{c} \cdot \mathbf{a} & \mathbf{c} \cdot \mathbf{b} & \mathbf{c} \cdot \mathbf{c} \end{bmatrix}$$

$$= (\mathbf{a} \cdot \mathbf{b})(\mathbf{b} \cdot \mathbf{b})(\mathbf{c} \cdot \mathbf{c}) + (\mathbf{a} \cdot \mathbf{b})(\mathbf{b} \cdot \mathbf{c})(\mathbf{c} \cdot \mathbf{a}) + (\mathbf{a} \cdot \mathbf{c})(\mathbf{b} \cdot \mathbf{a})(\mathbf{c} \cdot \mathbf{b})$$

$$- (\mathbf{b} \cdot \mathbf{b})(\mathbf{a} \cdot \mathbf{c})^2 - (\mathbf{c} \cdot \mathbf{c})(\mathbf{a} \cdot \mathbf{b})^2 - (\mathbf{a} \cdot \mathbf{a})(\mathbf{c} \cdot \mathbf{b})^2$$

$$= a^2 b^2 c^2 (1 + 2\cos \alpha \cos \beta \cos \gamma - \cos^2 \alpha - \cos^2 \beta - \cos^2 \gamma)$$

The two properties of determinants used are that rows and columns can be interchanged without altering the value and that they multiply as indicated. Although a Cartesian system was used for the components of **a**, **b**, and **c**, the resulting dot products are independent of the choice of coordinate system, and the result has general validity.

2.1.4   Reciprocal Vectors

Let $\mathbf{a}_1$, $\mathbf{a}_2$, and $\mathbf{a}_3$ be three noncoplanar vectors. Later they define the unit cell of a crystal. Define three related vectors, called *reciprocal vectors*, by the equations $\mathbf{a}_i \cdot \mathbf{b}_j = \delta_{ij}$, where $\delta_{ij} = 1$ if $i = j$ and $\delta_{ij} = 0$ if $i \neq j$. Later the vectors $\mathbf{b}_j$ are used to describe the diffraction pattern of the crystal whose unit cell is defined by the three $\mathbf{a}_i$ terms.

We measure the a_i and the angles α_i between them (α_1 is the angle between a_2 and a_3, and so on). Hence, it is useful to have explicit formulas for the b_j in terms of the a_i and the α_i. Consider, for example, b_1. Since $a_2 \cdot b_1 = 0$ and $a_3 \cdot b_1 = 0$, b_1 must be perpendicular to both a_2 and a_3. Therefore, b_1 must be parallel to $a_2 \times a_3$, or $b_1 = \varepsilon(a_2 \times a_3)$, where ε is a scalar to be determined. But on taking the dot product on both sides with a_1, we obtain $1 = \varepsilon(a_1 \cdot a_2 \times a_3) = \varepsilon v_a$, where v_a is the cell volume. Hence $\varepsilon = 1/v_a$, and

$$b_1 = \frac{a_2 \times a_3}{v_a}, \qquad b_2 = \frac{a_3 \times a_1}{v_a}, \qquad b_3 = \frac{a_1 \times a_2}{v_a} \tag{12}$$

$$a_1 = \frac{b_2 \times b_3}{v_b}, \qquad a_2 = \frac{b_3 \times b_1}{v_b}, \qquad a_3 = \frac{b_1 \times b_2}{v_b} \tag{13}$$

which follow because the relations are symmetrical to a cyclic permutation, and the definition is symmetrical in a_i and b_j. The volumes v_a and v_b are reciprocally related ($v_a = 1/v_b$):

$$v_a v_b = \begin{bmatrix} a_{1x} a_{1y} a_{1z} \\ a_{2x} a_{2y} a_{2z} \\ a_{3x} a_{3y} a_{3z} \end{bmatrix} \begin{bmatrix} b_{1x} b_{2x} b_{3x} \\ b_{1y} b_{2y} b_{3y} \\ b_{1z} b_{2z} b_{3z} \end{bmatrix} = \begin{bmatrix} a_1 \cdot b_1 & a_1 \cdot b_2 & a_1 \cdot b_3 \\ a_2 \cdot b_1 & a_2 \cdot b_2 & a_2 \cdot b_3 \\ a_3 \cdot b_1 & a_3 \cdot b_2 & a_3 \cdot b_3 \end{bmatrix} = \begin{bmatrix} 1 & 0 & 0 \\ 0 & 1 & 0 \\ 0 & 0 & 1 \end{bmatrix} = 1$$

A geometrical interpretation is very helpful. Let $\alpha_1 = \alpha_2 = \alpha_3 = 90°$, and $a_1 < a_2 < a_3$ (an orthorhombic system). Then $b_1 = 1/a_1$, $b_2 = 1/a_2$, $b_3 = 1/a_3$, $b_1 \| a_1$, $b_2 \| a_2$, $b_3 \| a_3$, $v_a = a_1 a_2 a_3$, and $v_b = 1/a_1 a_2 a_3$ (Figure 1.29). Now consider the monoclinic system, where $\alpha_1 = \alpha_3 = 90°$, and choose $\alpha_2 = 110°$. Again let $a_1 < a_2 < a_3$ (Figure 1.30). The equations between the a_i and the b_j are

$$b_1 = \frac{1}{a_1 \sin \alpha_2}, \qquad b_2 = \frac{1}{a_2}, \qquad b_3 = \frac{1}{a_3 \sin \alpha_2}$$

and $v_a = a_1 a_2 a_3 \sin \alpha_2$. Thus the lengths of the b_j are not always the simple reciprocals of the a_i.

We make use of these geometrical interpretations later. Meanwhile let us consider another application of reciprocal vectors—the evaluation of the product $a_1 \times (a_2 \times a_3)$. The result of this operation, by the definitions of the cross

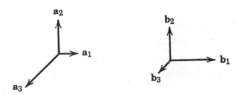

Figure 1.29 Relations between crystal vectors **a** and reciprocal vectors **b** in the orthorhombic system. Although the origins are separated here, it is sometimes useful to regard them as common. If the a_i are measured in angstroms, the b_j are measured in reciprocal angstroms (Å^{-1}).

Figure 1.30 Relations between crystal vectors a_i and reciprocal vectors b_j in the monoclinic system. $\beta_2 = 180° - \alpha_2$.

product, must be a vector in a_1, a_2, and a_3 space. Therefore,

$$a_1 \times (a_2 \times a_3) = la_1 + ma_2 + na_3$$

Now taking the dot product of both sides with b_1, b_2, and b_3, we obtain the three equations

$$l = b_1 \cdot a_1 \times (a_2 \times a_3)$$

$$m = b_2 \cdot a_1 \times (a_2 \times a_3)$$

$$n = b_3 \cdot a_1 \times (a_2 \times a_3)$$

Recognizing that $a_2 \times a_3 = v_a b_1$

$$l = (b_1 \cdot a_1 \times b_1)v_a$$

$$m = (b_2 \cdot a_1 \times b_1)v_a$$

$$n = (b_3 \cdot a_1 \times b_1)v_a$$

Rearranging,

$$l = (a_1 \cdot b_1 \times b_1)v_a = 0$$

$$m = (a_1 \cdot b_1 \times b_2)v_a = (a_1 \cdot a_3)v_a v_b = (a_1 \cdot a_3)$$

$$n = (a_1 \cdot b_1 \times b_3)v_a = -(a_1 \cdot a_2)$$

Therefore,

$$a_1 \times (a_2 \times a_3) = (a_1 \cdot a_3)a_2 - (a_1 \cdot a_2)a_3 \qquad (14)$$

A number of further relations are now easily proved. For example, $(A \times B) \cdot (C \times D) = (C \cdot A)(D \cdot B) - (C \cdot B)(A \cdot D)$ and $(A \times B) \times (C \times D) = C(D \cdot A \times B) - D(C \cdot A \times B)$. Interchange of the dot and the cross is possible: $A \cdot B \times C = A \times B \cdot C$. The useful relation

$$\cos \beta_3 = \frac{\cos \alpha_1 \cos \alpha_2 - \cos \alpha_3}{\sin \alpha_1 \sin \alpha_2}$$

may be proved by expressing all the $\mathbf{b}_j$ in $\mathbf{b}_1 \cdot \mathbf{b}_2 = (\mathbf{b}_1 \cdot \mathbf{b}_1)^{1/2}(\mathbf{b}_2 \cdot \mathbf{b}_2)^{1/2} \cos \beta_3$ in terms of the $\mathbf{a}_i$ and then making use of the preceding equations. Finally, the expression of a vector $\mathbf{U}$ in terms of either the $\mathbf{a}_i$ or $\mathbf{b}_j$ is sometimes useful. First write $\mathbf{U} = r\mathbf{a}_1 + s\mathbf{a}_2 + t\mathbf{a}_3$; then form the products $\mathbf{b}_1 \cdot \mathbf{U} = r$, and so on. The result is

$$\mathbf{U} = (\mathbf{b}_1 \cdot \mathbf{U})\mathbf{a}_1 + (\mathbf{b}_2 \cdot \mathbf{U})\mathbf{a}_2 + (\mathbf{b}_3 \cdot \mathbf{U})\mathbf{a}_3$$

$$= (\mathbf{a}_1 \cdot \mathbf{U})\mathbf{b}_1 + (\mathbf{a}_2 \cdot \mathbf{U})\mathbf{b}_2 + (\mathbf{a}_3 \cdot \mathbf{U})\mathbf{b}_3 \tag{15}$$

2.1.5   Complex Numbers

A complex number $Z = A + iB$, where $i = \sqrt{-1}$, is like a vector in two dimensions. The components A and B can be plotted in an orthogonal system, in which the unit vectors are 1 along the horizontal axis and i along the vertical axis (Figure 1.31). The real and the imaginary parts are treated like components of a vector in addition: $Z_1 + Z_2 = A_1 + A_2 + i(B_1 + B_2)$. The magnitude of a complex number is the length

$$|Z| = (ZZ^*)^{1/2} = [(A + iB)(A - iB)]^{1/2} = (A^2 + B^2)^{1/2}$$

The polar representation is also useful. Define $R = |Z|$, and Θ by the equations $\sin \Theta = B/R$ and $\cos \Theta = A/R$. Then $Z = A + iB = R(\cos \Theta + i \sin \Theta)$, which can be written more conveniently as $Re^{i\Theta}$ from Euler's formula. To prove this, define the function $f(\Theta)$ to be determined as $f(\Theta) = \cos \Theta + i \sin \Theta$. Differentiation then gives $df(\Theta)/d\Theta = -\sin \Theta + i \cos \Theta$. Comparison with $f(\Theta)$ then shows that $df(\Theta)/d\Theta = if(\Theta)$, or $df(\Theta)/f(\Theta) = id\Theta$, which on integration gives $\ln f(\Theta) = i\Theta + \ln K$, where K is a constant. But, when $\Theta = 0$, $f(\Theta) = 1$; hence $K = 1$, and $f(\Theta) = e^{i\Theta}$, which is $\cos \Theta + i \sin \Theta$. The complex conjugate is $e^{-i\Theta} = \cos \Theta - i \sin \Theta$. Note that the magnitude of $e^{i\Theta}$ is $e^{i\Theta}e^{-i\Theta} = 1 = \cos^2 \Theta + \sin^2 \Theta$, so that the points on the plane of Figure 1.31 representing $e^{i\Theta}$ lie on a circle of radius 1. Later it will be useful to remember this circle when expressions such as $\exp(2\pi ih/4)$ are encountered. When h is an integer, $0, 1, 2, \ldots$, this expression has the values $1, i, -i$, and so on.

2.1.6   Matrices

It is sometimes convenient to represent operators and other quantities in terms of matrices, rectangular arrays of dimension $m \times n$,

$$\begin{bmatrix} a_{11} & a_{12} & a_{13} & \cdots & a_{1n} \\ a_{21} & a_{22} & & \cdots & a_{2n} \\ a_{m1} & a_{m2} & & \cdots & a_{mn} \end{bmatrix}$$

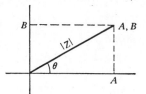

Figure 1.31   The complex number $Z = A + iB$.

For example, the twofold operation in the y direction (x, y, z transforms to $-x$, $y, -z$) can be represented by the matrix $(-1, 0, 0/0, 1, 0/0, 0, -1)$. Matrices can be multiplied if the number of columns in the first is equal to the number of rows in the second. If

$$\mathbf{A} = \begin{bmatrix} a_{11} & a_{12} & a_{13} \\ a_{21} & a_{22} & a_{23} \end{bmatrix}$$

and

$$\mathbf{B} = \begin{bmatrix} b_{11} & b_{12} & b_{13} \\ b_{21} & b_{22} & b_{23} \\ b_{31} & b_{32} & b_{33} \end{bmatrix}$$

then

$$\mathbf{AB} = \begin{bmatrix} a_{11}b_{11} + a_{12}b_{21} + a_{13}b_{31} & a_{11}b_{12} + a_{12}b_{22} + a_{13}b_{32} \\ a_{21}b_{11} + a_{22}b_{21} + a_{23}b_{31} & a_{21}b_{12} + a_{22}b_{22} + a_{23}b_{32} \end{bmatrix}$$
$$\begin{matrix} a_{11}b_{13} + a_{12}b_{23} + a_{13}b_3 \\ a_{21}b_{13} + a_{22}b_{23} + a_{23}b_3 \end{matrix}$$

In the final matrix the number of rows is equal to that found in the first matrix, and the number of columns is equal to that in the second. Also $\mathbf{AB} \neq \mathbf{BA}$; in fact multiplication is not defined in the latter case since the number of columns of $\mathbf{B}$ do not equal the number of rows of $\mathbf{A}$.

Square matrices can be inverted if their determinants are nonzero. Hence, if $\mathbf{A}$ were to represent an $n \times n$ coefficient matrix and $\mathbf{X}$ and $\mathbf{B}$ represent $1 \times n$ column matrices such that $\mathbf{AX} = \mathbf{B}$, then, if the determinant of $\mathbf{A}$, $|\mathbf{A}| \neq 0$, $\mathbf{X} = \mathbf{A}^{-1}\mathbf{B}$, where $\mathbf{A}^{-1}\mathbf{A} = 1$.

2.1.7   Fourier Series

A function $f(x)$, reasonably well-behaved (discontinuous at a finite number of points) and periodic with interval p, can be expressed as

$$f(x) = \sum_{n=-\infty}^{\infty} a_n \exp\left(\frac{2\pi i n x}{p}\right) \tag{16}$$

where

$$a_n = \frac{1}{p} \int_{-p/2}^{p/2} f(x) \exp\left(\frac{-2\pi i n x}{p}\right) dx$$

Extending p to infinity and considering functions that are small for numerically large values of x, we obtain

$$f(x) = \int_{-\infty}^{\infty} F(t) \exp(2\pi i t x)\, dt$$

$$F(t) = \int_{-\infty}^{\infty} f(x) \exp(-2\pi i t x)\, dx$$

The function $F(t)$ is called the Fourier transform of $f(x)$. If x is measured in units of length, then t must have units of 1/length, since tx must be a unitless quantity.
Generalizing to three dimensions, we obtain

$$f(\mathbf{r}) = \int F(\mathbf{h}) \exp(2\pi i \mathbf{h} \cdot \mathbf{r})\, dv_b \qquad \text{(17a)}$$

$$F(\mathbf{h}) = \int f(\mathbf{r}) \exp(-2\pi i \mathbf{h} \cdot \mathbf{r})\, dv_a \qquad \text{(17b)}$$

where $\mathbf{r} = x\mathbf{a}_1 + y\mathbf{a}_2 + z\mathbf{a}_3$, expressed in terms of three noncoplanar vectors $\mathbf{a}_1$, $\mathbf{a}_2$, and $\mathbf{a}_3$. Since $\mathbf{h}$ must be a vector of dimension (length)$^{-1}$, it can then be expressed in terms of components of the $\mathbf{b}_1$, $\mathbf{b}_2$, and $\mathbf{b}_3$ vectors as follows:

$$\mathbf{h} = h\mathbf{b}_1 + k\mathbf{b}_2 + l\mathbf{b}_3$$

Thus

$$f(\mathbf{r}) = \iiint F(\mathbf{h}) \exp[2\pi i(hx + ky + lz)]v_b\, dh\, dk\, dl \qquad \text{(18a)}$$

$$F(\mathbf{h}) = \iiint f(\mathbf{r}) \exp[-2\pi i(hx + ky + lz)]v_a\, dx\, dy\, dz \qquad \text{(18b)}$$

where v_a and v_b are the volumes in real and reciprocal space, respectively. If $f(\mathbf{r})$ is real, then $F(\mathbf{h}) = F^*(-\mathbf{h})$.
Another characteristic of Fourier series is worth noting. A convolution of two functions fg given by

$$s(y) = \int g(x)f(x + y)\, dx$$

can be thought of as a summation of products of the two functions with one shifted by an amount y relative to the other. The function f could, for example, be the experimental spectrum as obtained from an isolated peak and the function g, a function made up of deltalike contributions corresponding to the theoretical spectrum. Substituting into the expressions given above for Fourier series, one can readily obtain

$$s(y) = \int F(t)G(-t)\exp(2\pi ity)\,dt$$

Since it is possible to obtain Fourier coefficients directly from the experimental function representing the observed spectrum, that is, $s(y)$, it is often possible to factor out the aberrations caused by the experimental apparatus and obtain a better representation of the deltalike theoretical spectrum.

2.2   Superposition of Waves

Consider the function $E = E_o \cos 2\pi(\alpha x - \beta t + \delta)$, where E_o, α, β, and δ are constants. At $t = 0$, $E_{t=0} = E_o \cos 2\pi(\alpha x + \delta)$, which is a function that repeats itself in space when the argument $(\alpha x + \delta) = 0$, 1, 2, and so on; that is, when $x = -\delta/\alpha$, $-\delta/\alpha + 1/\alpha$, $-\delta/\alpha + 2/\alpha$, and so on. The distance between successive crests (Figure 1.32) is thus $1/\alpha$, and is defined as the wavelength $\lambda = 1/\alpha$. At a later time t_1 the wave looks just the same except that it has moved along the x axis by an amount $\beta t_1/\alpha$ (Figure 1.32b). In other words, maxima occur for $E_{t=t_1}$ when $\alpha x - \beta t_1 + \delta = 0$, 1, 2,...; that is, when $x = -\delta/\alpha + \beta t_1/\alpha$, $-\delta/\alpha + 1/\alpha + \beta t_1/\alpha$, $-\delta/\alpha + 2/\alpha + \beta t_1/\alpha$, and so on. The wave velocity u is the distance $\beta t_1/\alpha$ that any given crest has moved divided by the time t_1; hence, $u = \beta/\alpha$; and since $\lambda = 1/\alpha$, we find $\beta = u/\lambda$. But this is the frequency, $v = u/\lambda$, the number of waves per unit of time. We can then write the wave in the form

$$E = E_o \cos 2\pi(x/\lambda - vt + \delta)$$

where the velocity $u = \lambda v$. This idealization will do for the usual crystallographic studies in which refractive indices can be assumed to be unity within one part in

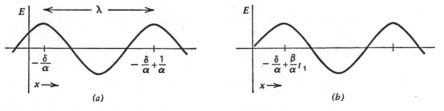

Figure 1.32   (a) The wave $E_{t=0} = E_o \cos 2\pi(\alpha x + \delta)$ at $t = 0$. (b) The wave $E_{t=t_1} = E_o \cos 2\pi(\alpha x - \beta t_1 + \delta)$ at a slightly later time t_1.

10^5, boundary effects are neglected, coherent attenuation of the wave in the crystal is neglected, and a strictly monochromatic wave is assumed.

The scattering between two different electrons at different points in space (different δ values) is then obtained by adding the amplitudes of scattering from each. The result is a simple trigonometric function. First expand

$$E = E_1 + E_2 = E_{o_1} \cos 2\pi(x/\lambda - vt + \delta_1) + E_{o_2} \cos 2\pi(x/\lambda - vt + \delta_2)$$

$$= (E_{o_1} \cos 2\pi\delta_1 + E_{o_2} \cos 2\pi\delta_2) \cos 2\pi(x/\lambda - vt)$$

$$- (E_{o_1} \sin 2\pi\delta_1 + E_{o_2} \sin 2\pi\delta_2) \sin 2\pi(x/\lambda - vt)$$

If we define $E_o' \cos 2\pi\delta_3 = E_{o_1} \cos 2\pi\delta_1 + E_{o_2} \cos 2\pi\delta_2$, and $E_o' \sin 2\pi\delta_3 = E_{o_1} \sin 2\pi\delta_1 + E_{o_2} \sin 2\pi\delta_2$, we find $E_o' \cos 2\pi(x/\lambda - vt + \delta_3)$. We can also find E_o' and δ_3 from the defining equations

$$(E_o')^2 = E_{o_1}^2 + E_{o_2}^2 + 2E_{o_1}E_{o_2} \cos 2\pi(\delta_1 - \delta_2)$$

Thus the amplitudes are additive, but the intensities, which are proportional to the square of the amplitudes, are not additive in general and depend in a sensitive and important way on the constants δ_1 and δ_2, which we later relate to the positions of the scattering centers in space. {Proof that the total energy P carried through an area A normal to the direction of propagation of a wave $E = E_0 \cos 2\pi(x/\lambda - vt + \delta)$ is proportional to E_o^2 follows. From electrodynamics we write (in vacuum) $P = (8\pi)^{-1} \int (E^2 + H^2) \, dv = (4\pi)^{-1} \int E^2 \, dv = (4\pi)^{-1} \int_0^{t_1} E_o^2 [\cos 2\pi(x/\lambda - vt + \delta)] A u \, dt = E_o^2 \, Aut_1/(8\pi)$, plus a trigonometric term, $-(1/4\pi v)[\sin 4\pi(x/\lambda - vt_1 + \delta) - \sin 4\pi(x/\lambda + \delta)]$, which is negligible if $t_1 \gg 1/v$.}

The equation that demonstrates the effect of interference on the intensities can be derived much more briefly with the equivalent use of complex waves. Represent the wave by the equation

$$E = E_o \exp[2\pi i(x/\lambda - vt + \delta)] \tag{19}$$

the real part of which is the same as we had previously. Multiplication by the complex conjugate gives E_o^2, the square of the amplitude constant. The two superimposed waves are

$$E = E_1 + E_2 = E_{o_1} \exp[2\pi i(x/\lambda - vt + \delta_1)] + E_{o_2} \exp[2\pi i(x/\lambda - vt + \delta_2)]$$

The intensity is then, aside from a proportionality constant,

$$EE^* = E_{o_1}^2 + E_{o_2}^2 + 2E_{o_1}E_{o_2} \cos 2\pi(\delta_1 - \delta_2)$$

where we have used the relations $2 \cos \Theta = e^{i\Theta} + e^{-i\Theta}$ and $\cos \Theta = \cos(-\Theta)$.

2.3   Scattering of X Rays by a Single Electron

The periodically varying electric field of a light wave accelerates the electron into a periodic motion; hence, it emits an electromagnetic wave in all directions, except precisely along the direction of its acceleration. The necessity for the electric vector of a light wave to be perpendicular to the direction of propagation causes the scattered wave to be polarized (Figure 1.33). Consider an initially unpolarized wave incident along s_o, and resolve all electric vectors into components along $E_\perp^i$ or $E_\parallel^i$. Since these two directions are equally probable in an unpolarized wave, $(E_\perp^i)^2 = (E_\parallel^i)^2$. In the scattering process the amplitude in any given direction is reduced. At zero angle of scattering the two directions ($\perp$ and $\parallel$) for E, the amplitude of the scattered wave, are equally probable; so, $(E_\perp)^2 = (E_\parallel)^2$. But when scattering at an angle ψ occurs, $E_\parallel$ is reduced to $E_\parallel \cos \psi$, the component of $E_\parallel$, which is perpendicular to the direction of propagation. Hence,

$$P = \frac{\text{intensity of scattering at angle } \psi}{\text{intensity of scattering at angle } 0°} = \frac{E_\perp^2 + (E_\parallel \cos \psi)^2}{E_\perp^2 + E_\parallel^2}$$

But $E_\perp^2 = E_\parallel^2$; therefore,

$$P = \frac{1 + \cos^2\psi}{2} \tag{20a}$$

is the polarization factor, the factor by which the density of scattering is reduced by polarization of part of the scattered radiation. Later we show that $\psi = 2\Theta$, where Θ is the Bragg angle. If the incident beam were first monochromatized by scattering off a monochromating crystal, then assuming this scattering were in the same plane as that of the sample, the incident $E_\parallel$ becomes $E_\parallel \cos \psi_M$, $\psi_M(=2\Theta_M)$ being the scattering angle of the monochromator, and

$$P = \frac{1 + \cos^2\psi \cos^2\psi_M}{1 + \cos^2\psi_M} \tag{20b}$$

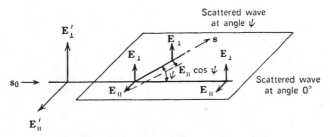

Figure 1.33   Polarization of wave scattered at angle ψ.

In the complete expression for the intensity I_e of a wave I scattered by a single electron at the origin,

$$I_e = I_o \frac{e^4}{r^2 m^2 c^4} \left(\frac{1 + \cos^2 \psi}{2} \right)$$

only the polarization factor is of use to us later. The expected inverse square dependence does not appear, since we integrate the total intensity over the area of a reflection in Section 2.4. The physical constants e, m, and c, the charge and mass of the electron and the velocity of light, are not usually needed because the scale of the diffracted maxima is ordinarily obtained from a comparison of observed and calculated amplitudes of scattering from the crystal, as is described later. In summary, we show that the integrated intensity of scattering I_{hkl} in a maximum characterized by the numbers h, k, and l is

$$I_{hkl} = kLPT^2 |F_{hkl}|^2 \tag{21}$$

where k is the scale factor; P is the polarization factor; L is the Lorentz factor, which arises because of the integration of the intensity over the maximum; T is the temperature factor, $\exp[-B(\sin^2\Theta/\lambda^2)]$, where B is determined empirically; and $|F_{hkl}|^2$ is calculated from the expression

$$F_{hkl} = \sum_{\substack{\text{unit} \\ \text{cell}}} f_j \exp[-2\pi i(hx_j + ky_j + lz_j)] \tag{22a}$$

We now derive these expressions and define the symbols.

2.4   The Atomic Scattering Factor f and the Structure Factor F_{hkl}

Define F_{hkl} as the ratio of the amplitude of scattering by the contents of the unit cell to the amplitude of scattering by a single electron at the origin of the unit cell.

Consider a wave (Figure 1.34a) incident upon an electron at the origin. The scattered wave has amplitude $E = E_o \exp[2\pi i(|R|/\lambda - vt + \delta)]$, where $|R| = \mathbf{R} \cdot \mathbf{s}$

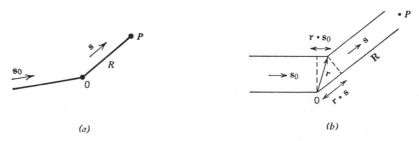

(a)   (b)

Figure 1.34   (a) The unit vectors s_0 and s represent directions of the incident and scattered waves, respectively. (b) Scattering from a point not at the origin. The point of observation P is very far away compared with the distance r.

is the distance from the origin to the point of observation P. If there are n electrons, then

$$E = E_o \exp[2\pi i(|R|/\lambda - vt)][\exp(2\pi i \delta_1)$$
$$+ \exp(2\pi i \delta_2) + \cdots + \exp[(2\pi i \delta_n)]$$

where $\delta_1, \delta_2, \delta_3, \ldots$ are the phases of the n-component waves. If $\mathbf{r}_j$ is drawn from electron 1 at the origin to electron j, then the difference in phase, $\delta_j - \delta_1$, is given by $(\mathbf{r}_j \cdot \mathbf{s}_o - \mathbf{r}_j \cdot \mathbf{s})/\lambda$, as shown in Figure 1.34b. Thus,

$$E = E_o \exp[2\pi i(|R|/\lambda_o - vt + \delta_1)]F'$$

where

$$F' = 1 + \exp[-2\pi i \mathbf{r}_2 \cdot (\mathbf{s} - \mathbf{s}_o)/\lambda] + \cdots + \exp[-2\pi i \mathbf{r}_n \cdot (\mathbf{s} - \mathbf{s}_o)/\lambda]$$

But this expression is just a special form of a Fourier series. More generally,

$$F' = \int_V \rho(\mathbf{r}) \exp[-2\pi i \mathbf{r} \cdot (\mathbf{s} - \mathbf{s}_o)/\lambda] \, dV \tag{23}$$

where $\rho(\mathbf{r})$ is the electron density.

Let Θ be any angle between $\mathbf{r}$ and $\mathbf{s} - \mathbf{s}_o$. Then the volume can be expressed in spherical coordinates as

$$dV = r^2 \sin \Theta \, dr \, d\Theta \, d\phi$$

Also,

$$(\mathbf{s} - \mathbf{s}_o)^2 = (\mathbf{s}^2 + \mathbf{s}_o^2 - 2|\mathbf{s}| \cdot |\mathbf{s}_o| \cos \psi) = 4(\sin \psi/2)^2$$

since both $\mathbf{s}$ and $\mathbf{s}_o$ are unit vectors, where ψ is the angle between $\mathbf{s}$ and $\mathbf{s}_o$. Therefore

$$F'(\psi) = \int_V \rho(\mathbf{r}) \exp(ikr \cos \Theta) r^2 \sin \Theta \, dr \, d\Theta \, d\phi \tag{24}$$

where

$$k = \frac{4\pi \sin \psi/2}{\lambda}$$

We make no assumptions yet as to the state of the material giving rise to the diffraction; hence, this expression is perfectly general.

Now, if $\rho(\mathbf{r})$ is a function of $\mathbf{r}$ only,

$$F'(\psi) = 4\pi \int_{-\infty}^{\infty} \rho(r)r^2 \frac{\sin kr}{kr}\, dr \tag{25}$$

and is the appropriate expression to use for diffraction of gases or most liquids.
For a crystal

$$\mathbf{r} = (x + p)\mathbf{a}_1 + (y + m)\mathbf{a}_2 + (z + n)\mathbf{a}_3$$

where p, m, and n are integers. Since the dot product of $\mathbf{r}$ and $(\mathbf{s} - \mathbf{s}_o)/\lambda$ must be a dimensionless quantity, the latter must be expressible in terms of any three basis vectors in the reciprocal space, hence, in terms of $\mathbf{b}_1$, $\mathbf{b}_2$, and $\mathbf{b}_3$:

$$(\mathbf{s} - \mathbf{s}_o)/\lambda = h\mathbf{b}_1 + k\mathbf{b}_2 + l\mathbf{b}_3 = \mathbf{h} \tag{26}$$

Therefore,

$$\mathbf{r} \cdot (\mathbf{s} - \mathbf{s}_o)/\lambda = hx + ky + lz + hp + km + ln$$

and

$$F'(\psi) = \sum_{p,m,n} \int_{\substack{\text{unit} \\ \text{cell}}} \rho(\mathbf{r}) \exp[-2\pi i(hx + ky + lz + hp + km + ln)]\, dV$$

$$= \sum_{p,m,n} \exp[-2\pi i(hp + km + ln)] \int_V \rho(\mathbf{r}) \exp[-2\pi i(hx + ky + lz)]\, dV$$

where the volume is now that of the unit cell. If the special case of one electron at the origin of each unit cell is considered, the equation reduces to

$$g(\psi) = \sum_p \sum_m \sum_n \exp[-2\pi i(hp + km + ln)]$$

$$= \left[\sum_{p=0}^{N_1} \exp(-2\pi ihp)\right]\left[\sum_{m=0}^{N_2} \exp(-2\pi ikm)\right]\left[\sum_{n=0}^{N_3} \exp(-2\pi iln)\right]$$

where N_1, N_2, and N_3 are the number of unit cells in each of the three directions, respectively.

Diffraction maxima occur whenever the phase difference is equal to an integer. Therefore,

$$hp + km + ln = \text{integer}$$

for all values of p, m, and n, but since p, m, and n must be integers, it follows that h, k, and l must also be integers.

We have shown that the function g can have maxima only at integer values of h, k, and l. Next, let us consider the width of these diffraction maxima. The sum of the series can be calculated as

$$\sum_{p=0}^{N_1} \exp(-2\pi i h p) = \frac{1 - \exp(-2\pi i N_1 h)}{1 - \exp(-2\pi i h)}$$

when $h \neq$ integer. However, when $h =$ integer, the series sum equals $N_1 + 1$.

Expressing nonintegral values of h as $h' + \Delta h$, where h' is an integer, the ratio of the intensity for a general value of h to that at diffraction maxima is

$$\frac{I(h)}{I(h')} = \frac{2}{(N_1 + 1)^2 (1 - \cos 2\pi\Delta h)}$$

Thus for $N_1 = 1000$, the ratio of the intensity at $\Delta h = 0.01$ is approximately 0.001 of the value at the maximum of the crystal peak. It is apparent that diffraction maxima for crystals are very sharp indeed. Because of the mosaic character of crystals and because of experimental conditions, maxima broaden slightly, and consequently integrated intensities are ordinarily used.

The general expression for $F'(\psi)$ can now be written as

$$F'(\psi) = g\left(\int_V \rho(\mathbf{r}) \exp[-2\pi i(hx + ky + lz)] \, dV \right)$$

and $F'(\psi)$ can have nonzero values only where h, k, and l are integers, since g has nonzero values only at these points. Consequently, if we divide $F'(\psi)$ by g, the scattering caused by one electron at the origin of each cell, we obtain

$$F_{hkl} = \frac{F'(\psi)}{g} = \int_V \rho(\mathbf{r}) \exp[-2\pi i(hx + ky + lz)] \, dV \qquad (22b)$$

where F_{hkl} is called the structure factor. Note that the intensities at these hkl points will be proportional to $|F_{hkl}|^2$ and thus depend on the arrangement of electrons, and hence atoms in the unit cell, but the positions of the maxima will be determined by $\mathbf{h}$ and therefore by the reciprocal vectors $\mathbf{b}_1$, $\mathbf{b}_2$, and $\mathbf{b}_3$.

The expression for the structure factor can be modified somewhat by substituting

$$\rho(\mathbf{r}) = \sum_{j=1}^{N} \rho_j(\mathbf{r}' + \mathbf{r}_j)$$

where ρ_j is the electron density associated with atom j, $\mathbf{r}'$ is measured from the atom center, and $\mathbf{r}_j$ denotes the position of atom j. Thus

$$F_{hkl} = \sum_{j=1}^{N} \int_{atom} \rho_j(\mathbf{r}') \exp\{-2\pi i[h(x_j + x') + k(y_j + y') + l(z_j + z')]\}\, dV'$$

$$= \sum_{j=1}^{N} \exp[-2\pi i(hx_j + ky_j + lz_j)] \int_{atom} \rho_j(\mathbf{r}') \exp[-2\pi i(hx' + ky' + lz')]\, dV'$$

Furthermore, if the assumption is made that in the molecules $\rho_j(\mathbf{r}')$ is essentially the same as for the free atom, then this integral can be evaluated once an approximate wave function ψ is known from quantum mechanics, since $\rho(\mathbf{r}') = \psi\psi^*$.

Defining f, the atomic scattering factor, as

$$f = \int_{atom} \rho(\mathbf{r}') \exp[-2\pi i(hx' + ky' + lz')]\, dV'$$

$$= \int_{atom} \rho(\mathbf{r}') \exp(ikr' \cos \Theta)\, dr'\, d\Theta\, d\phi r'^2 \sin \Theta \tag{27}$$

the structure can now be written as

$$F_{hkl} = \sum_{j=1}^{N} f_j \exp[-2\pi i(hx_j + ky_j + lz_j)] \tag{22a}$$

where x_j, y_j, and z_j are now the coordinates of the atom centers, and the summation is over the N atoms in the unit cell.

Let us assume that $\rho(x, y, z)$ is known from solutions of the Schröedinger wave equation for many electron atoms and that it is a function of $\mathbf{r}$ only; in other words, that the atoms are spherically symmetrical. Then $\rho(x, y, z) = \rho(\mathbf{r})$, and we can integrate over Θ and ϕ in polar coordinates in (27), to obtain

$$f = 4\pi \int_0^\infty r^2 \rho(\mathbf{r}) \frac{\sin kr}{kr}\, dr \tag{28}$$

As an example consider the evaluation of the scattering factor for the hydrogen atom. From the solution of the Schröedinger wave equation, $\psi = (\pi a^3)^{-1/2} e^{-r/a}$ for the lowest electronic state, where $a = 0.53$ Å is the Bohr radius. The value of $\rho(r)$ is then $(\pi a^3)^{-1} e^{-2r/a}$, which, when substituted into (28), integrates into

$$f = 1 + \left(\frac{k^2 a^2}{4}\right)^{-2}$$

A graph of f as a function of $k = (4\pi/\lambda) \sin \psi/2$ is shown in Figure 1.35. The shape of this curve is representative of atomic scattering factors in general. Note that at zero angle of scattering all parts of the electronic cloud scatter in phase;

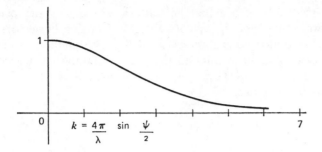

Figure 1.35 Atomic scattering factor for hydrogen.

therefore, for all atoms the value of f at $k = 0$ is just the number of electrons. As the angle of scattering increases, interference begins to occur between different parts of the electron cloud; hence, the atomic scattering factor falls off with angle. Values of f for atoms and many ions are listed in the *International Tables for X-Ray Crystallography* [61, 62].

2.5   The Reciprocal Lattice Vector *h*

We have stated that the vector $\mathbf{h}$ is defined by $\mathbf{h} = h\mathbf{b}_1 + k\mathbf{b}_2 + l\mathbf{b}_3$ and have shown that $\mathbf{h} = (\mathbf{s} - \mathbf{s}_o)/\lambda$. The latter equation is called the Laue equation and contains the geometrical information of X-ray diffraction photographs. Currently we know only that h, k, and l are the orders of the X-ray diffraction maxima. Proof of their relation to the planes into which a crystal can be resolved follows.

Assume that we have a plane passing through the unit cell with intercepts along the three axes given by $X_1 = \mathbf{a}_1/h$, $X_2 = \mathbf{a}_2/k$, $X_3 = \mathbf{a}_3/l$. We now prove that the vector $\mathbf{h}$ is perpendicular to this plane. The vectors $X_3 - X_1$ and $X_3 - X_2$ must be parallel to this plane; also the vector $(X_3 - X_1) \times (X_3 - X_2)$ must by definition of the cross product be perpendicular to the plane:

$$(X_3 - X_1) \times (X_3 - X_2) = (\mathbf{a}_3/l - \mathbf{a}_1/h) \times (\mathbf{a}_3/l - \mathbf{a}_2/k)$$

$$= -\mathbf{a}_3 \times \mathbf{a}_2/kl - \mathbf{a}_1 \times \mathbf{a}_3/hl + \mathbf{a}_1 \times \mathbf{a}_2/hk$$

$$= (v/hkl)(h\mathbf{b}_1 + k\mathbf{b}_2 + l\mathbf{b}_3)$$

$$= (v_a/hkl)(\mathbf{h})$$

Hence $\mathbf{h}$ is perpendicular to the plane with indices hkl obtained in the preceding manner. The indices so used are usually termed crystallographic indices, and they differ slightly from Miller indices in that Miller indices are defined with the auxiliary condition that the indices be three relatively prime numbers. Thus it is proper to refer to a 224 plane if crystallographic indices are being used, but the same plane in Miller index notation would be the second order of the 112 plane.

Now we shall also prove that $|\mathbf{h}| = (d)^{-1}$, where d is the perpendicular distance between planes of the type hkl. Select any one of the vectors describing the plane intercept along one of the axes, say $\mathbf{X}_1$. The component of this vector along a unit vector perpendicular to the plane of interest is just equal to the distance between planes d. However, we have just shown that $\mathbf{h}$ is a vector perpendicular to the plane; therefore,

$$d = \mathbf{X}_1 \cdot \mathbf{h}/|\mathbf{h}|$$
$$= (\mathbf{a}_1/h) \cdot (h\mathbf{b}_1 + k\mathbf{b}_2 + l\mathbf{b}_3)/|\mathbf{h}|$$
$$= 1/|\mathbf{h}|$$

We said earlier that $|(\mathbf{s} - \mathbf{s}_o)|/\lambda = 2(\sin\Theta)/\lambda$, where $\Theta = \psi/2$. Now by the Laue equation, $\mathbf{h} = (\mathbf{s} - \mathbf{s}_o)/\lambda$; hence, on taking magnitudes on both sides, this vector equation reduces to the scalar equation $(d)^{-1} = 2(\sin\Theta)/\lambda$, or

$$\lambda = 2d \sin\Theta \qquad (29)$$

which is Bragg's law. An alternate form of this equation involves Miller indices, in which case $n\lambda = 2d' \sin\Theta$, where $1/d = n/d'$. Figures 1.36a and b illustrate these equivalent considerations of diffraction.

The reciprocal lattice vector $\mathbf{h}$ is quite helpful for proof of certain useful relations. For example, the interplanar spacing $d = 1/|\mathbf{h}|$ in any crystallographic set of coordinates can be calculated by forming the product $|\mathbf{h}|^2 = \mathbf{h} \cdot \mathbf{h}$, substituting $\mathbf{h} = h\mathbf{b}_1 + k\mathbf{b}_2 + l\mathbf{b}_3$, writing the $\mathbf{b}$ values in terms of the $\mathbf{a}$ values, and then using the equations among the vectors discussed earlier. For the monoclinic system the resulting equation is

$$d = \left[\frac{1}{\sin^2\beta} \left(\frac{h^2}{a^2} + \frac{l^2}{c^2} - \frac{2hl}{ac}\cos\beta \right) + \frac{k^2}{b^2} \right]^{-1/2}$$

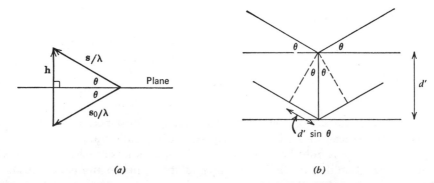

(a)   (b)

Figure 1.36   (a) Graphical representation of $\mathbf{s}/\lambda - \mathbf{s}_0/\lambda = \mathbf{h}$. (b) Path difference between successive planes a distance d' apart is $2d' \sin\theta$, which for a maximum is set equal to $n\lambda$.

Angles between plane normals are obtainable from such equations as $\cos(\mathbf{h}_p, \mathbf{h}_q) = \mathbf{h}_p \cdot \mathbf{h}_q / (|\mathbf{h}_p| |\mathbf{h}_q|)$. If the normals $\mathbf{h}$ for a set of planes are all perpendicular to an axis $\mathbf{T} = u_1\mathbf{a}_1 + u_2\mathbf{a}_2 + u_3\mathbf{a}_3$ of the crystal, then the equation $\mathbf{h} \cdot \mathbf{T} = 0$ holds. Therefore the indices hkl of these planes satisfy the equation $hu_1 + ku_2 + lu_3 = 0$. For example, if the axis in question is the monoclinic $\mathbf{b}$ axis [010], then all planes with indices $(h0l)$ satisfy this equation for all values of h and l. It is then said that the $(h0l)$ planes lie in the zone of the $\mathbf{b}$ axis.

The vectors $\mathbf{h} = h\mathbf{b}_1 + k\mathbf{b}_2 + l\mathbf{b}_3$ for the various hkl describe a three-dimensional lattice. With each point of this lattice we associate the three numbers hkl and the value of F_{hkl}. This weighted reciprocal lattice is a complete description of the information ordinarily obtainable in X-ray diffraction studies of single crystals. If we combine the reciprocal lattice with the graphical representation of the Laue equation, shown in Figure 1.36a, the result is the Ewald construction (Figure 1.37). The origin of the reciprocal lattice can be thought of as the point at which the crystal vectors and the reciprocal lattice vectors have a common origin and fixed relative orientation. Thus as the crystal rotates, the reciprocal lattice also rotates. The vector $\mathbf{s}_o/\lambda$ is defined by the direction of the incident X-ray beam. The length of this vector, $1/\lambda$, is measured in the same units as the lengths of the reciprocal lattice vectors. To satisfy the Laue equation we draw $\mathbf{s}_o/\lambda$ so that its head is at the common origin of all reciprocal lattice vectors. A sphere of radius $1/\lambda$ about its tail describes the locus of all possible vectors $\mathbf{s}/\lambda$ that can represent diffraction directions. (Note that the center of the sphere is not coincident with the reciprocal lattice origin.) When a point in reciprocal space touches the sphere, the Laue equation is satisfied for the particular set of parallel planes and a reflection results represented by the numbers hkl that identify the reciprocal lattice point. Then, if $F_{hkl} \neq 0$, a diffraction maxima will occur in the direction of $\mathbf{s}$.

2.6 The Lorentz and Temperature Factors

The Lorentz factor arises because different reflections are in a position to reflect for different lengths of time. If, for example, in Figure 1.37 the crystal is rotated about the $\mathbf{b}$ axis with uniform angular velocity ω, not all of the points in reciprocal space cross the circle of reflection with the same velocity. As an illustration, the Lorentz factor is derived for this case. Define the Lorentz factor

Figure 1.37 The Ewald construction. Let the reciprocal lattice net represent the $h0l$ zone of a monoclinic crystal. Then the $\mathbf{a}_2$ axis [010] is perpendicular to the plane of the paper. As the crystal is rotated about this axis, various points of the reciprocal lattice cross the circle of reflection; as they do so, the equation $\mathbf{s}/\lambda - \mathbf{s}_0/\lambda = \mathbf{h}$ is satisfied, and a diffracted beam occurs. Diffraction by the (101) plane is illustrated; $\mathbf{h}$ is normal to this plane.

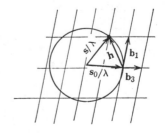

as à function inversely proportional to the linear velocity normal to the sphere $v_\perp$ of a reciprocal lattice point (Figure 1.38). Now the linear velocity of the end of the vector $\mathbf{h}$ is $\omega|\mathbf{h}|$, and its component normal to the surface of the sphere is $|\mathbf{h}| \cos \Theta = v_\perp$. But $\lambda = 2d \sin \Theta$ and $d = 1/|\mathbf{h}|$, so that $v_\perp = (\lambda/\omega)2 \sin \Theta \cos \Theta = (\omega/\lambda) \sin 2\Theta$. Hence, aside from a proportionality constant, $L = \lambda/(\omega \sin 2\Theta)$ is the Lorentz factor for a zero-level oscillation, rotation, or Weissenberg photograph, or for the four-circle diffractometer when the counter is constrained to move in the plane of $\mathbf{s}_o$ and $\mathbf{s}$. Graphical charts for Lorentz or combined Lorentz and polarization factors are available for other experimental methods. The principle of the calculation is the same for all experimental arrangements.

The thermal motion of atoms spreads out the electron density and hence reduces the amplitude of scattering more at the larger scattering angles. In this simplified treatment we assume that the atoms move independently and for convenience we assume orthogonal axes. The contribution of a single atom to the structure factor is $F_1 = f \exp[-2\pi i(hx + hy + lz)]$. Suppose the atom is displaced (in angstroms) by the distance $a\Delta x$ in the x direction, by $b\Delta y$ along y, and by $c\Delta z$ along z. Then F_1 is changed by ΔF_1, so that

$$F_1 + \Delta F_1 = f \exp\{-2\pi i[h(x + \Delta x) + k(y + \Delta y) + l(z + \Delta z)]\}$$
$$= F_1 \exp[-2\pi i(h\Delta x + k\Delta y + l\Delta z)]$$

Now assume that x, y, and z are independent variables, and average over all displacements.

$$\overline{F_1} = \overline{F_1 + \Delta F_1} = F_1 \overline{\exp(-2\pi i h\Delta x)}\, \overline{\exp(-2\pi i k\Delta y)}\, \overline{\exp(-2\pi i l\Delta z)}$$

A representative term is $e^{-i\alpha}$, where $\alpha = 2\pi h\Delta x$ is small. But the mean value of

$$e^{-i\alpha} = 1 - \frac{i\alpha}{1!} + \frac{(i\alpha)^2}{2!} - \frac{(i\alpha)^3}{3!} + \frac{(i\alpha)^4}{4!} + \cdots$$

over positive and negative values of α is

$$\overline{e^{-i\alpha}} = 1 - \frac{\overline{\alpha^2}}{2} + \frac{\overline{\alpha^4}}{24} - \cdots$$

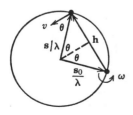

Figure 1.38 The Lorentz factor is proportional to $v_\perp$ along a radius of the sphere.

which is the same as the mean value of

$$e^{-\alpha^2/2} = 1 - \frac{\alpha^2}{2} + \frac{\alpha^4}{8}$$

within a negligible error. Therefore,

$$\bar{F}_1 = F_1 \exp\{-2\pi^2[h^2(\Delta x)^2 + k^2(\Delta y)^2 + l^2(\Delta z)^2]\}$$

This is the form of an anisotropic temperature factor in an orthogonal or nearly orthogonal system, when the largest or smallest amplitudes are primarily along x, y, or z. The constants $\overline{(\Delta x)^2}$, $\overline{(\Delta y)^2}$, and $\overline{(\Delta z)^2}$ are determined empirically when the atomic arrangement is known, but they can also be obtained approximately at the start of an X-ray diffraction study from mean values of F_{hkl}. Generally, however, the cross terms do not completely average to zero, and the more general form

$$\bar{F}_1 = F_1 \exp[-(\beta_{11}h^2 + \beta_{22}k^2 + \beta_{33}l^2 + \beta_{12}hk + \beta_{13}hl + \beta_{23}kl)]$$

is therefore required, where the β_{ij} are constants to be determined empirically.

Often an isotropic temperature factor is satisfactory if the crystal is not bound by forces of widely different strength in various directions. If so, we define the mean square amplitude of vibration as $\mu^2 = a^2\overline{(\Delta x)^2} = b^2\overline{(\Delta y)^2} = c^2\overline{(\Delta z)^2}$, and remember that $\lambda = 2d \sin \Theta$ and that $d = (h^2/a^2 + k^2/b^2 + l^2/c^2)^{-1/2}$ in the orthorhombic system. Then,

$$\bar{F}_1 = F_1 \exp(-8\pi^2\mu^2 \sin^2\Theta/\lambda^2) = F_1 \exp(-B \sin^2\Theta/\lambda^2)$$

where B is the temperature factor constant, to be determined empirically.

2.7   Electron Density and Patterson Functions

2.7.1   The Electron-Density Function

As we have said earlier, (22b) is a Fourier integral representation of the structure factor. Therefore the Fourier transform of this function should exist and, with proper scaling, should give the electron-density function, $\rho(\mathbf{r})$:

$$\rho = \sum_{\substack{\text{all} \\ hkl}} C_{hkl} \exp[2\pi i(hx + ky + lz)] \tag{30}$$

The coefficients are as yet unknown but can be determined by (22b), which we can rewrite as

$$F_{HKL} = \int_0^1 \int_0^1 \int_0^1 \rho \exp[-2\pi i(Hx + Ky + Lz)]\mathbf{a}\, dx \cdot \mathbf{b}\, dy \times \mathbf{c}\, dz$$

to distinguish the particular HKL from those in (30). Now substitute (30) into (22b). In each term of the sum the integrals over x, y, and z can be expressed as a product of three integrals, one of which is

$$\int_0^1 \exp[2\pi i(h - H)x\,dx = \frac{\exp[2\pi i(h - H)x]}{2\pi i(h - H)}\bigg|_0^1 = 0 \qquad \text{if } h \neq H$$

$$= \int_0^1 dx = 1 \qquad \text{if } h = H$$

There are similar integrals over y and z. Therefore all terms vanish except the term for which $h = H$, $k = K$, and $l = L$. The result is $F_{hkl} = \mathbf{a} \cdot \mathbf{b} \times \mathbf{c}\; C_{hkl}$, or $C_{hkl} = (1/V)F_{hkl}$. The three-dimensional electron-density function is thus

$$\rho(x, y, z) = \frac{1}{V} \sum_{h=-\infty}^{\infty} \sum_{k=-\infty}^{\infty} \sum_{l=-\infty}^{\infty} F_{hkl} \exp[2\pi i(hx + ky + lz)] \qquad (31)$$

The electron density can be projected along any axis, say the y axis, as

$$\rho(x, z) = \int_0^1 \rho(x, y, z)b\,dy$$

Again we can separate the exponential into products so that we need only to consider the factor

$$\int_0^1 \exp(2\pi i k y)\,dy$$

which equals 0 if $k \neq 0$ and equals 1 if $k = 0$. Therefore,

$$\rho(x, z) = \frac{b}{V} \sum_{h=-\infty}^{\infty} \sum_{l=-\infty}^{\infty} F_{h0l} \exp[2\pi i(hx + lz)] \qquad (32)$$

In general the structure factor $F_{hkl} = A_{hkl} - iB_{hkl}$, has both real and imaginary parts. We know that $\rho(x, y, z)$ is an observable quantity and hence must be real, and thus equal to its complex conjugate

$$\rho(x, y, z) = \frac{1}{V} \sum_h \sum_k \sum_l F_{hkl}^* \exp[-2\pi i(hx + ky + lz)]$$

If this equation is now added to (31) and the result is divided by 2, we obtain

$$\rho(x, y, z) = \frac{1}{V} \sum_h \sum_k \sum_l [A_{hkl} \cos 2\pi(hx + ky + lz)$$
$$+ B_{hkl} \sin 2\pi(hx + ky + lz)] \qquad (33)$$

For a centrosymmetric crystal in which the origin is chosen at a center of symmetry, for every atom at x_j, y_j, and z_j, there is another atom at $-x_j$, $-y_j$, and $-z_j$. Hence, F_{hkl} becomes

$$F_{hkl} = \sum_{j=1}^{N} f_j \exp[-2\pi i(hx_j + ky_j + lz_j)]$$

$$= \sum_{j=1}^{N/2} f_j \{\exp[-2\pi i(hx_j + ky_j + lz_j)] + \exp[2\pi i(hx_j + ky_j + lz_j)]\}$$

$$= \sum_{j=1}^{N/2} 2f_j \cos 2\pi(hx_j + ky_j + lz_j) \tag{34}$$

where the sum extends over the $N/2$ atoms not related by the center of symmetry. The electron-density function is then given by

$$\rho(x, y, z) = \frac{1}{V} \sum_h \sum_k \sum_l F_{hkl} \cos 2\pi(hx + ky + lz) \tag{35}$$

since $A_{hkl} = F_{hkl}$ and $B_{hkl} = 0$. The result for the acentric case can be expressed as

$$\rho(x, y, z) = \frac{1}{V} \sum_h \sum_k \sum_l \{|F_{hkl}| \cos[2\pi(hx + ky + lz) + \alpha_{hkl}]\} \tag{36}$$

where $|F_{hkl}| = (A_{hkl}^2 + B_{hkl}^2)^{1/2}$, and α_{hkl} (the so-called *phase angle*) is determined by the relations $\cos \alpha_{hkl} = A_{hkl}/|F_{hkl}|$ and $\sin \alpha_{hkl} = -B_{hkl}/|F_{hkl}|$. The reader can verify that the substitution of these relations into (36) gives (33).

The crystallographer must obtain the correct signs or phases of the F_{hkl} to calculate the complete three-dimensional map. The observed data give only $|F_{hkl}|^2$, and even in the centric case there are far too many intensities to attempt all possible sign combinations, even with the fastest computers. (Note that 20 terms would involve approximately 1 million electron density maps if all possible sign combinations were tried.)

However, (22a) and (36) are usually employed in the refinement of structures according to the following process:

1. Calculate $|F_{hkl}|$ by taking the positive square root of the observed $|F_{hkl}|^2$.
2. Calculate A_{hkl} and B_{hkl} from a trial structure obtained by methods described above.
3. If centric, assign the same sign to $|F_{hkl}|$ as found for the calculated $A_{hkl}(B_{hkl} = 0)$; or if acentric, compute α_{hkl}.
4. From the resultant electron-density map, obtain new atomic coordinates x_j, y_j, and z_j. Then repeat steps 2, 3, and 4 until the coordinates no longer change.

2.7.2   The Patterson Function

THE FUNCTION

After correction for the Lorentz and the polarization factors, each observed intensity yields $|F_{hkl}|^2$, which here includes the temperature factor. A Fourier series could be readily computed with these terms as coefficients, since no phase ambiguity is involved, but one may well want to know the physical meaning of such a function.

The electron density values $\rho(x, y, z)$ at x, y, z, and $\rho(x + u, y + v, z + w)$ at $x + u$, $y + v$, $z + w$ are both especially large if these coordinates are chosen at atomic positions. The product of the two functions is therefore large whenever the differences between these coordinates correspond to an interatomic distance; that is, when u, v, w represents the components of a vector between the two atoms. The totality of all such vectors, all drawn from a common origin, is obtained by summing this product over all space, here over the unit cell.

$$P(u, v, w) = \int_0^1 \int_0^1 \int_0^1 \rho(x, y, z)\rho(x + u, y + v, z + w)\mathbf{a}\,dx \cdot \mathbf{b}\,dy \times \mathbf{c}\,dz$$

Remembering that ρ is real and therefore equal to its complex conjugate, we substitute (31) for $\rho(x, y, z)$ and

$$\rho(x + u, y + v, z + w)$$

$$= \frac{1}{V}\sum_H \sum_K \sum_L F^*_{HKL} \exp\{-2\pi i[H(x + u) + K(y + v) + L(x + w)]\}$$

into $P(u, v, w)$. A representative integral is

$$\int_0^1 \exp(2\pi i h x)\exp[-2\pi i H(x+u)]\,dx = \exp(-2\pi i H u)\int_0^1 \exp[2\pi i(h-H)x\,dx = 0$$

unless $H = h$, where the value is $\exp(-2\pi i h u)$. Hence, only those terms remain for which $H = h$, $K = k$, $L = l$, and

$$P(u, v, w) = \frac{1}{V}\sum_h \sum_k \sum_l |F_{hkl}|^2 \exp[-2\pi i(hu + kv + lw)]$$

$$= \frac{1}{V}\sum_{h=-\infty}^{\infty} \sum_{k=-\infty}^{\infty} \sum_{l=-\infty}^{\infty} |F_{hkl}|^2 \cos 2\pi(hx + ky + lz)$$

(37)

This last result follows because $P(u, v, w)$ and $|F_{hkl}|^2 = F_{hkl}F^*_{hkl}$ are both real; therefore, the imaginary part of $P(u, v, w)$ vanishes. The Patterson function is always centrosymmetric whether the structure is or not; if we have a vector from atom 1 to atom 2, we also have the vector in the reverse direction. Projections can be made and are analogous to those for the electron-density map.

Methods based on the three-dimensional Patterson function can be very powerful for the solution of crystal structures if properly applied. Its power is greatly increased by establishing an approximate absolute scale for the observed $|F_{hkl}|$ and by sharpening the peaks. These two modifications are discussed next.

SCALE FACTOR AND TEMPERATURE FACTOR ESTIMATES

Let $|F_o|$ be the observed $|F^o_{hkl}|$, all of which are assumed to be on the same relative scale; they all need to be multiplied by a scale factor k to bring them into agreement with $|F_c|$, where $|F_c| = |F^c_{hkl}|$. We shall suppose that the $|F_c|$ do not contain the temperature factor; therefore,

$$k|F_o| = |F_c| \exp(-B \sin^2 \Theta / \lambda^2) \tag{38}$$

where

$$F_c = \sum_{j=1}^{N} f_j \exp(-2\pi i \mathbf{h} \cdot \mathbf{r}_j)$$

At the beginning of the structure investigation the atomic coordinates $\mathbf{r}_j$ are unknown, but the number of atoms N and their scattering factors are known. Now

$$|F_c|^2 = \left[\sum_{j=1}^{N} f_j \exp(-2\pi i \mathbf{h} \cdot \mathbf{r}_j) \right] \left[\sum_{k=1}^{N} f_k \exp(2\pi i \mathbf{h} \cdot \mathbf{r}_k) \right]$$

$$= \sum_{j=1}^{N} f_j^2 + \sum_{\substack{j \ k \\ j \neq k}} f_j f_k \exp[2\pi i \mathbf{h} \cdot (\mathbf{r}_k - \mathbf{r}_j)]$$

Suppose that the structure is sufficiently complex that the $\mathbf{r}_k - \mathbf{r}_j$ are roughly random and that a large fraction of the d spacings are not $\gg |\mathbf{r}_k - \mathbf{r}_j|$; these qualifications require modification in the discussion for simple structures, highly disordered structures, and proteins.

Aside from these cases, the average value of $|F_c|$ over many reflections is $\sum_{j=1}^{N} f_j^2$. Since both f_j and the temperature factor term vary with $\sin \Theta$, it is most reasonable to average over regions about 0.1 Å^{-1} in $\sin \Theta / \lambda$ and plot

$$\log \frac{\overline{|F_o|^2}}{|F_c|^2} = \log \frac{\overline{|F_o|^2}}{\sum_{j=1}^{N} f_j^2} = -\log k^2 - \left(\frac{2B}{2.303} \right) \frac{\sin^2 \Theta}{\lambda^2} \tag{39}$$

that is, plot $\log(\overline{|F_o|^2})/(\sum f_j^2)$ as ordinate against $\sin^2 \Theta / \lambda^2$ as abscissa. The slope, $-2B/2.303$, will give the value of B, and the intercept on the vertical axis, $-\log k^2$, will give the value of k. Usually the values of k are accurate to about 10%, which is sufficient to establish the identity of single interactions in the Patterson function if they are involved. If $|F_o|$ is placed on an absolute scale, the single vector between atoms i and j will have approximate height $Z_i Z_j$.

THE SHARPENED PATTERSON FUNCTION

As we have seen, peaks in the Patterson function result from the products of electron-density peaks. The electron-density peaks are appreciably wide, and thus the peaks in Patterson space are even wider. Since in the Patterson function there are $N(N-1)$ peaks in general, resolution becomes a real problem. It is advantageous therefore to sharpen these peaks; that is, to reduce the half-width as much as possible. To modify F_{hkl} so that the scattering corresponds to that of a point atom: (1) replace the atomic scattering factor f_j by the corresponding atomic number Z_j, and (2) multiply by a term of the type $\exp(B\sin^2\Theta/\lambda^2)$ to counteract the smearing out of the atom because of thermal motion effects. The shape of all scattering factors, although not exactly the same, is similar (excluding hydrogen), and therefore dividing $|F_{hkl}|$ by $\hat{f} = f/Z$ effectively replaces f by Z. If more than one type of atom is involved, then $\hat{f}$ can be computed by $\Sigma f_j/\Sigma Z_j$. Note that a Patterson function involving truly point atoms requires an infinite number of terms in the Fourier series. Therefore a factor of the type $\exp(-B'\sin^2\Theta/\lambda^2)$ is usually included to reduce the series termination effects. Thus the new coefficients for the Patterson function calculation are $(|F_{hkl}|^2/\hat{f}^2)\exp[(2B-B')\sin^2\Theta/\lambda^2$.

It is also possible to obtain a sharpened Patterson-like function by considering the product of two derivative electron-density functions in place of the electron-density functions in the usual formulation of the Patterson function [20]. Since a product is involved, the result is a function having large and sharper peaks at those positions corresponding to interatomic distances and small ripples around the peak, but these ripples can be completely removed by a slight modification of the coefficients. In general, both types of sharpening approaches can be used together. The effect of sharpening is illustrated in Figures 1.39a and b, which show, respectively, an unsharpened and a sharpened Patterson projection of the Ni(L)Br complex discussed earlier.

SYMMETRY IN THE PATTERSON FUNCTION

The effect of symmetry on the Patterson function is illustrated by a twofold screw axis, 2_1, parallel to **b**. We consider two pairs of atoms, 1 and 2, with coordinates $x_1, y_1, z_1; \bar{x}_1, y_1 + \frac{1}{2}, \bar{z}_1;$ and $x_2, y_2, z_2;$ and $\bar{x}_2, y_2 + \frac{1}{2}, \bar{z}_2$. First (Figure 1.40a) we assume that $v_1 \neq v_2$; hence, the vectors with component $\frac{1}{2}$ along **b** are $2x_1, \frac{1}{2}, 2z_1$ and $2x_2, \frac{1}{2}, 2z_2$. These vectors occur as single interactions in the section $v = \frac{1}{2}$ in Patterson space. The difficulty that often complicates this interpretation arises when two atoms by accident have the same y coordinate; this situation produces other vectors in the $v = \frac{1}{2}$ section that have twice the weight of the single interactions (Figure 1.40b). There are similar effects because of other symmetry elements, and all of these form part of the general analysis of the Patterson function. The very clearly written original paper on this subject by Patterson [63] is recommended to the reader.

2.7.3   The Least-Squares Method of Refinement

Let us assume that approximate positions were derived from the heavy atom method, direct methods, the Patterson function, or from some other method. We

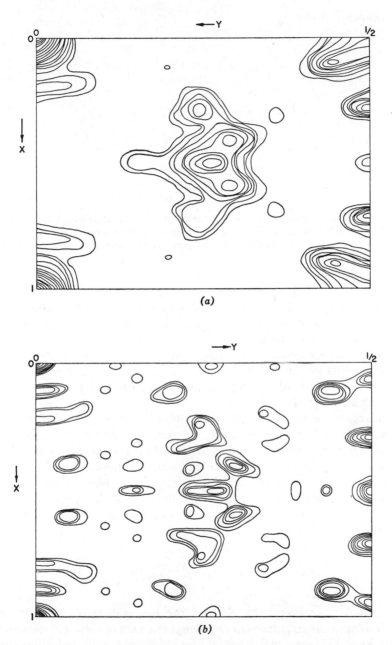

Figure 1.39 (*a*) An unsharpened Patterson map computed from the observed data for Ni(L)Br$_2$, where L is the tridentate ligand 6,6'-dimethyldi(2-pyridylmethyl)amine, and projected along the *c* axis. (*b*) A sharpened Patterson map for Ni(L)Br$_2$ projected along the *c* axis. A *B'* value of 3.0 Å^2 was used and the derivative sharpening function was also employed.

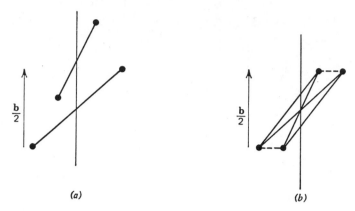

(a)   *(b)*

Figure 1.40 (*a*) Vectors with components **b**/2 when the two pairs of atoms do not have coincident *y* coordinates. (*b*) Additional interactions with components **b**/2, when *y* coordinates are coincident. The two interactions that occur in the $v = 0$ section can be used to help identify these additional interactions in the $v = \frac{1}{2}$ section.

wish now to refine these trial coordinates to obtain the best possible values consistent with the observed X-ray data. The most satisfactory method of refinement is the least-squares method [64]. Its advantages over methods based on Fourier series are that corrections for nonconvergence need not be made, and weighting factors for the observations can be related to the experimental errors in these observations. Also, errors in the final positional parameters can be readily calculated [65].

Let us regard the magnitude of the structure factor $|F_{hkl}|$ as a function of the scale factor $k' = 1/k$, multiplying the F_c; a function of the positional parameters $x_1, y_1, z_1, x_2, y_2, z_2$, and so on; and a function of the thermal parameters B_1, B_2, and so on; or β_{ij}, the corresponding anisotropic parameters. Let the symbol x_i represent a typical variable. Let $|F_j^c|$ represent for plane j the calculated magnitude of the structure factor, which is a function of all these variables; since $F_j^c = |F_j^c|e^{i\alpha}$, then $|F_j^c| = F_j^c e^{-i\alpha} = A_j^c \cos \alpha_j + B_j^c \sin \alpha$. Now $|F_j^c|$ can be developed in a Taylor's series neglecting terms higher than two as

$$|F_j^c|(x_1', x_2', \ldots, x_n') = |F_j^c|(x_1, x_2, \ldots, x_n) + \sum_i \left(\frac{\delta|F_j^c|}{\delta x_i}\right)\Delta x_i \qquad (40)$$

where x_i and x_i' represent the old and the new values of the parameters, respectively, and the Δx_i represent the changes in each of these parameters; that is, $\Delta k'$, Δx_1, Δy_1, Δz_1, ΔB_1, and so on. Now for each reflection we set $\Delta|F_j^c| = |F_j^o| - |F_j^c| - \varepsilon_j = |F_j^c(x')| - |F_j^c(x)|$, where ε_j is the experimental error associated with the reflection j and $|F_j^o|$ is the observed structure factor magnitude for plane j.

Assuming that the errors follow a Gaussian distribution, we obtain the "best values" of the parameter shifts, Δx_i, by a minimization of the sum of squares of the errors. Weighting factors, to be introduced later, are omitted here. Let

$$S = \sum_j \varepsilon_j^2$$

$$= \sum_j [(|F_j^o| - |F_j^c|) - \sum_i c_{ij} \Delta x_i]^2$$

where $c_{ij} = (\delta|F_j|)/(\delta x_i)$. The derivative of S with respect to each parameter change is set equal to 0,

$$\frac{\delta S}{\delta \Delta x_k} = 2 \sum_j [(|F_j^o| - |F_j^c|) - \sum_i c_{ij} \Delta x_i] c_{kj} = 0 \tag{41}$$

since this is the condition for a minimum in S relative to some particular Δx_k. This procedure reduces the large number of equations to the same number of equations as unknowns,

$$\sum_j c_{1j}^2 \Delta x_1 + \sum_j c_{1j} c_{2j} \Delta x_2 + \cdots + \sum_j c_{1j} c_{nj} \Delta x_n = \sum_j c_{1j}(|F_j^o| - |F_j^c|)$$

$$\sum_j c_{2j} c_{1j} \Delta x_1 + \sum_h c_j^2 \Delta x_2 + \cdots + \sum_j c_{2j} c_{nj} \Delta x_n = \sum_j c_{2j}(|F_j^o| - |F_j^c|) \tag{42}$$

$$\sum_j c_{nj} c_{1j} \Delta x_1 + \sum_j c_{nj} c_{2j} \Delta x_2 + \cdots + \sum_j c_{nj}^2 \Delta x_n = \sum_j c_{nj}(|F_j^o| - |F_j^c|)$$

which can be solved simultaneously. The c_{ij} values are assumed to be constant in this discussion, but their values are slightly different after the new parameters x_k' are obtained from the above equations. Hence, the whole process is repeated with these new values of x_k as starting values, until the changes in x_k are well within their probable errors.

Not all of the observations should contribute equally in the final set of equations. Because some observations might be more accurate than others, they should be weighted more heavily than those that were the less accurate. Therefore, we minimize the weighted sum of errors, $S' = \sum_j \omega_j \varepsilon_j^2$. This process has the net effect of multiplying the observational equations by $\omega_j^{1/2}$:

$$\omega_j^{1/2} \varepsilon_j = \omega_j^{1/2}(|F_j^o| - |F_j^c|) - \omega_j^{1/2} \sum_i c_{ij} \Delta x_i$$

The best choice of the $\omega_j^{1/2}$ is that they be inversely proportional to $\sigma_j(F)$, the standard deviation associated with the observed structure factor magnitude $|F_j^o|$. The $\sigma_j(F^o)$ can be obtained from the $\sigma_j(I)$ by $\sigma(I)/2|F|$, unless $|F|$ is quite small; then the finite difference method should be used [66].

However, many observations are required to determine $\sigma_j(I)$. Therefore, in practice, one of two schemes is usually followed, depending on the method used to take the data. For data recorded by countertechniques it is common to evaluate $\sigma_j(I)$ based on errors in counting statistics plus an attempt to include effects caused by systematic errors that were not completely eliminated.

Therefore an expression similar to

$$\sigma(I) = \left[T_c + \left(\frac{t_c}{2t_b} \right)^2 (B_1 + B_2) + (pT_c)^2 + (pB)^2 \right]^{1/2} \tag{43}$$

is used for the standard deviation in the net intensity, where T_c is the total count in a scan time t_c; B_1 and B_2 are the background counts each obtained in a time t_b; $I = T_c - (t_c/2t_b)(B_1 + B_2)$; and $B = B_1 + B_2$. The latter two terms in (43) are used to take into account effects caused by systematic errors, and a value of p of $0.03 - 0.05$ is usually selected.

For film data the usual choice is

$$\sigma^2(I) = (I_j^o/4I_{min}^o)^2 \qquad \text{for } I_j^o \geqslant 4I_{min}^o \tag{44}$$

$$= 1 \qquad \text{for } I_j^o \leqslant 4I_{min}^o$$

where I_{min}^o is the minimum observable intensity for the appropriate region of the film. The weighting factors arise because the larger intensities have a constant percentage error, whereas the smaller intensities have a constant error.

These least-squares equations are in a form readily handled on a digital computer. The weighting factors are quite different from those normally used in Fourier refinement, and hence it is reasonable to favor the results of least squares over those obtained from Fourier methods. In particular, the weighting factors inherent in the usual Fourier method give incorrect standard deviations as compared with the proper weighting factors.

The least-squares approach can, however, be quite consuming of computer time if the structure contains many atoms since a matrix of $P \times P$ in size must be inverted, where P is the number of parameters involved. For anisotropic refinement there are nine parameters per atom in the asymmetric unit. A block-diagonal approach is therefore sometimes used; in this approach it is assumed that the major nonzero elements correspond to those involving interactions between parameters of the same atom, and the other interactions involving parameters of different atoms have very small elements. This assumption is usually reasonably good; however, the scale factor–temperature factor interactions are more important, and unless properly handled they can lead to nonconvergence. A convenient way to circumvent such problems is to define the structure factor as

$$F_{\mathbf{h}} = \sum_{j=1}^{N} k_j f_j \exp[-2\pi i (\mathbf{h} \cdot \mathbf{r}_j)] T_j$$

where individual scale factors are assigned to each atom and T_j is the temperature factor for the atom. Using such an approach we can break down the large matrix into N small 5×5 or 10×10 matrices (isotropic or anisotropic) and handle each small matrix separately. Although individual scale-factor shifts are calculated, only one scale-factor shift is applied, which corresponds to that obtained by a weighted average, using atomic numbers for weights. In practice it was found that convergence is only slightly slower than with full matrix in terms of the number of cycles required, and the computational time is much less [67].

Least-squares approaches based on I_{hkl} instead of F_{hkl} can also be used.

3 SYMMETRY OF CRYSTALS

3.1 Symmetry Operations and Symmetry Elements

For our discussion of internal symmetry we assume that the structure extends indefinitely in three dimensions. (For further discussion of the topics in this section, the reader should consult [68].) The smallest unit that shows the full symmetry (aside from purely translation symmetry) and that is repeated by translational elements only is generally taken as the unit cell. Most unit cells contain additional symmetry elements; this means that only a fraction rather than all of the atoms in the unit cell need be located.

Consider an infinite two-dimensional array of points (which is termed a *two-dimensional lattice*), as shown in Figure 1.41, in which the pair of vectors $\mathbf{a}_1$ and $\mathbf{a}_2$ are perpendicular. The lattice is generated from this pair of vectors by the set of translations $\mathbf{T} = l_1\mathbf{a}_1 + l_2\mathbf{a}_2$, where l_1 and l_2 are positive, negative, or zero integers. A translation of this two-dimensional net by, for example, $\mathbf{a}_1, \mathbf{a}_2$, or $\mathbf{T}$ leaves this infinite net unchanged to an observer. This operation is called a *symmetry operation*, and the vectors $\mathbf{a}_1, l_1\mathbf{a}_1$, or $\mathbf{T}$ are *symmetry elements*. Similarly, this two-dimensional lattice is left unchanged by a rotation of $\pi, 2\pi, 3\pi$, and so on, or by reflection across a line along either $\mathbf{a}_1$, or $\mathbf{a}_2$, or by a center of inversion (replace $+x$ and $+y$ by $-x$ and $-y$, respectively) at any lattice point. These are symmetry operations; that is, the axis of rotation, the line of symmetry, and the center of inversion are symmetry elements. The symmetry operations considered here preserve distances and angles and involve no distortions.

If a symmetry element, such as a translation or a rotation, transforms an arbitrary figure into a congruent figure, it is called a *symmetry element of the first kind*. On the other hand, if a symmetry element, such as a line or plane of symmetry or a center of inversion, transforms an arbitrary figure into the mirror image of a congruent figure, it is called a *symmetry element of the second kind*.

Figure 1.41 A two-dimensional lattice generated by the orthogonal vectors $\mathbf{a}_1$ and $\mathbf{a}_2$.

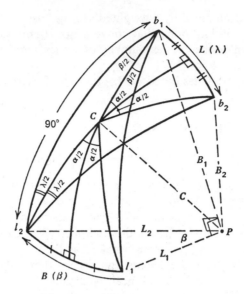

Figure 1.42 A sphere is drawn around point P that is common to S_1 and S_2. The general operation that superimposes l_1 on l_2 and then b_1 on b_2 is $B_1(\beta)$, a rotation of β about B_1 as an axis, followed by $L_2(\lambda)$, a rotation of λ about L_2 as an axis, where L_2 is $\perp B_1$. The axis $C(\alpha) = B_1(\beta)L_2(\lambda)$ is located from the intersection of the bisectors of β and λ. The equalities marked in the drawing then indicate the equivalence of $C(\alpha)$ and $B_1(\beta)L_2(\lambda)$ for superposing l_1 on l_2 and b_1 on b_2 in a single rotation by the angle α about the axis C.

This is an important classification, and it forms the basis, for example, of a discussion of optical activity. No molecule having a symmetry element of the second kind can show optical activity.

It can be shown that the general symmetry operation of the first kind is a screw rotation. The simple translation and the axis of rotation are special cases. Any figure S_1 can be superimposed on a congruent figure S_2 by a translation, which superimposes an equivalent point of S_1 and S_2, followed by two rotations. Suppose that we have carried out this translation, and now draw a sphere about the point P that is common to S_1 and S_2 (Figure 1.42). A careful inspection of this drawing, in which solid lines are drawn on the surface of the sphere and dotted lines toward the center at P, indicates that the superposition of three noncollinear points is now achieved by the operation $TC(\alpha)$. But this is not yet a simple screw rotation because the directions of $\mathbf{T}$ and $C(\alpha)$ are as yet unrelated. Resolve the vector $\mathbf{T}$ into components $\mathbf{T}_\parallel$ parallel to C and $\mathbf{T}_\perp$ normal to C, where $\mathbf{T} = \mathbf{T}_\parallel \mathbf{T}_\perp$. The general operation is now $TC(\alpha) = \mathbf{T}_\parallel \mathbf{T}_\perp C(\alpha)$. We can reduce $\mathbf{T}_\perp C(\alpha)$ to a single rotation $A(\alpha)$ as follows. Let $C(\alpha)$ be normal to the plane of the paper (Figure 1.43). The operation $\mathbf{T}_\perp$ carries the point A into A', while the rotation $C(\alpha)$ carries A' back to A; A is so located along the perpendicular bisector of $\mathbf{T}_\perp$ that $\mathbf{T}_\perp$ subtends the angle α at A. Then A is invariant, and a rotation $A(\alpha)$ is equivalent to $\mathbf{T}_\perp C(\alpha)$. Thus the general

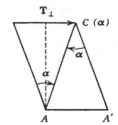

Figure 1.43 T is followed by the rotation α about the axis C extending perpendicular to the plane of the paper. We locate the point A on the perpendicular bisector of $\mathbf{T}_\perp$ so that the angle α occurs as indicated at A. Then the rotation $A(\alpha) = \mathbf{T}_\perp C(\alpha)$, and the point A is invariant.

operation of the first kind,

$$\mathbf{T}B_1(\beta)L_2(\lambda) = \mathbf{T}C(\alpha) = \mathbf{T}_\parallel\mathbf{T}_\perp C(\alpha) = \mathbf{T}_\parallel A(\alpha)$$

is a screw rotation.

The general operation of the second kind, which superimposes a figure S onto S' (which is the mirror image of S), can also be simplified. First we translate S so that point P is brought into coincidence with the equivalent point P' of S'. Then we carry out an inversion with this common point as a center, thus converting S into a figure congruent with S'. This figure can then be superimposed by the single rotation $A(\alpha)$. Thus the general operation of the second kind is $TIA(\alpha) = TA(\alpha)I$. The order of the last two operations is immaterial, and the center of inversion lies on the axis of rotation. The operation $A(\alpha)I$ can be seen (Figure 1.44) to be the same as $A(\alpha + \pi)M$, where M is a mirror plane perpendicular to the axis A. The general operation of the second kind is then $TA(\alpha)I = TA(\alpha')M$, where $\alpha' = \alpha + \pi$. However, a further simplification is possible. If no rotation is necessary, S and S', the object and its mirror image, are termed parallel; and the general operation reduces to TM. But T can be resolved into $\mathbf{T}_\parallel$ parallel to the plane M, and $\mathbf{T}_\perp$ perpendicular to the plane M. Then $\mathbf{T}_\perp M$ can be shown to be equivalent to M', a mirror plane parallel to M and halfway along $\mathbf{T}_\perp$ (Figure 1.45). Hence, the general operation reduces (when S and S' are parallel) to $TM = \mathbf{T}_\parallel\mathbf{T}_\perp M = \mathbf{T}_\parallel M'$, which is a gliding reflection, with the glide along a line parallel to the plane. If a rotation is necessary, we may resolve the translation $\mathbf{T}$ into $\mathbf{T}_\parallel$ parallel to the axis and $\mathbf{T}_\perp$ perpendicular to this axis and then the general operation becomes (when S and S' are not parallel)

$$TA(\alpha)I = TA(\alpha')M = TMA(\alpha') = \mathbf{T}_\perp\mathbf{T}_\parallel MA(\alpha') = \mathbf{T}_\perp M'A(\alpha') = M'A'(\alpha')$$

$$= A'(\alpha)M$$

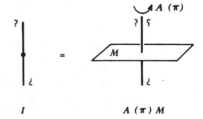

Figure 1.44 Proof of the equivalence $I = A(\pi)M$. This is easily generalized to $A(\alpha)I = A(\alpha + \pi)M$. The order of the operations, which is immaterial in this case, is not always so. Note that $M \perp A$ and that the center of inversion lies on A.

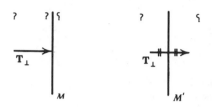

Figure 1.45 Proof of the equivalence of $T_\perp M = M'$, where M' is parallel to M and bisects $T_\perp$. (The pair of reversed question marks shown on each side of the mirror in the figure on the left should be considered as superimposed on the mirror. They are shown separated only for clarity.)

We have used our previous relations, remembering that $T_\parallel$ parallel to the axis is perpendicular to the mirror plane. Thus the general operation of the second kind is a rotatory reflection with the mirror plane perpendicular to the axis, or a rotatory inversion with the center of inversion on the axis. It is convenient to keep the gliding reflection as a special case.

Thus all symmetry operations are included in the *screw rotation*, the *gliding reflection*, or the *rotatory inversion*. The first is an operation of the first kind, the second two are operations of the second kind. A figure left invariant after a symmetry operation has been applied is said to be symmetric, and the symmetry elements are the vectors along which it is translated, planes across which it is reflected, centers through which it is inverted, or axes about which it is rotated.

The rotational axes of a crystal structure are restricted to one-, two-, three-, four-, and sixfold axes, which represent rotations by 0°, 180°, 120°, 90°, and 60°, respectively. This restriction arises because repeated application of the symmetry element must eventually restore the original position. The rotation is by $360°/n = 2\pi/n$, where n is an integer, and because only these types of axes combine with the lattice vectors, only certain values of n are possible. It is useful to include the axis 1, the rotation by 0° for completeness. Consider a rotation axis $A(\alpha)$, which is as yet arbitrary. Let T_1 be a vector of the lattice not parallel to $A(\alpha)$. The operation $A(\alpha)$ then generates the vectors $T_2, T_3, \ldots$ from T_1. The vectors $T_1 - T_2$, $T_2 - T_3$, and so on, are perpendicular to $A(\alpha)$, and these vectors or other similarly formed vectors can be chosen to describe a two-dimensional lattice perpendicular to $A(\alpha)$. It is a property of this lattice, as we have noted earlier, that if a vector $\mathbf{a}$ occurs, the vector $-\mathbf{a}$ also occurs (Figure 1.46); and if a rotation $A(\alpha)$ is possible, a rotation $A(-\alpha)$ is also possible. Since these rotations are symmetry operations, the two points at the end are lattice vectors and form a vector $\mathbf{b}$ parallel to $\mathbf{a}$; hence, $\mathbf{b} = m\mathbf{a}$, where m is an integer. But $\mathbf{b} = 2\mathbf{a} \cos \alpha$, and we find from these two equations that $\cos \alpha = m/2$. Now $\cos \alpha$ lies within the range from -1 to $+1$, so that $m = -2, -1, 0, 1, 2$ are the only permissible values. The resulting value of $\cos \alpha = -1, -\frac{1}{2}, 0, \frac{1}{2}, 1$ correspond,

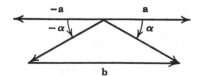

Figure 1.46 The rotation axis $A(\alpha)$ at a lattice point. Note that the vector $\mathbf{b}$ is parallel to $\mathbf{a}$. The reader should examine the situation when $A(\alpha)$ is halfway along the vector $\mathbf{a}$ and when $A(\alpha)$ is at some other position relative to $\mathbf{a}$.

respectively, to rotations of 180°, 120° and 240°, 90° and 270°, 60° and 300°, and 360°. Hence, only rotations by $2\pi/n$, where $n = 1, 2, 3, 4,$ or 6, are permitted. In particular, neither a fivefold axis nor axes where $n > 6$ can comprise a symmetry element of the lattice of a crystal structure.

Similarly, the inversion axes $A(\alpha)I$, which are permissible in crystal structures, can be shown to be $\bar{1}, \bar{2}, \bar{3}, \bar{4},$ and $\bar{6}$, where the bar indicates the inversion.

Symmetry operations are grouped together to form point or space groups, and examples follow. The assembly of symmetry operations form a group if (1) the successive operation R_1R_2, or operation R_2 followed by operation R_1, is equivalent to another operation R_3 of the group; (2) the associative law holds, $R_1R_2R_3 = R_1(R_2R_3) = (R_1R_2)R_3$, where those operations in parentheses are carried out and the result is combined with the remaining operation; (3) the identity operation E is a member of the group; and (4) the inverse R^{-1} of every operation R is also a member of the group. These group properties determine whether one has finished discovering the number of operations in a group or for our purposes the number of atoms equivalent by symmetry in any symmetric arrangement.

A space group occurs whenever a translation is involved. No single point is left invariant upon application of a symmetry element of which a translation is a part, such as a lattice translation, a screw axis, or a glide plane. Moreover, by virtue of the group properties, all translation groups are infinite; that is, property (1) leads to an infinite number of translational operations and to infinite lattices of points on which structures are based. Of course, no structure is infinite, but this idealization forms the basis of classification of three-dimensional periodic structures in Euclidean space and serves as a basis for discussing the symmetries of crystals.

A point group occurs when the set of symmetry operations, such as a rotation axis, a center of inversion, a mirror plane, or an inversion axis, leaves at least one point in space fixed in position. Point groups form the basis for discussion of the external forms of crystals, and they occur as subgroups in space groups. A subgroup is any set of symmetry operations that among themselves satisfy the preceding group properties but are part of a larger group. Because of this subgroup relation, a discussion of point groups, followed by their extension by translation groups, forms a simple geometrical basis for discovering the symmetry properties of crystals having no disorders, and hence for a discussion of space groups.

3.2   Point Groups

There are 32 point groups of importance to crystallography. Not all of these are developed here, but the methods employed for the first few illustrate the principle.

Cyclic groups of the first kind with a single axis are simply the axial groups 1, 2, 3, 4, and 6, or, in alternative notation, $C_1, C_2, C_3, C_4,$ and C_6. (The first notation, as we shall see later, is related to Hermann–Mauguin notation while

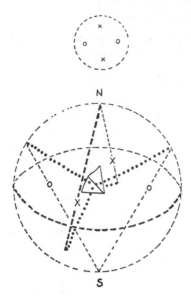

Figure 1.47 Stereographic projection of the faces of a tetrahedron. A sphere is drawn about the crystal, in this case a tetrahedron. The normals to the faces are drawn out to where they intersect the sphere. Now define an equatorial plane, and a north pole, N, and south pole, S. Connect intersections of face normals in the northern hemisphere by straight lines to the S pole, and face normals in the southern hemisphere with the north pole. The intersections of these connecting lines with the equator give the stereographic projection. Points above the plane are represented by o, and those below the plane by x.

the second resembles that of Schoenflies.) They can be represented in stereographic projections, the principle of which is illustrated in Figure 1.47. Stereographic projections of these five cyclic groups are shown in Figure 1.48. The order of the group is the total number of equivalent points (or objects) generated by the group when the first object does not lie on a symmetry element. The orders of C_1, C_2, C_3, C_4, and C_6, are 1, 2, 3, 4, and 6, respectively.

Rotation groups that have one n-fold axis but more than one twofold axis can be found by extension of the cyclic groups of the first kind with a twofold axis. If we are to have only one n-fold axis, the twofold axis must be perpendicular to the n-fold axis; otherwise, an additional n-fold axis appears. Let us extend each

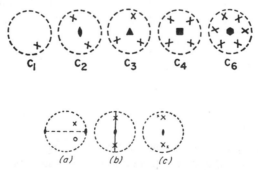

Figure 1.48 The five cyclic groups of the first kind. Note that C_2 could also be represented as shown in (a), where the twofold axis is horizontal. The points x and o should be regarded as asymmetric so that we should not be tempted to introduce a mirror plane into C_2 as shown in (b); if any doubt remains concerning the symmetry, add another independent point as shown in (c), which clearly shows that (b) is incorrect.

of the cyclic groups by one twofold axis, as indicated in Figure 1.49, and examine the symmetry of the resulting arrangement of equivalent objects, remembering that the objects may be asymmetric. The additional symmetry elements, all twofold axes, so generated are shown in the last line along with the group symbols.

Rotation groups with more than one axis of order greater than two form axial systems characteristic of regular polyhedra, as may be shown by starting with two intersecting axes of order greater than two and carrying out the rotations. A sphere drawn about the point of intersection of these axes is then covered by regular polyhedra when the symmetry operations of these axes are carried out. This result is not derived here; but the five regular polyhedra are the tetrahedron (4 faces, 4 vertices, 6 edges), the cube (6 faces, 8 vertices, 12 edges), the octahedron (8 faces, 6 vertices, 12 edges), the dodecahedron (12 faces, 20 vertices, 30 edges), and the icosahedron (20 faces, 12 vertices, 30 edges). Of these the last two have fivefold axes and hence are noncrystallographic; also, the cube and the octahedron have the same symmetry group. Thus there are only two new purely axial symmetry groups, T-23 and O-432, which are shown in Figure 1.50.

Now that the 11 axial point groups are derived using symmetry elements of the first kind, let us examine the cyclic groups of the second kind: $\bar{1}, \bar{2}, \bar{3}, \bar{4}$, and $\bar{6}$. Their stereographic projections are illustrated in Figure 1.51. The principle of

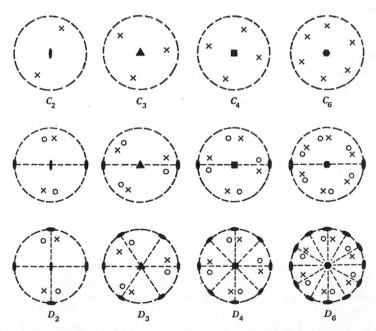

Figure 1.49   The cyclic groups in the first line are extended by a single twofold axis in the second line, and the additional twofold axes so generated are indicated in the third line. The order of the group is the number of equivalent points generated when the points do not lie on a symmetry element. The corresponding Hermann–Mauguin symbols are 2, 3, 4, 6, 222, 32, 422, and 622.

Figure 1.50   The cubic axial groups. Point groups having more than one threefold axis are described in terms of cubic axes, which are equal and orthogonal.

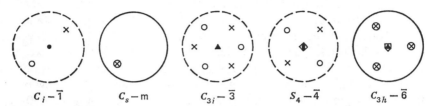

Figure 1.51   Cyclic groups of the second kind. The operation $\bar{N}$ is a rotation by $(2\pi/N)$ followed by an inversion through a center of symmetry. The reader should check the equivalence of $\bar{1}$ to i, of $\bar{2}$ to m, of $\bar{3}$ to C_3 extended by i, and of $\bar{6}$ to C_3 extended by m_h. A mirror plane is called m_h if perpendicular and m_v if parallel to the principal axis. A mirror plane in the plane of the stereographic projection is represented by a solid circle. Note that $C_s - m$, which is the same as $\bar{2}$, can also be represented, as shown in Figure 1.52.

extension is illustrated here by the equivalence of $\bar{3}$ with the group obtained by extending C_3-3 with a center of inversion; hence, the symbol C_{3i}-$\bar{3}$. The symbol S *is occasionally used (e.g., S_4),* where S_h denotes rotation coupled with reflection, while $\bar{N}$ denotes rotation coupled with inversion. For example, S_6, identical with C_{3i}, occurs, and it is left to the reader to show that the group generated by a rotation of 60° followed by reflection in a plane normal to this axis is identical with a rotation by 120° followed by an inversion through a center on the axis of rotation. The only group that is impossible to obtain from extension of the cyclic groups of the first kind by i or m is S_4-$\bar{4}$.

Sixteen of the 32 point groups are now derived. Since all of the axial groups are derived, the remainder are obtained by extending these 16 by mirror planes or centers of inversion added in such a way that the axial system remains invariant. It is a simple exercise to finish the list. In doing so, be careful to exclude the noncrystallographic groups; that is, those that have axes of the first or second kind of order other than 1, 2, 3, 4, or 6. The remaining 16 are included in the complete list, where the 32 are classified into reference systems of axes that prove convenient for their description (Figure 1.52).

These point groups are extended, by the use of translation groups and the introduction of screw axes and glide planes, to the space groups in the following section. In addition, the centrosymmetric point groups, which are C_i, C_{2h}, D_{2h}, C_{4h}, D_{4h}, T_h, O_h, C_{3i}, D_{3d}, C_{6h}, and D_{6h}, are the symmetries shown by the

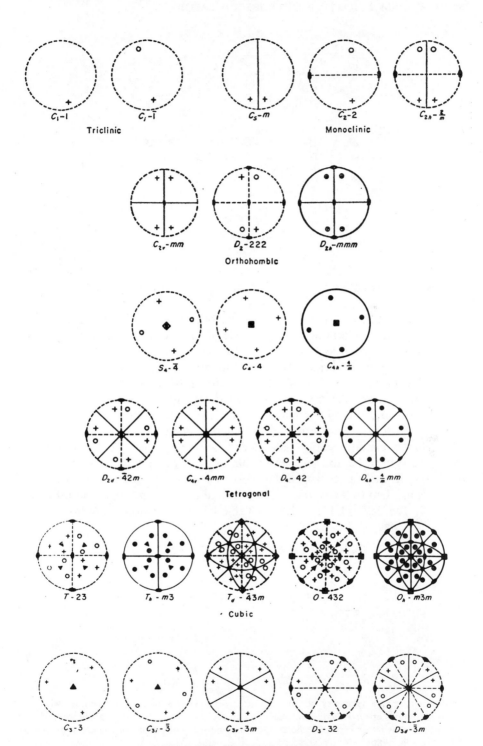

Figure 1.52 The 32-point groups.

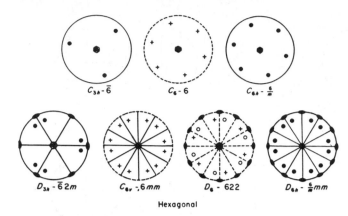

$C_{3h} - \bar{6}$ $C_6 - 6$ $C_{6h} - \frac{6}{m}$

$D_{3h} - \bar{6}2m$ $C_{6v} - 6mm$ $D_6 - 622$ $D_{6h} - \frac{6}{m}mm$

Hexagonal

Figure 1.52 (Cont.)

diffraction pattern of any crystal.[†] The identification of this symmetry is usually the first step in a structure determination.

The knowledge that point groups must obey the mathematical rules for a group is useful in determining the presence of all necessary symmetry elements; that is, the completeness of the group. Consider, for example, the group C_{2h}. From the notation it is evident that the group must consist of a twofold axis and a mirror plane whose normal is in the direction of the twofold axis. The identity operation, E, must also be a member of the group. Since the inverse of the twofold axis and the mirror plane are just the same twofold axis and mirror plane, respectively, there are three symmetry elements thus far. To test for completeness, a multiplication table, called the *Cayley table* is constructed. It is a square table in which the rows and columns are labeled with the operations of the group. In each space within the array is written the operation of the group that is equivalent to a row operation followed by a column operation. For the group C_{2h} the Cayley table is given in Table 1.3; the direction of the twofold axis

Table 1.3 Cayley Table for the Point Group C_{2h}

	E	2	m	i
E	E	2	m	i
2	2	E	i	m
m	m	i	E	2
i	i	m	2	E

[†]This symmetry, the symmetry of the reciprocal lattice, can be noncentrosymemtric when the structure contains optically active molecules and an X-ray wavelength near an absorption edge of one of the atoms in the crystal is chosen. This method forms the basis for the determination of absolute configuration.

is assumed to be the y direction, and this is also then the direction of the normal to the mirror plane. Thus to complete the table it is apparent that an inversion operation must be added.

3.3 Space Groups

3.3.1 Translational Groups in Two and Three Dimensions

The symmetry groups obtained by combining the screw rotation, the rotatory inversion, the gliding reflection, or some special case of these, with an infinite three-dimensional translation group produce what are termed the *space groups* [69]. Hence, our first consideration is of the translation groups only, and we begin by listing the two-dimensional translation groups (Figure 1.53). Any two-dimensional net can be expressed as a two-dimensional set of vectors $\mathbf{T} = l_1\mathbf{a}_1 + l_2\mathbf{a}_2$, where l_1 and l_2 are integers. A net so defined is called a *primitive net*. But such a choice of a primitive pair of vectors for the C-centered orthorhombic net (see Figure 1.53) does not define a unit cell that exhibits the D_{2h} symmetry of the net, and hence a nonprimitive description is preferred. The nonprimitive set for this C-centered cell is $\mathbf{T} = l_1\mathbf{a}_1 + l_2\mathbf{a}_2 + (m/2)(\mathbf{a}_1 + \mathbf{a}_2)$,

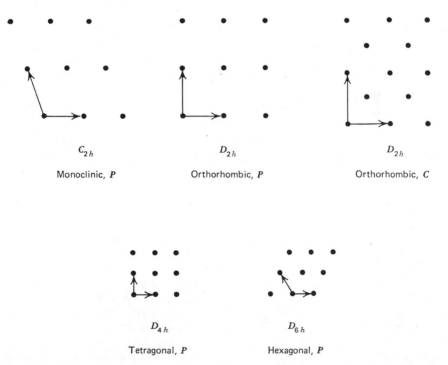

C_{2h}

Monoclinic, *P*

D_{2h}

Orthorhombic, *P*

D_{2h}

Orthorhombic, *C*

D_{4h}

Tetragonal, *P*

D_{6h}

Hexagonal, *P*

Figure 1.53 The two-dimensional nets. The symmetry at every net point and the system of axes into which these nets most naturally fall is indicated. All of these nets indicated by *P* are primitive; the *C*-centered net is indicated by *C*.

where $m = 1$ or 0, and l_1 and l_2 are integers. Although other-centered cells are possible, such as a centered cell in the case of the net with D_{4h} symmetry, it is easy to demonstrate that such a cell can also be described as a primitive cell by a new choice of axes less by a factor of $\sqrt{2}$ than the axes of the centered cell and oriented at $45°$ relative to them. Finally, recall that the only axial symmetries possible perpendicular to a net are 1, 2, 3, 4, and 6. Since the minimum symmetry at a point of a two-dimensional net is C_{2h}, such nets with axes 1 and 3 are not possible; therefore, the possible two-dimensional nets are those listed in Figure 1.53.

The three-dimensional translation groups can be found in an analogous way by a straightforward extension of the two-dimensional case. There are a greater variety of possible centered lattices as indicated in Table 1.4. These coupled with allowed symmetry elements yield 14 three-dimensional lattices called the *Bravais lattices* (Table 1.5). The coordinate systems that are most natural for descriptions of these lattices are indicated in the last column. The axes x, y, and z are measured along the translations $\mathbf{a}_1$, $\mathbf{a}_2$, $\mathbf{a}_3$, respectively, of the lattices. Restrictions on the lengths of these axes and on the angles between them (α_1 is the angle between $\mathbf{a}_2$ and $\mathbf{a}_3$, and so on) required by the symmetry of each lattice are indicated in this last column. It must be emphasized that the symmetry, not some accidental coincidences of axial lengths or angles, determines the choice of axes and the crystal system.

To acquire additional familiarity with these lattices, the reader may show that an orthorhombic cell centered on only two pairs of opposite faces is not possible, since such a set of vectors will generate equivalent points on the remaining pair of faces by virtue of the group properties, and may prove the equivalence of C and P tetragonal cells or of F and I tetragonal cells. The axis that is labeled $\mathbf{a}$, $\mathbf{b}$, or $\mathbf{c}$ ($\mathbf{a}_1$, $\mathbf{a}_2$, or $\mathbf{a}_3$) in the orthorhombic system is not determined by symmetry; hence, relabeling of axes converts an orthorhombic A- or B-centered cell into a C-centered cell. A similar statement holds for the equivalence of monoclinic A- and C-centered cells. A trigonal (rhombohedral) lattice can be indexed on

Table 1.4   General Types of Lattices

Type	Label	Cell Multiplicity	Description
Primitive	P	1	Primitive
End-centered	A	2	Centered on $\mathbf{b}$, $\mathbf{c}$ face
	B	2	Centered on $\mathbf{a}$, $\mathbf{c}$ face
	C	2	Centered on $\mathbf{a}$, $\mathbf{b}$ face
Body-centered	I	2	Body-centered
Face-centered	F	4	Centered on all faces
Rhombohedral	R	3	Two internal points $\frac{1}{3}$ and $\frac{2}{3}$ along the body diagonal (if described with the hexagonal axes)

Table 1.5   The 14 Bravais Lattices

Symmetry at a Lattice Point	Name	Points per Unit Cell	Elementary Cell Relations
$C_i - 1$	Triclinic	1	
$C_{2h} - \dfrac{2}{m}$	Simple monoclinic	1	$\alpha_1 = \alpha_3 = 90°$
	Side-centered monoclinic C	2	
D_{2h}	Simple orthorhombic	1	$\alpha_1 = \alpha_2 = \alpha_3 = 90°$
	End-centered C, A, B	2	
	Face-centered F	4	
	Body-centered I	2	
$D_{4h} - \dfrac{4}{m} mm$	Simple tetragonal	1	$\alpha_1 = \alpha_2 = \alpha_3 = 90°$
	(equivalent to end-centered)	2	$a_1 = a_2$
	Body-centered	2	
	(equivalent to face-centered I)	4	
$D_{3d} - \bar{3}m$	Trigonal	1	$\alpha_1 = \alpha_2 = \alpha_3$
			$a_1 = a_2 = a_3$ (Rhombohedral axes for hexagonal description, see below)
$D_{6h} - \dfrac{6}{m} mm$	Hexagonal	1	$\alpha_1 = \alpha_2 = 90°$
			$\alpha_3 = 120°$
			$a_1 = a_2$
$O_h - m3m$	Simple cubic	1	$\alpha_1 = \alpha_2 = \alpha_3 = 90°$
	Face-centered cubic F	4	$a_1 = a_2 = a_3$
	Body-centered cubic I	2	

hexagonal axes, and a hexagonal lattice can be indexed on rhombohedral axes [70], but the resulting transformations convert primitive cells to nonprimitive cells.

3.3.2   Space Operations

Screw axes permissible in three-dimensional crystallographic groups are limited to those involving rotations by 60°, 90°, 120°, 180°, and 360° by the same argument as that for ordinary rotation axes. The operation for an n-fold screw axis A, always taken as a right-handed screw, is $A(2\pi/n, \mathbf{t} = m\mathbf{T}/n)$ and is designated by n_m. The operation for various fourfold screw axes is illustrated in

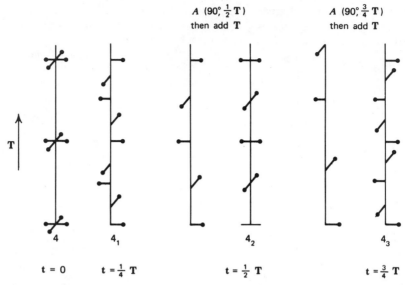

Figure 1.54 Operations of fourfold screw axes. Although the operation is defined as right-handed, the result of 4_3 is a left-handed rotation after the effect of the primitive translation **T** is taken into account.

Figure 1.54. Diagrams for the twofold screw axis 2_1, the threefold screw axes 3_1 and 3_2, and the sixfold screw axes 6_1, 6_2, 6_3, 6_4, and 6_5 are analogous. A onefold screw axis is simply a lattice translation. Note that four successive applications of operation $A(2\pi/4, \mathbf{t})$ of the screw axis 4_1 result in a lattice translation **T**; hence, as with any rotation axis the screw axis is always parallel to a row of the lattice and perpendicular to a net plane of the lattice.

The kinds of glide planes are also very limited in number. Let us represent a glide plane as $M(\mathbf{t})$, a reflection across a mirror plane m combined with a translation **t** parallel to the plane m (Figure 1.55). If the operation is repeated, the result is a translation of the original object by $2\mathbf{t}$, which must be a lattice translation **T**. The plane must therefore be parallel to a net plane of the lattice, and if this net plane is described by a primitive pair $\mathbf{a}_1$ and $\mathbf{a}_2$, the only possibilities are glides of $\frac{1}{2}\mathbf{a}_1$, $\frac{1}{2}\mathbf{a}_2$, or $\frac{1}{2}(\mathbf{a}_1 + \mathbf{a}_2)$. If the net is centered, then a glide by $\frac{1}{4}(\mathbf{a}_1 + \mathbf{a}_2)$ is also possible. The glide planes corresponding to translations of $\frac{1}{2}\mathbf{a}_1$, $\frac{1}{2}\mathbf{a}_2$, $\frac{1}{2}\mathbf{a}_3$, $\frac{1}{2}(\mathbf{a}_1 + \mathbf{a}_2)$, and $\frac{1}{4}(\mathbf{a}_1 + \mathbf{a}_2)$ are called a, b, c, n, and d glide planes, respectively. By combining these symmetry elements together the 230 space groups can be obtained.

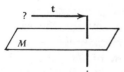

Figure 1.55 The glide plane.

Two designators for space groups are commonly used. The first is the Schoenflies symbol, which consists of a point-group symbol representing one of the 32 possible point groups from which space groups can be derived followed by a superscript indicating the particular space group derived from this point symmetry.

The second designator, the Hermann–Mauguin symbol, is the one now more commonly used and the one we use in naming space groups. The first letter, a capital, indicates the type of cell; that is, primitive, body-centered, and so on; whereas the remaining symbols indicate the symmetry elements present in the space group, written in an order that refers to the direction of each symmetry element relative to the crystallographic axes.

In the triclinic system the only possible symmetry element is 1 or $\bar{1}$. Thus only one symbol need be given after the cell-type designation.

In the monoclinic system there is one unique direction (usually the b direction); therefore, in the Hermann–Mauguin symbol it is necessary to give only the symmetry element in that direction. If two symmetry elements have the same direction they are written in fractional form, for example, $2/m$.

In the orthorhombic system, however, symmetry elements lie along all three directions. Therefore the symmetry element in the x direction is given first, followed by the one in the y direction, and then the one in the z direction.

In the tetragonal and hexagonal systems the axis of order higher than two is assumed to be in the c direction, and this symbol is listed first. The next symbol indicates the symmetry in the a direction, and the third symbol indicates the symmetry along the ab diagonal.

In the trigonal system rhombohedral description the threefold or threefold-inversion axis is along the body diagonal, and this symbol is listed following the cell type. The second symbol indicates the symmetry along a projection of a on the (111) plane.

In the cubic system the first symbol indicates the symmetry in the a direction, the second along the body diagonal, and the third along the face diagonal.

To illustrate, the orthorhombic space group *Pnma* indicates that the cell is primitive with a glide plane that reflects in the a direction and translates in the b and c directions, a mirror plane that reflects in the b direction, and a glide that reflects in the c direction and translates in the a direction. Just as in the related point group $(D_{2h}\text{-}mm)$, the mirror-type operations, noted above, when combined, yield a twofold screw parallel to the a axis, a twofold screw parallel to the b axis, a twofold screw parallel to the c axis, and an inversion center.

3.3.3 Monoclinic Space Groups With Axes Only

As first examples of three-dimensional space groups, consider the possible combinations of the only lattices permitted in the monoclinic system, P and C, with the only axes permissible, 2 and 2_1. The origin of the translation group can be placed anywhere as long as the axis is parallel to a row of the lattice and perpendicular to a net plane. We lose no generality if we choose the origin

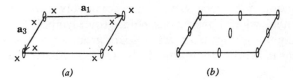

Figure 1.56 The array of points generated by a twofold axis at the origin and the primitive translation group: (a) incomplete, (b) complete. Note that additional twofold axes at $\frac{1}{2}\mathbf{a}_1$, $\frac{1}{2}\mathbf{a}_3$, and $\frac{1}{2}(\mathbf{a}_1 + \mathbf{a}_3)$ have been generated as shown in part (b) of the figure, which is the complete representation of the space group $C_2^1 - P2$.

(upper left corner) at the axis, which is perpendicular to the plane of the paper. The combination of a primitive translation group with a twofold axis is shown in Figure 1.56. This unit cell may contain any number of molecules or atoms; for example, if it contained only one molecule, this molecule would have to lie on a twofold axis and would have to have a twofold axis of symmetry (assuming no disorder). If there are two molecules in the unit cell, they could lie either on twofold axes or in the general position indicated by the "x" in Figure 1.56. Of course, more equivalent pairs of molecules in general positions may also occur. To illustrate one value of a knowledge of the space group, one needs to determine the coordinates of atoms in only one-half of this unit cell (Figure 1.56), since the remainder of the cell is determined by symmetry.

If we combine a twofold screw axis with a primitive translation group, the result is shown in Figure 1.57. This cell cannot have only one molecule; the translational element present in the screw axis prevents the reduction in the number of equivalent positions possible for a twofold axis. This space group, $P2_1$, frequently occurs in crystals of optically active molecules, but perhaps less frequently than the orthorhombic space group $P2_12_12_1$ following.

Now combine a C-centered lattice with a twofold axis. The set of equivalent points is shown in Figure 1.58. The coordinates of the set of equivalent points produced by this space group are $x, y, z; \frac{1}{2} + x, \frac{1}{2} + y, z; \bar{x}, y, \bar{z}; \frac{1}{2} - x, \frac{1}{2} + y, -z$, where $\bar{x}$ means $-x$. If an atom or molecule resides on the twofold axis, the coordinates of its center would be $0, y, 0; \frac{1}{2}, \frac{1}{2} + y, 0$.

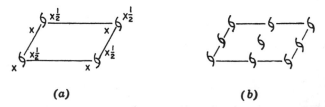

Figure 1.57 Combination of a P lattice with 2_1 (a) incomplete, (b) complete. The twofold screw axis carries the point x halfway along the translation out of the plane of the paper and rotates it 180° around the screw axis, as indicated by the symbol $x\frac{1}{2}$. Note that the set of equivalent points so generated also requires the presence of screw axes halfway along the axes and at the center of the cell, as shown in the complete drawing (b) of the space group $C_2^2 - P2_1$.

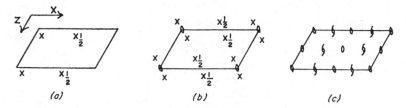

Figure 1.58 The space group $C_2^3 - C2$ is shown in part (c). The C centering is shown in (a); an atom at x, y, z has an equivalent atom at $\frac{1}{2} + x, \frac{1}{2} + y, z$. In (b) a twofold axis introduced at the origin produces the set of equivalent atoms shown, which gives the complete symmetry group shown in (c).

Finally, the combination of a C lattice with a 2_1 axis yields the same result as we have just obtained from C and 2. The choice of origin is arbitrary, and either the 2 or the 2_1 axis is a satisfactory choice.

3.3.4 Monoclinic Space Groups Related to $C_{2h} - 2/m$

Consider the extension of the space groups derived above by the possible kinds of symmetry planes, m, a, c, n, and d, perpendicular to the monoclinic axis 2 or 2_1. If we extend with a d-glide plane, the cell becomes B centered and can be reduced to a P cell. Hence, the d-glide plane cannot occur in either a P or a C cell. The directions of the a and c axes are not determined by symmetry in the monoclinic system, and hence the c axis of the unit cell can be chosen along the glide plane. If this is done we need only consider extension of $P2$ and $P2_1$ by m or c. (The extension of $C2$ is quite analogous, and it is not considered here.) These four possibilities are distinct and are summarized in Figure 1.59. It is left to the reader to show by careful study of Figure 1.59 how they arise from the extension of $P2$ and $P2_1$ by m or c planes.

A few details regarding the space group $P2_1/c$ may be helpful because this space group occurs very frequently, particularly for organic molecules. Let the coordinates of the point x in $P2_1$ (Figure 1.59) be x, y, z. The point o is at $-x$, $y + \frac{1}{2}$, $-z$. If the glide plane is introduced at $\frac{1}{2}b$, as indicated in $P2_1/c$, the additional points at $-x, \frac{1}{2} - y, \frac{1}{2} - z$ and $x, -y, \frac{1}{2} + z$ are generated. Let us refer these coordinates to the center of symmetry at $0\,\frac{1}{4}\,\frac{1}{4}$; that is, subtract $0\,\frac{1}{4}\,\frac{1}{4}$ from each set of coordinates. The result is $x, y - \frac{1}{4}, z - \frac{1}{4}$; $-x, y + \frac{1}{4}, -z - \frac{1}{4}$; $-x$, $-y + \frac{1}{4}, -z + \frac{1}{4}$; $x, -y - \frac{1}{4}, z + \frac{1}{4}$. Suppose we define $y_1 = y - \frac{1}{4}$ and also $z_1 = z - \frac{1}{4}$; then, the four equivalent points are x_1, y_1, z_1; $-x_1, -y_1, -z_1$; $x_1, \frac{1}{2} - y_1, \frac{1}{2} + z_1$; $-x_1, \frac{1}{2} + y_1, \frac{1}{2} - z_1$, where we remember that $+\frac{1}{2}$ is the same as $-\frac{1}{2}$ because of the lattice translation.

This result can also be obtained using a somewhat different approach. Define a general symmetry operator by the quantity $[\mathbf{S} + \mathbf{t}]$, where $\mathbf{S}$ is a point group matrix operator and $\mathbf{t}$ is the associated translation. For example, if the point x, y, z is transformed into the point $\frac{1}{2} - x, \frac{1}{2} + y, -z$ by a symmetry operation,

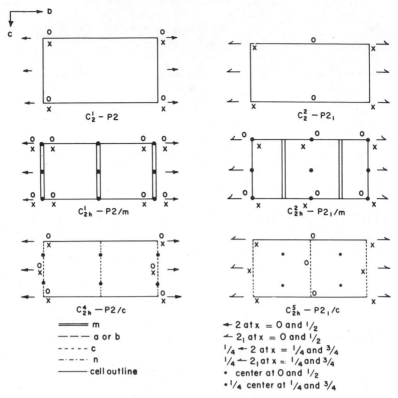

Figure 1.59 Monoclinic space groups $P2$ and $P2_1$, and groups obtained upon extending these by m or c. A center of symmetry is usually chosen as origin.

the associated **S** and **t** operators are

$$
\mathbf{S} = \begin{bmatrix} -1 & 0 & 0 \\ 0 & 1 & 0 \\ 0 & 0 & -1 \end{bmatrix} \qquad \mathbf{t} = \begin{bmatrix} \dfrac{1}{2} \\[4pt] \dfrac{1}{2} \\[4pt] \dfrac{1}{2} \end{bmatrix}
$$

The operation $[\mathbf{S}_1 + \mathbf{t}_1]$ followed by $[\mathbf{S}_2 + \mathbf{t}_2]$ is equivalent to the operation given by $[\mathbf{S}_2\mathbf{S}_1 + \mathbf{S}_2\mathbf{t}_1 + \mathbf{t}_2]$. Taking the space groups related to the point group C_{2h} we can write the Cayley table as shown in Table 1.6, where $[\mathbf{S}/\mathbf{t}] = [\mathbf{S} + \mathbf{t}]$. Consider the space group $P2_1/c$. Only the translations modulo the unit-cell translation are of interest. Since the cell is primitive, $\mathbf{t}_1 = 0$. Also from the space group symbol it is evident that t_{2y} and $t_{3z} = \frac{1}{2}$. Since this group possesses a center of symmetry, it is most convenient to place the center of symmetry at the origin; that is, $\mathbf{t}_4 = 0$; and from row 2, column 3, $2\mathbf{t}_3 + \mathbf{t}_2 = 0$. Solving for the

Table 1.6 Cayley Table for Space Groups Isomorphous With C_{2h}

	$[E/t_1]$	$[2/t_2]$	$[m/t_3]$	$[i/t_4]$
$[E]/t_1]$	$[E/2t_1]$	$[2/t_2 + t_1]$	$[m/t_3 + t_1]$	$[i/t_4 + t_1]$
$[2/t_2]$	$[2/2t_1 + t_2]$	$[E/2t_2 + t_2]$	$[i/2t_3 + t_2]$	$[m/2t_4 + t_2]$
$[m/t_3]$	$[m/mt_1 + t_3]$	$[i/mt_2 + t_3]$	$[E/mt_3 + t_3]$	$[2/mt_4 + t_3]$
$[i/t_4]$	$[i/it_1 + t_4]$	$[m/it_2 + t_4]$	$[2/it_3 + t_4]$	$[E/it_4 + t_4]$

individual components of these column vectors one obtains $t_{2x} = t_{3x} = 0$, $t_{2z} = \frac{1}{2}$, and $t_{3y} = \frac{1}{2}$, which produces the same set of coordinates described earlier.

Note that when the equivalent points lie on the screw axes or glide planes there are still four equivalent points in the unit cell, but if they lie on the symmetry centers, they reduce to only two: $0, 0, 0$ and $0, \frac{1}{2}, 0$. Very frequently the determination of the unit cell and the space group leads to the unique conclusion that there are two molecules in a unit cell of symmetry $P2_1/c$ (or the equivalent $P2_1/a$ or $P2_1/n$). Except for the rare case of orientational disorder one can then be sure that the molecule has a center of symmetry. This particular case arises so frequently that it is worth testing for if there is any possibility that a molecule has a center of symmetry, for the amount of time involved in such a determination is quite short.

3.3.5 The Space Group $P2_12_12_1$

In the orthorhombic system symmetry elements occur along the **a**, **b**, and **c** directions, and in particular the space group $P2_12_12_1$ is very frequently chosen by chiral organic molecules when they crystallize. Four molecules, or some multiple of four, occur in this unit cell.

Consider the extension of $P2_1$ by a screw axis along another direction perpendicular to the first. Introduction of the second screw axis so that it intersects the first leads to the perfectly good space group $P2_12_12$, but we introduce the second screw axis so that it does not intersect the first one (Figure 1.60). The points x, y, z and $-x, y + \frac{1}{2}, -z$ of $P2_1$ then create the additional points $\frac{1}{2} - x, -y, \frac{1}{2} + z$ and $\frac{1}{2} + x, \frac{1}{2} - y, \frac{1}{2} - z$, and these four points are equivalent in $P2_12_12_1$. If we list these points in a column, subtract $0\ 0\ \frac{1}{4}$ to change the origin by $\frac{1}{4}$ **c** in the second column, and define $z_1 = z - \frac{1}{4}$, we find

$$
\begin{array}{lll}
x, y, z & x, y, z - \frac{1}{4} & x, y, z_1 \\
-x, y + \frac{1}{2}, -z & -x, y + \frac{1}{2}, -\frac{1}{4} - z & -x, y + \frac{1}{2}, \frac{1}{2} - z_1 \\
\frac{1}{2} - x, -y, \frac{1}{2} + z & \frac{1}{2} - x, -y, \frac{1}{4} + z & \frac{1}{2} - x, -y, \frac{1}{2} + z_1 \\
\frac{1}{2} + x, \frac{1}{2} - y, \frac{1}{2} - z & \frac{1}{2} + x, \frac{1}{2} - y, \frac{1}{4} - z & \frac{1}{2} + x, \frac{1}{2} - y, -z_1
\end{array}
$$

It is worth noting that in the space group $P2_12_12_1$ there are no special positions that lead to less than four equivalent atoms.

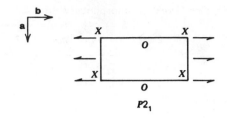

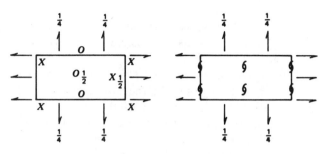

Figure 1.60 The space group $P2_12_12_1$, obtained by extension of $P2_1$ with another 2_1 not intersecting the first.

3.4   Effect of Symmetry Elements on X-Ray Intensities

3.4.1   Extinctions

Any symmetry element involving a translation produces systematic absences of certain types of X-ray diffraction maxima. These elements include centering operations in translation groups, glide planes, and screw axes. Three illustrations are given below.

EXTINCTIONS CAUSED BY A B-CENTERED LATTICE

If a lattice is B centered, then for every atom or molecule at x, y, z there exists an equivalent atom or molecule at $\frac{1}{2} + x, y, \frac{1}{2} + z$. The structure factor expression then becomes, for these two atoms only,

$$F_{hkl} = \sum_{j=1}^{N=2} f_j \exp[-2\pi i(hx_j + ky_j + lz_j)]$$

$$= f(\exp[-2\pi i(hx + ky + lz)] + \exp\{-2\pi i[h(\tfrac{1}{2} + x) + ky + l(\tfrac{1}{2} + z)]\})$$

$$= f \exp[-2\pi i(hx + ky + lz)](1 + \exp\{-2\pi i[(h + l)/2)]\})$$

Remember that h and l are always integers and that $e^{i\Theta} = \cos \Theta + i \sin \Theta$. The exponential $\exp\{-2\pi i[(h + l)/2]\}$ is -1 when $h + l$ is odd and is $+1$ when $h + l$ is even. Hence, $F_{hkl} = 0$ when $h + l$ is odd.

In general, if there are more than two atoms in the unit cell the reduction is

$$F_{hkl} = \sum_{j=1}^{N} f_j \exp[-2\pi i(hx_j + ky_j + lz_j)]$$

$$= \sum_{j=1}^{N/2} f_j \exp[-2\pi i(hx_j + ky_j + lz_j)](1 + \exp\{-2\pi i[(h+l)/2]\})$$

where the sum from $j = 1$ to $N/2$ extends over all atoms not related by the B centering.

EXTINCTION CAUSED BY A b-GLIDE PLANE PERPENDICULAR TO THE c AXIS

A b glide perpendicular to c, with z coordinate $z = 0$, means that if there is an atom at x, y, z, then there is an equivalent atom $x, \frac{1}{2} + y, \bar{z}$. The structure factor then becomes

$$F_{hkl} = \sum_{j=1}^{N/2} f_j(\exp[-2\pi i(hx_j + ky_j + lz_j)] + \exp\{-2\pi i[hx_j + k(\frac{1}{2} + y_j) - lz_j]\})$$

where the sum extends over the $N/2$ atoms not related by this b glide. For the general hkl reflections no systematic extinction results, but for the $hk0$ reflections,

$$F_{hk0} = \sum_{j=1}^{N} f_j \exp[-2\pi i(hx_j + ky_j)]\{1 + \exp[-2\pi i(k/2)]\}$$

Hence, all reflections of the type $hk0$ vanish when k is odd.

EXTINCTION CAUSED BY A TWOFOLD SCREW AXIS ALONG $\mathbf{b}$

The symmetry operation of a 2_1 axis along $\mathbf{b}$, with x and z coordinates zero, transforms an atom at x, y, z into an equivalent atom at $\bar{x}, y + \frac{1}{2}, \bar{z}$. The structure-factor expression becomes

$$F_{hkl} = \sum_{j=1}^{N/2} f_j(\exp[-2\pi i(hx_j + ky_j + lz_j)] + \exp\{-2\pi i[-hx_j + k(\frac{1}{2} + y_j) - lz_j]\})$$

Systematic absences occur for the $0k0$ reflections. The expression reduces to

$$F_{0k0} = \sum_{j=1}^{N/2} f_j \exp[-2\pi i(ky_j)]\{1 + \exp[-2\pi i(k/2)]\}$$

where the sum extends over the $N/2$ atoms not related by the 2_1 axis. Thus the $0k0$ reflections are absent when k is odd.

EXTINCTIONS CAUSED BY OTHER SYMMETRY ELEMENTS

A complete list of extinctions caused by symmetry are given in Table 1.7 [71].

Table 1.7 Extinctions Caused by Symmetry Elements[a]

Class of Reflection	Condition for Nonextinction (n, an Integer)	Interpretation of Extinction	Symbol for Symmetry Element
hkl	$h + k + l = 2n$	Body-centered lattice	I
	$h + k = 2n$	C-centered lattice	C
	$h + l = 2n$	B-centered lattice	B
	$k + l = 2n$	A-centered lattice	A
	$h + k, k + l$	Face-centered lattice	F
	$h + l = 2n$		
	$-h + k + l = 3n$	Rhombohedral lattice indexed on hexagonal lattice	R
	$h + k + l = 3n$	Hexagonal lattice indexed on rhombohedral lattice	H
$0kl$	$k = 2n$	Glide plane $\perp \mathbf{a}$, translation $\mathbf{b}/2$	b
	$l = 2n$	Glide plane $\perp \mathbf{a}$, translation $\mathbf{c}/2$	c
	$k + l = 2n$	Glide plane $\perp \mathbf{a}$, translation $(\mathbf{b} + \mathbf{c})/2$	n
	$k + l = 4n$	Glide plane $\perp \mathbf{a}$, translation $(\mathbf{b} + \mathbf{c})/4$	d
$h0l$	$h = 2n$	Glide plane $\perp \mathbf{b}$, translation $\mathbf{a}/2$	a
	$l = 2n$	Glide plane $\perp \mathbf{b}$, translation $\mathbf{c}/2$	c
	$h + l = 2n$	Glide plane $\perp \mathbf{b}$, translation $(\mathbf{a} + \mathbf{c})/2$	n
	$h + l = 4n$	Glide plane $\perp \mathbf{b}$, translation $(\mathbf{a} + \mathbf{c})/4$	d

hk0	$h = 2n$	Glide plane $\perp \mathbf{c}$, translation $\mathbf{a}/2$	a
	$k = 2n$	Glide plane $\perp \mathbf{c}$, translation $\mathbf{b}/2$	b
	$h + k = 2n$	Glide plane $\perp \mathbf{c}$, translation $(\mathbf{a} + \mathbf{b})/2$	n
	$h + k = 4n$	Glide plane $\perp \mathbf{c}$, translation $(\mathbf{a} + \mathbf{b})/4$	d
hhl	$l = 2n$	Glide plane $\perp(\mathbf{a} - \mathbf{b})$, translation $\mathbf{c}/2$	c
	$2h + l = 2n$	Glide plane $\perp(\mathbf{a} - \mathbf{b})$, translation $(\mathbf{a} + \mathbf{b} + \mathbf{c})/2$	n
	$2h + l = 4n$	Glide plane $\perp(\mathbf{a} - \mathbf{b})$, translation $(\mathbf{a} + \mathbf{b} + \mathbf{c})/4$	d
h00	$h = 2n$	Screw axis $\parallel \mathbf{a}$, translation $\mathbf{a}/2$	$2_1, 4_2$
	$h = 4n$	Screw axis $\parallel \mathbf{a}$, translation $\mathbf{a}/4$	$4_1, 4_3$
00l	$l = 2n$	Screw axis $\parallel \mathbf{c}$, translation $\mathbf{c}/2$	$2_1, 4_2, 6_3$
	$l = 4n$	Screw axis $\parallel \mathbf{c}$, translation $\mathbf{c}/4$	$4_1, 4_3$
	$l = 3n$	Screw axis $\parallel \mathbf{c}$, translation $\mathbf{c}/3$	$3_1, 3_2, 6_2, 6_4$
	$l = 6n$	Screw axis $\parallel \mathbf{c}$, translation $\mathbf{c}/6$	$6_1, 6_5$
hh0	$h = 2n$	Screw axis $\parallel(\mathbf{a} + \mathbf{b})$, translation $(\mathbf{a} + \mathbf{b})/2$	2_1

[a]Adapted from M. J. Buerger, *X-Ray Crystallography*, Wiley, New York, 1942, p. 83; *International Tables for X-Ray Crystallography*, Vol. I, The Kynoch Press, Birmingham, England, 1952, p. 54.

3.4.2   Statistical Relations

While symmetry elements without translations do not produce systematic extinctions, they do influence the intensity distributions. For example, Wilson [72] has shown that for a crystal containing a reasonably large number of atoms of about the same atomic number in the unit cell, the ratio of the square $\langle F \rangle^2$ of the mean value $\langle F \rangle$ to the mean value $\langle I \rangle$, where $I = |F|^2$, is significantly different for centrosymmetric unit cells (0.785) than for noncentrosymmetric unit cells (0.637). A more detailed examination of this and related, more powerful criteria [73], and a comparison with experimental data do indicate that this method is useful in deciding between centrosymmetric and noncentrosymmetric space groups for complex structures. However, some caution is necessary, for pseudosymmetry may exist; that is, the structure may closely resemble a centrosymmetric arrangement without actually having this symmetry, and the statistical method may then give an incorrect result. Similar statistical relations exist for other symmetry elements, and they are usually most apparent from examination of the limited amount of data most strongly influenced by these symmetry elements.

4   PRINCIPLES OF EXPERIMENTAL METHODS

4.1   Introduction

In Section 1 only a few of the many methods available for the solution of structures were discussed in any detail. We follow the same philosophy in this, the experimental section. (For more detailed discussion of the experimental methods of this section consult [74].) In recent years, counterdata collection techniques have almost completely replaced film techniques for nonprotein studies, and often film techniques receive minimal use even in the preliminary examination of crystals. While film techniques lack the efficiency and linearity of countertechniques, they do have the advantage of affording a two-dimensional view of reciprocal space. Hence, we begin this section with the details of a structure determination by film techniques of a simple compound, hydrogen fluoride. The major reason for the selection of the compound is the relatively small amount of data involved. We only consider the location of the fluorine atoms and do not consider the hydrogen atom positions, which are, however, discussed in the original publication [75].

Some of the difficulties in this study somewhat resemble difficulties in studies of very complex structures. Low-temperature techniques [76] must be used here because of the low melting point of the crystal, and they must also be used in proteins to decrease the destruction rate of the crystal by the products of X radiation. Also, the material must be sealed into capillaries, here in fluorothene (polytrifluorochloroethylene) capillaries to place the liquid in an inert container, and in glass capillaries for compounds unstable in air, for proteins, or for other crystals containing solvent of crystallization to prevent destruction of the single crystal by evaporation of the solvent.

Several experimental methods are available from which to obtain space-group and unit-cell information and from which to collect intensity data, but the simplest of the film techniques is the *precession method* [77], which was developed by M. J. Buerger. The precession method gives an undistorted picture of the reciprocal lattice. In fact it is useful to think of all other film methods for the study of single crystals as giving photographs of distorted reciprocal space.

4.2 The Principles of the Precession Method

Undistorted photographs of layers of the reciprocal lattice may be obtained in experiments in which the crystal and X-ray film are always maintained in the same relative orientation as they are moved (Figure 1.61). Then, as described in Section 2, as a reciprocal lattice point crosses the sphere of reflection of radius $1/\lambda$, a diffraction maximum occurs in the direction of s, the scattered ray. Since in this method the reciprocal lattice plane shown in Figure 1.61 is always maintained parallel to the film, the resulting scattered rays merely project the reciprocal lattice plane onto the film.

In the precession method the normal N to the reciprocal lattice plane and the normal N' to the X-ray film are maintained parallel as they are rotated slowly, about once a minute, about the line OO' (Figure 1.62). Remember from Section 2 that when a reciprocal lattice plane is being photographed, the direction of N coincides with an axis of the crystal. Usually Mo K_α radiation ($\lambda = 0.711$ Å) is employed with which one can obtain a total number of reflections at least comparable with other methods now in use. A usual crystal-to-film distance is 6.00 cm for zero-level photographs. If we recall from Section 2 that $d_{hkl} = l/|h|$

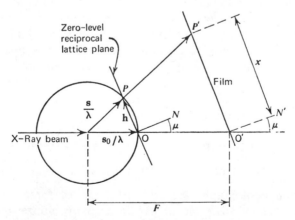

Figure 1.61 Geometry of the de Jong and Bouman and the precession methods. The plane in reciprocal space extends above and below the plane of this figure. A reciprocal lattice point at the end of the vector h satisfies the conditions $s/\lambda - s_0/\lambda = h$ for reflection. The position x at which the reflection occurs on the film is, from similar triangles, $x/F = |h|/|s_0/\lambda| = \lambda/d_{hkl}$. Hence, $d_{hkl} = \lambda F/x$. In this construction, note that the crystal-film distance F is measured from the point at which the diffraction vector s originates; it is helpful to think of the conversion of the crystal into a reciprocal lattice in a volume that is comparable with the size of the crystal.

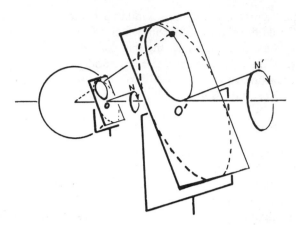

Figure 1.62   As N and N' rotate about the line OO', the portion of the reciprocal lattice in the dotted circle about O is projected undistorted onto the film in the dotted circle about O' as a center.

and that $\mathbf{s}$ and $\mathbf{s_o}$ are unit vectors, the interplanar spacing is, from similar triangles, $d_{hkl} = \lambda F/x$, as derived in the legend of Figure 1.61. During the complete precession motion of N and N' about OO', a circle of radius OP is projected onto the X-ray film in the zero-level photograph illustrated in Figure 1.62.

Recording of n levels is also possible but this is not described in detail here. Because the n level is displaced from the zero level it becomes necessary to move the film to a different crystal-film distance so that the n-level photograph is also undistorted. In addition, because the n level does not pass through the origin, the central region is missing in its photograph. Finally, in both the zero and n levels, an annular screen is required between the crystal and the film to isolate only those reflections that are in the level being photographed. The geometry for an n level is illustrated in Figure 1.63, and the details can be found in [77, 78].

The precession instrument is still useful for the rapid survey of reciprocal space especially when superlattices, modulated structures, or anisotropic-unit-cell distortions are involved, and for its multidetector characteristics in protein and other very large molecule investigations.

4.3   Example: The Structure of Crystalline Hydrogen Fluoride

4.3.1   The X-Ray Diffraction Photographs

Many diffraction photographs are normally required for a structure determination; it is usually advisable to obtain as many nonzero reflections as possible to obtain certainty of the structure and to keep standard deviations of parameters as low as possible. Of the 20 or so diffraction photographs taken of a single hydrogen fluoride crystal, three are shown in Figure 1.64.

Some of the features of these photographs are common to all X-ray diffraction photographs of single crystals. The finite size of the maxima is mostly

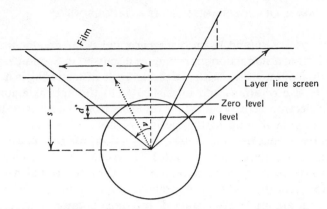

Figure 1.63   The geometry of an n-level photograph, and the position of the layer line screen, which excludes reflections from other levels, such as the zero-level reflection shown by a dotted line. The relation $\tan v = r/s$, which determines the setting of the layer line screen, may be expressed in terms of d^* and the precession angle.

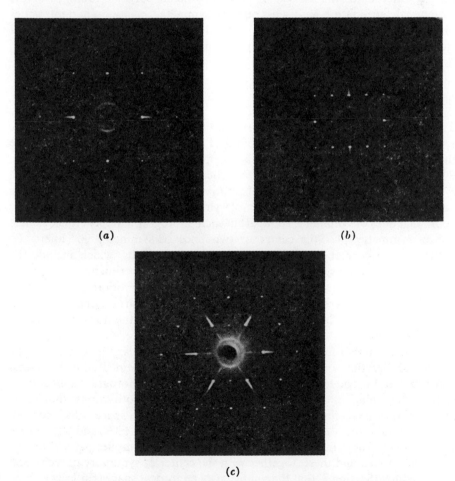

(a)

(b)

(c)

Figure 1.64   Diffraction photographs of a single crystal of hydrogen fluoride.

caused by crystal size and beam divergence and is far greater than the usual intrinsic width of a diffraction maximum as discussed in Section 2. The average overall decline of intensities with increasing angle of scattering is mostly caused by the finite size of atoms and in particular to the approximately Gaussian decrease of electron density as one moves away from the nucleus; but there are other factors contributing to this decline of intensities, such as the thermal motions of the atoms, the polarization of radiation, and the time each reflection is in a position to reflect (Lorentz factor). These effects were given a preliminary quantitative description in Section 2. In addition, the radial white radiation streaks (*wheaks* [79]) actually appear on both sides of the most intense reflections, but are much less just inside the spots because of absorption by the zirconium filter that was introduced to remove the K_β radiation. The black area in the center of the photograph was shielded from the direct X-ray beam by a circular piece of lead. A white ring around this dark area and the much less intense outer halos are caused by the fluorothene polymer tubing. For the structure determination we discuss only the symmetry, positions, and intensities of the diffraction maxima, which occur as white spots on these prints and as dark spots on the original negatives, which were exposed to the X-ray scattering from the crystal.

4.3.2   Symmetry of the Reciprocal Lattice

The symmetry of the reciprocal lattice is the point group obtained from the symmetry of the various levels of reciprocal space on these photographs. Since the substance under study does not show appreciable X-ray absorption near the wavelength of Mo K_α radiation, the reciprocal lattice necessarily has a center of symmetry. These photographs are of zero levels of reciprocal space, and hence contain the origin, so that a center of symmetry is necessarily present. The additional lines of symmetry, 90° apart in each photograph, imply either a single line of symmetry plus the center or two lines of symmetry 90° apart. To determine the correct possibility additional photographs, which include the higher level reflections on the zero levels, are taken in experiments in which the annular layer line screen is omitted. The additional photographs, not shown here, indicate that there are two lines of symmetry 90° apart on photographs (*a*) and (*c*), but only one line of symmetry perpendicular to the horizontal line on photograph (*b*).

Now it is known from the experiment that the capillary axis is along the horizontal direction and that these photographs are obtained with the crystal reoriented at 33° and 90° about the horizontal axis for photographs (*b*) and (*c*), respectively, relative to photograph (*a*). Thus the photographs should be regarded as three sections through the origin of reciprocal space, which contain the horizontal direction in common and are oriented at 0°, 33°, and 90° relative to each other. The symmetry of photographs (*a*) and (*c*) implies pairs of mirror planes 90° apart; and these, together with the center of symmetry of reciprocal space, require (Section 3) that the symmetry of reciprocal space is at least as high as $D_{2h} - mmm$. The single mirror plane of photograph (b) is consistent with these observations. The directions from the origin to the first reflection along the vertical directions of photographs (a) and (c) and along the horizontal direction of all three photographs represent the directions of special symmetry, and hence will be chosen as the axial directions. That the distances to these first reflections in these three directions are not related by an integer or some multiple of $\sqrt{2}$ or $\sqrt{3}$ is a strong indication, later confirmed by many additional photographs, that the symmetry of reciprocal space is not higher than $D_{2h} - mmm$. Therefore, we conclude that the crystal structure is orthorhombic. Note that this conclusion is based on symmetry requirements and not on any special axial lengths or angles between axes over and above those required by the symmetry requirements. Finally, the symmetry of reciprocal space; that is, the diffraction symmetry, is not usually the same as the symmetry of the crystal. All orthorhombic crystals show $D_{2h} - mmm$ diffraction symmetry, but there are many different orthorhombic crystal symmetries.

4.3.3 The Unit Cell

The simplest approach to the determination of the unit cell is to choose axes according to symmetry directions, find the unit cell in reciprocal space, and convert this unit cell by the methods of Section 2 into the unit cell in crystal space. In the orthorhombic system the directions of all axes are determined by symmetry, and the directions of the reciprocal and crystal axes coincide. In reciprocal space these directions are perpendicular to the mirror planes; that is, along the twofold axes of the $D_{2h} - mmm$ symmetry group. Hence, we are required to choose one axis, say b_1, along the vertical direction of photograph (a); the second axis, say b_2, along the horizontal direction of photograph (a), (b), or (c); and the third axis, b_3, along the vertical direction of photograph (c). The distances of the closest reflections along b_1 and b_2, the relative orientations of the three photographs about b_2, and the smallest reciprocal cell in the b_1 and b_3 directions are summarized in Figure 1.65. We must now choose a cell in three-dimensional reciprocal space in such a way that we account for all observed reflections in a three-dimensional array. The distances of the closest reflections to the center along b_1, b_2, and b_3 are, respectively, 2.49, 1.97, and 1.58 cm. But a unit cell of this size would not include the first reflection along the vertical direction of photograph (b), and hence we are required to choose a cell half as big along b_1 and b_3, as illustrated in Figure 1.65. Similarly we are required to halve the 1.97 direction along b_2, the horizontal direction on the photographs, if we are to avoid omitting about half of the reflections, which have components along b_2 of $(\frac{1}{2})$ 1.97 cm. Thus the reciprocal unit cell has dimensions of $b_1 = 1.245$ cm, $b_2 = 0.985$ cm, and $b_3 = 0.79$ cm. This cell accounts for all observed reflections, and there are a number of absent reflections that are discussed below.

Having chosen a suitable cell in reciprocal space, we can now "index" the photographs; that is, assign three integers to each reflection. Along the

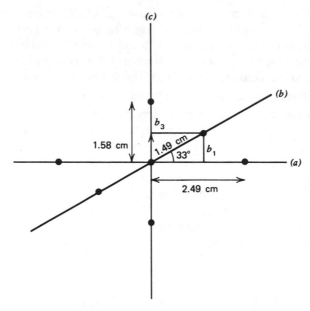

Figure 1.65 Relative orientations of photographs (a), (b), and (c) about the horizontal axis common to all three. Measurements are those from the center to the first reflection in the vertical direction on each photograph. The smallest unit cell of reciprocal space that will account for all reflections has dimensions $b_1 = 1.245$ cm, $b_2 = 0.985$ cm ($\perp b_1$ and b_3), and $b_3 = 0.79$ cm.

horizontal direction of the photographs we find the 020 and 040 reflections. Along the three vertical directions of the photographs we find 200 on (a), 101 on (b), and a weak 002 and a strong 004 both on (c). The strongest reflection on (c) is 012. As proven in Section 2, reflections on (a) should be of the type $hk0$, those on (b) of the type hkh, and those on (c) of the type $0kl$. The axes about which these photographs were taken are, respectively, [001] for (a), [10$\bar{1}$] for (b), and [100] for (c), and reflections (hkl) lying in the zero level about an axis [uvw] satisfy the relation $hu + kv + lw = 0$, as also proven in Section 2.

We have identified the vectors $\mathbf{h}_1 = \mathbf{b}_1 + \mathbf{b}_3$ and $\mathbf{h}_2 = \mathbf{b}_2$ in photograph (b) from our choice of unit cell. Let $\mathbf{T} = l_1\mathbf{a}_1 + l_2\mathbf{a}_2 + l_3\mathbf{a}_3$ be any vector in crystal space. Then, if $\mathbf{h}_1$ and $\mathbf{h}_2$ are perpendicular to $\mathbf{T}$, $\mathbf{T} \cdot \mathbf{h}_1 = 0 = \mathbf{T} \cdot \mathbf{h}_2$, so that $l_1 + l_3 = 0$ and $l_2 = 0$. Hence $l_1 = -l_3$ and $\mathbf{T} = l_1\mathbf{a}_1 - l_1\mathbf{a}_3 = l_1(\mathbf{a}_1 - \mathbf{a}_3)$, and the axis [101] is therefore perpendicular to $\mathbf{h}_1$ and $\mathbf{h}_2$. Finally, the general reciprocal lattice vector is $\mathbf{h} = h_1\mathbf{b}_1 + h_2\mathbf{b}_2 + h_3\mathbf{b}_3$. Those $\mathbf{h} \perp \mathbf{T} = \mathbf{a}_1 - \mathbf{a}_3$ must satisfy the relation $\mathbf{h} \cdot \mathbf{T} = 0 = h_1 - h_3$. Hence $h_1 = h_3$, but no restriction occurs for h_2, and planes with indices $h_1 h_2 h_1 = hkh$ lie in the zone of the [10$\bar{1}$] axis. This general procedure is valuable for indexing photographs.

In general, the safest procedure to be followed to obtain the unit cell is first to calculate the dimensions of the reciprocal axes in reciprocal angstroms (Å^{-1}) and the angles among them, and then to convert to the real cell by the equations of Section 2. The geometry of the preceding precession method was shown to give the relation $|\mathbf{h}| = 1/d = x/(\lambda F)$. Hence, with the knowledge that the crystal-to-film distance is 6.00 cm and the wavelength of the X rays is 0.711 Å, we can then convert these film measurements into reciprocal axes in reciprocal angstrom units. Thus $b_1 = |\mathbf{h}_{100}| = 1/d_{100} = 1.245$ cm$/[(0.711$ Å$)$ $(6.00$ cm$)] = 0.292$ Å^{-1}. Then, in the orthorhombic system, $a_1 = 1/b_1 = 1/(0.292$ Å$^{-1}) = 3.42$ Å. Similarly, we find $a_2 = 4.32$ Å and $a_3 = 5.41$ Å. The volume V of the unit cell is 80 Å^3.

4.3.4   Number of Molecules in the Unit Cell

The measured density of solid hydrogen fluoride is 1.658 g $\cdot$ cm^{-3}. If in the unit cell there are Z molecules of hydrogen fluoride and if each molecule weighs $M/N = 20.008/6.023 \times 10^{23} = 3.32 \times 10^{-23}$ g, then the density is $D = (ZM)/(NV)$. Therefore, there are

$$Z = \frac{DV}{(M/N)} = \frac{(1.658)(80 \times 10^{-24})}{3.32 \times 10^{23}} = 4.0$$

molecules in the unit cell.

Sometimes an experimental value of the density of a substance is not available; however, it is usually possible to obtain an unambiguous answer for the number of molecules in the unit cell. One estimates either the density from densities of known related compounds or the volume per molecule from related molecules. Moreover, symmetry elements are usually present that may require multiples of two, three, four, and more, for the number of molecules in the unit cell. We shall see that for the symmetry of the hydrogen fluoride crystal there must be multiples of four; that is, the unit cell of hydrogen fluoride may have only 4, 8, 12,... molecules from symmetry arguments alone. Thus the experimental value of the density need not be known with accuracy, and hence rough estimates often suffice.

4.3.5   Symmetry Elements from Extinctions

If we use the symbols s, m, w, and v for *strong, medium, weak,* and *very,* we then find on photograph (*a*) the reflections 020 (s), 040 (m), 200 (s), 400 (vvw), 220 (m), 240 (vw); on photograph (*b*) the reflections 020 (s), 040 (m), 101 (vs), 111 (s), 121 (m), 131 (w), 141 (vw), 212 (w), 232 (vw); and on photograph (*c*) the reflections 020 (s), 040 (m), 002 (w), 012 (vs), 032 (m), 052 (w), 004 (m), 024 (w), 044 (vw), 016 (vw), 036 (w). In all, 42 reflections, listed in Table 1.8, were observed on the films. Systematic absences were observed for all reflections *hkl* when $h + l$ was odd, and a further systematic absence of *hk*0 reflections was noted when *k* was odd. Many well-exposed photographs with low background are required for one to be reasonably sure of these systematic absences by film techniques, or these reflections have to be measured by countertechniques. Erroneous structure interpretations have occurred because of lack of care at this stage of the investigation.

Table 1.8 Comparison of Observed and Calculated Values[a] of F_{hkl}

hkl	F_o	$F_{0.115}$	$F_{0.135}$	hkl	F_o	$F_{0.115}$	$F_{0.135}$
200	14.6	16.2	16.2	303	2.4	−2.0	−3.0
400	1.8	3.2	3.2	204	4.4	−4.6	−4.6
020	20.8	20.9	−20.9	103	6.6	−6.6	−9.7
040	8.5	8.6	8.6	206	<1.1	−0.7	0.7
002	3.4	2.8	−2.8	111	14.3	−14.0	−15.9
004	8.9	−8.4	−8.4	121	11.0	−11.5	−10.1
006	1.0	−1.1	1.1	131	7.4	7.0	7.9
220	10.0	−9.7	−9.7	141	5.0	5.4	4.8
240	4.4	5.2	5.2	151	3.8	−3.2	−3.6
420	1.5	−2.1	−2.1	212	9.0	−9.1	−9.1
012	18.5	−20.1	−20.1	222	<1.3	−1.0	1.0
022	1.2	−1.9	1.9	232	5.4	5.8	5.8
032	9.0	10.1	10.1	242	<1.3	0.5	−0.5
042	<1.0	0.9	−0.9	313	2.6	−2.9	−2.0
052	4.9	−4.6	−4.6	323	1.8	1.7	2.5
014	2.1	−2.0	2.0	333	1.9	2.1	1.4
024	7.0	6.6	6.6	214	<1.4	−1.1	1.1
034	1.0	1.3	−1.3	224	3.7	3.5	3.5
044	4.3	−3.8	−3.8	234	<1.4	0.7	−0.7
016	3.4	2.6	2.6	113	8.9	−9.0	−6.1
026	1.0	0.9	−0.9	123	5.3	5.0	7.3
036	3.8	−2.0	−2.0	133	4.9	5.5	3.8
101	18.4	17.8	15.7	143	2.8	−2.7	−4.0
202	1.2	1.2	−1.2	216	2.1	1.6	1.6

[a]Values of $\dfrac{\Sigma \, \|F_o| - |F_c\|}{\Sigma |F_o|}$:

	All Observed F values	F values for $l = 1,3$	F values for $l = 3$
for $z = 0.115$ are	0.08	0.06	0.05
for $z = 0.135$ are	0.13	0.21	0.34

Thus $z = 0.115$.

The absences shown in Section 3 indicate a *B*-centered cell and a *b*-glide plane perpendicular to the *c* axis. Thus the space group symbol is $B\_\_b$, where a symmetry element such as *m*, 2, or 2_1 is present in each of the blank spaces. Note that the extinction of 0*k*0, when *k* is odd, is already contained in the glide-plane extinction, so the possible presence of 2_1 along *b* is obscured.

A word of caution about the space group is added. The hydrogen atom scatters so little compared to fluorine that the true space group can be of different symmetry. This point is discussed more fully in the published paper [75], and in what follows we discuss the fluorine positions only. While this is a general comment that applies to organic crystals, too, the hydrogen atoms often are important in determining the packing of molecules; hence, they help determine the positions of other heavier atoms. Thus it is unlikely that many organic structures have a true symmetry different from that established by the atoms heavier than hydrogen.

Of course quantitative intensity measurements are required for the structure determination. For the measurement of intensities, scintillation or proportional counters are commonly used and give quite accurate intensity values. However, in some cases experimental factors dictate the use of intensities estimated from films. One technique involves using a microdensitometer to evaluate the amount of film blackening; another employs a standard scale of intensities and measurements by visual comparison of intensities. Relative intensities are then corrected for Lorentz and polarization factors, as described in Section 2, usually on a computer. Equivalent data are averaged and one thus obtains a complete set of all observed $|F_o|^2$ on the same relative scale. The usual $|F_o|$ is obtained by taking the positive square root of each $|F_o|^2$.

4.3.6   Solution of the Structure

There are only four fluorine atoms in the unit cell; since both the B-centering and b-glide planes perpendicular to $\mathbf{c}$ have translational components, an atom placed anywhere in the unit cell will have three others generated by these symmetry elements. Hence, we do not need to investigate possible space groups further to solve the structure as given by the fluorine atoms. If an atom is placed at x, y, z, the b glide perpendicular to $\mathbf{c}$ produces an equivalent atom at $x, \frac{1}{2} + y, \bar{z}$; and the B centering produces equivalent atoms at $\frac{1}{2} + x, y; \frac{1}{2} + z$; and $\frac{1}{2} + x$; $\frac{1}{2} + y; \frac{1}{2} - z$ from these two. Note that x and y can have any values, provided that there are only 4F in the unit cell. Hence, for convenience, we choose $x = 0$, $y = \frac{1}{4}$, to place the origin at a center of symmetry. Accordingly, the 4F are at the positions

$$0, \tfrac{1}{4}, z \qquad 0, \tfrac{3}{4}, \bar{z} \qquad \tfrac{1}{2}, \tfrac{1}{4}, \tfrac{1}{2} + z \qquad \tfrac{1}{2}, \tfrac{3}{4}, \tfrac{1}{2} - z$$

and thus there is only one positional parameter z to be fixed by detailed consideration of the intensities. The structure factor becomes

$$
\begin{aligned}
F_{hkl} &= \sum_{j=1}^{4} f_j \exp(-2\pi i \mathbf{h} \cdot \mathbf{r}_j) \\
&= f_F(\exp[-2\pi i(k/4 + lz)] + \exp[2\pi i(k/4 + lz)] \\
&\quad + \exp\{-2\pi i[h/2 + k/4 + l(\tfrac{1}{2} + z)]\} + \exp\{2\pi i[h/2 + k/4 + l(\tfrac{1}{2} + z)]\}) \\
&= f_F\{\exp[-2\pi i(k/4 + lz)] + \exp[2\pi i(k/4 + lz)]\}(1 + \exp\{-2\pi i[(h + l)/2]\}) \\
&= f_F 2 \cos 2\pi(k/4 + lz)(1 + \exp\{-2\pi i[(h + l)/2]\}) \qquad\qquad (45)
\end{aligned}
$$

We have used the substitution of $-\frac{1}{2} - z$ for $\frac{1}{2} - z$, which is a cell translation. Now the final bracket is 0 if $h + l$ is odd and 2 if $h + l$ is even, and hence

$$F_{hkl}(h + l \text{ even}) = f_F 4 \cos 2\pi(k/4 + lz)$$

An approximate value for z can be found from the weak 002 reflection. If 002 were zero, then $z = \frac{1}{8}, \frac{3}{8}, \frac{5}{8}$, or $\frac{7}{8}$. These choices all lead to identical structures (Figure 1.66) except for the choice of origin, and hence, we discuss values of z near $\frac{1}{8}$. Now we consider a slight displacement of the fluorine atom from $z = \frac{1}{8}$ to $z = \frac{1}{8} \pm \delta$. Then

$$F_{hkl}(h + l \text{ even}) = 4f_F \left[\cos 2\pi\frac{k}{4} \cos 2\pi l \left(\frac{1}{8} \pm \delta\right) - \sin 2\pi\frac{k}{4} \sin 2\pi l \left(\frac{1}{8} \pm \delta\right) \right]$$

which for 002 becomes $\pm \sin 4\pi\delta$. Hence, the magnitude of F_{002} is the same whether the displacement δ from $\frac{1}{8}$ is positive or negative. Only those reflections for which l is odd distinguish between these two possible structures; but only one

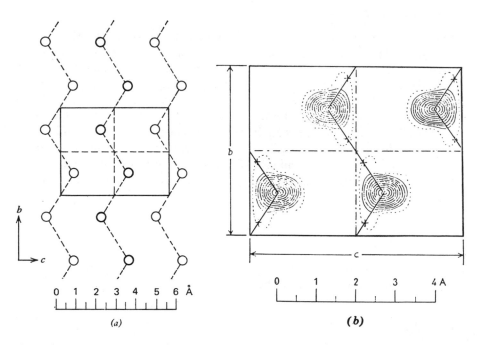

(a)   (b)

Figure 1.66 (a) Arrangements of the fluorine atoms in the a-axis projection. Atoms represented by heavy and light circles are at $x = \frac{1}{2}$ and 0, respectively. Hydrogen bonds are indicated by broken lines. (b) Electron-density projection along the a axis. Each contour line represents a density increment of $1e/\text{Å}^2$, the two-electron line being broken. The cross symbols, showing the possible positions of the hydrogen atoms, are based on the hydrogen fluoride distance of 0.92 Å in the gas phase and suggest some randomness with a half-hydrogen at each cross.

of these structures is correct. The detailed calculations are carried out by substituting the values for f_F at the approximate value of $\sin \Theta/\lambda$ from the *International Tables for X-Ray Crystallography* [61] into (45), including temperature factors (see the original publication [75]), and computing all F_{hkl}. The results are summarized in Table 1.8.

It is unusual that an incorrect structure can give such good agreement, but the error in the fluorine position is only $0.02b$, about 0.1 Å. These results indicate that caution is necessary in deriving structures and that one must demand equally good agreement among all classes of reflections. Usually poor agreement for one projection, for example, as compared with the agreement for other projections, or for all reflections for $h + l$ odd as compared with $h + l$ even, or for the weighted agreement for small $\sin \Theta$ as compared with those for large $\sin \Theta$ indicates that the structure is either incorrect or insufficiently refined.

4.4   The Weissenberg Method

The Weissenberg method [80] is another commonly used film method for the investigation of the reciprocal lattice; we discuss it only briefly here. In this method, in the most general case, the crystal is rotated about an axis through a $> 180°$ range. Referring to the Ewald construction shown in Figure 1.37, if we assume the rotation axis is perpendicular to the plane of the paper, then all the points in that zero-layer reciprocal lattice plane within the radius $2/\lambda$ pass through the sphere and hence give rise to diffraction maxima. The crystal is surrounded by a cylindrical film, the axis of the cylinder being coincident with the rotation axis. A cylindrical screen is employed to allow only reflections from the desired layer to reach the film. To resolve and index the resulting pattern of intensities the film carriage is translated while the crystal is being rotated ($2°/\text{mm}$). Because of the translational motion of the film, any radial line in reciprocal space is represented by a line making an angle of $\tan^{-1} (2) = 63°26'$ with the direct beam trace (the horizontal line of Figure 1.67). Copper radiation is usually employed in the Weissenberg technique, and this method gives similar information to that obtained in the precession technique; that is, space group, unit cell, and intensity information. Note, however, that for a particular orientation of the crystal these two techniques give complementary information because of the different way reciprocal space is sampled. More details on the Weissenberg method, including a discussion of upper-level photography, can be found in Buerger's *X-Ray Crystallography* [80].

4.5   The Four-Circle Single-Crystal Diffractometer

Single-crystal diffractometers are used to acquire some of the most accurate intensity measurements. They are commonly used for data collection of all but the largest unit cells where the inherent multidetector capability of film and its versatility become overriding factors, as in the case of proteins. One of the most

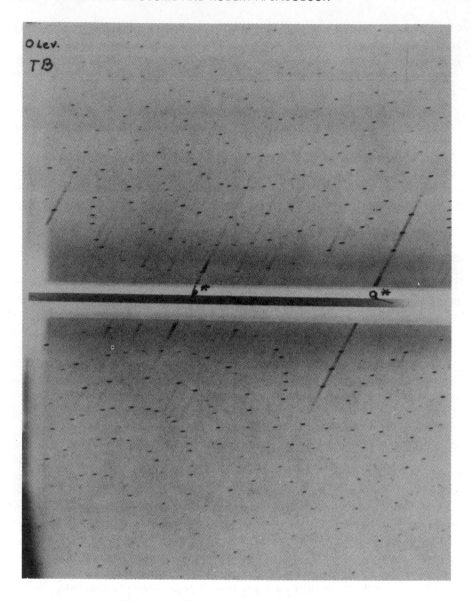

Figure 1.67   Zero-level Weissenberg photograph.

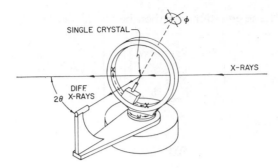

Figure 1.68   Schematic of a four-circle diffractometer.

common diffractometers of this type is the four-circle diffractometer illustrated in Figure 1.68.

The principles of operation of this type of diffractometer follow. Using a Bragg picture, we can visualize the crystal as consisting of numerous sets of planes. To observe a particular diffraction maximum, the appropriate set of planes must be oriented in space such that the angle between the incident beam, which is fixed in space, and the set of planes in the crystal is made equal to the angle Θ given by the Bragg equation, $\Theta = \sin^{-1}(\lambda/2d)$. The counter, usually a scintillation or proportional type, is constrained to move in the horizontal plane in the usual system, and therefore to observe the diffraction maximum the crystal is rotated until the normal to the diffraction plane also lies in the horizontal plane. Then all that remains is to position the counter at an angle Θ to the diffraction plane, and hence at an angle 2Θ to the incident beam.

The crystal orientation axes are shown in Figure 1.68: ω (also Θ and 2Θ), rotation about the vertical axis of the diffractometer; χ, angle of rotation about a horizontal axis through the center of the circle and perpendicular to the plane of the circle; and Φ rotation about the axis of the goniometer head (see also [81]).

Mathematically, these angles can be determined for a particular reflection corresponding to the lattice vector $\mathbf{h}$ as follows. From Bragg's law Θ can be readily determined as we have already indicated, since $|\mathbf{h}| = d^{-1}$. The usual solution for ω is to set it equal to Θ. The angles χ and ϕ are somewhat more difficult to determine. Three unit vectors can be thought of as superimposed on the four-circle diffractometer such that the $\mathbf{k}$ direction is along the ϕ axis at $\chi = 0$; that is, the position at which the ϕ and ω axis are colinear and hence the $\mathbf{k}$ axis is vertical; the $\mathbf{i}$ vector is in the horizontal plane parallel to a vector from the source to the crystal; and the $\mathbf{j}$ vector is a unit vector that also lies in the horizontal plane and forms an orthogonal right-handed system with $\mathbf{i}$ and $\mathbf{k}$. Each of the three principal reciprocal lattice vectors can be resolved into components along the three unit vectors, obtaining b_{1x}, b_{1y}, b_{1z}, b_{2x}, and so on.

Since any reciprocal lattice vector $\mathbf{h}$ can be written by components along $\mathbf{b}_1$,

b_2, and b_3, the Laue equation can then be written as

$$\mathbf{h} = \begin{bmatrix} b_{1x} & b_{2x} & b_{3x} \\ b_{1y} & b_{2y} & b_{3y} \\ b_{1z} & b_{2z} & b_{3z} \end{bmatrix} \begin{bmatrix} h \\ k \\ l \end{bmatrix} = \mathbf{s}/\lambda - \mathbf{s}_o/\lambda \tag{46}$$

or,

$$\mathbf{BH} = \mathbf{s}/\lambda - \mathbf{s}_o/\lambda$$

where the **B** matrix will usually be referred to as the orientation matrix; that is, it supplies information on the orientation on the three principal lattice vectors relative to the laboratory coordinate system, as represented by $\mathbf{i}, \mathbf{j}, \mathbf{k}$. The orientation matrix so described is appropriate for χ and ϕ equal to 0 and $\omega = \Theta$. At any other angular value a new orientation matrix $\mathbf{B}'$ replaces $\mathbf{B}$, where $\mathbf{B}' = \mathbf{F}(\Delta\omega, \chi, \phi)\mathbf{B}$, and $\mathbf{F}$ represents the product of three rotation matrices evaluated as follows:

$$\mathbf{F} = \begin{bmatrix} \cos \Delta\omega & -\sin \Delta\omega & 0 \\ -\sin \Delta\omega & \cos \Delta\omega & 0 \\ 0 & 0 & 1 \end{bmatrix} \begin{bmatrix} 1 & 0 & 0 \\ 0 & \cos \chi & -\sin \chi \\ 0 & -\sin \chi & \cos \chi \end{bmatrix}$$

$$\times \begin{bmatrix} \cos \phi & -\sin \phi & 0 \\ -\sin \phi & \cos \phi & 0 \\ 0 & 0 & 1 \end{bmatrix}$$

In the case of the ordinary four-circle diffractometer, the χ, ϕ, and $\Delta\omega$ angles are selected such that the diffraction vector is brought into coincidence with the unit vector $\mathbf{j}$,

$$\mathbf{B}'\mathbf{H} = \begin{bmatrix} 0 \\ |\mathbf{h}| \\ 0 \end{bmatrix} \quad \text{or} \quad \mathbf{BH} = \mathbf{F}^{-1} \begin{bmatrix} 0 \\ |\mathbf{h}| \\ 0 \end{bmatrix}$$

When $\Delta\omega = 0$, a general description is given by the following:

$$\begin{bmatrix} b_{1x} & b_{2x} & b_{3x} \\ b_{1y} & b_{2y} & b_{3y} \\ b_{1z} & b_{2z} & b_{3z} \end{bmatrix} \begin{bmatrix} h \\ k \\ l \end{bmatrix} = \begin{bmatrix} |\mathbf{h}| & \sin \phi & \cos \chi \\ |\mathbf{h}| & \cos \phi & \cos \chi \\ |\mathbf{h}| & \sin \chi & \end{bmatrix} = \begin{bmatrix} X \\ Y \\ Z \end{bmatrix} \tag{47}$$

and $\tan \phi = (X/Y)$, $\tan \chi = -Z/(X^2 + Y^2)^{1/2}$.

X-RAYS

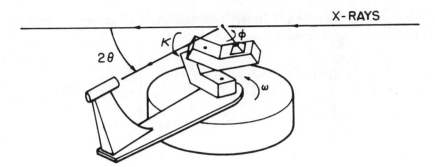

Figure 1.69   Schematic of a kappa diffractometer.

An alternate axes arrangement for a four-circle diffractometer is shown schematically in Figure 1.69 and is termed a *kappa goniometer*. The χ motion in the full-circle diffractometer is eliminated; instead a new axis is introduced that is termed the κ axis, and it intersects the axis at an angle of less than 90°, typically 50°. As noted in Figure 1.69 the ω position of the ϕ axis in space changes as κ changes—remember that ϕ changes with χ in a full-circle diffractometer. The ϕ axis is supported such that at $\kappa = 0$, the ϕ axis and the ω axis are coincident. At $\kappa = 180°$, the ϕ axis is positioned at the equivalent of $\chi = 100°$ (assuming the 50° angle noted previously). In general the foregoing ω, χ, and ϕ rotation matrices need merely to be replaced by ω, κ, and ϕ matrices. The design of the κ goniometer allows access to a greater region of reciprocal space; that is, it has less "blind" areas than the full-circle diffractometer.

4.6   Indexing and Unit Cell Determination on a Diffractometer

Procedures were developed to enable one to rapidly and automatically index an arbitrary subset of reflections and determine the orientation matrix, and hence the unit cell dimensions. In this section we give the reader an outline of the details of one such procedure [82].

Assume that a crystal whose unit cell is unknown is centered on a four-circle diffractometer. Let us next assume that the positions (χ, ω, and 2Θ) of approximately a dozen reflections were determined either with the help of a few quick oscillation photographs using a Polaroid cassette attachment directly on the diffractometer or from a random search. As noted earlier, the angular values can be converted to X, Y, and Z values. Hence, one can write

$$\begin{bmatrix} b_{1x} & b_{2x} & b_{3x} \\ b_{1y} & b_{2y} & b_{3y} \\ b_{1z} & b_{2z} & b_{3z} \end{bmatrix} \begin{bmatrix} h_1 & h_2 & \cdots & h_n \\ k_1 & k_2 & \cdots & k_n \\ l_1 & l_2 & \cdots & l_n \end{bmatrix} = \begin{bmatrix} X_1 & X_2 & \cdots & X_n \\ Y_1 & Y_2 & \cdots & Y_n \\ Z_1 & Z_2 & \cdots & Z_n \end{bmatrix}$$

and since n is greater than 3, in fact approximately 12 in our example, this equation can be subdivided into two equations

$$\begin{bmatrix} b_{1x} & b_{2x} & b_{3x} \\ b_{1y} & b_{2y} & b_{3y} \\ b_{1z} & b_{2z} & b_{3z} \end{bmatrix} \begin{bmatrix} h_1 & h_2 & h_3 \\ k_1 & k_2 & k_3 \\ l_1 & l_2 & l_3 \end{bmatrix} = \begin{bmatrix} X_1 & X_2 & X_3 \\ Y_1 & Y_2 & Y_3 \\ Z_1 & Z_2 & Z_3 \end{bmatrix} \qquad \text{or} \qquad \mathbf{BH_1 = X_1}$$

and

$$\begin{bmatrix} b_{1x} & b_{2x} & b_{3x} \\ b_{1y} & b_{2y} & b_{3y} \\ b_{1z} & b_{2z} & b_{3z} \end{bmatrix} \begin{bmatrix} h_4 & \cdots & h_n \\ k_4 & \cdots & k_n \\ l_4 & \cdots & l_n \end{bmatrix} = \begin{bmatrix} X_4 & \cdots & X_n \\ Y_4 & \cdots & Y_n \\ Z_4 & \cdots & Z_n \end{bmatrix} \qquad \text{or} \qquad \mathbf{BH_2 = X_2}$$

where the $\mathbf{X_1}$ matrix is selected such as to have a nonzero determinant.

Solving for $\mathbf{B}$ from the first equation of the pair and substituting into the second, we obtain

$$\mathbf{B = X_1 H_1^{-1}}$$

$$\mathbf{(X_1 H_1^{-1})H_2 = X_2}$$

$$\mathbf{H_2 = H_1 X_1^{-1} X_2} = \begin{bmatrix} h_1 & h_2 & h_3 \\ k_1 & k_2 & k_3 \\ l_1 & l_2 & l_3 \end{bmatrix} \begin{bmatrix} XX_4 & XX_5 & \cdots & XX_n \\ YY_4 & YY_5 & \cdots & YY_n \\ ZZ_4 & ZZ_5 & \cdots & ZZ_n \end{bmatrix}$$

where the $\mathbf{XX}$ matrix is $\mathbf{X_1^{-1} X_2}$. Note that $\mathbf{XX}$ can be directly evaluated from experimental data. This matrix equation yields scalar equations of the type

$$h_j = h_1 XX_j + h_2 YY_j + h_3 ZZ_j \qquad (j = 4, n) \qquad (48)$$

This equation is now in a form that is readily amenable to rapid testing and elimination procedures on a computer. A valid h_1, h_2, h_3 triplet must not only be integer, but also yield $h_j(j = 4, n)$, which are integers. Since there are many ways to describe a unit cell corresponding to different choices of axes, there are many equally valid solutions. Once one is obtained, the resultant orientation matrix can be "reduced"; that is, converted to one whose angles are closest to 90° and whose lengths are a minimum without altering the volume [83]. This reduction procedure also has the advantage of providing indication of centered-cell possibilities in a higher symmetry system.

Usually, if a good random selection of initial reflections is made, this procedure yields unit cell information and an orientation matrix that can then be used to begin the actual data collection procedure. Moreover, this process is quite rapid, often taking less than 90 min between the time the crystal is first put on the diffractometer, the reflections are found, the crystal is accurately centered, the reflections are indexed, the unit cell is determined, and actual data collection is begun.

5   APPENDIX: STRUCTURE DETERMINATION AND REFINEMENT OF THE TRITHIOCARBAMATE COMPLEX $Fe(C_5H_5)(CO)(CS_3CH_3)$

A brief description of the structure determination and refinement techniques used in the X-ray crystal structure determination of the trithiocarbamate complex $Fe(C_5H_5)(CO)(CS_3CH_3)$ [84] is presented in this appendix to illustrate some of these procedures and the methods used to display the results of a structure determination.

5.1   Experimental Crystal Data

$$FeS_3OC_8H_8, \quad MW = 274.75, \quad \text{Triclinic } P\bar{1}$$

$$a = 8.070(3), \qquad b = 9.331(3), \qquad c = 7.644(3) \text{ Å}$$

$$\alpha = 99.26(3), \qquad \beta = 100.00(3), \qquad \gamma = 102.08(3)°$$

$$V = 542.6(3) \text{ Å}^3, \qquad D_c = 1.68 \text{ g} \cdot \text{cm}^{-3}, \quad Z = 2,$$

$$\text{Mo } K_\alpha (\lambda = 0.7107 \text{ Å}), \qquad \mu_{abs} = 41.00 \text{ cm}^{-1}.$$

A crystal with approximate dimensions $0.20 \times 0.20 \times 0.30$ mm was mounted on a glass fiber for data collection. Intensity data were taken at room temperature (22°C) using a fully automated Nicolet diffractometer equipped with a scintillation counter and controlled by a Nova minicomputer. Initial indexing of the cell from the tuned 2Θ, ω, χ, and ϕ angular settings of 12 reflections was accomplished with the automatic indexing procedure BLIND [82]. Data were taken from the four octants h, k, l; $h, k, -l$; $h, -k, l$; and $h, -k, -l$ using ziconium-filtered Mo K_α radiation within a Θ sphere of 25° (sin $\Theta/\lambda = 0.596$). A Θ-2Θ, stepscan, with 0.01° per step and a variable counting time per step, was employed. Individual backgrounds were obtained from stationary counter-measurements for one-half the total scan time at each end of the scan.

As a general check of the electronic and crystal stability, the intensity of one of the standard reflections was measured periodically during the data collection. Monitoring options based on these standard counts were employed to maintain crystal alignment and to stop the data collection process if the standard count fell below statistically allowed fluctuations. A total of 2687 reflections were recorded in this manner.

At the completion of data collection, the final cell parameters and their standard deviations were obtained from a least-squares fit to the tuned 2Θ values for 12 independent, near-axial reflections.

The intensity data were corrected for Lorentz-polarization effects. Because of the near sphericity of the crystal, intensity variations caused by absorption effects were calculated to be negligible. Furthermore, the consistency of equivalent data was checked and equivalent reflections were averaged ($R_{av} = 0.015$). Standard deviations were estimated from the average intensity and background values by (43). Of the 2513 independent reflections 1162 were observed, having $I > 3.0*\sigma_I$. These reflections were used in all refinement and map calculations.

5.2   Solution and Refinement of the Structure

The Patterson map was determined using sharpened coefficients calculated as

$$|F^s_{hkl}| = \left[\frac{|F^o_{hkl}|}{\hat{f}}\right] \exp[(2B - B')\sin^2\Theta/\lambda^2]$$

where $\hat{f} = \Sigma f_j/\Sigma Z_j$, with the summations over the assumed stoichiometric unit; B is the overall isotropic temperature factor and B' is a variable used to minimize rippling from sharpening. Estimates of the overall temperature and scale factors were obtained from a Wilson plot. The map was printed with a resolution of approximately 0.25 Å per grid point. A list of the positions and heights of all peak maxima was obtained from least-squares fits of the map values to individual three-dimensional Gaussian functions.

Instead of the use of the heavy-atom method, which relies on the iron atom to adequately phase the structure factor magnitudes, we employed an automated Patterson analysis technique to obtain the majority of the structure, with the Patterson-superposition method playing an integral role.

From the list of Patterson peak positions, a vector corresponding to a double Fe—S vector was chosen as the shift vector for a minimum-function Patterson superposition. The resultant map of this superposition should contain, in theory, only two images: an iron image and a sulfur image. The iron image was identified by imposing a center of symmetry halfway along the assumed $(-2x, -2y, -2z)$ vector relating the iron atoms. Only peaks consistent with this imposed symmetry were accepted as part of the image. Atomic positions were then calculated by shifting the averaged positions of symmetry-equivalent peaks by the vector $(-x, -y, -z)$.

Interatomic distances and angles, calculated using these atomic positions (derived directly from the superposition map), were then used to identify the other atoms. This procedure provided positions for 12 of the 13 non-hydrogen atoms in the molecule. Least-squares refinement of these positions resulted in a conventional residual index of $R = \Sigma ||F_o| - |F_c||/\Sigma |F_o| = 0.178$ and a weighted residual index of $Rw = [\Sigma w(|F_o| - |F_c|)^2/\Sigma w|F_o|^2]^{1/2} = 0.198$. The remaining carbon atom was readily located in the subsequent electron-density map.

Three cycles of block-diagonal matrix least-squares refinement, on $|F|$, of all positional and isotropic thermal parameters for the non-hydrogen atoms gave residual indices of $R = 0.093$ and $Rw = 0.107$ for the 1162 observed reflections. The scattering factors used were those of Cromer and Waber [61], with the real and imaginary contributions to the anomalous scattering of the iron atom from Templeton [62]. A difference electron-density map at this stage showed that all non-hydrogen atoms had been accounted for, but that some anisotropic motion was evident. Anisotropic refinement of all heavy-atom thermal parameters lowered the residual indices to $R = 0.044$ and $Rw = 0.058$. The resulting difference electron-density map indicated the positions of all cyclopentadienyl and methyl hydrogen atoms. The refinement proceeded smoothly, with hydrogen positions and isotropic thermal parameters converging to reasonable values. The final values of $R = 0.037$ and $Rw = 0.044$ were obtained from a full-matrix refinement of 150 variables including all positions, anisotropic thermal parameters for non-hydrogen atoms, and isotropic thermal parameters for hydrogen atoms. A final difference electron-density map showed no peaks greater than $0.4e\,\text{Å}^{-3}$ in magnitude.

Tables 1.9 and 1.10 list the final positional and thermal parameters of the non-hydrogen atoms along with their standard deviations. In Table 1.11 are the refined positional and isotropic thermal parameters and their standard deviations for the hydrogen atoms. Standard deviations for these parameters were obtained from the inverse matrix of the final least-squares refinement cycle. Figures 1.70 and 1.71 show the molecular and crystal structures, respectively. An indication of the directions and root-mean-squares amplitudes of vibration, for the atoms refined anisotropically, is provided by these figures. Selected bond

Table 1.9 Final Non-Hydrogen Atom Coordinates

Atom[a]	x^b	y^b	z^b
Fe	0.2421(1)	0.1855(1)	0.2562(1)
S(1)	0.1912(2)	0.2280(2)	−0.0311(2)
S(2)	0.3202(2)	0.4402(2)	0.2908(2)
S(3)	0.2481(3)	0.5599(2)	−0.0443(3)
O(1)	−0.1069(7)	0.1749(7)	0.3018(8)
C(1)	0.0300(10)	0.1808(7)	0.2803(10)
C(2)	0.2530(8)	0.4118(7)	0.0637(8)
C(3)	0.1632(17)	0.4659(13)	−0.2778(12)
C(4)	0.3393(16)	0.1102(13)	0.4828(11)
C(5)	0.2382(12)	−0.0121(10)	0.3440(16)
C(6)	0.3070(14)	−0.0164(9)	0.1975(13)
C(7)	0.4588(13)	0.0981(13)	0.2392(17)
C(8)	0.4766(12)	0.1728(9)	0.4126(15)

[a]Numbers in parentheses denote atom position, for example, S(1) = sulfur atom number 1 position.
[b]Numbers in parentheses correspond to standard deviations in the least significant figures.

Table 1.10 Final Non-Hydrogen Atom Thermal Parameters[a]

Atom[b]	B_{11}[c]	B_{22}[c]	B_{33}[c]	B_{12}[c]	B_{13}[c]	B_{23}[c]
Fe	3.75(5)	3.50(4)	3.50(5)	0.94(3)	0.47(3)	0.72(3)
S(1)	6.09(10)	3.94(8)	3.46(8)	1.10(7)	0.56(7)	0.56(6)
S(2)	6.21(10)	3.53(8)	3.75(8)	0.71(7)	0.30(7)	0.14(6)
S(3)	8.51(13)	4.13(9)	5.48(10)	1.52(8)	1.56(9)	1.88(8)
O(1)	5.1(3)	10.5(4)	11.6(4)	3.7(3)	4.4(3)	5.6(3)
C(1)	5.0(4)	4.4(3)	6.3(4)	1.2(3)	1.0(3)	2.4(3)
C(2)	3.9(3)	3.8(3)	4.7(3)	0.9(2)	1.4(3)	1.6(2)
C(3)	10.4(8)	7.8(6)	4.8(4)	4.9(6)	2.5(4)	3.2(4)
C(4)	9.8(7)	9.8(6)	4.1(4)	7.3(5)	2.3(5)	2.7(4)
C(5)	4.8(4)	5.5(5)	9.2(7)	1.8(4)	1.4(5)	4.4(5)
C(6)	7.7(6)	4.4(4)	6.3(5)	3.3(4)	−0.2(4)	0.1(4)
C(7)	4.8(5)	9.2(7)	8.1(7)	3.9(5)	2.6(5)	3.0(6)
C(8)	5.5(5)	4.7(4)	7.6(6)	2.1(4)	−2.8(4)	−0.2(4)

[a] $T = \exp[-\frac{1}{4}(B_{11}h^2a^{*2} + B_{22}k^2b^{*2} + B_{33}l^2c^{*2} + 2B_{12}hka^*b^* + 2B_{13}hla^*c^* + 2B_{23}klb^*c^*)]$.
[b] Numbers in parentheses denote atom position, for example, S(1) = sulfur atom in 1 position.
[c] Numbers in parentheses correspond to standard deviations in the least significant figures.

lengths and bond angles with standard deviations are given in Table 1.12. These standard deviations were estimated using the variance–covariance matrix.

The best least-squares plane through the cyclopentadienyl was calculated from the final positional parameters. The equation of the least-squares plane and the displacements of the non-hydrogen atoms from this plane are given in Table 1.13.

Table 1.11 Refined Hydrogen Atom Parameters

Atom[a]	x[b]	y[b]	z[b]	B[b]
H(31)	0.972(9)	0.570(8)	0.254(9)	4(2)
H(32)	0.772(11)	0.594(10)	0.273(11)	5(3)
H(33)	0.857(10)	0.478(8)	0.330(10)	3(2)
H(4)	0.313[c]	0.141	0.605	5.5
H(5)	0.134(12)	−0.054(10)	0.397(12)	7(2)
H(6)	0.294(11)	−0.083(10)	0.063(13)	8(3)
H(7)	0.507(13)	0.100(11)	0.159(13)	7(3)
H(8)	0.558(9)	0.278(8)	0.484(9)	4(2)

[a] Numbers in parentheses denote atom position, for example, H(31) = hydrogen atom in position 31.
[b] Numbers in parentheses correspond to standard deviations in the least significant figures.
[c] Atom H(4) was not refined.

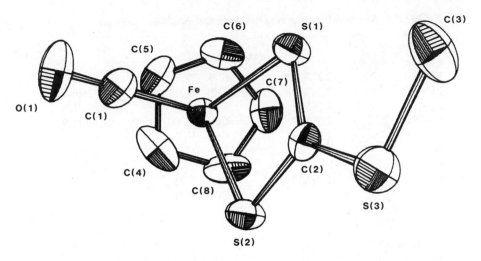

Figure 1.70 Molecular structure of the trithiocarbamate complex $Fe(C_5H_5)(CO)(CS_3CH_3)$.

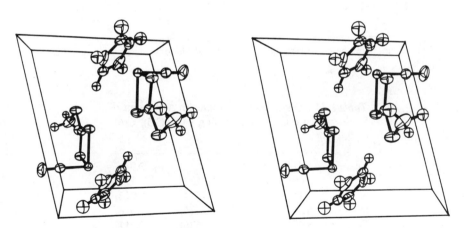

Figure 1.71 Stereographic drawing showing the molecular packing in the unit cell. Note: Distance should be 55–60 mm between image centers.

5.3 Computational Considerations

The programs used in this structure determination include: SUPR [85], for superposition map computations; FOUR [86], for the Patterson, electron-density and difference electron-density map calculations; ALCAMPS [27], for the automatic interpretation of the superposition map; ALLS [67], for the least-

Table 1.12 Selected Interatomic Distances and Angles

Distances (Å)[a]

Fe—S(1)	2.277(2)	S(2)—C(2)	1.686(6)
Fe—S(2)	2.287(2)	C(2)—S(3)	1.724(6)
Fe—C(1)	1.745(8)	S(3)—C(3)	1.801(10)
Fe—C(4)	2.060(14)	C(1)—O(1)	1.136(10)
Fe—C(5)	2.057(10)	C(4)—C(5)	1.401(18)
Fe—C(6)	2.062(9)	C(5)—C(6)	1.328(15)
Fe—C(7)	2.094(10)	C(6)—C(7)	1.397(14)
Fe—C(8)	2.098(10)	C(7)—C(8)	1.355(16)
S(1)—C(2)	1.682(6)	C(8)—C(4)	1.404(18)

Angles (°)[a]

Fe—S(1)—C(2)	86.9(2)	C(4)—C(5)—C(6)	109.8(10)
Fe—S(2)—C(2)	86.5(2)	C(5)—C(6)—C(7)	107.9(9)
S(1)—C(2)—S(2)	111.4(4)	C(6)—C(7)—C(8)	108.5(9)
S(1)—C(2)—S(3)	127.4(4)	C(7)—C(8)—C(4)	108.1(10)
S(2)—C(2)—S(3)	121.2(4)	C(8)—C(4)—C(5)	105.7(11)
C(2)—S(3)—C(3)	102.1(4)	S(1)—Fe—S(2)	75.1(1)
Fe—C(1)—O(1)	177.2(7)	S(1)—Fe—C(1)	93.6(2)
		S(2)—Fe—C(1)	94.1(2)

[a]Numbers in parentheses correspond to standard deviations in the least significant figures.

Table 1.13 Distances from Plane (in Å) for Least-Squares Plane Calculated Using C(4), C(5), C(6), C(7), C(8)[a]

Fe	1.711	C(2)	3.911
S(1)	3.092	C(3)	5.728
S(2)	3.154	C(4)	0.013
S(3)	5.362	C(5)	0.016
O(1)	3.269	C(6)	0.011
C(1)	2.681	C(7)	0.002
		C(8)	0.007

[a]Equation relative to a, b, c: $0.6720X - 0.6985Y + 0.3449Z = 2.2399$.

squares refinements; and ORTEP [87], for generating Figures 1.70 and 1.71. All programs were run on a VAX 11/780 computer. Approximately 40 s of computer time were required for each cycle of block-diagonal refinement, nearly $1\frac{1}{2}$ min for full-matrix refinement, and 40 s for each electron-density map calculation.

References

1. J. C. Kendrew, *Science*, **139**, 1259 (1963).

2. M. F. Perutz, M. G. Rossmann, A. F. Cullis, H. Muirhead, G. Will, and A. T. C. North, *Nature (London)*, **185**, 416 (1960); M. F. Peutz, H. Muirhead, J. M. Cox, L. C. G. Goaman, F. S. Mathews, E. L. McGandy, and L. E. Webb, *Nature (London)*, **219**, 29 (1968).

3. T. Imoto, L. M. Johnson, A. C. T. North, D.C. Phillips, and J. A. Rupley, "Vertebrate Lysozymes," in P. D. Boyer, Ed., *The Enzymes*, Vol. 7, Academic, New York, 1972, pp. 665–868.

4. D. C. Rees, M. Lewis, and W. N. Lipscomb, *J. Mol. Biol.*, **168**, 367 (1983); W. N. Lipscomb, J. C. Coppola, J. A. Hartsuck, M. L. Ludwig, H. Muirhead, J. Searl, and T. A. Steitz, *J. Mol. Biol.*, **19**, 423 (1966).

5. G. Kartha, J. Bello, and D. Harker, *Nature (London)*, **213**, 862 (1967); H. P. Avey, M. O. Boles, C. H. Carlisle, S. A. Evans, S. J. Morris, R. A. Palmer, and B. A. Woodhouse, *Nature (London)*, **213**, 557 (1967).

6. R. B. Honzatko, J. L. Crawford, H. L. Monaco, J. E. Ladner, B. F. P. Edwards, D. R. Evans, S. G. Warren, D. C. Wiley, R. C. Ladner, and W. N. Lipscomb, *J. Mol. Biol.*, **160**, 219 (1982).

7. C. Abad-Zapatero, S. S. Abdel-Meguid, J. E. Johnson, A. G. W. Leslie, I. Raymant, M. G. Rossmann, D. Suck, and T. Tsukihara, *Nature (London)*, **286**, 33 (1980).

8. J. S. Richardson, *Adv. Protein Chem.*, **34**, 167 (1981); for other examples, see E. R. Kantrowitz, and W. N. Lipscomb, *Science*, **241**, 669 (1988); M. S. Chapman, S. W. Suh, P. M. G. Curmi, D. Cascio, W. W. Smith, and D. S. Eisenberg, *Science*, **241**, 71 (1988).

9. L. Pauling and M. D. Shappell, *Z. Kristallogr.*, **75**, 128 (1930).

10. D. L. Plummer, B. A. Karcher, R. A. Jacobson, and R. J. Angelici, *J. Organomet. Chem.*, **260**, 347 (1984).

11. C. W. Bunn, *Chemical Crystallography*, 2nd ed., Oxford University Press, Oxford, 1961; J. D. Dunitz, *X-Ray Analysis and the Structure of Organic Molecules*, Cornell University Press, Ithaca, NY, 1979; M. F. C. Ladd and R. A. Palmer, *Structure Determination by X-Ray Crystallography*, Plenum, New York, 1977; K. Lonsdale, *Crystals and X-rays*, Van Nostrand, New York, 1949; J. M. Robertson, *Organic Crystals and Molecules*, Cornell University Press, Ithaca, NY, 1953; D. Sherwood, *Crystals, X-Rays and Proteins*, Longman, London, 1976; G. H. Stout and L. H. Jenson, *X-Ray Structure Determination*, Macmillan, New York, 1968.

12. R. A. Young and D. B. Wiles, *Adv. X-Ray Anal.*, **24**, 1 (1981).

13. J. Waser, *J. Am. Chem. Soc.*, **65**, 1451 (1943); **66**, 2035 (1944).

14. D. C. Hodgkin, J. Pickworth, J. H. Robertson, K. N. Trueblood, R. J. Posen, and J. G. White, *Nature (London)*, **176**, 325 (1955).

15. A. L. Patterson, *Z. Kristallogr.*, **A90**, 517, 543 (1935).

16. D. Wrinch, *Philos. Mag.* (7)**27**, 98 (1939); M. J. Buerger, *Acta Crystallogr.*, **3**, 87 (1950); and **4**, 531 (1951); J. H. Robertson and C. A. Beevers, *Acta Crystallogr.*, **4**, 270 (1951); J. Clastre and R. Gray, *C. R. Acad. Sci. Ser. C*, **230**, 1876 (1950); J. Garrido, *C. R. Acad. Sci. Ser. C*, **230**, 1878 (1950); *C. R. Acad. Sci. Ser. C*, **231**, 297 (1951); D. McLachlan, *Proc. Natl. Acad. Sci. USA*, **37**, 115 (1951).

17. J. Rodgers and R. A. Jacobson, *J. Chem. Soc. (A)*, 1826 (1970).

18. R. A. Pasternak, *Acta Crystallogr.*, **9**, 341 (1956).

19. R. A. Jacobson and W. N. Lipscomb, *J. Chem. Phys.*, **31**, 605 (1959).

20. R. A. Jacobson, J. A. Wunderlich, and W. N. Lipscomb, *Acta Crystallogr.*, **14**, 598 (1961).

21. S. H. Kim and R. D. Rosenstein, *Acta Crystallogr.*, **22**, 648 (1967).

22. J. H. Derissen, H. J. Endeman, and A. F. Peerdeman, *Acta Crystallogr.*, **B24**, 1349 (1968).

23. A. W. Schlueter and R. A. Jacobson, *J. Chem. Soc.*, **26A**, 2317 (1968).

24. D. E. Beckman and R. A. Jacobson, *J. Organomet. Chem.*, **179**, 187 (1979); R. A. Jacobson and D. E. Beckman, *Acta Crystallogr.*, **A35**, 339 (1979).

25. S. H. Kim, G. A. Jeffrey, R. D. Rosenstein, and P. W. R. Cornfield, *Acta Crystallogr.*, **22**, 733 (1967); T. A. Beineke and L. L. Martin, Jr., *J. Organomet. Chem.*, **20**, 65 (1969).

26. A. D. Mighell and R. A. Jacobson, *Acta Crystallogr.*, **16**, 443 (1963); P. G. Simpson, R. D. Dobrott, and W. N. Lipscomb, *Acta Crystallogr.*, **18**, 169 (1965); W. C. Hamilton, *Acta Crystallogr.*, **18**, 866 (1965).

27. J. W. Richardson, Jr., and R. A. Jacobson, "Computer-Aided Analysis of Multi-solution Patterson Superpositions," in J. P. Glusker, B. K. Patterson and M. Rossi, Eds., *Patterson and Pattersons*, Oxford University Press, Oxford, 1987, p. 311.

28. M. J. Buerger, *Vector Space*, Wiley, New York, 1959; R. A. Jacobson, *Trans. Am. Crystallogr. Assoc.*, **2**, 39 (1966); R. A. Jacobson and L. J. Guggenberger, *Acta Crystallogr.*, **20**, 592 (1966).

29. D. Sayre, *Acta Crystallogr.*, **5**, 60 (1952).

30. J. Karle and I. L. Karle, *Acta Crystallogr.*, **21**, 849 (1966).

31. W. Cochran and M. M. Woolfson, *Acta Crystallogr.*, **8**, 1 (1955).

32. F. A. Muller and R. A. Jacobson, *Cryst. Struct. Commun.*, **9**, 325 (1980).

33. H. Hauptman and J. Karle, *Solution of the Phase Problem I. Centrosymmetric Crystal*, American Crystallographic Association, Monograph No. 3, Polycrystal Book Service, Pittsburgh, PA, 1953.

34. J. Karle and H. Hauptman, *Acta Crystallogr.*, **9**, 635 (1956).

35. I. L. Karle and J. Karle, *Acta Crystallogr.*, **21**, 860 (1966).

36. Y. Kainuma and W. N. Lipscomb, *Z. Kristallogr.*, **113**, 44 (1960).

37. J. Karle, *Acta Crystallogr.*, **B24**, 182 (1968).

38. I. L. Karle and J. Karle, *Acta Crystallogr.*, **17**, 835 (1964).

39. J. Karle and I. L. Karle, *Acta Crystallogr.*, **B24**, 81 (1968).

40. G. N. Reeke, Jr. and W. N. Lipscomb, *J. Am. Chem. Soc.*, **90**, 1663 (1968).

41. D. Harker and J. S. Kasper, *Acta Crystallogr.*, **1**, 70 (1948).

42. M. F. C. Ladd and R. A. Palmer, Eds., *Theory and Practice of Direct Methods in Crystallography*, Plenum, New York, 1980; G. German and M. M. Woolfson, *Acta Crystallogr.*, **B24**, 91 (1968).

43. J. D. Bernal and D. Crowfoot, *Nature (London)*, **133**, 794 (1934).

44. C. Bokhoven, J. C. Schoone, and J. M. Bijvoet, *Proc. Acad. Sci. Amsterdam*, **52**, 120 (1949); *Acta Crystallogr.*, **4**, 275 (1951).

45. D. W. Green, V. M. Ingraham, and M. F. Perutz, *Proc. R. Soc. London Ser. A*, **A225**, 287 (1954).

46. J. C. Kendrew, G. Bodo, H. M. Dintzis, R. G. Parrish, and H. Wyckoff, *Nature* (*London*), **181**, 662 (1958).

47. J. C. Kendrew, R. E. Dickerson, B. E. Strandberg, R. G. Hart, D. R. Davies, D. C. Phillips, and V. C. Shore, *Nature* (*London*), **185**, 4711 (1960).

48. W. N. Lipscomb, "Structure of an Enzyme," in A. Rich and N. Davidson, Eds., *Structural Chemistry and Molecular Biology*, Freeman, San Francisco, 1968, p. 38.

49. W. N. Lipscomb, J. A. Hastruck, G. N. Reeke, Jr., F. A. Quiocho, P. H. Bethge, M. L. Ludwig, T. A. Steitz, H. Muirhead, and J. C. Coppola, "The Structure of Carboxypeptidase A. VII. The 2.0 Å Resolution Studies of the Enzyme and of Its Complex with Glycylityrosine, and Mechanistic Deductions," *Brookhaven Symposia in Biology No. 21*, Brookhaven National Laboratory, Upton, NY, 1968, p. 24.

50. H. W. Wyckoff, K. D. Hardman, N. M. Allewell, T. Imagami, L. N. Johnson, and F. M. Richards, *J. Biol. Chem.*, **242**, 3984 (1967).

51. B. W. Matthews, P. B. Sigler, R. Henderson, and D. M. Blow, *Nature* (*London*), **214**, 652 (1967).

52. J. Drenth, J. N. Jansonius, R. Koekoek, H. M. Swen, and B. G. Wolthers, *Nature* (*London*), **218**, 929 (1968).

53. C. S. Wright, R. A. Alden, and J. Kraut, *Nature* (*London*), **221**, 235 (1969).

54. D. Rabinovich and G. M. J. Schmidt, *Nature* (*London*), **211**, 1391 (1966).

55. D. E. Williams, *Acta Crystallogr.*, **A25**, 464 (1969); C. P. Brock, *Acta Crystallogr.*, **A33**, 898 (1977).

56. C. E. Nordman and K. Nakatsu, *J. Am. Chem. Soc.*, **85**, 353 (1963); C. E. Nordman and S. K. Kumra, *J. Am. Chem. Soc.*, **87**, 2059 (1965); C. E. Nordman, *Trans. Am. Crystallogr. Assoc.*, **2**, 29 (1966).

57. W. Hoppe and E. F. Paulus, *Acta Crystallogr.*, **23**, 339 (1967); E. Egert, *Acta Crystallogr.*, **A39**, 936 (1983).

58. M. G. Rossmann and D. M. Blow, *Acta Crystallogr.*, **15**, 24 (1962); **16**, 39 (1963).

59. M. G. Rossmann and W. N. Lipscomb, *J. Am. Chem. Soc.*, **80**, 2592 (1958).

60. L. V. Aza'roff, *Elements of X-Ray Crystallography*, McGraw-Hill, New York, 1968; M. J. Buerger, *Crystal-Structure Analysis*, Wiley, New York, 1960; J. D. Dunitz, *X-Ray Analysis and the Structure of Organic Molecules*, Cornell University Press, Ithaca, NY, 1979. H. P. Klug and L. E. Alexander, *X-Ray Diffraction Procedures*, 2nd ed., Wiley, New York, 1974. D. Sherwood, *Crystals, X-Rays and Proteins*, Longmans, London, 1976. M. M. Woolfson, *An Introduction to X-Ray Crystallography*, Cambridge University Press, Cambridge, UK, 1970. W. H. Zachariasen, *Theory of X-Ray Diffraction in Crystals*, Wiley, New York, 1945.

61. D. T. Cromer and J. T. Waber, *International Tables for X-ray Crystallography*, The Kynoch Press, Birmingham, England, 1974, Vol. IV, Table 2.2a, p. 71.

62. D. H. Templeton, *International Tables for X-ray Crystallography*, The Kynoch Press, Birmingham, England, 1974, Vol. III, Table 3.3.2c, pp. 215–216.

63. A. L. Patterson, *Z. Krist.*, **90**, 517 (1935); D. Harker, *J. Chem. Phys.*, **4**, 381 (1935); G. Albrecht and R. B. Corey, *J. Am. Chem. Soc.*, **61**, 1087 (1939); H. Klug and L. Alexander, *J. Am. Chem. Soc.*, **66**, 1056 (1944); A. J. C. Wilson, *Nature* (*London*), **150**, 151 (1942).

64. E. W. Hughes, *J. Am. Chem. Soc.*, **63**, 1737 (1941).

65. E. W. Hughes and W. N. Lipscomb, *J. Am. Chem. Soc.*, **68**, 1970 (1964).

66. S. L. Lawton and R. A. Jacobson, *Inorg. Chem.*, **7**, 2124 (1968).

67. R. L. Lapp and R. A. Jacobson, *ALLS, A Generalized Crystallographic Least Squares Program*, United States Department of Energy Report IS-4708, Iowa State University, Ames, IA, 1979.

68. L. V. Aza'roff, *Elements of X-ray Crystallography*, McGraw-Hill, New York, 1968; M. J. Buerger, *Elementary Crystallography*, rev. ed., Wiley, New York, 1963; H. D. Megaw, *Crystal Structures: A Working Approach*, Saunders, Philadelphia, PA, 1973.

69. *International Tables for X-ray Crystallography*, Vol. I, The Kynoch Press, Birmingham, England, 1952.

70. M. J. Buerger, *X-Ray Crystallography*, Wiley, New York, 1942, pp. 69–70.

71. M. J. Buerger, *X-Ray Crystallography*, Wiley, New York, 1942, p. 83; *International Tables for X-Ray Crystallography*, Vol. I, The Kynoch Press, Birmingham, England, 1952, p. 54.

72. A. J. C. Wilson, *Acta Crystallogr.*, **2**, 318 (1944).

73. E. R. Howells, D. C. Phillips, and D. Rogers, *Acta Crystallogr.*, **3**, 210 (1950).

74. U. W. Arndt and b. T. M. Willis, *Single Crystal Diffractometery*, Cambridge University Press, Cambridge, UK, 1966; L. V. Aza'roff, *Elements of X-ray Crystallography*, McGraw-Hill, New York, 1968; M. J. Buerger, *X-ray Crystallography*, Wiley, New York, 1942; E. W. Nuffield, *X-ray Diffraction Methods*, Wiley, New York, 1966.

75. M. Atoji and W. N. Lipscomb, *Acta Crystallogr.*, **7**, 173 (1954).

76. T. B. Reed and W. N. Lipscomb, *Acta Crystallogr.*, **6**, 45 (1953); W. N. Lipscomb, *Norelco Rep.*, **4**, 54 (1957); C. Altona, *Acta Crystallogr.*, **17**, 1282 (1964); G. Abowitz and P. G. Cath, *J. Sci. Instrum.*, **44**, 156 (1967); G. Abowitz and J. Ladell, *J. Sci. Instrum.*, Series 2, **1**, 113 (1968); R. Rudman, *Low-Temperature X-Ray Diffraction*, Plenum, New York, 1976.

77. M. J. Buerger, *X-Ray Crystallography*, Wiley, New York, 1942, pp. 331ff; G. H. Stout and L. H. Jensen, *X-Ray Structure Determination*, Macmillan, New York, 1968, pp. 122ff; E. W. Nuffield, *X-Ray Diffraction Methods*, Wiley, New York, 1966, Chapter 9.

78. M. J. Buerger, *The Procession Method*, Wiley, New York, 1964.

79. R. Wass and J. Donohue, *Acta Crystallogr.*, **10**, 375 (1957).

80. M. J. Buerger, *X-Ray Crystallography*, Wiley, New York, 1942, Chapter 12; E. W. Nuffield, *X-Ray Diffraction Methods*, Wiley, New York, 1966, Chapter 11; G. H. Stout and L. H. Jensen, *X-Ray Structure Determination*, Macmillan, New York, 1968, pp. 96ff.

81. W. C. Hamilton, "Angle Settings for Four-Circle Diffractometers," *International Tables for X-ray Crystallography*, Vol. IV, The Kynoch Press, Birmingham, England, 1974, pp. 273–284.

82. R. A. Jacobson, *J. Appl. Crystallogr.*, **9**, 115 (1976).

83. P. Niggli, *Haubuch der Experimentalphysike* 7 Part 1. Leipzig: Akademische Verlagsgesellschaft, 1928; M. J. Buerger, *Z. Kristallogr.*, **109**, 42 (1957); A. Santoro and A. D. Mighell, *Acta Crystallogr.*, **A26**, 124 (1970); V. Balashov, *Acta Crystallogr.*, **9**, 319 (1956); S. L. Lawton and R. A. Jacobson, *The Reduced Cell and its Crystallographic Applications*, USAEC Report IS-1141, Ames Laboratory, Ames, IA, 1965.

84. J. R. Matachek, J. W. Richardson, Jr., R. J. Angelici, and R. A. Jacobson, unpublished.

85. C. R. Hubbard and R. A. Jacobson, *A Fortran IV Crystallographic System of Programs for Generalized Superpositions*, USAEC Report IS-2210, Iowa State University, Ames, IA, 1969.

86. D. R. Powell and R. A. Jacobson, *FOUR: A General Crystallographic Fourier Program*, US DOE Report IS-4737, Iowa State University, Ames, IA 1980.

87. C. K. Johnson, *ORTEP, A Fortran Thermal-Ellipsoid Plot Program for the Crystal Structure Illustrations*, Report ORNL-3794, Oak Ridge National Laboratory, Oak Ridge, TN, 1965.

Chapter **2**

THE APPLICATION OF NEUTRON DIFFRACTION TECHNIQUES TO STRUCTURAL STUDIES

Chuck F. Majkrzak, Mögens S. Lehmann, and David E. Cox

1   INTRODUCTION

Since the pioneering work of Shull and Wollan [1] more than 35 years ago, neutron diffraction has become a well-established method for determining the microscopic chemical and magnetic structures of condensed matter; and this method is now widely used by chemists, physicists, and biologists [2]. The methodology is well documented, and there are many excellent general textbooks on neutron scattering (see, e.g., [3–12]). This chapter gives a brief introduction to some fundamental principles and experimental techniques, describes techniques of measurement and data analysis of particular interest to chemists for structure determination, and provides some illustrative examples.

Although the theoretical descriptions of neutron and X-ray diffraction are similar, the principal interaction of X rays with matter is electronic; whereas two interactions predominate for neutrons: one is nuclear and the other is magnetic in nature. Hence, neutrons are particularly sensitive to magnetic architectures. In fact, neutron scattering data constitute the most significant body of experimental evidence regarding long-range magnetic order in solids. The strength of nuclear interaction varies erratically with the isotope, although it is typically of the same order of magnitude, unlike the interaction of an X-ray photon with an atom, which is proportional to the number of electrons and so increases regularly with atomic number. Therefore, neutrons can be useful in distinguishing between atoms of neighboring atomic number or different isotopes of the same element, and they are sensitive to the positions of light elements, most notably hydrogen. Furthermore, because the neutron scattering lengths for hydrogen and deuterium are markedly different, many small-angle scattering studies in biology, chemistry, and polymer science are possible with neutrons since the contrast between particulate samples and the homogeneous matrix in which they are suspended can be varied by adjusting deuterium/hydrogen ratios. Hydrogen–deuterium substitution can also be used to label specific parts of a structure. Other secondary, though not inconsequential, advantageous properties of neutrons include relatively high penetration depths and low radiation damage levels.

For the wavelengths that are required for diffraction studies, which are of the order of the interatomic spacings in crystals, the corresponding neutron and X-ray energies differ greatly, the former being of millivolts and the latter thousands of electronvolts. Therefore, the neutron's energy is comparable to thermal fluctuations and magnetic excitations in solids; consequently, the neutron is an ideal probe of the dynamics of condensed matter. Inelastic neutron scattering also has extensive applications to chemical problems; however, this topic is not covered in this chapter.

Since the chapter by Hastings and Hamilton [13] on neutron scattering written in this series in 1972, there has been a rapid growth in neutron facilities throughout the world and a tremendous increase in the number of users from outside institutions for whom neutron diffraction represents an important technique that complements many others used in their fields of research. In this chapter we focus on an area of primary importance to chemists, physicists, biologists, and materials scientists; namely, the application of neutron diffraction to structural studies by powder and single-crystal techniques. Both techniques have large and active user communities; and although the objectives are by and large similar, the many varied applications warrant separate consideration.

In this chapter we summarize modern developments in instrumentation, measurement techniques, and data analysis in these areas; and we describe some examples that illustrate the power of neutron diffraction for structural studies. One important development, which has had considerable impact over the past few years, is the application of white beam techniques at pulsed neutron sources to structural problems. This topic is covered in some detail. The chapter is aimed at those who are unfamiliar with neutron techniques and have some knowledge of X-ray powder and single-crystal diffraction, but who do not regard themselves as crystallographers. Constraints on length prevent consideration of related topics such as small-angle neutron scattering, surface diffraction, or scattering from noncrystalline materials. For these and other related areas of interest the reader is referred to the general neutron scattering textbooks [3–12].

2 COHERENT ELASTIC SCATTERING

2.1 Physical Properties of the Neutron

The neutron has a mass of 1.675×10^{-27} kg, which is slightly larger than that of a proton. In fact, a free neutron decays into a proton that is accompanied by the emission of an electron and an antineutrino with a half-life of about 1000 s. The neutron is electrically neutral and is a Fermi particle with spin $\frac{1}{2}$ and a magnetic dipole moment of -1.913 nuclear magnetons (the negative sign indicates that the magnetic moment is in the opposite direction to that of the spin). The neutron wavelength λ is related to its mass m and velocity v by the well-known quantum mechanical relation $\lambda = h/mv$, where h is Planck's constant. The principle interactions of thermal or slow neutrons with matter are nuclear and magnetic. Other interaction potentials are several orders of magnitude weaker, and they are ordinarily of no consequence in conventional diffraction experiments [14].

2.2 Nuclear Scattering

2.2.1   Scattering by a Single Nucleus

Consider the scattering of a neutron by a single fixed nucleus. The incident neutron can be described by the plane wave function

$$\psi_1 = \exp(ikz) \tag{1}$$

where the wave vector $k = 2\pi/\lambda$ and z is the displacement. For a rigidly bound nucleus, the scattering of neutrons is an elastic process; and although the neutron–nucleus interaction is relatively strong, it is of sufficiently short range compared to thermal or slow neutron wavelengths (about 0.1 Å or greater) that the scattering is practically isotropic. This is different from the scattering of X rays from the electron distribution of an atom, which is considerably more extended in space than the volume occupied by the atomic nucleus. The spherically symmetric wave function ψ_s of a scattered neutron at a distance r from a fixed nucleus can be written as

$$\psi_s = \frac{-b}{r} \exp(ikr) \tag{2}$$

where b is the nuclear scattering amplitude or length. A similar expression can be written for the wave function of an X-ray photon scattered by atomic electrons except that the scattering length $f = f(\theta)$ is not a constant but is dependent on the scattering angle 2θ. The scattering process is depicted schematically in Figure 2.1 in two dimensions. The scattering vector $\mathbf{Q} \equiv \mathbf{k}_f - \mathbf{k}_i$, where $\mathbf{k}_f$ and $\mathbf{k}_i$ are the final and incident neutron wave vectors, respectively. Because $|k_f| = |k_i|$, $|Q| = 2k \sin \theta = 4\pi \sin \theta/\lambda$. The cross section σ_s for the scattering of neutrons by a fixed nucleus is defined as

$$\sigma_s = \frac{\text{current of scattered neutrons (neutrons per second)}}{\text{incident neutron flux (neutrons per square centimeter per second)}}$$

$$= 4\pi r^2 v \frac{|-(b/r)\exp(ikr)|^2}{v|\exp(ikz)|^2} = 4\pi b^2 \tag{3}$$

where v is the velocity of the neutron.

Generally, the nuclear scattering length for neutrons is different for different isotopes; and for a nucleus with spin I, there are two values of the scattering

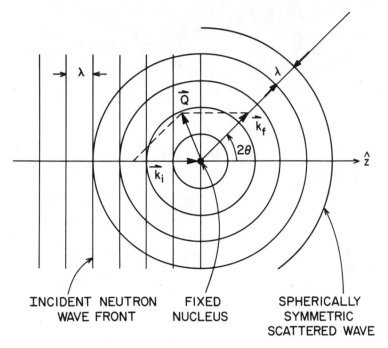

INCIDENT NEUTRON   FIXED   SPHERICALLY
WAVE FRONT   NUCLEUS   SYMMETRIC
                                SCATTERED WAVE

Figure 2.1 Two-dimensional projection of the scattering geometry for a neutron–nucleus inter-action. The scattering vector $\mathbf{Q} \equiv \mathbf{k}_f - \mathbf{k}_i$; $|Q| = 2k \sin \theta = 4\pi \sin \theta / \lambda$, since $|\mathbf{k}_i| = |\mathbf{k}_f| \equiv k = 2\pi/\lambda$.

length—one corresponding to a parallel, the other to an antiparallel alignment of neutron and nuclear spins. The nuclear scattering length has no regular relationship to nuclear mass or charge. This is a result of the resonance scattering that occurs in addition to the *hard-sphere* scattering (e.g., see [15]). Furthermore, the nuclear scattering length is usually complex. The imaginary part is related directly to the absorption cross section, and the real part can be either positive or negative. Absorption is most often negligible (exceptions include certain isotopes of Cd, B, and Li and several rare earths such as Gd). Usually the real part of the scattering length is practically constant over the thermal or slow neutron energy range, whereas the absorption cross section is proportional to the wavelength (except in the neighborhood of a resonance). Table 2.1 lists a few selected values of neutron scattering lengths and absorption cross sections together with corresponding X-ray values. More extensive tables can be found in [3] and [16]. Crystalline materials are usually composed of elements that are a mixture of several different isotopes and/or of a single isotope with nonzero nuclear spin. The result is that an atom of a given element at a particular position in a unit cell is in effect randomly selected from a distribution of possible isotopes and neutron–nucleus spin states so that atoms in equivalent positions in different unit cells have nonuniform scattering lengths. The result is a component of the scattering cross section, which is incoherent.

Table 2.1 Selected Neutron and X-Ray Scattering Lengths and Absorption Cross Sections[a]

| Atomic | | | Neutrons | | X Rays $f(10^{-12}$ cm) | | |
Element	Number	Isotope	$b(10^{-12}$ cm)	$\sigma_{abs}(10^{-24}$ cm^2)	$\sin\theta/\lambda = 0$ Å^{-1}	$\sin\theta/\lambda = 0.5$ Å^{-1}	$\sigma_{abs}(10^{-24}$ cm^2)
H	1	^{1}H	−0.37423	0.3326	0.28	0.02	0.7
		^{2}H	0.6674	0.000519	0.28	0.02	0.7
C	6	^{12}C	0.66535	0.00353	1.69	0.48	92
N	7	^{14}N	0.937	1.91	1.97	0.53	175
O	8	^{16}O	0.5805	0.00010	2.25	0.62	305
Al	13	^{27}Al	0.3449	0.231	3.65	1.55	2,180
Si	14	b	0.4149	0.171	3.95	1.72	2,830
V	23	^{51}V	−0.04024	4.9	6.5	2.8	19,700
Fe	26	b	0.954	2.56	7.3	3.3	28,600
Ni	28	^{58}Ni	1.44	4.6	7.9	3.6	4,450
		^{62}Ni	−0.87	14.5	7.9	3.6	4,450
Nb	41	^{93}Nb	0.7054	1.15	11.5	5.7	23,600
Cd	48	b	0.51	2520	13.6	6.9	43,000
Pb	82	b	0.94003	0.171	23.1	12.9	79,700

[a]Selected values of the real part of the nuclear scattering lengths b and absorption cross sections σ_{abs} for neutrons, together with corresponding X-ray values (at $\lambda = 1.8$ Å for neutrons and $\lambda = 1.54$ Å for X rays). More extensive tables are found in [3] and [16].
[b]Values for the naturally occurring isotopic mixture.

Only the average value of the different individual scattering lengths contributes to coherent scattering processes, such as the diffraction from ordered arrays of atoms to be discussed below. For the majority of the elements the incoherent scattering cross sections are relatively small, although hydrogen and vanadium are important exceptions. However, note that the spin-incoherent scattering cross section can in principle be eliminated by polarizing the nuclear magnetic moments. For hydrogen, this was accomplished dynamically at low temperatures using microwave pumping techniques [17], and it may prove to be of value in determining the structures of complicated hydrogenous materials.

Finally, the scattering lengths a for free nuclei or atoms (e.g., in the gaseous state) are related to the fixed scattering lengths b by the equation

$$a = \left(\frac{A}{A+1}\right) b \qquad (4)$$

where A is the atomic weight.

2.2.2   Diffraction by a Collection of Nuclei

KINEMATIC APPROXIMATION

Consider now the coherent elastic scattering of neutrons by the ordered assemblage of atoms comprising a single crystal. Each atom irradiated by the incident neutron wave becomes a source of spherical Huyghens wavelets that subsequently interfere with each other. For the present discussion we neglect refraction effects (which arise from the interference between the incident and scattered neutron waves, which is small since the neutron refractive index is usually close to unity) and assume that the atomic nuclei are rigidly fixed in space. We first ask what is the amplitude of a neutron plane wave of infinite lateral extent if it is specularly reflected by a *single* infinite plane sheet of atoms, such as the first plane from the left depicted in Figure 2.2. The incident and final neutron wave vectors, k_i and k_f, respectively, are normal to the propagating wave fronts and are of equal magnitude. Using the well-known Fresnel zone construction for wave optics, it can be shown (see, e.g., [18]) that the ratio of reflected amplitude A_R to incident amplitude A_I is given by

$$\frac{A_R}{A_I} = nb \frac{\lambda}{\sin \theta} = \rho \frac{4\pi}{Q} \qquad (5)$$

where n is the number of atoms per unit area, ρ is the planar scattering length density, and $Q = 4\pi \sin \theta/\lambda$. Suppose now that there is a set of N planes with, in general, different atoms and/or scattering length densities ρ_j and that this set or unit cell is repeated periodically in space M times as shown schematically in Figure 2.2. In adding the amplitudes contributed by successive individual planes, account must be taken of the phase differences. If we neglect extinction; that is, the depletion of the incident beam intensity caused by the relatively small

N·M ATOMIC PLANES

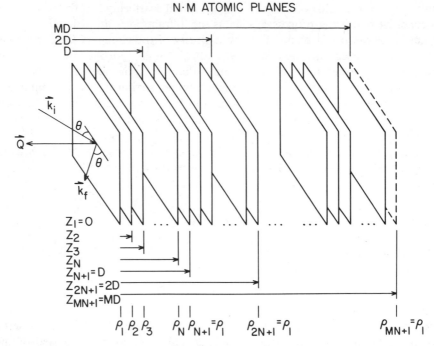

Figure 2.2 Schematic representation of the coherent elastic scattering from an ordered array of $N \cdot M$ parallel planes. A group of N planes (one unit cell) is repeated regularly M times.

fraction reflected by each successive plane (this is the so-called *kinematic approximation*), then the resultant amplitude q scattered coherently by $N \cdot M$ planes is

$$q = \frac{4\pi}{Q} \sum_{l}^{NM} \rho_l \exp(iQz_l) \tag{6}$$

where z_l is the position of the lth plane from the origin, as depicted in Figure 2.2. Because $\rho_{jN+l} = \rho_l$ and $z_{jN+l} = z_l$, where j and l are integers, the summation in (6) can be expanded and recombined to give for the reflectivity $R \equiv |q|^2$ the expression

$$R = \left(\frac{4\pi}{Q}\right)^2 \left| \sum_{j=1}^{N} \rho_j \exp(iQz_j) \right|^2 \frac{\sin^2(MQD/2)}{\sin^2(QD/2)} \tag{7}$$

with D as defined in Figure 2.2. The summation in the expression for the preceding reflectivity is commonly referred to in both X-ray and neutron diffraction as the *structure factor* F. The function $G(Q) \equiv [\sin^2(MQD/2)/\sin^2(QD/2)]$ has a maximum value of M^2 at $Q = 2\pi/D$ and for large M approaches a δ function. Thus, the diffracted radiation has well-defined maxima at values of Q equal to integer multiples of $2\pi/D$. Since $|\mathbf{Q}| = 2k \sin \theta$, $= (4\pi/\lambda) \sin \theta$ substitution into the expression $Q = 2\pi/D$ gives

$$\lambda = 2d \sin \theta \tag{8}$$

the well-known Bragg condition. The principal maximum of the function $G(Q)$ has a full width at half-maximum (fwhm) Γ approximately equal to $2\pi/MD$. Thus, Γ is a measure of the coherence length of the crystalline planes contributing to the diffraction peak. The coherence of a set of atomic planes can be limited by finite size and strain, for example.

In practice it is often more useful to define an integrated reflectivity

$$\mathscr{R}(Q_0) \equiv \int R(Q) \, dQ \tag{9}$$

where the integration limits include the range of appreciable scattered intensity centered about a particular Bragg reflection maximum (at Q_0). This is done because a Bragg reflection always has a measured width that is a convolution of an intrinsic width (e.g., $\propto Q_0/M$) and a finite instrumental resolution width corresponding to distributions in the magnitudes and directions of $\mathbf{k}_i$ and $\mathbf{k}_f$ that are in general Q dependent. To calculate $\mathscr{R}$ it is convenient to change the variable of integration from Q to the angle θ since a spectrometer is frequently stepped in uniform angular increments. Therefore, we write

$$\mathscr{R}(\theta_0) = \int R(\theta) \, d\theta = \int \left(\frac{4\pi}{Q}\right)^2 |F|^2 \frac{\sin^2(MQD/2)}{\sin^2(QD/2)} \, d\theta \tag{10a}$$

or, using $Q = 4\pi \sin \theta/\lambda$,

$$\mathscr{R}(\theta_0) = \int \frac{\lambda^2}{\sin^2\theta} |F|^2 \frac{\sin^2[(MD/2)(4\pi/\lambda) \sin \theta]}{\sin^2[(D/2)(4\pi/\lambda) \sin \theta]} \, d\theta \tag{10b}$$

Because we are interested in a relatively narrow range of angles about a particular Bragg reflection centered at θ_0, we can write $\theta = \theta_0 + \varepsilon$ and replace θ with the constant value θ_0 everywhere except in the rapidly varying function $G(Q)$ where an expansion for small ε is appropriate. Performing these substitutions and expansions gives

$$\mathscr{R}(\theta_0) \simeq \frac{\lambda^2}{\sin^2\theta_0} |F|^2 \int \frac{\sin^2[(MD/2)(4\pi/\lambda)(\cos \theta_0) \cdot \varepsilon]}{\sin^2[(D/2)(4\pi/\lambda)(\cos \theta_0) \cdot \varepsilon]} \, d\varepsilon \tag{11a}$$

For small angles ε, the denominator of the integrand can be replaced by the square of its argument. Making the substitution $x = MB\varepsilon$, where

$B \equiv 2\pi D \cos \theta_0 / \lambda$, we obtain

$$\mathscr{R}(\theta_0) = \frac{\lambda^2}{\sin^2 \theta_0} |F|^2 \frac{M\lambda}{2\pi D \cos \theta_0} \int \frac{\sin^2 x}{x^2} \, dx \qquad (11b)$$

Because the value of the integrand is significant only for values of ε close to θ_0, the limits of integration can be arbitrarily extended to $\pm \infty$ with the result that

$$\mathscr{R}(\theta_0) = \frac{\lambda^3}{\sin \theta_0 \sin 2\theta_0} |F|^2 \frac{M}{D} \qquad (11c)$$

If we replace the structure factor F as defined in (7) with the more general three-dimensional form

$$F_c = \sum_{j=1}^{N} b_j \exp(i\mathbf{Q} \cdot \mathbf{r}_j) \qquad (12)$$

where now the summation is over all of the atoms in a unit cell at positions $\mathbf{r}_j$, and write the density of atoms in the reflecting plane n(atoms per unit area) as $\mathscr{N}$(atoms per unit volume)$\cdot D$, where $\mathscr{N}$ is the density of atoms in the unit cell, then

$$\mathscr{R}(\theta_0) = \frac{\mathscr{N}^2 D^2 |F_c|^2 \lambda^3}{\sin \theta_0 \sin 2\theta_0} \frac{M}{D} \qquad (13)$$

Recall that this derivation assumes that the crystal planes and wave fronts of the incident beam are of infinite lateral extent. In practice an incident beam of finite cross-sectional area A illuminates a volume of crystal $V = MDA/\sin \theta$ (e.g., see [18]). Multiplying the integrated reflectivity by A we obtain the integrated reflecting power $\mathscr{P}(\theta_0)$

$$\mathscr{P}(\theta_0) \equiv \int P(\theta) \, d\theta = \mathscr{R}(\theta_0) A \qquad (14)$$

Thus

$$\mathscr{P}(\theta_0) = \frac{\mathscr{N}^2 |F_c|^2 \lambda^3}{\sin 2\theta_0} \frac{MDA}{\sin \theta_0}$$

or

$$\mathscr{P}(\theta_0) = \frac{\mathscr{N}^2 |F_c|^2 \lambda^3}{\sin 2\theta_0} V \qquad (15)$$

Equation (15) is in fact valid for any crystal shape and size provided the cross-sectional area of the beam is sufficient to illuminate the crystal completely and absorption and extinction can be neglected. These effects are considered in more detail in the next section.

DYNAMICAL THEORY

Thus far neither absorption nor extinction effects were taken into account. Clearly, the reflectivity R cannot be greater than unity. The expression for R given by (7) becomes arbitrarily large for sufficiently large M at $Q = 2\pi/D$. For a perfect single crystal, the process through which the incident beam intensity is depleted by reflection from successive atomic planes is called *primary extinction*. This effect is taken into account in *dynamical* theories of diffraction such as the one developed by Zachariasen [19], which gives

$$R = \tanh^2|q| \tag{16}$$

Another type of extinction can occur in mosaic crystals composed of an angular distribution of perfect microcrystallites centered about some mean direction. As the incident radiation propagates through a mosaic crystal, a reduction in intensity occurs by primary extinction within the perfect mosaic blocks but only when a block with an orientation sufficiently close to the appropriate Bragg angle is encountered. This process is referred to as *secondary extinction*. Extinction effects can be significant in perfect and mosaic single crystals, but they are usually negligible in powder or polycrystalline samples consisting of a random distribution of microcrystallites.

Another possibility is that multiple or simultaneous scattering effects will occur, involving more than one set of atomic planes satisfying conditions for Bragg reflection. These and other dynamical effects including refraction are discussed at length in the review article by Sears [20].

POWDER DIFFRACTION

For a three-dimensional crystal the structure factor F_c is given by (12)

$$F_c = \sum_j b_j \exp(i\mathbf{Q} \cdot \mathbf{r}_j)$$

The summation is over all of the atoms in a unit cell, $\mathbf{r}_j$ being the position of the jth atom within that cell. For a given crystal lattice with lattice parameters $\mathbf{a}$, $\mathbf{b}$, and $\mathbf{c}$, reciprocal lattice vectors $\mathbf{a}^*$, $\mathbf{b}^*$, and $\mathbf{c}^*$ can be defined such that the value of $\mathbf{Q}$, which is perpendicular to and which satisfies the condition for Bragg reflection from the (hkl) atomic plane (where the integers h, k, and l are the Miller indices), is given by $\mathbf{Q} = h\mathbf{a}^* + k\mathbf{b}^* + l\mathbf{c}^*$ (see, e.g., [21]). Writing the atomic positions $\mathbf{r}_j$ in terms of the lattice vectors we obtain

$$\mathbf{Q} \cdot \mathbf{r}_j = (h\mathbf{a}^* + k\mathbf{b}^* + l\mathbf{c}^*) \cdot \left(\frac{x_j}{a}\mathbf{a} + \frac{y_j}{b}\mathbf{b} + \frac{z_j}{c}\mathbf{c} \right)$$

$$= 2\pi \left(\frac{x_j}{a}h + \frac{y_j}{b}k + \frac{z_j}{c}l \right) \tag{17}$$

The diffracted radiation from a polycrystalline sample composed of a random angular distribution of microcrystallites, given a monochromatic and collimated incident beam, occurs at values of Q or scattering angles 2θ corresponding to the Bragg conditions for reflection from the atomic planes (hkl). Furthermore, the scattered intensity is uniform in angle about an axis parallel to the incident beam direction, giving rise to Debye–Scherrer rings or cones as shown in Figure 2.3.

One of the standard geometrical configurations for powder diffraction is shown in Figure 2.4. The sample is either in the form of a pellet or it is enclosed in a cylindrical container with the cylinder axis perpendicular to the scattering plane defined by $\mathbf{k_i}$ and $\mathbf{k_f}$. The *integrated* diffracted intensity (integrated over solid angle in uniform angular increments) I_{hkl} (number of neutrons per unit time) is given in [3]

$$I_{hkl} = \frac{I_0}{A} \frac{lV}{8\pi R} jN_c^2 |F_{hkl}|^2 e^{-2W} \frac{\lambda^3}{\sin\theta_0 \sin 2\theta_0} A_{hkl} \tag{18}$$

where I_0 is the incident beam intensity; A is the beam area; l is the detector height; V is the volume of specimen in the beam; R is the specimen-to-detector distance; j is the multiplicity, or number of cooperating planes for the particular (hkl) reflection being measured (reflecting planes with different Miller indices that have the same interplanar spacing contribute at the same value of Q or 2θ); N_c is the number of unit cells per unit volume; e^{-2W} is the Debye–Waller factor correction for thermal motion of the atoms; and A_{hkl} is the absorption or self-shielding factor.

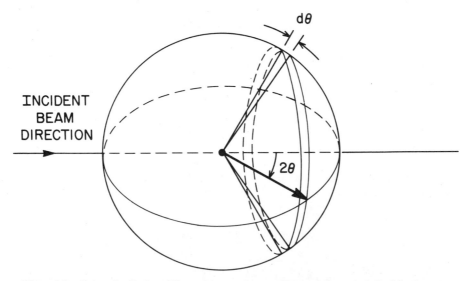

Figure 2.3 Cone of radiation diffracted by a polycrystalline sample as described in the text.

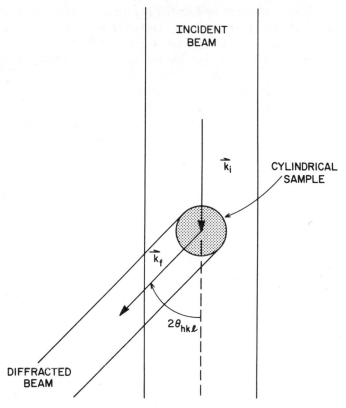

Figure 2.4 Cylindrical sample container with cylinder axis perpendicular to the scattering plane defined by $\mathbf{k}_i$ and $\mathbf{k}_f$.

The Debye–Waller factor takes into account the thermal motion of the atoms, which to this point has been neglected. It can be shown (see, e.g., [21]) that if the instantaneous position of the jth atom in a unit cell is $\mathbf{r}_j + \mathbf{u}_j$, where $\mathbf{u}_j$ is the atom's displacement from its equilibrium (or lattice) position $\mathbf{r}_j$, then the argument of the Debye–Waller factor is given by

$$2W_j = \langle (\mathbf{Q} \cdot \mathbf{u}_j)^2 \rangle \tag{19}$$

where the brackets $\langle\ \rangle$ indicate an average. For a cubic crystal with one atom per unit cell, it can be shown [13, 22] that

$$W = \frac{6h^2}{mk_B T_D} \left[\frac{\phi(T_D/T)}{(T_D/T)} + \frac{1}{4} \right] \frac{\sin^2\theta}{\lambda^2} \tag{20}$$

where T_D is the Debye characteristic temperature, m is the mass of the atom, k_B is the Boltzmann constant, and T is the absolute temperature of the sample. The function $\phi(T_D/T)$ was tabulated by Debye [23]. For more complicated crystals each atom has an anisotropic Debye–Waller factor, and the relationship to the characteristic temperature for the crystal is not simple. An expression for the intensity diffracted by a powder sample in the form of a flat slab, which is another geometrical configuration sometimes used, is given in [3].

The general expression for the absorption factor A_{hkl} depends in a complicated manner on both the Bragg angle θ_0 (for a given hkl) and the absorption cross section. In practice, however, the absorption is relatively small compared to X rays; consequently, A_{hkl} does not vary appreciably with angle θ_0 [3, 24].

2.3 Magnetic Scattering

The magnitude of the magnetic scattering length p for neutrons is given by

$$p = \left(\frac{e^2\gamma}{2mc^2}\right)gJf = (0.2695 \times 10^{-12}\text{ cm})\,gJf \qquad (21)$$

where e, m, and c are the electron charge, electron mass, and speed of light, respectively; γ is 1.9 nuclear magnetons, g is the Landé splitting factor, and the quantum number J corresponds to the total angular momentum of the atom. The quantity f is an angle-dependent form factor (usually normalized to unity in the forward direction) analogous to the electronic form factor for X rays. However, for neutrons, f corresponds to the Fourier transform of the *unpaired* electron density, which is responsible for the net magnetic moment and which extends over a volume of space with linear dimensions comparable to the

Table 2.2 Representative Compilation of Magnetic Scattering Lengths[a]

	$p(10^{-12}\text{ cm})$	
Ion	$\dfrac{\sin\theta}{\lambda} = 0\text{ Å}^{-1}$	$\dfrac{\sin\theta}{\lambda} = 0.25\text{ Å}^{-1}$
Cr^{2+}	1.08	0.45
Mn^{2+}	1.35	0.57
Fe (metal)	0.60	0.35
Fe^{2+}	1.08	0.45
Fe^{3+}	1.35	0.57
Co (metal)	0.47	0.27
Co^{2+}	1.21	0.51
Ni (metal)	0.16	0.10
Ni^{2+}	0.54	0.23

[a]Adapted, with permission, from Table 16 in G. E. Bacon, *Neutron Diffraction*, 3rd ed., Oxford University Press, London, 1975.

neutron wavelength, unlike the nucleus, which is effectively a point scatterer. Table 2.2 is a representative compilation of magnetic scattering length magnitudes. Thus, to measure accurately the structure factors corresponding to a magnetic crystal, it is necessary to know the angular dependence of the form factor. The measured form factor for Gd [25] is shown in Figure 2.5. On the other hand, by accurately measuring the integrated Bragg intensities for a single crystal with a known structure, it is in principle possible to deduce the magnetization distribution within the unit cell. An example is described in Section 5.3.

In general, the more a component of the unpaired electron distribution is extended in space, the less it contributes to the intensity of magnetically scattered neutrons at higher values of Q or scattering angle. It should also be noted that for some atoms; for example, most of the rare earths, the magnetic moment arises from both spin and orbital momenta of the electrons; consequently, the form factors are actually a combination of two components, although only the resultant can be measured experimentally.

The interaction between the magnetic moments of the neutron and atom depends not only on the magnitude of the moments, but also on their orientations relative to one another and to the scattering vector $\mathbf{Q}$. To take into account the vectorial nature of the interaction it is useful to define a vector $\mathbf{q}$ so that

$$\mathbf{q} = \hat{\mathbf{Q}}(\hat{\mathbf{Q}} \cdot \hat{\mathbf{S}}) - \hat{\mathbf{S}} \qquad (22)$$

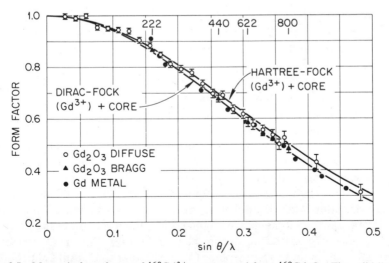

Figure 2.5 Magnetic form factor of $^{160}Gd^{3+}$ as measured from $^{160}Gd_2O_3$. The solid line is the Blume–Freeman–Watson form factor. The closed circles show the experimental $4f$ form factor measured on the metal. Except for very small scattering angles the calculated form factor fits the experimental oxide form factor remarkably well. Adapted, with permission, from Figure 5 in R. M. Moon and W. C. Koehler, *Phys. Rev. Sect. B*, **11**, 1609 (1975).

where $\hat{\mathbf{Q}}$ and $\hat{\mathbf{S}}$ are unit vectors along the directions of the scattering vector and atomic magnetic moment, respectively. Then a magnetic structure factor analogous to the nuclear structure factor in (12) is defined as

$$F_M = \sum_j^N q_j p_j \exp(i\mathbf{Q}\cdot\mathbf{r}_j) \tag{23}$$

For unpolarized neutrons, the scattering formulas given by (15) and (18) are now modified to contain $|F_c|^2 + |F_M|^2$.

For a polarized beam, it is possible to obtain additional information by polarization analysis of the scattered neutrons by the spin-dependent scattering processes, which are discussed in detail in [4, 26]; we simply give the appropriate structure factors

$$F^{++} = \sum_j^N (b_j + p_j q_{z_j}) \exp(i\mathbf{Q}\cdot\mathbf{r}_j) \tag{24a}$$

$$F^{--} = \sum_j^N (b_j - p_j q_{z_j}) \exp(i\mathbf{Q}\cdot\mathbf{r}_j) \tag{24b}$$

$$F^{+-} = \sum_j^N p_j(q_{x_j} + i q_{y_j}) \exp(i\mathbf{Q}\cdot\mathbf{r}_j) \tag{24c}$$

$$F^{-+} = \sum_j^N p_j(q_{x_j} - i q_{y_j}) \exp(i\mathbf{Q}\cdot\mathbf{r}_j) \tag{24d}$$

where $(++)$, $(--)$, $(+-)$, and $(-+)$ refer to the spin directions [$(+)$ parallel, $(-)$ antiparallel] of the incident and scattered neutron, respectively, with respect to the quantization axis; and the z axis is stipulated as the direction of the neutron polarization.

Consider now some general, as well as several important special, consequences of the spin-dependent expressions for the structure factors.

Note first that coherent nuclear scattering is always nonspin-flip (NSF). It can also be shown that nuclear spin-incoherent scattering is $\frac{2}{3}$ spin-flip (SF) and $\frac{1}{3}$ NSF for unaligned nuclear moments regardless of the relative orientation of the neutron polarization and scattering vector. On the other hand, isotopic incoherent scattering is always NSF.

If the sample is ferromagnetic, a magnetic field of sufficient strength to align individual domains must be applied to prevent depolarization of the neutron beam. If the magnetic moments are aligned along the direction of the scattering vector, no magnetic scattering occurs since it is only the projection of the magnetic moment in the reflecting plane (perpendicular to $\mathbf{Q}$) that contributes to the scattering. If, however, the parallel moments are all aligned perpendicular to $\mathbf{Q}$, then there is interference between the nuclear and magnetic scattering so that F^{++} and F^{--} are proportional to $b_j + p_j$ and $b_j - p_j$, and F^{+-} and F^{-+} are zero. By *selecting* $b_j = p_j$ in a monochromating crystal, only one spin state is reflected. This is one of the principal means by which a polarized beam can be obtained.

On the other hand, if the sample is an antiferromagnet and $\mathbf{Q}$ is parallel to the neutron polarization axis (the z axis), all magnetic scattering is SF.

The use of polarized neutrons for separating nuclear spin-incoherent scattering and magnetic scattering from coherent nuclear scattering was demonstrated in several papers, for example, [27].

3 EXPERIMENTAL METHODS

3.1 Neutron Sources

There are two principal sources of neutrons: the nuclear reactor and the spallation source. In the former the high-energy neutrons that are produced during the fissioning of uranium atoms within the reactor core are subsequently moderated by inelastic collisions with a moderating material such as heavy or light water. The continuous spectrum of neutron energies emerging from a beam hole is Maxwellian with a peak at a wavelength given by [3]

$$\lambda_p = \frac{h}{\sqrt{5mk_{\mathrm{B}}T}} \tag{25}$$

where k_{B} is Boltzmann's constant. For a typical moderator temperature of 300 K, λ_p is 1.125 Å. Useful intensities are generally obtained from approximately 0.5 to 5.0 Å. With secondary *hot* or *cold* moderators, this wavelength range can be extended considerably. For example, with a liquid hydrogen moderator at 20 K, the Maxwellian peak can be shifted to longer wavelengths to give a usable flux of neutrons with wavelengths of 20 Å or more.

Spallation sources on the other hand are pulsed in nature, providing bursts of neutrons at regular intervals. These accelerator-based devices produce high-energy neutrons when, for example, a pulsed proton beam impinges on a target such as uranium. The high-energy neutrons that result from the nuclear reaction are then moderated as in the nuclear reactor source. However, the inherent time dependence of the neutron spectrum allows for energy analysis to be performed by time-of-flight (TOF) techniques, which are discussed in detail in the textbook by Windsor [28].

3.2 General Features of a Neutron Triple-Axis Diffractometer

Space does not allow for complete descriptions of the various types of neutron diffractometers currently in use. We first describe some of the general features of a triple-axis spectrometer, which is one of the more commonly used and versatile instruments for neutron scattering studies, and then we consider more specifically the diffractometers used for structural studies of polycrystalline and single-crystal samples.

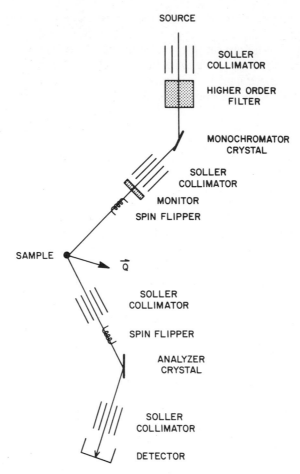

Figure 2.6 Schematic representation of a triple-axis spectrometer as described in detail in the text. The quantity B_{GF} represents the magnetic guide field along the neutron flight path that is necessary to maintain an axis of quantization for a polarized neutron beam. The scattering vector **Q** lies in the plane of the paper, which in practice usually corresponds to the horizontal plane.

Let us consider sequentially the essential individual components of a triple-axis spectrometer, which is depicted schematically in Figure 2.6, beginning with the monochromator. The monochromator is a single crystal that selects out of a collimated Maxwellian distribution of neutron wavelengths a nearly monochromatic beam by the Bragg diffraction process. Because of the finite angular divergence of the beam and the mosaic or natural width of the reflection, a relatively narrow but nonzero energy bandwidth is reflected. One of the most efficient nonpolarizing monochromators available is pyrolytic graphite. Other important monochromators include Cu, Ge, and Be crystals. The properties of crystal monochromators for neutrons are discussed at length in an article by Freund and Forsyth [29]. To obtain a polarized monochromatic beam a crystal such as the Heusler alloy Cu_2MnAl is used, which has a nearly vanishing structure factor for one spin state. Thin-film multibilayer devices can also be used to polarize neutrons, and they are reviewed in [30].

One problem with crystal monochromators is that the higher order wavelengths $\lambda/2, \lambda/3,...$ are also reflected, in general, thereby contaminating the beam incident on the sample. These higher order components can be suppressed by using either a polycrystalline Be filter for primary wavelengths longer than 4 Å or pyrolytic graphite filters, which have several convenient primary wavelength *windows* in the thermal energy range, a particularly useful one being at about 2.4 Å.

Additional collimators between monochromator and sample, sample and analyzer, and analyzer and detector further define the angular divergences of incident and scattered beams and ultimately affect the overall instrumental resolution.

At one of the larger research reactors such as the high flux beam reactor (HFBR) at Brookhaven National Laboratory (BNL), the high flux isotope reactor (HFIR) at Oak Ridge National Laboratory (ORNL), or the high flux reactor (HFR) at the Institute Laue–Langevin (ILL) in Grenoble, typical monochromatic fluxes incident upon the sample are from 10^6 to 10^9 neutrons per square centimeter per second. Because the scattering and absorption cross sections of aluminum and other structural materials for thermal neutrons are relatively low, cryostats and other sample environments are readily constructed. An incident beam monitor preceding the sample is used to correct for temporal intensity fluctuations of the source.

Neutrons scattered by the sample with the proper k_f are subsequently reflected by the analyzer crystal into the detector. Because coherent elastic scattering cross sections are usually orders of magnitude larger than those for inelastic processes, an analyzer is often unnecessary.

If it is desired to measure the four spin-dependent scattered intensities and corresponding structure factors [given by (24) in Section 2.3], polarizing monochromator and analyzer crystals are required as well as magnetic guide fields (of the order of 10–100 Oe) over the flight path of the neutrons (Figure 2.6). Neutrons are first simultaneously monochromated and polarized, then scattered by the sample (a flat coil flipper preceding the sample permits selection of either of the two neutron spin states). The second polarizing crystal or analyzer (along with another flipper) distinguishes the SF from the NSF scattering.

3.3 Detectors

3.3.1 Gas-Filled Devices

The detection of radiation most often relies on some ionization process that can be amplified to produce a measurable electrical impulse. Because neutrons have no charge and the energies used in scattering studies are from 0.001 to 1 eV, the neutron interactions with matter do not normally produce charged particles. Therefore, to a large extent the detection of neutrons has been based on nuclear capture reactions that produce electrons, protons, α particles, and other heavier fission fragments. The most commonly used reactions are [31]: ^{3}He(n,p)^{3}H + energy, ^{6}Li(n,α)^{3}H + energy, and ^{10}B(n,α)^{7}Li + energy + γ rays. The energies are in the million-electronvolt range, and the two particles produced are emitted in opposite directions, producing numerous ion pairs along the track. The cross sections for these reactions are sufficiently high to make the detection process very efficient. In a typical gas-filled ^{3}He detector, a pressure of 3 atm and a pathlength of 10 cm ensures more than 90% detection efficiency for 1-Å neutrons. For the neutron wavelengths of interest, the cross sections are proportional to the wavelength, so that for 0.5-Å neutrons the pressure or the pathlength must be doubled to retain 90% detection, but this is still feasible. It is important to notice that the number of charged particles produced in a neutron-capture incident is independent of the neutron kinetic energy. The electric impulse produced by the detection process is therefore independent of the neutron wavelength.

The most common detectors contain ^{3}He or ^{10}BF$_3$ as the neutron-capture medium. To detect the ionized particles an electric field is set up in the detector, so that electrons and positive ions drift toward the positive and negative electrode, respectively. By suitable choice of the field, the particles can be accelerated sufficiently to ionize further atoms, and a gas multiplication takes place. With proper design the final pulse is proportional to the number of primary ions produced in the capture reaction. The simplest of these is a gas-filled cylinder with the anode being a thin wire coincident with the cylinder axis. The neutrons enter along the axis through a ceramic window at one end of the cylinder. Typical dimensions are a length of 10–15 cm and a diameter of 5 cm with an operating voltage in the kilovolt region, although many different geometries are available, as summarized in [31]. Detector shielding typically includes some lead (for γ rays), borated paraffin or polyethylene (to moderate and absorb fast neutrons), and cadmium (to absorb the slow or thermal neutrons).

If we want additional spatial information, we must add further electrodes or make those we have position sensitive. Linear position-sensitive detectors (PSD) can be made using a cylindrical detector with one wire in the middle, but the neutrons now impinge orthogonal to the axis. For a captured neutron an electric impulse occurs at both ends of the wire, and using the two pulses we can determine the location of the incidence. This is done either by comparing the times of arrival or the pulse sizes. Two-dimensional information can be obtained by combining a series of one-dimensional detectors, or more commonly, by having several parallel electrodes inside the same gas volume. Each electrode may then be position sensitive [32], or there may be two orthogonal sets of wires to allow $X-Y$ detection of the point of incidence [33].

A summary of the many different available geometries is given by Convert and Forsyth [31]. A typical one-dimensional curved detector for studies of powders or liquids contains between 64 and 800 cells with an angular resolution of between 0.05° and 0.2°. For detection of two-dimensional scattering patterns the range is between 32×32 pixels and perhaps 200×200 pixels, and resolution on the detector surface is normally between 0.2 and 1 cm. The spatial resolution is defined by the length of the tracks of the primary particles. These are in the range of millimeters, but they can be reduced by adding stopping molecules to the gas mixture.

The detection efficiency is good for wavelengths above 1 Å. The detectors normally have a homogeneous response over the surface and good long-term stability. Their γ sensitivity is low, which makes them very suitable for use at steady-state reactor sources, where γ radiation is appreciable, and their count-rate capacity is adequate in this case. For TOF measurements, they are used mainly for longer wavelengths, as here the detector thickness must be kept small, which leads to low efficiency for shorter wavelengths.

3.3.2 Scintillation Devices

The first step in the detection process is again the conversion of neutrons to other particles by nuclear reactions. The converter is in direct contact with the scintillator material, and the particles excite electrons in the scintillator to higher energy states. The decay of these states produces emission of light in the ultraviolet and visible spectrum, which is then detected by a photomultiplier system. Most commonly a $^6LiO_2$-containing silicate glass doped with Ce^{3+} is used, and typically the scintillator thickness is a few millimeters.

A single piece of scintillator connected optically to a photomultiplier produces a detector with no space resolution. To introduce position sensitivity two approaches are used at present. One creates a mosaic of small optically independent scintillator elements, which are coupled by fiber optics to individual photomultipliers. By connecting each element to several photomultipliers and by decoding the combined output of all photomultipliers, the number of photomultipliers can be much smaller than the number of scintillator elements [34]. The size of the individual elements can be as small as a few millimeters squared; the elements can be arranged in any geometry suitable for the measurement; and the number of elements is normally comparable to the previously quoted values for gas-containing detectors.

Another approach uses one large continuous piece of scintillator connected to an array of photomultipliers. A neutron incident creates a flash of light, which is detected by several of the photomultipliers. The location of the neutron incidence is then deduced from the size of the photomultiplier signals. This detector type, which was originally invented by Anger [35] for use in nuclear medicine employing γ radiation, is now in routine use for neutron diffraction [36]. The resolution is in the millimeter range.

Because the conversion part of the detector is a solid of high density, the detection efficiency can be high even for short wavelengths. Therefore, these detectors are well suited for work with the more energetic part of the spallation-source spectrum. Likewise, they are well suited for TOF measurements, because the scintillator thickness contributes little uncertainty to the neutron pathlength.

However, their sensitivity toward γ radiation can be considerable. This makes the detectors more suitable for pulsed sources, as the γ and neutron pulse can be separated in time. The light produced in the scintillator can be recorded equally well on photographic plates. The nuclear converter–scintillator is normally $^6LiF/ZnS(Ag)$ in a plastic binder, the efficiency is around 20% at 1 Å, and the resolution is around 0.1 mm [37]. Intensity measurements are possible, but the accuracy is limited to 7–8%. Nevertheless, film techniques are often essential for a general exploration of reciprocal space.

Another possibility for neutron detection is to use the nuclear conversion in a foil of one of the materials quoted above followed by detection of the particles escaping from the foil. The major inconvenience of this method is that the average range of particles inside the foil is only from 10^{-2} to 10^{-3} mm. This limits the thickness of the foil and thus its efficiency. The best results are obtained for ^{157}Gd, where an efficiency of 40% can be obtained at 1 Å [31] if particles escaping from both sides of the foil are used. Other foils, such as ^{235}U, give much lower efficiencies, but they are used as low-efficiency detectors to monitor the neutron beam impinging on the sample.

3.4   Instrumental Resolution

As mentioned previously, the monochromatic beam of neutrons incident upon the sample is actually composed of a distribution of magnitudes and directions of wave vectors k_i. The instrumental resolution depends on: (1) the interatomic plane spacing for a particular Bragg reflection from monochromator, sample, and analyzer; (2) the corresponding widths of the angular distributions of microcrystallites (for mosaic single crystals); (3) the angular divergences of the incident and scattered beams as defined by the collimators; and (4) the relative orientation of the scattering vectors of monochromator, sample, and analyzer. The resolution properties (e.g., energy width, widths in reciprocal space parallel and perpendicular to the mean scattering vector Q_0, etc.) corresponding to a particular spectrometer configuration can, in general, be calculated in a straightforward manner [38]. A typical resolution of about 0.5 meV in energy and $0.015\,Å^{-1}$ parallel to Q_0 is obtained over a significant range of sample scattering angles at $\lambda = 2.35\,Å$ using a pyrolytic graphite monochromator and analyzer [(0002) reflection and mosaics of 25 min of arc] and 20-min horizontal collimations in the "W" configuration depicted in Figure 2.6. Proper consideration of spectrometer resolution is one of the most important factors in an experimental study, but a general discussion is beyond the scope of this chapter. However, remember that inevitably there is a trade-off between intensity and resolution that places constraints on the experimental configuration.

3.5   Powder and Single-Crystal Diffractometers

We mentioned previously how a monochromatic beam is produced for reactor neutrons, or how a pulsed white beam is produced in a spallation source. We now consider some of the special features of modern diffractometers for neutron powder measurements first with a constant-wavelength beam and then with a white beam using TOF techniques.

3.5.1   Powder Diffractometers

The essential features of a constant-wavelength instrument are a sample table large enough to support heavy accessory equipment such as cryostats and furnaces and some kind of detector assembly. The latter may be a stationary PSD covering a suitably wide angular region such as that used at the instrument D1B at ILL [39], but it is more often a set of individual BF_3 or ^3He counters mounted on a scattering arm, as used at the instrument D1A at ILL [40] and several other reactors. Both the sample table and the scattering arm should be capable of being driven with stepping motors under computer control to an accuracy of at least 0.01°. A PSD covering a limited range of 2θ may be mounted on the scattering arm instead and moved periodically to cover as many successive regions as desired. Detectors of this type are in operation at Kernforschungsanlage, Jülich [41] and at the University of Missouri, Columbia, Missouri [42].

The success of the profile technique of structure refinement developed over the past decade by Rietveld [43], which is discussed in more detail in Section 4.1.2, has resulted in a trend toward the construction of dedicated high-resolution powder diffractometers. The design features necessary to optimize intensity and resolution for this purpose were first set out by Hewat [44] and embodied into D1A [40] at ILL, which has served as a prototype for constant-wavelength machines throughout the world. Hewat's design was based on some simple expressions derived by Caglioti and co-workers [45] many years previously for the diffraction peak fwhm, Γ, as a function of 2θ. For a two-axis diffractometer

$$\Gamma = [U \tan^2\theta + V \tan\theta + W]^{1/2} \qquad (26)$$

where U, V, and W are instrumental parameters that are functions of the collimator divergences α_1, α_2, and α_3; the mosaic spread of the monochromator β; and the Bragg angle of the latter θ_M. The geometry is shown schematically in Figure 2.7. For $\theta_M \approx 60°$, and appropriate choices of the other variables, the minimum value Γ_{min} in this quadratic expression can be made to fall in the 2θ region where the peaks in the diffraction pattern are most closely spaced; that is, about 90°–110°, giving a resolution function that is fairly well optimized for Rietveld analysis. For D1A and several other instruments of this type, Γ_{min} is about 0.3°–0.4°, corresponding to a resolution expressed as $\Delta Q/Q$ or $\Delta d/d$ of about 2–3×10^{-3}. This quantity is a more generally useful specification of resolution since it enables a qualitative comparison of constant-wavelength and TOF diffractometers to be made [46]. Still better resolution is possible if the collimation is tightened further, but only at the expense of considerable intensity. The recently commissioned machine D2B at ILL [47] has 64 counters

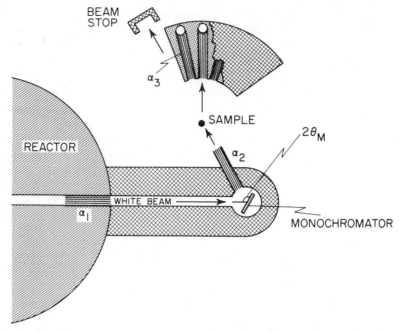

Figure 2.7 Schematic diagram of the geometry of a high-resolution constant-wavelength neutron powder diffractometer. The quantities α_1, α_2, and α_3 are, respectively, the collimation in-pile; monochromator sample; and sample detector, defined as d/l, where d is the spacing between the vertical blades and l is the length of the collimator; β is the mosaic spread of the monochromator (fwhm of the rocking curve); and $2\theta_M$ is the take-off angle.

and a resolution minimum $\Delta d/d$ of about 5×10^{-4}, which probably represents the practical limit at presently operating reactors.

Neutron powder techniques are primarily exploited for structure refinement, where fast counting rates are sacrificed for high resolution and data collection times are typically a few hours. However, for studies of reaction kinetics, phase transformations, and crystallization processes, emphasis is on high counting rates with only modest resolution. The D1B at ILL is an instrument built with these features in mind [39], and it incorporates a curved PSD covering an angular range of 80° with $\Delta d/d$ around 10^{-2}. Data collection can be accomplished in as little as a few minutes. A new instrument, D20, has recently commenced operations [48], and this should allow an order of magnitude increase in counting rates.

Although TOF techniques were used for a few powder-diffraction experiments at reactors, their application to high-resolution powder diffractometry was coupled mainly with the development of pulsed neutron spallation sources [28, 49]. Here a white beam is incident upon the sample, the detector bank is at a fixed scattering angle, and a diffraction pattern is recorded as a function of the flight time of a short pulse of neutrons of various energies along the total flight path. For data analysis the flight time is converted to the d spacing. The geometry is shown schematically in Figure 2.8.

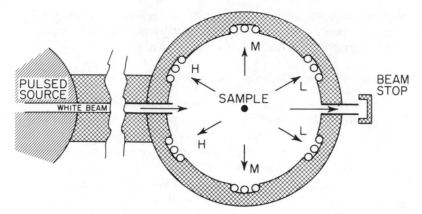

Figure 2.8 Schematic diagram of the geometry of a high-resolution TOF neutron powder diffractometer. Total flight path is typically 15–20 m (100 at the instrument at the Rutherford–Appleton Laboratory). Three pairs of detector banks are shown, with low (L), medium (M), and high (H) resolution depending on the scattering angle.

The resolution of such an instrument is given approximately by the expression

$$\Delta d/d = [(\Delta t/t)^2 + (\Delta L/L)^2 + (\cot \theta \cdot \Delta \theta)^2]^{1/2} \tag{27}$$

where Δt, ΔL, and $\Delta \theta$ are the uncertainties in the flight time t; the flight path L; and the Bragg angle θ; respectively. The main contribution to Δt arises from the finite pulse width of neutrons emanating from the moderator, and it is proportional to wavelength, as is the flight time. Thus $\Delta d/d$ is approximately constant, and it decreases with increasing L and θ.

These principles were embodied in the high-resolution TOF diffractometer described by Jorgensen and Rotella [50] and in two subsequent powder instruments in current operation at the Argonne intense pulsed neutron source (IPNS) [51]. These have flight paths of 14 and 20 m and resolutions $\Delta d/d$ of about 3.5 and 2.2 × 10^{-3}, respectively, which are comparable to those of many constant-wavelength instruments. Other diffractometers of this type have recently become operational at Los Alamos National Laboratory [52] and at the Japanese National Laboratory for High Energy Physics [53].

Considerably better resolution can be obtained by extending the flight path, and this was done at the recently commissioned powder diffractometer at the pulsed-neutron source at the Rutherford–Appleton Laboratory [54]. In this case, L is about 100 m and a resolution of 5 × 10^{-4} can be achieved; but the question of *frame overlap*, in which faster neutrons from one pulse overtake slower ones from the previous pulse, must be considered carefully.

Because of the much higher epithermal flux from a spallation source, it is possible to collect data at d spacings down to 0.2 Å, compared to 0.75–1 Å for constant-wavelength instruments. This may be an advantage for certain types of experiment where high-Q data are necessary, as is the fixed-angle geometry for studies in constrained environments such as high-pressure cells. Generally, structure refinements based on data collected by the two different techniques have yielded comparable results at a given level of resolution.

3.5.2   Single-Crystal Diffractometers

The detailed considerations of collimation and focusing effects that are necessary for the design of high-resolution neutron powder diffractometers are not so important for single-crystal instruments. These are usually four-circle diffractometers basically similar to their more familiar X-ray counterparts but with a χ-circle large enough for a closed-cycle He cryostat to be mounted, since many single-crystal investigations are carried out at low temperatures.

It is important to remember that because of the relatively low neutron flux, compared to that of X-ray sources, fairly large crystals are needed, typically 1–2 mm in average dimension. Although absorption is usually negligible, extinction effects can be extremely large, particularly at longer wavelengths, and it is common practice to work around 1 Å, which also has the advantage that d spacings down to 0.5 Å are accessible.

Most constant-wavelength diffractometers are equipped with a conventional detector, but the greatly enhanced rates of data collection possible with position-sensitive area detectors make these a very attractive, although expensive, alternative. Such a detector has been used for protein crystallography at the Brookhaven high flux beam reactor for several years [32].

Single-crystal diffractometers at pulsed neutron sources are based on the Laue white beam technique that is familiar to many X-ray diffractionists for crystal orientation. In this case the diffraction pattern is recorded on an area detector while the crystal is held stationary in the white beam. A particularly elegant feature of this technique is that TOF analysis allows all orders of any given set of planes to be measured. Unlike the corresponding high-resolution TOF powder diffractometers, the required time resolution for this purpose is usually not so critical, so that these instruments can be located on short flight paths. A diffractometer of this type has been operating at the Argonne intense pulsed neutron source for several years [55].

4   DATA COLLECTION AND ANALYSIS FOR STRUCTURAL STUDIES

4.1   Powder Diffraction

4.1.1   Experimental Considerations

Until very recently, neutron powder diffraction was used predominantly as a technique for structure refinement and hardly at all for the ab initio solution of unknown structures. Thus one of the requirements was at least partial knowledge of the basic crystal architecture. At most instruments, this continues to be important, but at the new high-resolution diffractometers at the ILL and Rutherford–Appleton Laboratory, it should be possible to extract enough individual integrated intensity data for ab initio structure solution by traditional methods of single-crystal analysis such as Patterson synthesis and direct methods, as illustrated by the recently reported structure determination for $FeAsO_4$ [56]. We can expect much more emphasis on experiments of this type.

An important consideration is the complexity of the structure and hence the level of resolution required. If one extends the qualitative arguments of Christensen and co-workers [57], it is possible to estimate from the density of inequivalent peaks in reciprocal space and an average atomic volume of 12 Å^3 that one should be able to successfully refine a centrosymmetric structure containing some 50 positional parameters at a resolution level $\Delta d/d = 2 \times 10^{-3}$, or perhaps 200 at the new powder machines at ILL and Rutherford–Appleton Laboratory. Clearly there are a great many structural problems, including time-dependence studies, that can be undertaken at much more modest levels of resolution, and this should be considered when the experiment is planned.

Since most neutron powder diffractometers are oversubscribed, it is important first to characterize the sample carefully by conventional X-ray powder techniques. The presence of impurity phases will generally detract from the quality of the subsequent refinement, although several Rietveld-type programs permit simultaneous refinement of two or more phases. Particle-size and strain broadening can also affect the line widths. Particle-size broadening takes the form $\Delta d/d \approx d/L$, where L is the apparent crystallite size; therefore, it is significant on most instruments if $L \leqslant 0.1 \mu m$. Similar considerations apply to strain broadening, which takes the form $\Delta d/d = $ constant. In favorable cases useful information can be obtained about these effects with appropriate modifications to the profile model.

Sample composition is another important consideration. Highly absorbing elements such as B, Cd, Gd, and Eu are generally to be avoided unless suitable isotopes are available. High concentrations of H are also undesirable because of the large incoherent scattering cross section and correspondingly high background. Although it is possible to refine structures containing significant amounts of H, substitution of as much of the latter as possible by D is strongly recommended.

Flat plate specimens can be used in transmission, but it is generally more convenient to use cylindrical samples, which can either be stacked on a pedestal in pellet form or loaded into a sample holder. To avoid extraneous Bragg peaks, thin-walled V holders are frequently used for this purpose, but at the expense of an appreciable amount of incoherent scattering. If low background is essential, thin-walled Al holders may be better in this respect.

For most samples, attenuation losses are not severe, since the effective absorption (self-shielding) coefficient μ_{eff}, which includes contributions from both true absorption and secondary extinction, is seldom greater than 1 cm^{-1} at 1.5 Å. Thus optimum sample dimensions may be 1–2 cm in diameter and 2–4 cm high, corresponding to perhaps 5–20 g of sample. For constant-wavelength experiments it is useful to make a direct determination of μ_{eff} by means of a transmission measurement in the forward direction with simple pinhole geometry. For TOF experiments, μ_{eff} must be treated as a parameter in the refinement process.

Since most powder diffractometers are already fairly well optimized for a particular function, the actual process of data collection is usually straightforward and involves only choices such as the counting time necessary for adequate statistics, the 2θ range to be covered, or the angular step interval [58].

One final point to note is that many elements have a significant neutron-activation cross section and may become appreciably radioactive after a few hours' exposure to the incident beam. Appropriate precautions should be taken when storing and handling these samples subsequently.

4.1.2   Data Analysis

A typical X-ray or neutron powder-diffraction pattern contains a series of peaks that become increasingly overlapped with increasing 2θ or Q. In the early days of neutron diffraction, many simple structures were determined from the relatively few integrated intensities that could be extracted manually, and this procedure is still used widely for solving magnetic structures. However, in the last decade, the emphasis has shifted almost entirely to the application of the profile analysis technique for structure refinement introduced in 1969 by Rietveld for constant-wavelength neutron data [43]. This involves least-squares fitting of individual data points comprising the overall profile of the pattern to some structural model so that all information contained in the pattern is utilized irrespective of the degree of overlap. This process, widely recognized as one of the major advances in powder diffraction methodology in recent years, is now generally known as *Rietveld refinement*.

The expression for the calculated intensity at the ith point on the pattern was formulated by Rietveld on the assumption that the peaks were Gaussian in shape; that is,

$$y_i(\text{calcd}) = \sum K \left[\frac{t2(\ln 2)^{1/2}}{\Gamma \pi^{1/2}} \right] \exp\left[-4\ln 2 \frac{(\Delta 2\theta)^2}{\Gamma} \right] \frac{F^2 j}{\sin\theta \sin 2\theta} \tag{28}$$

where K is an instrumental scale factor, t is the 2θ step interval, Γ is the fwhm, $\Delta 2\theta$ is the displacement from the peak maximum, and the other quantities have their usual crystallographic significance. Summation is taken over all peaks that can contribute to the ith point within some previously defined range about the peak maximum, usually $1.5\,\Gamma$ on either side. The dependence of Γ on 2θ is given by the expression:

$$\Gamma = (U \tan^2\theta + V \tan\theta + W)^{1/2} \tag{29}$$

which is already described in Section 3.5.1. The terms U, V, W, and the unit cell parameters are treated as refinable *profile* parameters in the least-squares fit, together with the usual *structural* parameters; that is, the positional parameters, thermal factors, and site occupancies.

The observed intensity y_i(obsd) is obtained from the raw intensity Y_i by subtraction of a background intensity Y_B, which is determined by linear interpolation between values estimated in regions of the pattern where there are no peaks. The variance of Y_i is

$$\text{var}(y_i) = \text{var}(Y_i) + \text{var}(Y_B) \tag{30}$$

Since it is difficult to make an estimate of var(Y_B), this is usually set equal to zero, and the weights w_i are taken as $1/Y_i$. To avoid this approximation, and because there may be no discernible background regions in heavily overlapped parts of the pattern, most Rietveld-type codes provide an option for background refinement in the form of some simple empirical function.

Rietveld also made provision for the refinement of magnetic structures, in which case the components of the moment along different axes are included as refinable parameters and the magnetic form factors are a necessary part of the input.

The Rietveld refinement technique can be applied to pulsed-source TOF neutron data in an analogous way [59]. In this case, the intensity spectrum must first be measured and input in an appropriate form, a more complex asymmetric peak-shape function must be used, and the peak-shape TOF dependence must be determined. In addition, a refinable parameter is needed to allow for extinction effects at longer d spacings. When TOF data are analyzed, background is always refined because it is difficult to make estimates at shorter d spacings where the pattern is usually heavily overlapped. Despite these additional complications, the fits are of comparable quality to those obtained with constant-wavelength data at similar levels of resolution.

Several systematic errors may need to be considered at this stage. Among these are (1) peak asymmetry caused by vertical divergence effects, which may be particularly severe in constant-wavelength patterns at low angles; (2) particle-size and strain broadening, which can contribute a Lorentzian-like component to the peak shape that is visible as extended tails; and (3) preferred orientation effects because of plate- or needlelike crystallites. Most modern versions of the Rietveld code include an empirical correction for these systematic errors.

The quality of the fit is indicated by several R factors, which are defined as:

$$R_I = 100 \sum |I(\text{obsd}) - I(\text{calcd})|/\sum I(\text{obsd}) \tag{31a}$$

$$R_p = 100 \sum |y(\text{obsd}) - y(\text{calcd})|/\sum y(\text{obsd}) \tag{31b}$$

$$R_{wp} = 100\{\sum w[y(\text{obsd}) - y(\text{calcd})]^2|/\sum w[y(\text{obsd})]^2\}^{1/2} \tag{31c}$$

$$R_e = 100\{(N_O - N_P)/\sum w[y(\text{obsd})]^2\}^{1/2} \tag{31d}$$

where the calculated values contain the overall scale factor, $1/c$. The factor R_I represents an unweighted integrated intensity index based on an approximation for I(obsd) in which individual values of y_i(obsd) are partitioned in the ratio of their calculated values. Therefore, R_I provides a useful measure of the quality of the fit that can be compared directly to the familiar single-crystal R factor; R_p and R_{wp} are unweighted and weighted indices, respectively, for the complete profile; and R_e is the R factor expected if the residuals were purely statistical. Both R_p and R_{wp} can be affected by factors such as the peak shape and the background as well as inadequacies in the structural model; therefore, they are not as easy to interpret as the R factors in a single-crystal refinement. The ratio $(R_{wp}/R_e)^2$ corresponds to the χ-squared goodness-of-fit index for the profile; that is,

$$S_p^2 = \sum w[y(\text{obsd}) - y(\text{calcd})]^2/(N_O - N_P) \tag{31e}$$

where N_O and N_P are the number of profile observations and refined parameters, respectively. Values of S_p^2 between 1.5 and 2.0 are generally regarded as perfectly respectable.

The extent to which the estimated standard deviations given by the Rietveld technique reflect the real accuracy of the refined parameters, particularly in comparison with those obtained from integrated intensity data, is a source of some controversy [60]. It is generally accepted that if the only source of error is statistical, the two techniques give essentially identical results. However, systematic errors usually predominate, as indicated by values of S_p^2 significantly greater than unity. Most versions of the Rietveld code conform to normal crystallographic practice in that the variances are scaled by S_p^2 [61], but whereas this generally works quite well for integrated intensities, giving estimated standard deviations within a factor of perhaps 1.5 of the true error, there is considerable evidence that profile estimated standard deviations for positional parameters sometimes underestimate the true error by a factor of 2–3 and even more for thermal and occupation factors. A detailed discussion is beyond the scope of this chapter, but it is recommended that considerable caution be exercised in interpreting, for example, the significance of small changes in bond lengths or deviations from stoichiometry.

For relatively complex structures, a generally useful refinement strategy is to apply selected constraints in the initial stages. For example, a single isotropic temperature factor may be assigned to the cations and another to the anions, or the half-width parameters may be fixed at some typical value. As the refinement progresses, one would probably want to assign individual isotropic temperature factors to nonequivalent atoms, and perhaps allow for the possibility of fractional occupancy or substitutional disorder in some sites. Unexpectedly high or negative temperature factors may be an indication of disorder. In certain cases, refinement of anisotropic temperature factors for some or all of the atoms may be justified, but the results should be interpreted with much caution. In cases of pseudosymmetry, where atoms are shifted slightly away from *ideal* positions and two of the unit-cell parameters may differ very little, it is important to check for false minima by using several different sets of starting parameters. Additional constraints such as rigid-body rotation may be useful in these situations.

Although the usual type of significance tests [62] can in principle be applied to R_{wp}, it is important to recognize that there is a distinction between the total *profile* model, which may contain many thousands of individual observations, and the *structural* model, for which the information is contained in perhaps a few hundred values of F^2. There is at present no consensus about whether a *significant* improvement in the former necessarily represents an improvement in the latter.

A modified Rietveld-type approach to profile analysis was developed by Cooper and co-workers [63] and Pawley [64], in which no structural model is assumed, but instead the profile parameters and integrated intensities are treated as refinable variables. This has the advantage of extracting information that can be used directly for ab initio structure solution and also gives estimated standard deviations that more directly reflect the fit to the profile parameters independent of a structural model. The disadvantage is that the amount of useful extracted information can be severely limited because of large correlations between intensities in heavily overlapped regions of the pattern, particularly in cases of pseudosymmetry.

4.2 Single-Crystal Diffraction

4.2.1 Experimental Considerations

Once again, many of the considerations involved in planning powder-diffraction experiments do not apply for single-crystal measurements. In this case, the goal is usually to collect as many integrated intensities as possible. Just as with X rays, the crystal gives rise to diffraction over a range of orientations because it is mosaic in nature and because of the instrumental angular beam divergence and wavelength spread. Therefore, it is necessary to integrate over these variables to obtain the intensity, and hence the observed structure amplitude $|F_{hkl}^{obsd}|$.

Figure 2.9 illustrates how this integration differs for constant-wavelength and white-beam techniques. In the former, the condition for Bragg scattering is shown in Figure 2.9a, where the Ewald sphere has a certain thickness depending on the wavelength spread and divergence of the incident beam. In addition, the reciprocal lattice points are smeared by the crystal mosaic. To record the total integrated intensity, the crystal is step scanned about an axis orthogonal to the scattering triangle, usually an ω scan about the vertical axis. It is important at this stage to check the crystal mosaic along the principal directions for possible twins or low-angle boundaries, as these can greatly complicate the process of data collection. Once the crystal is suitably oriented, the strategy for data collection depends on the type of detector available. For a single detector, the reflections are first scanned one after another by zigzagging through reciprocal space fairly quickly. As Q increases, a point is reached where the signal-to-noise ratio becomes unacceptably low. If the ultimate goal is the precise analysis of some structural feature, the first data set can be used for preliminary analysis to

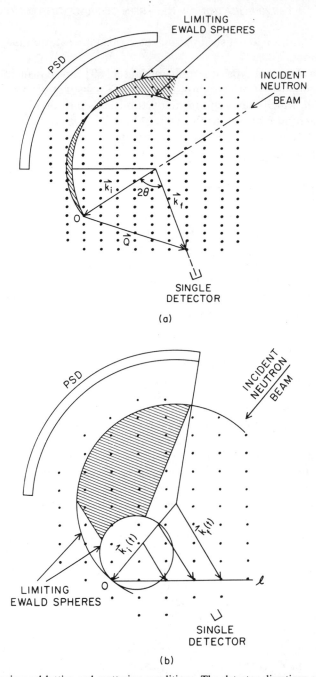

Figure 2.9 Reciprocal lattice and scattering conditions. The detector directions or positions are marked symbolically: (*a*) corresponds to a typical steady-state reactor measurement with a very limited wavelength range. The lower portion indicates a single scattering incident, while the upper portion shows the part of the reciprocal lattice that can be observed by a PSD in a given setting; (*b*) shows the situation for a pulsed-source employing a large range of wavelengths. This will lead to TOF dependent sets of K_0, K_1. The lower part indicates the location of a single detector for measurements of reflections corresponding to lattice points lying on line *l*. The shaded region in the upper part shows the area viewed by a PSD.

154

predict reflections beyond this point that are worth measuring with better counting statistics and thus to increase the number of useful observations. Conversely, if a less precise analysis suffices, reflections beyond this point can probably be ignored. For measurements at higher Q values, it is important to ensure that the detector aperture is sufficiently large to accept all of the diffracted beam.

If a PSD is used, the question of skipping particular reflections does not arise, since the detector sees all of these within a given Q range. However, it may be more difficult to ensure that all reflections are measured under similar conditions. It is worth noting that here the measurement strategy is less dependent on the crystal system under study. For example, if a hemisphere of reciprocal space is explored to some limiting Q value, then the settings with an area detector are independent of the crystal system.

The situation for white-beam measurements is shown in Figure 2.9b, from which it is evident that all reciprocal lattice points on a line extending from the origin lead to scattering in the same direction. As mentioned in Section 3.5.2, these different orders of reflection can be distinguished readily by TOF analysis at a pulsed neutron source. The upper part of Figure 2.9b indicates the area of reciprocal space that is accessible to a PSD. The crystal is fixed and integration of the intensity is over wavelength. Other areas of reciprocal space can be explored by changing the orientation angles in a predetermined way independently of the sample. Figure 2.10 shows an example of the intensity distribution measured over one such region [65].

It can be inferred from Figure 2.9a that for constant-wavelength measurements on a crystal with a small unit cell, the reciprocal lattice points are sufficiently well separated for the wavelength spread and collimation to be relaxed without risk of overlapping reflections, while the converse is true for large unit cells, such as those found in proteins. In addition, overlap can be reduced by use of a longer wavelength.

Under these circumstances, the decrease in flux at the sample may necessitate the use of an area detector. In these cases, wavelengths of 1–1.6 Å are generally used, and the data-collection time may be a few weeks. For small unit cells most measurements are made at about 1 Å with a single detector; but if a hot source is available, it is preferable to use shorter wavelengths of 0.5–0.7 Å. For these measurements, 300–1000 reflections can be measured in a day.

Rather similar considerations apply to TOF measurements, where the resolution depends on the flight path, pulse width, and wavelength range. Data-collection times for one setting may be about $\frac{1}{2}$ day [65]. The number of settings depends on the crystal symmetry, but it may exceed 20.

4.2.2   Data Analysis

The raw data from a diffractometer consists of one-, two-, or three-dimensional data arrays containing neutron counts. If the data are from a single detector, the intensity of a Bragg reflection is recorded as a function of the rotation of the crystal as the reflection passes through the Bragg position. The integrated intensity is then obtained by first identifying the peak position [66]. The

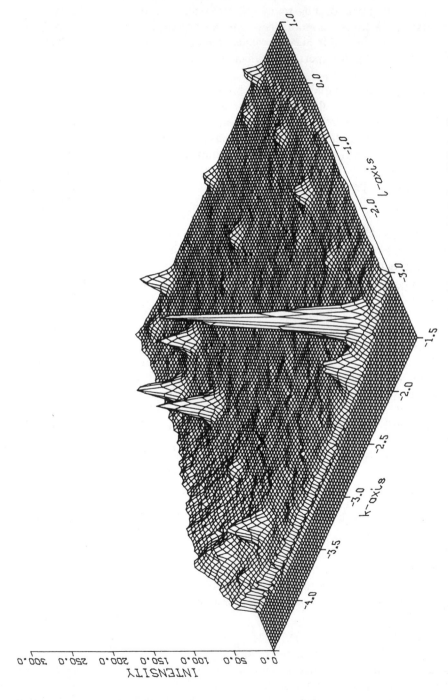

Figure 2.10 The intensity in the plane 5, k, l, observed in a single-crystal TOF study of the compound $(ET)_2I_3$ at 20 K. Satellites are seen next to some of the main reflections; they are especially pronounced near $(5, \bar{2}, \bar{3})$, $(5, \bar{2}, 1)$, and $(5, \bar{4}, 0)$. Adapted, with permission, from Figure 4 in T. J. Emge and coworkers, *Phys. Rev. Sect. B,* **30,** 6780 (1984).

remaining points are used to define the background, a straight line is assumed, the intensity is obtained by trapezoidal integration over the peak, and the counting error is estimated by employing standard statistical methods.

Intensities for multidimensional data are derived in a similar way. Figure 2.10 represents a section through a three-dimensional recording of part of reciprocal space that was obtained with white-beam techniques as described in Section 4.2.1. The coordinates of the three-dimensional raw data array are the X and Y coordinates of the detector, and the neutron flight time. From the known detector position and the crystal orientation these are then converted into reciprocal space coordinates. Again data reduction consists first in defining the outline of the individual Bragg spot. Many methods were developed [67–72] using a priori determination of the peak location, adjustable ellipsoidal enclosures, and image-recognition techniques. After the peak location is identified, the intensity can be extracted by integration over the volume of the peak followed by a background correction.

The observed intensity I_{hkl} can be related to the observed structure amplitude F_{hkl}^{obsd} by the following general expression:

$$I_{hkl} = K \Phi \varepsilon L A y (1 + \alpha) |F_{hkl}^{obsd}|^2 \tag{32}$$

where K is an instrumental scale factor that takes into account neutron flux, crystal size, and measurement time and is normally treated as an adjustable parameter relating $|F_{hkl}^{obsd}|^2$ to the calculated structure amplitudes in the refinement; Φ is the incoming intensity distribution as a function of λ; ε is the detector efficiency as a function of λ; L is the product of the λ dependence and the Lorentz correction, which takes into account the velocity with which the reciprocal lattice point moves through the Ewald sphere; A is the absorption correction; y is the extinction correction, which is caused by secondary scattering processes within the crystal; and α is the proportion of the observed intensity that results from inelastic scattering processes.

For constant-wavelength measurements the wavelength range is very limited, and Φ and ε may be incorporated into K. For white-beam measurements these functions must be determined independently. The wavelength dependence is similar for A and y. The absorption factor is calculated from the crystal geometry in the usual fashion, but the absorption coefficients are wavelength dependent. For all practical purposes this dependence is linear, so a satisfactory correction can be made. The situation is much more complicated for extinction corrections for white-beam data. Theories do exist [73] that account for the wavelength dependence, assuming some model for the crystal mosaic distribution. Unfortunately, however, there is a good deal of evidence [74] that most crystals do not have a simple mosaic behavior. At best the correction works, but with questionable significance; at worst it fails altogether. As the extinction effect is dependent on the structure amplitude and the crystal size, it can be avoided experimentally by the use of small crystals and by excluding reflections that are seriously affected.

The inelastic contribution α can be calculated if the elastic constants for the material are known. As the inelastic scattering depends on whether the neutron velocity is faster or slower than the velocity of sound in the crystal [75], it is best to work in a range where only one of these two conditions applies.

The factor L is different for the constant-wavelength and white-beam cases. In the first, where the integration is performed by rotating around an axis orthogonal to the scattering plane, we obtain $L = \lambda^3/\sin 2\theta$ [76], while for the white-beam measurements, where the integration is performed over λ, the corresponding expression is $L = \lambda^4/\sin 2\theta$ [77]. The strong dependence on λ underlines the advantages of measuring at long wavelengths. Unfortunately, as mentioned previously, the longer the wavelength, the fewer are the number of reflections that are observed, and the more numerous are the secondary scattering processes within the crystal, leading to increasing extinction effects. Thus the choice of wavelength is a delicate balance that depends very much on the aim of the study.

In principle all the quantities mentioned earlier can be accounted for, so after appropriate corrections are made, the observed structure factors should be independent of the instrument on which they were measured. This is an important point, because all further analysis can then be carried out using standard crystallographic analysis techniques. These have been described in several standard crystallographic textbooks [78] and are familiar to most X-ray crystallographers, so we do not further elaborate on this aspect.

Differences in details between measurements made with X rays and neutrons are pointed out as they arise in some of the subsequent examples.

5   SOME EXPERIMENTAL RESULTS

We now give a few recent representative examples of structural studies that illustrate the power and scope of modern neutron powder and single-crystal diffraction techniques.

5.1   Powder-Diffraction Studies

The main emphasis in the last decade has been on the application of the Rietveld technique to the refinement of inorganic structures, with typically 10–50 positional parameters, including many materials in areas that are technologically important, such as superconductivity, catalysis, fast-ion conductivity, ceramics, magnetism, and hydrogen storage. Neutron powder techniques are particularly well suited for temperature-dependence studies since precise temperature control is easy to achieve, the orientation is unimportant, the sample will not shatter because of stresses introduced through a crystallographic transition, and perhaps most important of all a large region of reciprocal space is automatically surveyed. In addition, Rietveld refinement yields precise values of lattice parameters that can signal the occurrence of subtle symmetry changes.

Other areas where powder techniques proved useful are magnetic structure determination, studies of chemical kinetics, and stress analysis.

5.1.1   Materials Research

Neutron diffraction has played an important role in the structural characterization of the new classes of Cu-based mixed oxides with high superconducting transition temperatures. The discovery of $YBa_2Cu_3O_7$, with a T_c of about 90 K, was quickly followed by several X-ray determinations of the cation positions. However, because of twinning problems and the relatively weak scattering from oxygen, X rays were unable to establish with any certainty the oxygen positions. The latter were determined in many independent neutron powder studies. For a review, see [79]. The structure is orthorhombic and can be viewed as a heavily distorted, oxygen-defective, ordered derivative of a perovskite-type structure, as illustrated in Figure 2.11a and b, taken from reference [80], which are representative of the many independent structural determinations referenced in [79]. Two-thirds of the Cu atoms have four near-neighbor oxygen atoms between 1.93 and 1.96 Å distant and one more distant neighbor at 2.28 Å in a tetragonal pyramidal arrangement, and they form a roughly two-dimensional network with corner-shared oxygen atoms in the ab plane. The remaining one-third have fourfold planar coordination in the bc planes with two sets of oxygen atoms 1.85 and 1.94 Å distant, and they form chains linked through corner-shared O4 atoms along the b axis (Figure 2.11b).

Detailed studies carried out as a function of temperature in different atmospheres by TOF methods at the Argonne intense pulsed neutron source [81] revealed that there is an order–disorder transition in pure O_2 at about 700°C to a tetragonal structure, at which point the equilibrium composition is about $YBa_3Cu_3O_{6.5}$. The basal plane sites are now disordered and only partially occupied (Figure 2.11c). By appropriate heat treatment in an inert atmosphere, the oxygen content can be reduced still further to $YBa_2Cu_3O_6$. As shown by several studies [79], this is nonsuperconducting and has a similar tetragonal structure in which the basal plane sites are completely vacant.

A series of Rietveld refinements were made on constant-wavelength data collected at ILL at temperatures between 5 and 300 K by Capponi and co-workers [82] to check for possible changes in structure associated with the superconducting transition; but apart from a few minor changes in bond lengths, the results were negative. However, subsequent high-resolution TOF data obtained at the Rutherford–Appleton Laboratory by David and co-workers [83] revealed evidence for a subtle change in volume expansivity at T_c.

Over the past few years, some very elegant structural studies have been carried out on zeolites, which have considerable commercial importance as ion exchangers, molecular sieves, and heterogeneous catalysts for hydrocarbon reactions [84]. Their general formula is $M_{x/m}Al_xSi_{2-x}O_4$, where AlO_4 and SiO_4 form a framework of tetrahedra and M is an extra-framework cation of valency m. The framework structure contains pores or cages of various shapes and sizes

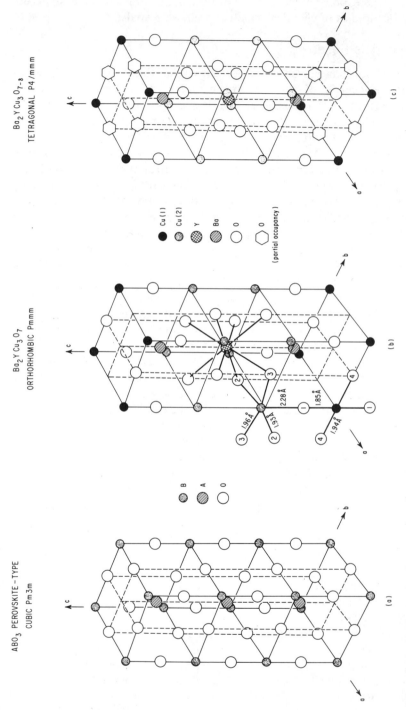

Figure 2.11 (a) Three unit cells of the prototype cubic ABO_3 perovskite-type structure. (b) Orthorhombic structure of $YBa_2Cu_3O_7$ with $Pmmm$ symmetry. Heavy lines show the fourfold arrangement of oxygen atoms around Cu(1), the tetragonal pyramidal arrangement around Cu(2), and the near-cubic coordination of Y. (c) Tetragonal structure of $YBa_2Cu_3O_{7-\delta}$ showing partially occupied oxygen sites in basal planes. Adapted, with permission, from Figure 3 in D. E. Cox and co-workers, *J. Phys. Chem. Solids*, **49**, 47 (1988).

into which water or organic molecules can be sorbed. Silicon and aluminum can be replaced by other ions, as in the large family of analogous aluminophosphates, for example, [85]. Rietveld refinement is particularly appropriate for these materials, since an approximate framework structure can very often be inferred, and it is possible to use difference Fourier techniques to locate extra-framework atoms or molecules.

A good example of this is provided by Fitch and co-workers [86] who recently determined the localization of benzene in sodium-Y zeolite. This compound has the approximate composition $Na_{0.58}Al_{0.58}Si_{1.42}O_4$, space group $Fd3m$, and $a \approx 24.8$ Å. There are eight atoms in the asymmetric unit, with Na in three types of sites (two with fractional occupancies) for a total of 12 positional parameters. The unit cell contains eight so-called *supercages* in which benzene molecules can be accommodated. A satisfactory refinement was first obtained for the base zeolite, and this information was used to establish the positions of the absorbed benzene molecules from difference Fourier maps. This is illustrated in Figure 2.12 for coverages of about 1.1 and 2.6 molecules per supercage. For the former coverage the benzenes are distributed fairly evenly among the supercages and are bonded to some of the Na ions. For the high-coverage material additional benzene molecules are accommodated in the windows between adjacent supercages.

Another study of this type was recently reported by Wright and co-workers [87]; it involved the location of pyridine in the gallo-silicate $K_{0.57}Ga_{0.57}Si_{1.43}O_4$, space group $P6/mmm$, with $a \approx 18.6$ Å and $c \approx 7.5$ Å, containing about 1.5 pyridine molecules per unit cell. The position found for the pyridine molecule was in excellent agreement with that based on calculations of the interaction energy, which take into account short-range and electrostatic terms.

Neutron powder techniques have been very useful for investigating the structures of a wide range of materials with high ionic conductivities. One important group constitutes the so-called *Nasicon* materials, of which the prototype system is $Na_{1+x}Zr_2Si_xP_{3-x}O_4$ [88]. In this system the conductivity changes by four orders of magnitude to reach a maximum at $x = 2$ ($\approx 0.2 \, \Omega^{-1} \, cm^{-1}$ at 300°C). Phases near the end members are rhombohedral, space group $R\bar{3}c$, while the high-conductivity phases have a closely related monoclinic structure, space group $C2/c$, which transforms to the rhombohedral form at about 150°C. The conductivity properties are quite sensitive to the structural changes, and the mechanism is of considerable interest.

Didisheim and co-workers [89] made detailed structural studies of several compounds in this series at room temperature, and also above the transition for materials with $x = 1.6$ and 2.0. The transformation involves relatively small shifts in atomic positions and no detectable change in the Na site occupancies. An interesting feature is the fact that the temperature factors of the two types of Na ions at 320°C are extremely large (about 35 and 15 Å^2, respectively), and in particular the scattering density for the sodium atom around the origin is highly delocalized. The maximum does not occur at the origin in fact, but in a toroidal volume surrounding this location.

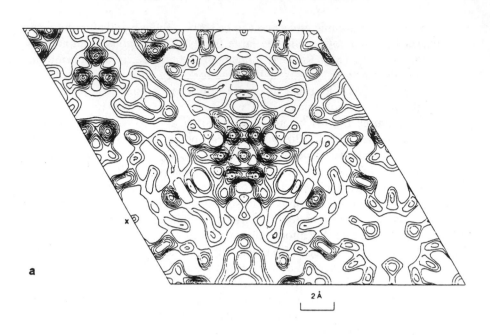

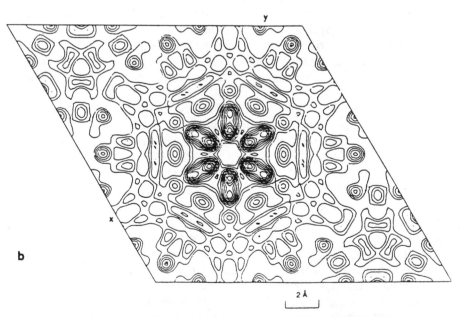

Figure 2.12 Fourier difference maps perpendicular to [111] for sodium-Y zeolite $Na_{56}Al_{56}$-$Si_{136}O_{384}$ containing chemisorbed benzene; x is parallel to $[\bar{1}, 1, 0]$, and y is parallel to $[\bar{1}, 0, 1]$. The sections are (*a*) about (0.295, 0.295, 0.295) showing the first benzene site for the sample with lower benzene coverage, and (*b*) about $(\frac{1}{2}, \frac{1}{2}, \frac{1}{2})$ showing the second benzene site for the sample with higher benzene coverage. Reprinted with permission from A. N. Fitch and co-workers, *J. Phys. Chem.*, **90**, 1311 (1986)]. Copyright 1986 American Chemical Society.

162

5.1.2   Structures Containing Deuterium

The application of neutron powder techniques to temperature-dependence studies of bond lengths and hydrogen bonding is very nicely demonstrated in a recent study by Hathaway and Hewat [90] of the deuterated Tutton's salt, $Cu(OD_2)_6(ND_4)_2(SO_4)_2$. Octahedral Cu^{2+} compounds are of particular structural interest because of Jahn–Teller distortions, and in this compound the usual type of elongated octahedron with four short (≈ 2.0 Å) and two long (≈ 2.3 Å) Cu—O bonds becomes progressively more distorted as the temperature is raised, approaching a compressed octahedron with four long and two short bonds. From the behavior of the lattice constants and the Cu—O bond lengths, a structural transition is predicted to occur at about 340 K, which probably corresponds to the onset of hindered rotation of the $(ND_4)^+$ ions. This study is an excellent example of the power of the Rietveld technique, for the structure is quite complex with a monoclinic cell of about 700 Å^3 volume, 62 structural parameters, and 10 profile parameters. At higher temperatures the fit is considerably improved by the introduction of anisotropic temperature factors— a total of 122 structural parameters. As illustrated in Figure 2.13, physically reasonable results were obtained for the corresponding thermal ellipsoids, with the largest amplitudes perpendicular to adjacent N—D and O—D bond directions, as required for librational motion.

5.1.3   Structure Solution by ab Initio Techniques

The examples in Sections 5.1.1 and 5.1.2 demonstrate that Rietveld analysis, particularly when coupled with Fourier techniques, is a very powerful tool for structure analysis when an approximate model is known. In contrast to this, there have been relatively few successful attempts to apply neutron powder data to the solution of unknown structures [91]. This is mainly because the resolution of most present-day diffractometers is not good enough to allow determination of an adequate set of unique integrated intensities. The limitation appears likely to be overcome as the new high-resolution instruments at the Rutherford–Appleton Laboratory and ILL become fully operational, which is illustrated by one of the first studies to be made at the former source, the determination of the structure of $FeAsO_4$ by Cheetham and co-workers [56] using direct methods. The compound $FeAsO_4$ is monoclinic with space group $P2_1/n$; integrated intensities were obtained for 139 reflections with d spacings between 1 and 3 Å. After the peaks in the output from the direct methods solution were identified, the structure shown in Figure 2.14 was obtained by refining the integrated intensities in the usual way. The most striking feature is the presence of five-coordinated Fe in approximately trigonal bipyramidal stereochemistry, with pairs of FeO_5 units sharing a common edge.

5.1.4   Studies at High Pressures

As noted in Section 3.5.1, the fixed-angle geometry of the TOF technique is especially well suited to structural studies at high pressures. This was illustrated

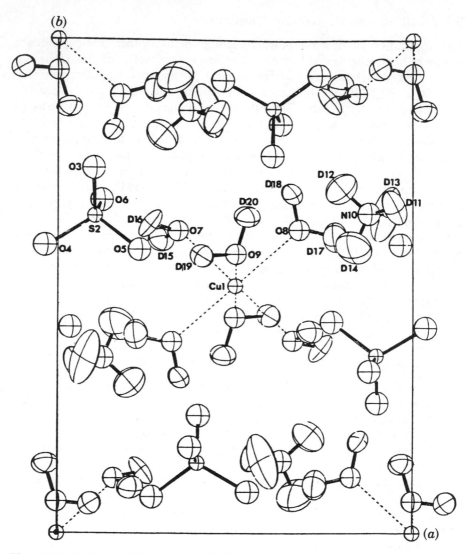

Figure 2.13 Projection of the structure of Cu(II) Tutton's salt, $Cu(OD_2)_6(ND_4)_2(SO_4)_2$ down the c axis. Notice the four ND_4, tetrahedra, two pointing down the c axis, labeled N10-D13, and two pointing up. Flipping of these tetrahedra is apparently responsible for the predicted phase transition near 340 K. Adapted, with permission, from Figure 4 in B. J. Hathaway and A. W. Hewat, *J. Solid State Chem.*, **51**, 373 (1984).

by a recent study of the pressure-induced phase transitions in ReO_3 by Jorgensen and co-workers [92]. At ambient pressure, ReO_3 has a cubic perovskite-like ABO_3 lattice, space group $Pm3m$, with $a_0 \approx 3.75$ Å, in which the A sites are vacant. At 5.2 kbar, a slightly distorted tetragonal structure is observed, with space group $P4/mbm$ and lattice parameters $a \approx a_0\sqrt{2}$, $c \approx a_0$. This arises from condensation of one set of so-called M_3 phonons whose

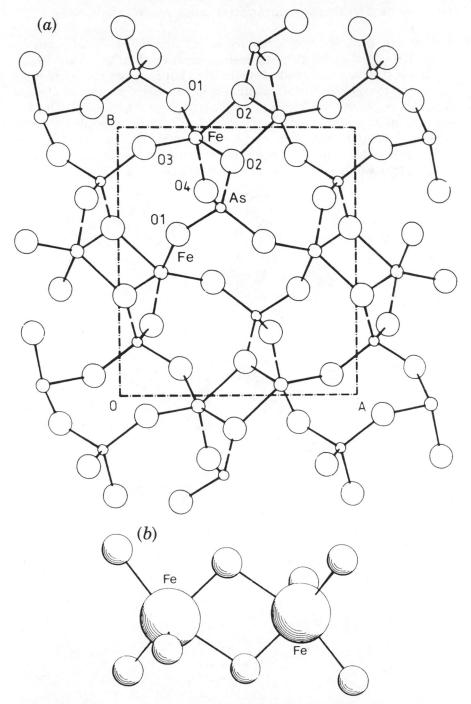

Figure 2.14 (*a*) Ab initio structure determination of $FeAsO_4$ from TOF neutron powder data collected at the Rutherford–Appleton Laboratory. Projection is shown down the *c* axis. Broken bonds indicate connections to atoms that are obscured in the projection. (*b*) The edge-sharing FeO_5 units in $FeAsO_4$; the principal axis of the trigonal bipyramid is $O(1)$—Fe—$O(2)$. Reprinted by permission from A. K. Cheetham and co-workers, *Nature (London)*, **320**, 46. Copyright © 1986 Macmillan Journals Limited.

displacements correspond to a coordinated rotation of the ReO_6 octahedra about $\langle 100 \rangle$ axes. At 7.3 kbar and higher pressures up to 27.4 kbar, a cubic structure is observed, with space group $Im3$ and lattice constant $a \approx 2a_0$, which corresponds to condensation of the three degenerate sets of M_3 phonons (Figure 2.15a). Although the transition between the tetragonal $P4/mbm$ and cubic $Im3$ structure must necessarily be first order, the pressure dependence of the octrahedral rotation angle ϕ appears to be a continuous function of pressure above 5 kbar with a critical exponent β of about one-third over the whole range (Figure 2.15b), which suggests the possibility of interesting tricritical behavior.

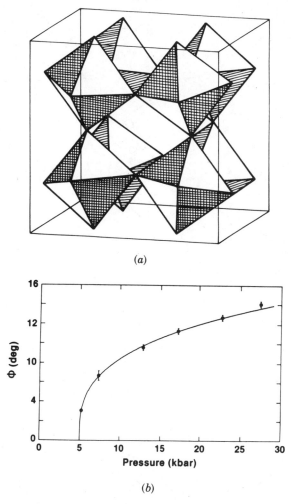

(a)

(b)

Figure 2.15 (a) The $Im3$ unit cell for ReO_3. The ReO_6 octahedra are rotated around the [111] direction, which corresponds to a condensation of all three M_3 phonons. (b) Pressure dependence of the angle of rotation of the ReO_6 octahedra. The solid line shows the curve $\phi \propto (P-P_c)^\beta$ with $\beta = 0.322(5)$. Adapted, with permission, from Figures 1 and 4 in J. E. Jorgensen and co-workers, *Phys. Rev. Sect. B*, **33**, 4795, 4797 (1986).

5.1.5 Time-Resolved Studies

Until fairly recently, neutron powder diffraction techniques focused heavily on structural studies of the type described in Sections 5.1.1–5.1.4, which were carried out on instruments optimized for high resolution at the expense of intensity. Typically, it takes several hours to collect a set of data, which rules out kinetic studies of chemical reactions or phase transformations. However, with an appropriately designed diffractometer incorporating a PSD such as D1B at ILL, many such studies are possible [39].

Much of this work was pioneered by Riekel and co-workers [93] in studies of chemical or electrochemical intercalation reactions, but for a simple illustration we describe a recent kinetic study by Christensen and co-workers [94] of a liquid–solid reaction, the hydration of $CaSO_4 \cdot 0.5H_2O$. This compound is formed by heating gypsum, $CaSO_4 \cdot 2H_2O$, and because it is the main ingredient of plaster of paris and related products it is a commercially important material. The hydration process was studied on the D1B with a sample of deuterated material that was mixed with D_2O. Data sets were collected at 3–5-min intervals at several temperatures between 293 and 335 K; the evolution of the diffraction patterns with time at 307 K is shown in Figure 2.16. At first there is a rapid drop in the intensities of peaks from the starting material; later these intensities decrease more slowly while peaks from gypsum begin to appear and increase in intensity with time. The observations are consistent with a process of hydration involving an intermediate gel state from which gypsum crystals then grow.

5.1.6 Measurement of Residual Stresses

A fairly new application of neutron powder diffraction that utilizes the high penetrating power of the neutron is the measurement of bulk residual stresses in

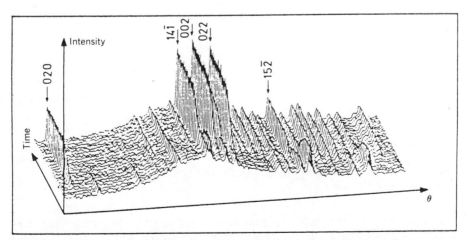

Figure 2.16 Time-resolved diffraction data showing the deuteration of $CaSO_4 \cdot 0.5D_2O$ at 307 K. The patterns show increasing intensities of $CaSO_4 \cdot 2D_2O$ and decreasing intensities of $CaSO_4 \cdot \frac{1}{2}D_2O$. Each pattern in the diagram represents a time period of 5 min. Adapted, with permission, from Figure 1 in A. N. Christensen and co-workers, *J. Appl. Crystallogr.*, **18**, 170 (1985).

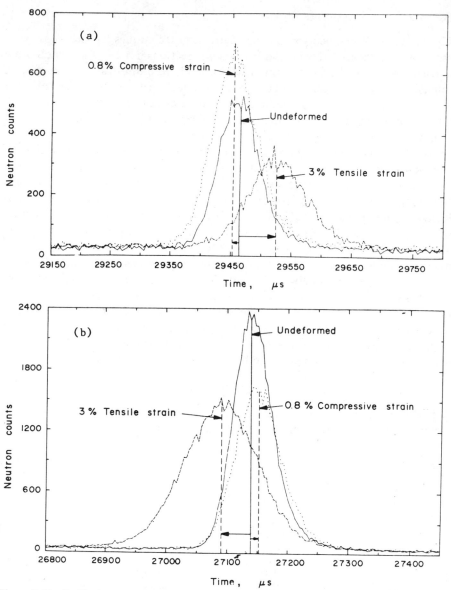

Figure 2.17 Residual stress measurements by TOF neutron techniques in deformed zircaloy-2 showing the effect of plastic deformation on the position and shape of (a) the ($10\bar{1}0$) line and (b) the (0002) line. Adapted, with permission, from Figures 8 and 9 in S. R. MacEwen and co-workers, *Acta Metall.*, **31**, 657 (1983).

alloys, which has considerable practical significance. This technique was recently put to good use by MacEwen and co-workers [95] to study residual stresses in deformed polycrystalline Zircaloy-2, which is a Zr-rich alloy used extensively for core components and fuel cladding in reactors. These measurements were made by using the TOF technique at the Argonne intense pulsed neutron source, and the effect of plastic deformation on the position and shape of the (10$\bar{1}$0) and (0002) peaks is seen in Figure 2.17 for a 3% tensile strain and an 0.8% compressive strain, respectively. The tensile test leaves the prism planes in residual tension, while the compressive test produces a net compressive strain normal to the (10$\bar{1}$0) planes. The effect is reversed for the basal planes. Line broadening is also observed, and these results can be used to analyze the residual stresses in terms of an idealized texture representative of that found in Zircaloy-2 rods.

5.2   Single-Crystal Diffraction Studies

As emphasized in Section 2.2.1, neutron-scattering amplitudes do not vary systematically with atomic number as do their X-ray counterparts. This leads to some very useful applications of neutron diffraction for the determination of light atom positions when heavy atoms are present. Another application of single-crystal techniques is in studies of atomic distribution functions, where high accuracy is required out to small values of d spacing. White-beam methods find particular application to studies of phase transitions in controlled environments where it is important to survey a large segment of reciprocal space as a function of temperature or pressure. Some examples illustrating these applications are now described.

5.2.1   Determination of Hydrogen and Other Light Atoms

The average contribution of an atom to the intensity of a reflection is proportional to the square of the scattering amplitude. Therefore, in a hypothetical organic compound containing 50% carbon and 50% hydrogen the relative contribution of hydrogen to a Bragg X-ray intensity is $1^2/(1^2 + 6^2) = \frac{1}{37}$; that is, 2.7%. The neutron scattering lengths of carbon and hydrogen are 0.665 and -0.374×10^{-12} cm, respectively, giving a ratio $0.374^2/(0.374^2 + 0.665^2)$; that is, 24%. If we now replace hydrogen with deuterium, which has a scattering length of 0.667×10^{-12} cm, the contribution from the light atom is 50%.

Although it should be possible to find hydrogen positions by X-ray diffraction analysis, the precision is clearly much superior when neutron data are employed. Many neutron diffraction studies were used to map out the relationships in various hydrogen bonds X—H$\cdots$Y [96]. The X and Y are valence-electron rich atoms such as nitrogen, oxygen, and fluorine; and the range of variation in the covalent bond X—H is as much as 0.2 Å, while the spread in the H$\cdots$Y contact is even larger. Typically the hydrogen atom can be located with a precision of around 0.003 Å; and the errors in the positions of the other atoms are even smaller, so if a sufficient number of studies are available, precise relationships can be established.

When hydrogen is bound to a metal atom it is even more difficult to locate it with X-ray diffraction methods; and as with hydrogen bonding, neutron diffraction is used routinely to solve problems involving such contacts [97]. As an example, we consider a study [98] of the cationic bis(hydrido)-bridged platinum–iridium complex $[(PEt_3)_2Pt(\mu\text{-}H)_2IrH_2(PEt_3)_2]^+[BPh_4]^-$ carried out at 22 K, which represents the first reported neutron diffraction analysis of a transition metal hydride complex with hydrogen bridging between two different metals. The asymmetry of the Ir(μ-H)$_2$Pt bridge is quite marked, with mean distances of Ir—H = 1.88 Å and Pt—H = 1.73 Å. The direction of this asymmetry is in accord with the position of platinum to the right of iridium in the periodic table and with the expected relative trans effect on the bridging hydrides caused by the terminal hydride and phosphine ligands (Figure 2.18).

As neutron diffraction techniques evolved, larger and larger molecules were investigated. The first protein crystal to be studied with neutrons was sperm whale myoglobin [99]. The goal of these studies was to a large extent to obtain information about hydrogen atoms, but because of large differences in the scattering lengths of nitrogen (0.921×10^{-12} cm), oxygen (0.581×10^{-12} cm), and carbon, the correct orientation of groups such as histidine and glutamine is more easily determined with neutrons than with X rays [71, 99].

The hydrogen atoms are studied for a multitude of reasons. As enzyme reactions often involve protonation processes, the location of the protons in the active site of an enzyme at a given step of the process helps to indicate the chemical pathway. Neutron data were used in this manner [100] to show that the histidine 57 group rather than the aspartic acid 102 group was protonated, by selecting one of several proposed enzyme mechanisms. Because of differences

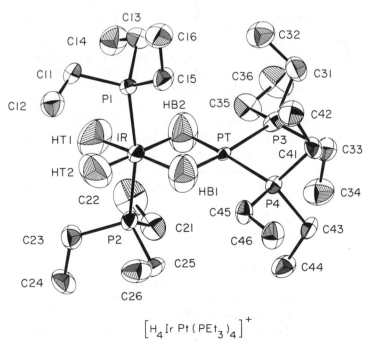

$$\left[H_4 \, \text{Ir Pt} \, (\text{PEt}_3)_4 \right]^+$$

Figure 2.18 View of the [(PEt$_3$)$_2$Pt(μ-H)$_2$IrH$_2$(PEt$_3$)$_2$]$^+$ cation. Reprinted, with permission, from A. Albinati and co-workers, *Inorg. Chem.*, **25**, 4821 (1986). Copyright 1986 American Chemical Society.

between the scattering length of hydrogen and deuterium, neutron data can be used to identify the hydrogen positions that are exchanged, when the protein or the protein crystal is soaked in D_2O. This can give indications of both the flexibility and the accessibility of various parts of the macromolecule, and a series of studies was done with these objectives [101, 102]. Finally, of course, the hydrogen-bond pattern can be studied, and a good example of such an analysis is the determination of pentagonal clusters of water along a hydrophobic part of the plant protein Crambin [103].

Hydrogen–deuterium exchange can be used equally well to mark parts of the molecule when specific interactions are studied. The power of this method is exemplified by a study of ethanol interactions with hen egg-white lysozyme [104]. Only the methyl hydrogen atoms of ethanol were deuterated. The average scattering density of this molecule, CD_3CD_2OH, is then 0.050×10^{-12} cm $\cdot$ Å^{-3}, while the protein molecule has an average scattering density 0.019×10^{-12} cm $\cdot$ Å^{-3}, and the water surrounding the protein has a scattering density of only -0.006×10^{-12} cm $\cdot$ Å^{-3}. An ethanol molecule will scatter more than one residue of the protein; and as it is found in the water region, it should be clearly observable. It is likely that the ethanol molecules will be somewhat disordered, so one cannot expect to observe these molecules at atomic resolution. Therefore, data were recorded only to a resolution of around 2.0 Å (i.e., the smallest d spacing of a Bragg reflection was 2 Å), which means that only a limited number of reflections were needed. The crystal used was soaked in a solution of 25% CD_2CD_2OH, and a total of 7060 reflection intensities were recorded. The crystal structure of lysozyme was previously determined with neutron data [105], so a scattering density can be calculated using the measured structure amplitudes and phases from the known structure. From this, after comparison with a scattering density of the normal structure, the possible ethanol sites can be identified, and constrained protein crystallographic refinements [106] can then be carried out. Figure 2.19 shows the shape of one of the ethanol molecules in the atomic site. The small molecule is only an elongated cloud of scattering density, but this is sufficient to get a good impression of the interaction with the protein molecule.

5.2.2   Measurements of Atomic Distribution Functions

To study atomic distribution functions in detail, Bragg reflections with very small d spacings must be measured, and as a consequence the number of data will be large. In addition, high precision and accuracy are required, so the measurements must be done with small counting errors, and the dynamical behavior of the crystals must be known to correct for inelastic scattering effects. Generally, therefore, only small systems are studied in this manner.

The distribution function can be represented either by the electron distribution or by the distribution of the nuclei. The former is observed with X rays, and it will be the convolution of the electron distribution with the distribution function of atoms resulting from thermal motion and disorder. Although a large

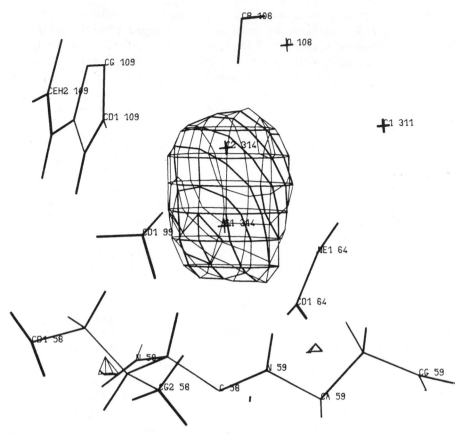

Figure 2.19   Scattering density showing ethanol interacting with part of the surface of a lysozyme molecule. The density of the lysozyme molecule was subtracted.

proportion of the electrons retain spherical symmetry when covalent bonds are formed, the nonspherical part is not negligible and makes deconvolution extremely difficult. The distribution of the nuclei is observed with neutrons, and it leads to distribution functions that are much simpler to understand. This difference between the two techniques has stimulated two main directions of research; namely, the interpretation of atomic thermal motion and disorder using neutron data and the effect of chemical bonding on the electron distribution using a combination of X-ray and neutron data.

The detailed determination of disorder is often used in the study of structural phase transitions [107], while the analysis of thermal motion can be used, for example, to determine the barrier to torsional motion in methyl groups [13]. Here we describe an analysis of common ice (Ice Ih), which combines the two aspects [108]. Ice Ih has a hexagonal structure with the hydrogen atoms disordered on two sites [109, 110], but the observed covalent oxygen–hydrogen bond length of 1.003 Å is far too long, both when compared to similar structures and when estimated from spectroscopic and thermodynamic information [111]. A diffractometer on a hot source with a relatively short neutron wavelength of 0.7 Å was used to collect data from H_2O ice at 60, 123, and 232 K, and from D_2O ice at 60 and 123 K. Typically, 250 symmetry-independent Bragg reflections were recorded. After the necessary corrections, structure refinements were made, and from the result of these the mean-square displacements of the atoms were derived. The relative mean-square displacement of hydrogen along the oxygen–hydrogen bond with respect to oxygen was then obtained by taking the difference between the atomic values, and from this the stretching frequency was obtained assuming harmonic oscillator behavior. Frequencies derived this way were in the range of 4200 to 4500 cm^{-1}—very much higher than those observed by spectroscopy [112], which are 3275 and 2421 cm^{-1} for H_2O and D_2O, respectively. The only plausible explanation for this difference is that the observed motion of oxygen is too large; that is, the motion derived from the crystallographic study contains two components, one from true thermal motion and one from multisite disorder. The disorder component of oxygen along the bond can be estimated to be 0.04 Å, and oxygen sites can be proposed that lead to a better oxygen–hydrogen bond length of 0.985 Å. Finally, it remains to show that the detailed atomic distribution function proposed agrees with such a model. Figure 2.20 shows a section through the water molecule. The density displayed is the difference between the density for harmonic motion and the

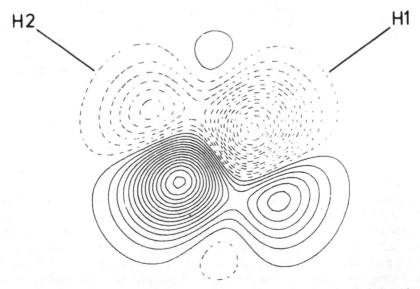

Figure 2.20 Section through the water molecule in ice Ih showing the deviations from a distribution corresponding to a harmonic oscillator. The O—H_1 bond is along the hexagonal axis of the structure. Adapted, with permission, from Figure 4 in W. F. Kuhs and M. S. Lehmann, "The Crystal Structure of Ice Ih," in F. Franks, Ed., *Water Science Reviews*, Cambridge University Press, New York, 1987.

density obtained assuming higher order terms in the atomic distribution description. This indicates that there is oxygen multisite disorder, and the sites are indeed in agreement with the simplest model proposed.

Although the electron distribution in crystals can be obtained directly from X-ray data, structural information obtained using neutron data is often important in extracting details of this distribution because a large portion of the electron density of the atoms retains spherical symmetry when the molecules are formed. Therefore, the total electron density, obtained as a Fourier transform of the phased diffraction data, is dominated completely by the spherical part of the atomic electron distributions, and it reveals none of the changes that occur as a consequence of the chemical bonding. These can normally be seen only from difference calculations, and one common way to obtain such differences is by subtracting from the electron density a superposition of neutral spherical atoms located at the positions of the atoms in the molecule. This requires accurate atomic locations and thermal parameters, and the very fact that the electron density is not exactly a superposition of spherical atoms can lead to difficulties if only X-ray data are used. One possibility is to derive atomic parameters from the part of the data that is least affected by the asphericity; namely, the data at high scattering angles; another is to include the atomic deformations in the model that is adjusted to fit the observations. In both cases the atomic positions and thermal parameters depend mainly on the high-order data, which have the lowest Bragg intensities and which are most likely to be in error. Therefore, at an early stage it was suggested that neutron data be used for this purpose [113]. This has since become common practice and leads to deformation electron densities based on both X-ray and neutron data (X–N maps). Of course this approach is not without problems, because the two sets of data must be recorded under identical conditions and be corrected for all instrument errors. The main advantage is that it nearly doubles the number of data, and thus it increases the number of checks for internal consistency.

An example of such an electron-density analysis is a study of the electron density in oxalic acid dihydrate. This compound was studied by several groups [114] to estimate the accuracy of such determinations. Four X-ray and five neutron data sets were involved. The measurement temperatures were around 100 K, the number of symmetry-independent X-ray data were around 3000, and there were about 2000 neutron reflections. Intercomparison showed that the atomic positions can be reproducibly determined with a precision of 0.001 Å or better, but there were large deviations among the thermal motion parameters. Nevertheless, it is possible to extract interesting chemical information, and Figure 2.21 shows an example of a deformation density map. In this case, the thermal parameters for hydrogen are taken from the neutron data, while for the other atoms the values are from the high-order X-ray data. The density shows the region of the short hydrogen bond between the acid group O(1)—H(1) and the oxygen O(3) of the water molecule. The section is orthogonal to the water molecule and contains the bisector O(3)—H' of the H—O—H angle. The map shows the lone pair of the hydrogen-bond acceptor oxygen, and there is clear evidence that this is polarized into the hydrogen-bond direction.

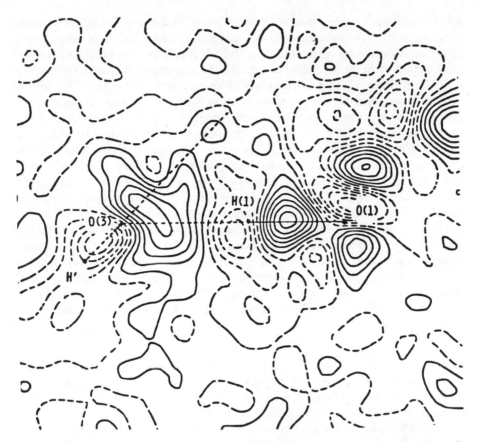

Figure 2.21 Deformation electron density in the plane orthogonal to the water molecule in oxalic acid dihydrate; O(1)—H(1) is part of the oxalic acid molecule, and H′—O(3) is the bisector of the H—O—H angle. Contours at $0.05e\ \text{Å}^{-3}$, with zero and negative contours dashed. Adapted, with permission, from Figure 4 in J. Dam and co-workers, *Acta Crystallogr. Sect. A*, **40**, 184 (1984).

5.2.3 Studies of Phase Transitions in Controlled Environments

In Section 2.2.1 it was mentioned that the neutron absorption cross section for most materials is very low, and this facilitates investigations in controlled environments. The crystal can be surrounded by millimeter-thick walls without a major effect on the quality of the data; therefore, studies at extreme temperatures are commonplace. In addition, very precise temperature control can be achieved, which allows studies to be made at well-defined points very near the critical temperature of structural phase transitions, for example. Under these conditions it is important to survey a larger part of reciprocal space, as new lattice points associated with some change in symmetry may occur at unexpected locations. White-beam techniques are well suited for this.

The most common variable is temperature, as cooling of crystals often leads to changes in their symmetry and physical behavior. One class of extensively suited materials are the so-called *organic superconductors*. These are organic salts, which at very low temperature—sometimes under pressure—undergo a phase transition to a superconducting state. It is clearly of interest to correlate this behavior with the crystal structure, and the first step in such an analysis is to study the reciprocal lattice as a function of temperature. This can be done with X rays, but for reasons mentioned in Section 4.1.2, it can also be done very efficiently with neutron white-beam techniques. The I_3^- salt of bis(ethylenedithio)tetrathiafulvalene, called $(ET)_2I_3$, becomes superconducting at 1.5 K. The room temperature structure of $(ET)_2I_3$ consists of a loosely packed àrrangement of ET molecules with I_3^- molecules incorporated in cavities with a relatively short intermolecular I—I distance. Below 200 K satellite reflections were found by X-ray film measurements [77], and these were followed in greater detail with neutrons down to 20 K. Figure 2.10 shows a section through reciprocal space. When transitions occur there is often a change in the size of the unit cell, and the simplest to understand are the commensurate transitions; that is, transitions where the new unit cell is some integer multiple of the original cell. In this instance the transition seems to be incommensurate. New lattice points occur at positions with h, k, $l \pm \tau$, where $\tau = (0.08, 0.27, 0.205)$, indicating a modulation with wavelength of 23.1 Å propagating through the crystal in an apparently general direction.

5.3   Magnetic Structures

As emphasized in Section 2.3, neutron diffraction is a unique technique for magnetic structure determination, and prior to the Rietveld era, powder methods were exploited mainly for this purpose. The need to use integrated intensities was not as severe a limitation as for chemical structure determination because there are usually relatively few magnetic atoms in the unit cell and the essential features of the magnetic structure can often be obtained from a few magnetic intensities. Hundreds of magnetic structures were solved in this way [115, 116] including many incommensurate ones where the periodicity of the magnetic wave vector is not a simple multiple of one of the unit cell dimensions.

Nevertheless, even in simple cases the Rietveld technique should yield the maximum information about magnetic moments and form factors, and different models can be explored systematically; for example, whether chemically similar but inequivalent atoms have different moments, or a noncollinear model gives a better fit.

The application of the Rietveld technique to a magnetic problem is illustrated by a recent study of $FeClMoO_4$ [117]. This has a simple layered tetragonal structure and undergoes an antiferromagnetic ordering transition at about 70 K, as can be seen by the appearance of additional peaks of magnetic origin (Figure 2.22a). These can be indexed in terms of an enlarged cell where both the a and c lattice parameters are doubled, which means that neighboring Fe moments separated by one translation along a or c are coupled antiparallel. However, as is frequently the case with powder experiments, two models fit the data equally well: one collinear, with orthorhombic magnetic symmetry, and one non-

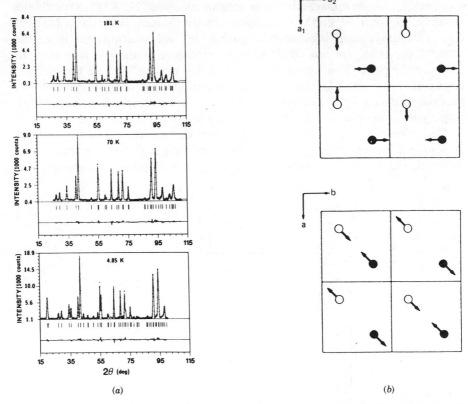

(a) (b)

Figure 2.22 (a) Neutron powder profile fits and difference plots for $FeClMoO_4$ at 181, 70, and 4.8 K. Note the extra magnetic peaks in the lower pattern. (b) Projection of the two possible doubled magnetic cells for $FeClMoO_4$ down the c axis: (top) tetragonal and (bottom) orthorhombic. Full and open circles represent Fe atoms at heights of $0.27\,c$ and $-0.27\,c$, respectively, in the chemical cell. Adapted, with permission, from Figures 8 and 11 in C. C. Torardi and co-workers, *J. Solid State Chem.*, **66**, 112, 113 (1987).

collinear, with tetragonal symmetry (Figure 2.22b). The profile fits and difference plots above and below the transition are shown in Figure 2.22a.

This kind of degeneracy arises because it is impossible to distinguish between magnetically inequivalent but chemically equivalent reflections in a powder pattern, and it can only be resolved with a single crystal (which consists of at least predominantly one magnetic domain). It is important to remember that the proper powder multiplicities depend on the magnetic symmetry of the particular model used, and they should be checked for a few general reflections before Rietveld refinement is attempted. A more detailed discussion of magnetic symmetry is in [3].

The possible relationship between magnetism and superconductivity in the new high-T_c Cu-based oxides has stimulated many investigations of magnetic ordering in these materials. For slightly oxygen-deficient samples of

La_2CuO_{4-y}, antiferromagnetic order sets in at about 220 K [118] with the structure shown in Figure 2.23a. Because the Cu moment is very small (about 0.5 μ_B), the Bragg scattering is weak (Figure 2.23b), and counting times of several minutes per point are needed. It is very important in these experiments to optimize the peak-to-background ratio, and it is desirable to use an analyzer crystal after the sample for this purpose. Furthermore, to unequivocally confirm that the scattering is truly magnetic and not nuclear in origin, it is necessary to employ neutron polarization analysis techniques [119] (Figure 2.23c).

The value of polarized beam techniques was demonstrated in a very elegant way in a much earlier single-crystal study of Fe [120]. In a ferromagnetic material, the magnetic scattering is always superimposed on the nuclear

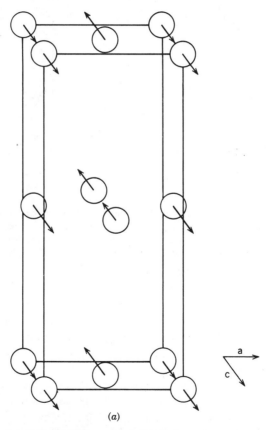

(a)

Figure 2.23 (a) Spin structure of antiferromagnetic La_2CuO_{4-y}. Only Cu sites in the orthorhombic unit cell are shown for clarity. (b) Intensity versus scattering angle 2θ for neutron powder scans for the (100) peak region at 15 K and room temperature. (c) Magnetic scattering from the (100) superlattice peak measured in SF (HF–VF) mode at 100 and 300 K. The intensities shown are normalized to a given monitor counting time. Adapted, with permission, from Figures 3 and 2(a) in D. Vaknin and co-workers, *Phys. Rev. Lett.*, **58**, 2803, 2804 (1987), and Figure 2(b) in S. Mitsuda and co-workers, *Phys. Rev. Sect. B*, **36**, 822 (1987).

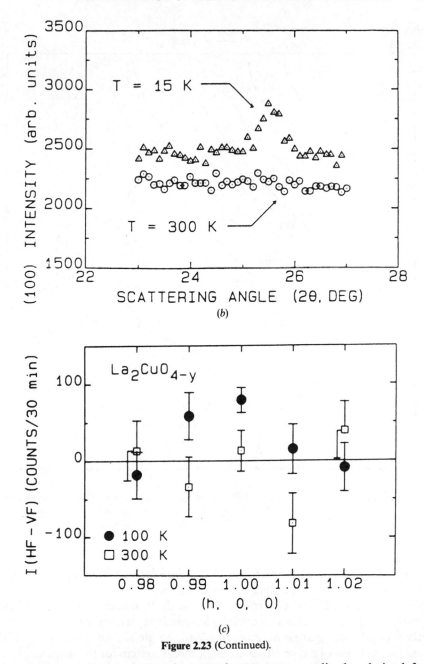

Figure 2.23 (Continued).

scattering, so that the observed magnetic structure amplitudes derived from unpolarized neutron measurements by subtraction of F_N^2 from the total scattering are subject to large errors at higher values of Q. In a polarized beam experiment of this type, one measures the ratio of $(F_M + F_N)^2 : (F_M - F_N)^2$, and hence, F_M can be determined with good accuracy. The magnetic moment density

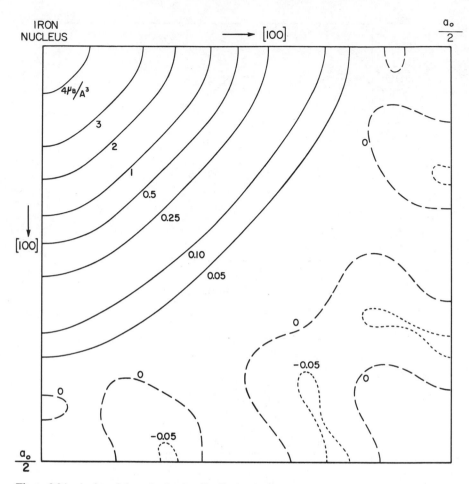

Figure 2.24 A plot of the spin-density distribution in the (100) plane of iron, indicating an excess along the [100] axis relative to that along [110]. Contours are in units of μ_β Å^{-3}. Adapted, with permission, from Figure 2 in C. G. Shull and Y. Yamada, *J. Phys. Soc. Jpn.*, **17**, Suppl. B-III, 1 (1962).

obtained by Fourier transformation of the F_M data for Fe is illustrated in Figure 2.24, and it shows the detail that can be extracted.

More recently, considerable efforts were made to understand the magnetic structures of synthetic layered materials. Magnetic layers comprising a discrete number of atomic planes of moments can be deposited alternately with nonmagnetic layers of a given thickness as a model system for investigating both the interfacial magnetism and the effects of reduced dimensionality. Analysis of polarized neutron diffraction data from ferromagnetic superlattices can in principle give a microscopic profile of the magnetization across the thickness of the film [121].

It is also possible to study interlayer magnetic coupling in synthetic superlattices. With metallic, magnetic rare earths (RE), the magnetic moments are well localized and the indirect exchange interaction is by the conduction electrons. The long-range nature of this Ruderman–Kittel–Kasuya–Yosida (RKKY) interaction can be expected to give rise to a modulation of the magnetic properties in an artificially layered magnetic RE–nonmagnetic RE structure. In a polarized neutron diffraction study [122] of $[Gd_{N_{Gd}}-Y_{N_Y}]_M$ superlattices composed of M successive bilayers of N_{Gd} basal planes of hexagonal closed-packed (hcp) Gd followed by N_Y such planes of Y it was found that below the

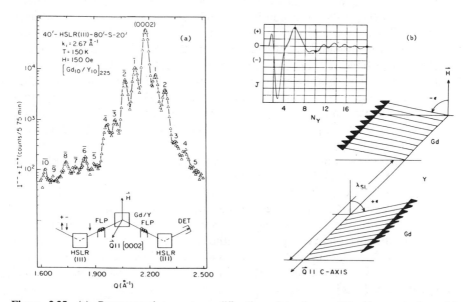

Figure 2.25 (a) Representative neutron diffraction data for a Gd—Y superlattice with $N_{Gd} = N_Y = 10$ (the spectrometer configuration including beam collimations is given in the upper left). Satellites occur at integer multiples of $\frac{1}{2}$ the superlattice wave vector with odd- and even-numbered satellites corresponding to SF and NSF scattering, respectively, as described in the text. The lower part of the figure is a diagram of the spectrometer configuration used to separate the four possible spin-dependent scattered intensities for polarized neutrons. The Heusler (depicted as HSLR) crystals reflect a narrow wavelength band of predominantly one neutron spin state when magnetized perpendicular to the scattering vector. The flat coil flippers (FLP) rotate the $+(-)$ to the $-(+)$ neutron spin eigenstate thereby allowing one or the other spin state to be selected. Adapted, with permission, from Figure 1 in C. F. Majkrzak and co-workers, *Phys. Rev. Lett.*, **56**, 2701 (1986). (b) Schematic representation of the antiphase domain configuration of the Gd magnetic moments in the $[Gd_{10}-Y_{10}]_{225}$ superlattice; ε is approximately $80°$ at 75 K and 150 Oe. The inset in the upper left-hand corner shows the results (solid curve) of the theoretical calculation of the relative strength and sign of the RKKY interaction as a function of N_Y that is described in [122]. The points indicate those values of N_Y for which the magnetic structures were determined experimentally. The interaction is positive or ferromagnetic for $N_Y = 6$ and negative or antiferromagnetic for $N_Y = 10$, in agreement with experiment. Adapted, with permission, from Figure 3 in C. F. Majkrzak and co-workers, *Phys. Rev. Lett.*, **56**, 2702 (1986).

Curie temperature and in low fields the ferromagnetic Gd layers tend to align antiferromagnetically relative to one another for $N_{Gd} = N_Y = 10$ in a microscopic antiphase domain structure that is coherent over many superlattice periods. For $N_Y = 6$ or 20, however, simple long-range ferromagnetic order is observed. These results are consistent with a simple theoretical calculation that indicates the RKKY coupling between Gd layers through the nonmagnetic Y layers can be either ferromagnetic or antiferromagnetic in sign, depending on the Y layer thickness, and of sufficient strength to produce long-range order.

The upper part of Figure 2.25a shows typical polarized neutron diffraction data for a Gd—Y superlattice at 150 K in a field of approximately 150 Oe perpendicular to $\mathbf{Q}$ with $\mathbf{Q}$ parallel to the [002] direction in reciprocal space. The positions Q_m of the even-numbered satellites measured from the primary (002) reflection correspond to integer multiples of $2\pi/\lambda_{SL}$, where $\lambda_{SL} = 58.4\text{Å}$ is the chemical modulation wavelength or bilayer thickness [i.e., $Q_m = (m/2)2\pi/\lambda_{SL}$]. Above T_c (ca. 290 K) the intensities of the even-numbered satellites arise solely from the nuclear scattering associated with the superlattice chemical modulation so that $I_m^{++} = I_m^{--}$, whereas below T_c it is found that $I_m^{++} \neq I_m^{--}$, which indicates the presence of an additional ferromagnetic component. To within experimental accuracy, no SF scattering is found to occur at the even-numbered satellite positions. The odd-numbered satellites, on the other hand, begin to appear below T_c in low fields at integer multiples of the wave vector for a *doubled* bilayer thickness [e.g., for $m = 1$, $Q_1 = \frac{1}{2}2\pi/\lambda_{SL}$; for $m = 3$, $Q_3 = \frac{3}{2}2\pi/\lambda_{SL}$; etc.] with intensities corresponding to SF scattering only.

The diffraction pattern calculated for the antiphase domain model depicted in Figure 2.25b with the Gd moments confined to the basal planes is in good agreement with the observed pattern, although better agreement is obtained with a more refined version of this basic model, which includes a more accurate description of the chemical composition and atomic plane spacing modulations along the c axis [122].

6   CONCLUDING REMARKS

Over the past few years there has been a dramatic growth in the application of neutron diffraction techniques to structural studies, both numerically and in diversity. With the availability of new high-resolution instruments and sophisticated position-sensitive detectors, continued progress and development is certain to occur over the next few years. At the same time, the outside community is being increasingly encouraged to view neutron centers as national resources and to make the widest possible use of these facilities. This can involve anything from a simple experiment lasting a day or two to active collaboration in developing highly specialized instrumentation; for example, for experiments in controlled environments.

We hope that readers have gained some insight from this chapter into the power and versatility of present-day neutron techniques and how these may be able to play a role in their own research in the future.

Acknowledgment

The work at Brookhaven National Laboratory was supported by the Division of Materials Sciences, United States Department of Energy, under contract DE-AC02-76CH00016.

References

1. E. O. Wollan and C. G. Shull, *Phys. Rev.*, **73**, 830 (1948); C. G. Shull and E. O. Wollan, *Phys. Rev.*, **81**, 527 (1951).
2. R. Mason, E. W. J. Mitchell, and J. W. White, Eds., *Neutron Scattering in Biology, Chemistry and Physics*, The Royal Society, London, 1980.
3. G. E. Bacon, *Neutron Diffraction*, 3rd ed., Oxford University Press, London, 1975.
4. S. W. Lovesey, *Theory of Neutron Scattering from Condensed Matter*, Vols. 1, 2, Oxford University Press, London, 1984.
5. B. T. M. Willis, Ed., *Chemical Applications of Thermal Neutron Scattering*, Oxford University Press, London, 1973.
6. G. Kostorz, Ed., *Neutron Scattering*, Academic, New York, 1979.
7. G. L. Squires, *Thermal Neutron Scattering*, Cambridge University Press, London, 1978.
8. H. Dachs, Ed., *Neutron Diffraction*, Springer-Verlag, Berlin, 1978.
9. B. T. M. Willis, Ed., *Thermal Neutron Diffraction*, Oxford University Press, London, 1970.
10. I. I. Gurevich and L. V. Tarasov, *Low-Energy Neutron Physics*, Elsevier, New York, 1968.
11. P. A. Egelstaff, Ed., *Thermal Neutron Scattering*, Academic, New York, 1965.
12. V. F. Turchin, *Slow Neutrons*, Sivan, Jerusalem, 1965.
13. J. M. Hastings and W. C. Hamilton, "Neutron Scattering," in A. Weissberger and B. W. Rossiter, Eds., *Physical Methods of Chemistry*, Wiley, New York, 1972.
14. L. Koester and A. Steyerl, *Neutron Physics*, Springer Tracts in Modern Physics, Vol. 80, Springer-Verlag, Berlin, 1977.
15. N. F. Mott and H. S. W. Massey, *The Theory of Atomic Collisions*, 2nd ed., Clarendon, Oxford, 1949.
16. V. F. Sears, *Thermal Neutron Scattering Lengths and Cross Sections for Condensed Matter Research*, Atomic Energy of Canada Report AECL-8490, Chalk River, Ontario, Canada, 1984.
17. J. B. Hayter, G. T. Jenkin, and J. M. White, *Phys. Rev. Lett.*, **33**, 696 (1974).
18. R. W. James, *The Optical Principles of the Diffraction of X-rays*, Cornell University Press, Ithaca, NY, 1965.
19. W. H. Zachariasen, *Theory of X-ray Diffraction in Crystals*, Wiley, New York, 1945.
20. V. F. Sears, *Can. J. Phys.*, **56**, 1261 (1978).
21. C. Kittel, *Introduction to Solid State Physics*, 5th ed., Wiley, New York, 1976.
22. R. Weinstock, *Phys. Rev.*, **65**, 1 (1944).

23. *International Tables for X-ray Crystallography*, Vol. 2, The Kynoch Press, Birmingham, England, 1959.

24. K. D. Rouse and M. J. Cooper, *Acta Crystallogr. Sect. A*, **26**, 682 (1970).

25. R. M. Moon and W. C. Koehler, *Phys. Rev. Sect. B.*, **11**, 1609 (1975).

26. O. Halpern and M. H. Johnson, *Phys. Rev.*, **55**, 898 (1939).

27. R. M. Moon, T. Riste, and W. C. Koehler, *Phys. Rev.*, **181**, 920 (1969).

28. C. G. Windsor, *Pulsed Neutron Scattering*, Halsted, New York, 1981.

29. A. Freund and J. B. Forsyth, "Materials Problems in Neutron Devices," in G. Kostorz, Ed., *Neutron Scattering*, Academic, New York, 1979, p. 462.

30. C. F. Majkrzak, *Appl. Opt.*, **23**, 3524 (1984).

31. P. Convert and J. B. Forsyth, Eds., *Position-Sensitive Detection of Thermal Neutrons*, Academic, New York, 1983.

32. J. Alberi, J. Fischer, V. Radeka, L. C. Rogers, and B. Schoenborn, *Nucl. Instrum. Methods*, **127**, 507 (1975).

33. R. Allemand, J. Bourdel, E. Roudaut, P. Convert, K. Ibel, J. Jacobe, J. P. Cotton, and B. Farnoux, *Nucl. Instrum. Methods*, **126**, 29 (1975).

34. P. L. Davidson and H. Wroe, *Proceedings of the Fourth Meeting of the International Collaboration on Advanced Neutron Sources*, *KENS Report II*, Argonne National Laboratory, Argonne, IL, 1981, p. 642.

35. H. O. Anger, *Rev. Sci. Instrum.*, **29**, 27 (1958).

36. M. G. Strauss, R. Brenner, F. J. Lynch, and C. B. Morgan, *IEEE Trans. Nucl. Sci.*, NS-28 (1981).

37. D. Hohlwein, "Photographic Methods in Neutron Scattering," in P. Convert and J. B. Forsyth, Eds., *Position-Sensitive Detection of Thermal Neutrons*, Academic, New York, 1983, p. 379.

38. M. J. Cooper and R. Nathans, *Acta Crystallogr.*, **23**, 357 (1967); *Acta Crystallogr. Sect. A*, **24**, 481 (1968).

39. J. Pannetier, *Chem. Scr.*, **26A**, 131 (1986).

40. A. W. Hewat and I. Bailey, *Nucl. Instrum. Methods*, **137**, 463 (1976).

41. W. Schäfer, E. Jansen, F. Elf, and G. Will, *J. Appl. Crystallogr.*, **17**, 159 (1984).

42. C. W. Tompson, D. F. R. Mildner, M. Mehregany, J. Sudol, R. Berliner, and W. B. Yelon, *J. Appl. Crystallogr.*, **17**, 385 (1984).

43. H. M. Rietveld, *J. Appl. Crystallogr.*, **2**, 65 (1969).

44. A. W. Hewat, *Nucl. Instrum. Methods*, **12**, 361 (1975).

45. G. Caglioti, A. Paoletti, and F. P. Ricci, *Nucl. Instrum.*, **3**, 223 (1958).

46. G. H. Lander and V. J. Emery, *Nucl. Instrum., Methods*, **B12**, 525 (1985).

47. A. W. Hewat, *Mater. Sci. Forum*, **9**, 69 (1986).

48. J. Pannetier, private communication.

49. J. M. Carpenter, G. H. Lander, and C. G. Windsor, *Rev. Sci. Instrum.*, **55**, 1019 (1984).

50. J. D. Jorgensen and F. J. Rotella, *J. Appl. Crystallogr.*, **15**, 27 (1982).

51. J. D. Jorgensen and J. Faber, *Proceedings of the Sixth Meeting of the International Collaboration on Advanced Neutron Sources*, Argonne National Laboratory, Argonne, IL, 1982.

52. J. A. Goldstone, R. B. Von Dreele, and A. T. Ortiz, "Design Proposal for a Variable Resolution Neutron Powder Diffractometer at LANSCE," Technical Report LA-UR-87-2384, Los Alamos National Laboratory, Los Alamos, NM, 1987.

53. N. Watanabe, H. Asano, H. Iwasa, S. Satoh, H. Murata, K. Karahashi, S. Tomiyoshi, F. Izumi, and K. Inoue, *Jpn. J. Appl. Phys.*, **26**, 1164 (1987).

54. W. I. David, *Mater. Sci. Forum*, **9**, 89 (1986).

55. A. J. Schultz, *Trans. Am. Crystallogr. Assoc.*, **23**, 61 (1987).

56. A. K. Cheetham, W. I. F. David, M. M. Eddy, R. J. B. Jakeman, M. W. Johnson, and C. C. Torardi, *Nature (London)*, **320**, 46 (1986).

57. A. N. Christensen, M. S. Lehmann, and M. Nielsen, *Aust. J. Phys.*, **38**, 497 (1985).

58. R. J. Hill and I. C. Madsen, *Powder Diffr.*, **2**, 146 (1987).

59. R. B. Von Dreele, J. D. Jorgensen, and C. G. Windsor, *J. Appl. Crystallogr.*, **15**, 581 (1982).

60. H. G. Scott, *J. Appl. Crystallogr.*, **16**, 159 (1983) and references cited therein.

61. W. C. Hamilton, *Statistics in Physical Science*, Ronald, New York, 1964.

62. W. C. Hamilton, *Acta Crystallogr.*, **18**, 502 (1965).

63. M. J. Cooper, K. D. Rouse, and M. Sakata, *Z. Kristallogr.*, **157**, 101 (1981).

64. G. S. Pawley, *J. Appl. Crystallogr.*, **14**, 357 (1981).

65. T. J. Emge, P. C. W. Leung, M. A. Beno, A. J. Schultz, H. H. Wang, L. M. Sowa, and J. M. Williams, *Phys. Rev. Sect. B*, **30**, 6780 (1984).

66. M. S. Lehmann and F. K. Larsen, *Acta Crystallogr. Sect. A.*, **30**, 580 (1974).

67. S. A. Spencer and A. A. Kossiakoff, *J. Appl. Crystallogr.*, **13**, 563 (1980).

68. A. Wlodawer and L. Sjölin, *Nucl. Instrum. Methods*, **201**, 117 (1982).

69. M. Roth and A. Lewit-Bentley, *Acta Crystallogr. Sect. A.*, **38**, 670 (1982).

70. C. Wilkinson and H. W. Khamis, "Location and Integration of Single Reflections Using Area PSDs," in P. Convert and J. B. Forsyth, Eds., *Position-Sensitive Detection of Thermal Neutrons*, Academic, New York, 1983, p. 358.

71. B. P. Schoenborn, "Data Processing in Neutron Protein Crystallography Using Position-Sensitive Detectors," in P. Convert and J. B. Forsyth, Eds., *Position-Sensitive Detection of Thermal Neutrons*, Academic, New York, 1983, p. 321.

72. A. Filhol, M. Thomas, G. Greenwood, and A. Barthelemy, "On-line Data Reduction Software for a 4-Circle Neutron Diffractometer Equipped With a Fly's-Eye Area PSD," in P. Convert and J. B. Forsyth, Eds., *Position-Sensitive Detection of Thermal Neutrons*, Academic, New York, 1983, p. 351.

73. P. J. Becker and P. Coppens, *Acta Crystallogr. Sect. A*, **30**, 129 (1974).

74. J. R. Schneider, *Nucl. Sci. Appl.*, 1, 227 (1981).

75. B. T. M. Willis, *Acta Crystallogr. Sect. A.*, **26**, 396 (1970).

76. U. W. Arndt and B. T. M. Willis, *Single Crystal Diffractometry*, Cambridge University Press, Cambridge, UK, 1966.

77. A. J. Schultz, K. Srinivasan, R. G. Teller, J. M. Williams, and C. M. Lukehart, *J. Am. Chem. Soc.*, **106**, 999 (1984).

78. J. P. Glusker and K. N. Trueblood, *Crystal Structure Analysis. A Primer*, Oxford University Press, Oxford, 1985; M. F. C. Ladd and R. A. Palmer, *Structure Determination of X-Ray Crystallography*, Plenum, New York, 1985.

79. J. D. Jorgensen, *Jpn. J. Appl. Phys. Suppl.*, **26**, 2018 (1987), and references therein.

80. D. E. Cox, A. R. Moodenbaugh, J. J. Hurst, and R. H. Jones, *J. Phys. Chem. Solids*, **49**, 47 (1988).

81. J. D. Jorgensen, M. A. Beno, D. G. Hinks, L. Soderholm, K. J. Volin, R. L. Hitterman, J. D. Grace, I. K. Schuller, C. U. Segre, K. Zhang, and M. S. Kleefisch, *Phys. Rev. Sect. B*, **36**, 3608 (1987).

82. J. J. Capponi, C. Chaillout, A. W. Hewat, P. Lejay, M. Marezio, N. Nguyen, B. Raveau, J. L. Soubeyroux, J. L. Tholence, and R. Tournier, *Europhys. Lett.*, **3**, 1301 (1987).

83. W. I. F. David, P. P. Edwards, M. R. Harrison, R. Jones, and C. C. Wilson, *Nature (London)*, **331**, 245 (1988).

84. J. Newsam, *Science*, **231**, 1093 (1986).

85. S. T. Wilson, *J. Am. Chem. Soc.*, **104**, 1146 (1982).

86. A. N. Fitch, H. Jobic, and A. Renouprez, *J. Phys. Chem.*, **90**, 1311 (1986).

87. P. A. Wright, J. M. Thomas, A. K. Cheetham, and A. K. Nowak, *Nature (London)*, **318**, 6047 (1985).

88. J. B. Goodenough, H. Y.-P. Hong, and J. A. Kafalas, *Mater. Res. Bull.*, **11**, 173 (1976).

89. J. J. Didisheim, E. Prince, and B. J. Wuensch, *Solid State Ionics*, **18, 19**, 944 (1986).

90. B. J. Hathaway and A. W. Hewat, *J. Solid State Chem.*, **51**, 364 (1984).

91. A. K. Cheetham, *Mater. Res. Forum*, **9**, 103 (1986).

92. J. E. Jorgensen, J. D. Jorgensen, B. Batlogg, J. P. Remeika, and J. D. Axe, *Phys. Rev. Sect. B*, **33**, 4793 (1986).

93. C. Riekel, in G. M. Rosenblatt and W. L. Worrell, Eds., *Progress in Solid-State Chemistry*, Vol. 13, Pergamon, New York, 1980, Chap. IV, p. 89.

94. A. N. Christensen, M. S. Lehmann, and J. Pannetier, *J. Appl. Crystallogr.*, **18**, 170 (1985).

95. S. R. MacEwen, J. Faber, and A. P. L. Turner, *Acta Metall.*, **31**, 657 (1983).

96. P. Schuster, G. Zundel, and C. Sandorfy, *The Hydrogen Bond*, North-Holland, Amsterdam, 1976.

97. R. Bau and R. G. Teller, *Struct. Bonding (Berlin)*, **64**, 1 (1981).

98. A. Albinati, T. J. Emge, T. F. Koetzle, S. V. Meille, A. Musco, and L. M. Venanzi, *Inorg. Chem.*, **25**, 4821 (1986).

99. B. P. Schoenborn, *Nature (London)*, **224**, 143 (1969).

100. A. A. Kossiakoff and S. A. Spencer, *Nature (London)*, **288**, 414 (1980).

101. A. A. Kossiakoff, *Nature (London)*, **296**, 713 (1982).

102. A. Wlodawer and L. Sjölin, *Proc. Natl. Acad. Sci. USA*, **79**, 1418 (1982).

103. M. M. Teeter, *Proc. Natl. Acad. Sci. USA*, **81**, 6014 (1984).

104. M. S. Lehmann, S. A. Mason, and G. J. McIntyre, *Biochemistry*, **24**, 5862 (1985).

105. S. A. Mason, G. A. Bentley, and G. J. McIntyre, "Kinetic Studies of Chemical Reactions: Inferences from Structural and Related Observations of Intercalation Compounds," in B. P. Schoenborn, Ed., *Neutrons in Biology*, Academic, New York, 1984, p. 323.

106. A. Wlodawer and W. A. Hendrickson, *Acta Crystallogr. Sect. A*, **38**, 239 (1982).

107. R. J. Nelmes, W. F. Kuhs, C. J. Howard, J. E. Tibballs, and T. W. Ryan, *J. Phys. C*, **18**, L711, L1023 (1985).

108. W. F. Kuhs and M. S. Lehmann, "The Crystal Structure of Ice Ih," in F. Franks, Ed., *Water Science Reviews*, Cambridge University Press, New York 1987.

109. L. Pauling, *J. Am. Chem. Soc.*, **57**, 2680 (1935).

110. S. W. Peterson and H. A. Levy, *Acta. Crystallogr.*, **10**, 70 (1957).

111. E. Whalley, *Mol. Phys.*, **28**, 1105 (1974).

112. C. Andreani, B. C. Boland, F. Sacchetti, and C. G. Windsor, *J. Phys. C*, **16**, L513 (1983).

113. P. Coppens, *Science*, **158**, 1577 (1967).

114. J. Dam, S. Harkema, D. Feil, R. Feld, M. S. Lehmann, R. Goddard, C. Krüger, E. Hellner, H. Johansen, F. K. Larsen, T. F. Koetzle, R. K. McMullan, E. N. Maslen, E. D. Stevens, and P. Coppens, *Acta Crystallogr. Sect. A*, **40**, 184 (1984).

115. D. E. Cox, *IEEE Trans. Magn.*, **MAG-8**, No. 1, 161 (1972).

116. Yu. A. Izyumov and R. P. Ozerov, *Magnetic Neutron Diffraction*, Plenum, New York, 1970.

117. C. C. Torardi, W. M. Reiff, K. Lazar, J. H. Zhang, and D. E. Cox, *J. Solid State Chem.*, **66**, 105 (1987).

118. D. Vaknin, S. K. Sinha, D. E. Moncton, D. C. Johnston, J. M. Newsam, C. R. Safinya, and H. E. King, *Phys. Rev. Lett.*, **58**, 2802 (1987).

119. S. Mitsuda, G. Shirane, S. K. Sinha, D. C. Johnston, M. S. Alvarez, D. Vaknin, and D. E. Moncton, *Phys. Rev. Sect. B*, **36**, 822 (1987).

120. C. G. Shull and Y. Yamada, *J. Phys. Soc. Jpn.*, **17**, Suppl. B-III, 1 (1962).

121. C. F. Majkrzak, *Physica*, **136B**, 69 (1986).

122. C. F. Majkrzak, J. W. Cable, J. Kwo, M. Hong, D. B. McWhan, Y. Yafet, J. V. Waszczak, and C. Vettier, *Phys. Rev. Lett.*, **56**, 2700 (1986).

Chapter **3**

X-RAY ABSORPTION SPECTROSCOPY: EXAFS AND XANES

Steve M. Heald and John M. Tranquada

1 INTRODUCTION

The interaction of X rays with matter occurs by two processes: scattering and absorption. When scattered the X-ray photons are deflected with little or no energy loss. If the X rays are scattered by a regular array, that is, the atoms in a crystal, striking interference patterns are observed that are characteristic of the crystal structure. The discovery and analysis of these patterns began in 1912 [1, 2]—soon after X rays were discovered, and X-ray scattering techniques are of enormous importance in many areas of science.

At typical X-ray energies absorption is a much more likely event than scattering. It occurs when an X-ray photon is absorbed by an electron in an atom causing the electron to be ejected with an energy given by the incoming photon energy minus the electron-binding energy. Since the electron can only be ejected if the incoming photon has more energy than its binding energy, an X-ray absorption spectrum shows sharp increases in absorption as the various electron-binding energies are crossed. The sharp steps are usually called *X-ray edges,* and the edge energy has a characteristic value for each element. Edges were first observed on film [3] at energy values that later turned out to be characteristic of Ag and Br in the emulsion.

Fricke [4] and Hertz [5] first observed that there is structure past the edge. These early measurements were only of structure very near the edge, and they were first explained by Kossel [6]. Later measurements [7, 8] showed that the structure continued well past the edge, which was not explained by Kossel's theory. Kronig [9] was the first to attempt an explanation, and he relied explicitly on the long-range order of the material by considering the scattering of the photoelectron from the periodic lattice. This theory was quite successful in qualitatively explaining many observations, such as the similar fine structure observed for similar crystal structures; the dependence of structure on the photoelectron wave vector rather than on its energy; and the dependence of the structure on the temperature, which was first observed by Hanawalt [10]. However, Hanawalt [10] also observed fine structure in some molecular gases that could not be explained by a long-range order theory.

Kronig also began development of a short-range order theory [11], which we now know is the correct approach. In this theory the fine structure is explained by modulations in the final state of the photoelectron that are caused by backscattering from the surrounding atoms. The process is shown in Figure 3.1 for a simple two-dimensional case. The details of this in short-range order theory were then developed over the years by several workers including

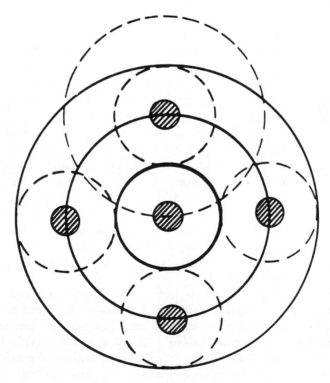

Figure 3.1   Illustration of the backscattering (–––) of the outgoing photoelectron wave (——) from the neighboring atoms.

Peterson [11], Kostarev [12], Sawada [13], and Shmidt [14]. The reader is referred to the original papers and the review by Azaroff and Pease [15] for these early developments. Although correct, the short-range order theory was difficult to calculate, and agreement with experiment was again often only qualitative. Thus, the two theories coexisted until 1970, and a quantitative description of the fine structure now known as the extended X-ray absorption fine structure (EXAFS) remained to be found.

A simpler way of comparing theory and experiment was needed that emphasized the underlying structural content rather than the details of the photoelectron scattering. X-ray diffraction developed rapidly because there was a clear correspondence between the symmetry of the diffraction pattern and the atomic order. A detailed description of the diffracted intensities thus had a firm base on which to develop. Similarly, the modern use of the EXAFS technique began when Sayers and co-workers [16] introduced Fourier analysis of the EXAFS signal. As we discuss in Section 2 each near-neighbor distance appears as a peak in the transform. Thus, the structural information is highlighted, and the difficult-to-calculate quantities affect the details of the transform peaks. Thus useful information is obtained from the EXAFS signal without having to make a complex calculation, and the difficult-to-calculate parameters can often be calibrated experimentally. The result brought a tremendous growth in EXAFS experiments; and spurred by the resulting data, the ability to calculate EXAFS spectra accurately progressed steadily.

The growth in EXAFS experiments was also accelerated greatly by the development of synchrotron radiation sources. Since the EXAFS signal is a small modulation of the absorption and X-ray tubes have a high intensity only at a few sharp lines, the early measurements were extremely tedious. The state of the art before 1970 was probably the automated apparatus of Lytle [17], which required a week or more to obtain high quality data. This was changed with the advent of synchrotron radiation sources where the intense continuum of X rays allowed similar quality data to be obtained in approximately 10 min.

Synchrotron radiation comes from charged particles traveling in circular motion, a typical situation in particle accelerators or storage rings. The development of such accelerators in the 1940s led to detailed calculations of the properties of synchrotron radiation independently by Ivanenko and Pomeranchuk [18] and by Schwinger [19]. The name *synchrotron radiation* comes from the fact that it was first observed at the 70-MeV General Electric Company synchrotron in 1947 [20].

However, the use of synchrotron radiation at X-ray energies required the development of more powerful machines. X rays were first used briefly at the Cambridge Electron Accelerator in 1972 and then on a permanent basis at SPEAR (located at the Stanford Linear Accelerator Laboratory) in 1974. Since then many synchrotron radiation facilities that are capable of providing X-ray energies have opened, including several that are designed solely as synchrotron radiation sources. These are summarized in Table 3.1. Interestingly enough, the competition of synchrotron sources spurred continued development of labora-

Table 3.1 Synchrotron Radiation Sources in the X-Ray Energy Range[a]

Machine	Location	E_c (keV)	I (mA)	E (GeV)
Synchrotrons				
LUSY	Lund, Sweden	1.1	10	1.2
SIRIUS	Tomsk, USSR	1.2	20	1.3
INS-ES	Tokyo, Japan	1.2	60	1.3
Pachra	Moscow, USSR	1.2	100	1.3
Bonn II	Bonn, FRG	4.5	30	2.5
ARUS	Yerevan, USSR	19.5	1.5	5.0
Storage Rings				
ALADDIN	Stoughton, Wisconsin	1.1	100	1.1
ADONE	Frascati, Italy	1.5	35	1.5
DCI	Orsay, France	3.0		1.72
SRS	Daresbury, UK	3.2		2.0
VEPP-3	Novosibirsk, USSR	3.8	100	2.2
Photon Factory	Tsukuba, Japan	3.8	100	2.2
SPEAR	Stanford, California	4.7	100	3.0
NSLS	Upton, New York	5.2	200	2.5
DORIS	Hamburg, FRG	9.3	70	3.7
CHESS	Ithaca, New York	11.5	40	5.5
VEPP-4	Novosibirsk, USSR	14.5	10	6.0

[a]E_c is the critical energy (see Section 5.1), and I and E are the electron beam current and energy. This table includes only machines with critical energies greater than 1 keV and does not include the approximately eight sources under construction or planned.

tory machines so that today good spectra on concentrated samples can be obtained with a laboratory system in about 1 h.

This chapter reviews the modern application of X-ray absorption measurements beginning with the general theory and data analysis techniques. Also discussed are important experimental topics, including measurement techniques, and a comparison of synchrotron and laboratory facilities. The chapter concludes with a summary of important applications. Since it is difficult to cover these topics fully in a single chapter, an attempt was made to include an extensive reference list so that the original sources can be consulted.

2 THEORY OF X-RAY ABSORPTION

2.1 General Description

In X-ray absorption spectroscopy X rays are directed onto a sample, and the fraction absorbed is detected as a function of the X-ray energy. In the simplest case, one measures the incident and transmitted X-ray fluxes, I_0 and I, respectively, which are related by the equation

$$I = I_0 e^{-\mu x} \qquad (1)$$

where $\mu = \mu(E)$ is the energy-dependent X-ray absorption coefficient and x is the sample thickness.

An X ray can be removed from the incident beam either by scattering or by the photoelectric effect, and in X-ray absorption spectroscopy we are concerned primarily with the latter process. The cross section for absorbing a photon is greatest when its energy is equal to the ionization energy of a bound electron. As the X-ray energy is increased through the binding energy of a given core electron level, an abrupt increase in absorption, called an *absorption edge*, occurs. The photoelectron resulting from an absorption process is in an unbound state and has an energy E equal to the difference between that of the incident photon $\hbar\omega$ and the binding energy of the excited core level E_c:

$$E = \hbar\omega - E_c \qquad (2)$$

Viewed as an outwardly traveling wave, the free electron has a wave vector $k = 2\pi/\lambda$ determined by the relation

$$E = \frac{\hbar^2 k^2}{2m_e} \qquad (3)$$

where $\hbar$ is Planck's constant divided by 2π, m_e is the electron's mass, and λ is its wavelength.

If the absorbing atom is closely surrounded by other atoms as in a molecule, liquid, or solid, the photoelectron will scatter from the neighboring atoms, as shown in Figure 3.1. The scattered part of the electron's wave function interferes with the outgoing part, and the phase of the interference varies with the electron (and hence photon) energy. Because the final state of the electron affects the absorption process, the interference effects cause a modulation of the X-ray absorption coefficient and result in fine structure near and above the absorption edge. The features occurring within a few tens of electron volts (eV) of the edge are commonly referred to as the *X-ray absorption near edge structure* (XANES), while EXAFS refers to the oscillations extending above the edge as much as 1000 eV.

The initial core state excited in EXAFS measurements is typically a $1s$, or K-shell, electron. It is convenient to isolate the fine structure above a K absorption edge and to define the corresponding EXAFS interference function $\chi(k)$ by

$$\chi(k) = \frac{\mu(E) - \mu_0(E)}{\mu_K(E)} \qquad (4)$$

where $\mu_0(E)$ is the smoothly varying part of the absorption due to all absorption processes, while $\mu_K(E)$ is that part of $\mu_0(E)$ due only to K-shell absorption. The interference function is described fairly well by the formula [21–23]

$$\chi(k) = \sum_j \frac{N_j}{kR_j^2} A_j(k) p_j(\hat{\varepsilon}) \sin[2kR_j + \psi_j(k)] \tag{5}$$

where the sum is over shells of neighboring atoms with shell j containing N_j atoms at a distance R_j from the origin (where the absorbing atom is located), and $p_j(\hat{\varepsilon})$ is a factor that depends on the X-ray polarization $\hat{\varepsilon}$. The amplitude factor $A_j(k)$ is given by

$$A_j(k) = S_0^2 F_j(k) Q_j(k) e^{-2R_j/\lambda} \tag{6}$$

where $F_j(k)$ is the magnitude of the complex atomic backscattering amplitude $f_j(k, \pi)$, $Q_j(k)$ is a term that accounts for thermal and structural disorder within a shell, S_0^2 is an amplitude reduction factor resulting from multielectron effects in the central atom, and $\exp(-2R_j/\lambda)$ is a mean free path factor included to account for inelastic scattering and core hole lifetime effects. The extra phase term $\psi_j(k)$ is described by

$$\psi_j(k) = 2\delta_1^c(k) + \theta_j(k) + \phi_j(k) \tag{7}$$

where $\theta_j(k)$ is the phase of $f_j(k, \pi)$, $\delta_1^c(k)$ is the p-wave phase shift caused by the potential of the absorbing atom, and $\phi_j(k)$ is a phase factor related to $Q_j(k)$ and due to disorder within a shell.

The various factors in (5)–(7) are discussed in Sections 2.2–2.7. The general form of (5) is derived using a simple one-electron, single-scattering model. The disorder factors enter when we consider the way an X-ray absorption measurement averages over a very large number of absorption sites in a sample. Multielectron effects tend to reduce the EXAFS amplitude and must be included to obtain a quantitatively accurate description of $\chi(k)$. Multiple scattering contributions are not properly included in the formulas presented so far, but in those cases where they are important the appropriate modifications can be made in a straightforward way. Finally, when a polarized X-ray source such as a synchrotron is used, the polarization dependence, which can be quite significant for samples lacking cubic symmetry, must be considered.

2.2   The One-Electron, Single-Scattering Model

To develop the EXAFS formula consider a one-electron model in which all electrons other than the photoelectron are replaced by a set of effective atomic potentials. We assume initially that the absorption process involves a single electron that is excited from a $1s$ level into an unbound state. The X-ray absorption coefficient for such an event is proportional to the absorption cross section, which according to Fermi's golden rule is given by

$$\sigma_K = 4\pi^2 \alpha \hbar \omega \sum_f |\langle \psi_f | \hat{\varepsilon} \cdot \mathbf{r} | \psi_K \rangle|^2 \delta(E_f + E_K - \hbar\omega) \tag{8}$$

where α is the fine structure constant, $\hbar\omega$ is the X-ray photon energy, $\hat{\varepsilon}$ is the X-ray polarization vector, and E_K is the K-shell binding energy. The sum is over all final states of the photoelectron, and the delta function ensures that energy is conserved.

Since the K-shell wave function is localized at the core of an atom, the magnitude of the dipole matrix element in (8) depends only on the amplitude of the final state wave function at the core of the absorbing atom. As required by the dipole selection rule, the photoelectron travels outward from the origin as a p wave. The total final state of the electron is the sum of the outgoing wave plus the part that is scattered by neighboring atoms. At the origin, the interference between scattered and outgoing parts is expected to result in a sinusoidal oscillation of the wave function amplitude, which depends on the photoelectron's wave vector and the distance to the scattering atoms.

To properly calculate the function $\chi(k)$ in this one-electron model, the curvature of the outgoing wave front must be considered [21, 24, 25]; however, the general result is fairly complicated and not very illuminating. A more useful expression is obtained by making the following approximations: (1) the scattering atoms are sufficiently small or far from the origin that the curved wave front at a scattering site can be replaced by a plane wave, and (2) the photoelectron's wavelength is sufficiently small that the asymptotic form of the radial part of the wave function can be used. With these approximations, the EXAFS resulting from scattering by a single neighbor can be written as

$$\chi'(k) = -\frac{1}{kR^2} 3(\hat{\varepsilon} \cdot \hat{\mathbf{R}})^2 F(k) \sin[2kR + 2\delta_1^c(k) + \theta(k)] \tag{9}$$

where $\mathbf{R}$ is the position vector of the scatterer relative to the central atom and $\hat{\mathbf{R}} = \mathbf{R}/R$. The quantities $F(k)$ and $\theta(k)$ are the magnitude and phase, respectively, of the complex atomic backscattering amplitude $f(k, \pi)$, while $\delta_1^c(k)$ is the phase shift caused by the potential of the central atom.

The function $f(k, \theta)$ is the complex amplitude for the scattering from an atom of an electron that is incident as a plane wave, where θ is the angle between the directions of motion of the incident and scattered electron. For backscattering, $\theta = \pi$. The backscattering amplitude can be calculated [21, 26] from the atomic electron density in terms of partial-wave phase shifts $\delta_l(k)$. Teo and Lee [27] have calculated and tabulated the backscattering amplitudes and phase shifts for a wide range of atomic numbers covering most of the atoms in the periodic table; examples of their calculations for the magnitude of the backscattering amplitude are shown in Figure 3.2. In general, the function $F(k)$ is peaked at low k and falls off roughly as k^{-2} at higher k. The overall magnitude of $F(k)$ increases

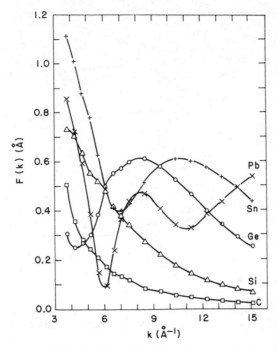

Figure 3.2 Backscattering amplitude $F(k) = f(k, \pi)$ for column IV elements calculated with the plane wave approximation. Reprinted with permission from B.-K. Teo and P. A. Lee, *J. Am. Chem. Soc.*, **101**, 2815 (1979). Copyright 1979 American Chemical Society.

with the atomic number of the scattering atom. The backscattering amplitude is fairly structureless for low Z atoms. For heavier atoms, a resonance can occur that is similar in origin to the Ramsauer–Townsend effect [28] in which, for a particular energy, the electron scattering is greatly reduced, thus causing a dip in $F(k)$. As a result, $F(k)$ tends to have a double-peaked or, for the heaviest atoms, a triple-peaked structure.

The backscattering of an electron changes the phase of its wave function as well as its magnitude. Away from any resonances, the backscattering phase shift $\theta(k)$ tends to have a roughly linear k dependence with a negative slope, as illustrated for several different atoms in Figure 3.3. The phase shifts for two different elements are separated by a constant offset that varies significantly with atomic number. The phase shift deviates sharply from its linear behavior in the region of a resonance (amplitude minimum) where it tends to bend so as to have a positive slope.

The potential of the absorbing atom also changes the phase of the electron from that of a free particle. The central atom phase shift (Figure 3.4) is structureless for all Z, and it decreases monotonically as k increases. As for the backscattering phase shifts, there is a significant constant offset between central atom phase shifts for elements with different atomic numbers.

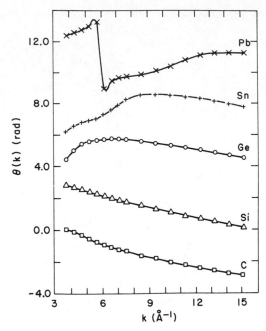

Figure 3.3 Phase of the backscattering amplitude $\theta(k)$. Reprinted with permission from B.-K. Teo and P. A. Lee, *J. Am. Chem. Soc.*, **101**, 2815 (1979). Copyright 1979 American Chemical Society.

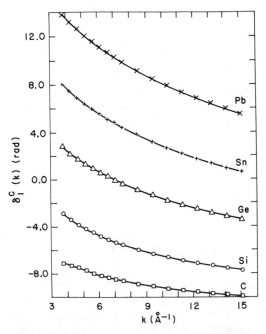

Figure 3.4 Central atom *p*-wave phase shift $\delta_1^c(k)$. Reprinted with permission from B.-K. Teo and P. A. Lee, *J. Am. Chem. Soc.*, **101**, 2815 (1979). Copyright 1979 American Chemical Society.

The atom-dependent information in the backscattering amplitudes and phase shifts can be both useful and troublesome in the analysis of EXAFS data. On the useful side, the structure of the backscattering amplitude can be helpful in identifying the species of atoms surrounding a given absorber, when their types are unknown. On the negative side, one must correct for these factors in order to extract from the data such quantities as coordination numbers and interatomic distances. Ways in which one can deal with this problem are discussed in Section 3.

The small-atom/plane-wave and short-wavelength approximations used in the derivation of (9) are frequently inappropriate for nearest-neighbor shells, so that if one calculates the EXAFS using the plane-wave backscattering amplitude and compares it with experimental results, significant discrepancies may be observed, especially at low k [29, 30]. Much better agreement can be obtained when the full curved wave formalism is applied [29–31]. However, despite any limitations they may have for theoretical calculations, the approximations still yield a useful parameterization of the EXAFS for empirical analysis. Equations (5)–(7) define the quantities $F(k)$, $\theta(k)$, and $\delta_1^c(k)$. The quantities so defined then have a weak R dependence; that is, as the interatomic distances within a system are changed, the functions $F(k)$, $\theta(k)$, and $\delta_1^c(k)$ also change. The fact that the approximate formulas provide a valid description of the EXAFS is quite important for the interpretation and analysis of experimental data.

2.3 Multiple Scattering

So far only single-scattering events have been considered. To be complete, one must also consider the contributions from multiple scatterings. The amplitude for a multiple-scattering process is proportional to a factor $f(k, \theta)$ for each scattering event involved. Because the scattering amplitude is small over the EXAFS energy range for angles θ away from the forward direction ($\theta = 0$), most multiple-scattering events make a negligible contribution to the EXAFS. Those events that cannot be ignored are discussed here. Fortunately, multiple-scattering effects can be isolated and treated in a straightforward manner when necessary.

The EXAFS data are commonly analyzed by the Fourier transform method (see Section 3), which separates different scattering contributions according to path length. Because they have longer path lengths, multiple-scattering contributions are easily separated from single-scattering peaks corresponding to nearest-neighbor coordination shells. Multiple-scattering effects need not interfere with the analysis of single-scattering EXAFS for the shortest length paths. Furthermore, besides the weakness of the scattering amplitude, the magnitudes of multiple-scattering terms are also reduced by inelastic effects (see Section 2.4). The photoelectron state can decay by various processes with a probability that increases with the distance traveled. The distance-dependent decay factor reduces the significance of multiple-scattering contributions to EXAFS.

However, there is one case in which multiple scattering is important, and that is when two neighboring atoms are collinear with the absorbing atom [32]. The electron–atom scattering amplitude is strongly peaked in the forward direction, so that the forward-scattering amplitude $f(k, 0)$ is always quite strong relative to $f(k, \pi)$ or $f(k, \theta)$ for finite angles. As a matter of fact, the electron-wave amplitude seen by a shadowed atom due to forward scattering by an intervening atom can be stronger than that of the direct, unscattered wave [21, 33]. (This is known as the *focusing effect*.) For such a shadowed atom, the multiple-scattering EXAFS can be much larger than the single-scattering term. A good example of this situation is the fourth shell in the face-centered cubic lattice, which is shadowed by the first shell, as was discussed for Cu [21, 22] and Pt [34]. The focusing effect can be significant even when backscattering by the shadowing atom is negligible, as has been demonstrated with hydrogen atoms in Ni and Cr hydrides [35].

When a pair of neighboring atoms are perfectly collinear with the central atom, the EXAFS from the shadowed atom can still be described by an equation such as (9) provided that $F(k)$ and $\theta(k)$ are replaced by renormalized quantities [36, 37]. The renormalized $F'(k)$, which is proportional to the single-atom backscattering amplitude, includes a term that describes the transmission of the electron wave through the shadowing atom.

In principle, multiple-scattering contributions contain structural information about bond angles that is not present in single-scattering EXAFS. By considering a triangular arrangement of an absorbing atom and two neighbors, one can derive formulas to describe a general multiple-scattering process [38, 39]. The Debye–Waller factors (which describe the effects of thermal vibrations) must be generalized and treated appropriately for the various scattering paths [40]. For a slightly noncollinear geometry, the analysis becomes complicated. Because the scattering amplitude is strongly peaked in the forward direction, thermal fluctuations in the bond angle cause large variations in the scattering amplitude for the intervening atom and result in an asymmetric effective radial distribution function [41]. The effects of thermal vibrations in such a case must be treated carefully [42].

2.4   Many-Body Effects

In the one-electron model, all of the X-ray photon's energy is used to excite a single core level, and all electrons other than the photoelectron are assumed to form a potential from which the photoelectron can scatter elastically. In reality, while elastic scattering tends to dominate, the other electrons can absorb energy from the photoelectron, and in doing so reduce the EXAFS amplitude. The many-body effects important to EXAFS can be divided into two categories [30]: (1) relaxation of the remaining electrons about the core hole in the absorbing atom, and (2) inelastic scattering of the photoelectron from neighboring atoms.

The relaxation effects are discussed first. In applying the golden rule (8) it was assumed that the dipole matrix element can be written in terms of single-electron states. The situation is more properly described in terms of a transition between N-electron states $|\Psi_i^N\rangle$ and $|\Psi_f^N\rangle$, where N is the number of electrons in the absorbing atom. Factoring the initial state as $|\Psi_i^N\rangle = |\psi_K\rangle|\Phi_i^{N-1}\rangle$, and treating the final state similarly, we obtain the dipole matrix element

$$\langle \Psi_f^N | \hat{\varepsilon} \cdot \mathbf{r} | \Psi_i^N \rangle = \langle \psi_f | \hat{\varepsilon} \cdot \mathbf{r} | \psi_k \rangle \langle \Phi_f^{N-1} | \Phi_i^{N-1} \rangle \qquad (10)$$

From (10) we see that the single-electron absorption coefficient is modified by an overlap factor [43] $S_0^2 = |\langle \Phi_f^{N-1} | \Phi_i^{N-1} \rangle|^2$. If the *passive* electrons do not adjust during the transition, then $S_0^2 = 1$, while relaxation about the core hole results in $S_0^2 < 1$. At low photoelectron energies the passive electrons may only be slightly perturbed from their initial states, but at high energies ($\gtrsim 200\,\mathrm{eV}$) the $N - 1$ electron final state should be fully relaxed. The excess probability for a transition, $1 - S_0^2$, corresponds to *shake-up* and *shake-off* processes [44], where one or more of the "passive" electrons jumps into a different unoccupied bound level or is knocked free from the atom. Events involving shake-up and shake-off transitions also contribute to the EXAFS interference function, but are smeared out because of the partitioning of the photon energy between particles; hence, the reduction of the amplitude of $\chi(k)$ by a factor S_0^2 relative to that of the one-electron model results [43].

At the photoabsorption threshold, there is insufficient energy to allow multielectron excitations, so that S_0^2 must equal 1, while for energies well above those required to excite shake-up or shake-off transitions, S_0^2 is expected to asymptotically approach a constant value less than 1 [43, 45]. A calculation [46] of the central atom amplitude reduction factor for the molecule Br_2 using an interpolation between the adiabatic and sudden approximation limits has found that the reduction factor decreases slowly throughout the EXAFS region and approaches its asymptotic value near $16\,\text{Å}^{-1}$. Calculations [45, 47] of the overlap factors for a large number of atoms and molecules indicated that significant chemical effects can be seen that are caused by charge transfer from neighboring atoms [47]. Studies of experimental EXAFS amplitudes find the reduction of the experimental relative to theoretical amplitudes to be in reasonable agreement with calculated S_0^2 values [23, 48].

Because of inelastic scattering and decay of the core hole, the photoelectron has a finite lifetime. According to the uncertainty principle, the lifetime τ is related to an uncertainty $\eta \approx \hbar/\tau$ in the photoelectron's energy. Treating η as the imaginary part of the photoelectron energy, the definition of the wave vector becomes [21]

$$[2m(E + i\eta)/\hbar^2]^{1/2} \approx k + (i\eta'/k) \qquad (11)$$

so that

$$e^{i2kR} \rightarrow e^{i2kR} e^{-2R\eta'/k} \qquad (12)$$

The factor $\exp(-2R\eta'/k)$ represents a reduction in the EXAFS amplitude because of finite lifetime effects.

There are two main physical effects responsible for the limited lifetime, and the net decay factor should be a convolution of them. One is the core-hole lifetime $\tau_h = \hbar/\Gamma_h$; the other is inelastic scattering of the photoelectron from electrons, plasmons, phonons, and so on. The effects of inelastic scattering can be described in terms of a mean free path $\lambda_i \approx v\tau_i$. Mean free path measurements made with various electron spectroscopic techniques were analyzed in terms of a *universal curve* and found to behave as

$$\lambda_i = k/\eta' \tag{13}$$

for $k \gtrsim 5\,\text{Å}^{-1}$, with $\eta' \approx 1\,\text{Å}^{-2}$ for elements and $\eta' \approx 0.5\,\text{Å}^{-2}$ for inorganic compounds [49]. By comparison, the effective mean free path corresponding to the core-hole lifetime of a Cu atom, using $v = \hbar k/m$ and $\Gamma_h = 1.55\,\text{eV}$ [50], is of the form of (13) with $\eta' \approx 0.2\,\text{Å}^{-2}$. The total mean free path is given by

$$\frac{1}{\lambda} = \frac{1}{\lambda_i} + \frac{1}{\lambda_h} \tag{14}$$

Evidently loss of coherence from inelastic scattering tends to dominate, although core-hole lifetime effects cannot be ignored. Assuming that the net mean free path has the form given in (13), the amplitude decay term can be written $\exp(-2R/\lambda)$.

Measurements of the EXAFS mean free path for tetrahedral semiconductors were made by Stern and co-workers [23]. By comparing the reduction in amplitude of the EXAFS contribution due to next-nearest neighbors relative to that from nearest neighbors, they found that the k dependence of λ is in agreement with mean free path behavior determined from Auger electron measurements. However, inelastic energy losses in the absorbing atom are already accounted for by the factor S_0^2. They suggest that with the inclusion of the S_0^2 factor, the exponential decay term should be modified to the form $\exp[-2(R-\Delta)/\lambda]$, where Δ is a distance corresponding to the radius of the absorbing atom. Nevertheless, the inelastic processes covered by S_0^2 are not entirely the same as those described by λ so that the unmodified mean free path factor $\exp(-2R/\lambda)$ may still be appropriate in the EXAFS formula.

Theoretical calculations of backscattering amplitudes sometimes take inelastic scattering effects into account by using a complex scattering potential [26, 27, 30]. When such theoretical backscattering amplitudes are used to calculate EXAFS, a distance factor Δ should be included in the mean free path term to avoid overcounting inelastic effects. Despite such corrections, the mean free path term may be overly simplified for an accurate description of amplitude reduction for higher shells in EXAFS. A recent experimental study of mean free paths in covalent zincblende structure materials finds significant discrepancies between equivalent analyses of the data that may be caused by anisotropic effects [51].

2.5   Structure and Temperature Dependence

The expression for $\chi'(k)$ given by (9), together with the mean free path term $\exp(-2R/\lambda)$ and the many-body overlap factor S_0^2 discussed earlier, give a quantitative description of the EXAFS due to absorption by an atom at a unique site in a sample with scattering by a single neighbor. In an actual X-ray absorption experiment, the photon beam samples many sites, and to describe such a result one must, in general, average the single site result over all local equilibrium configurations in the sample and then perform a thermal average to account for temperature-dependent fluctuations in interatomic distances. Assuming that neighboring atoms around the absorption sites can be grouped into well-defined coordination shells j containing an average of N_j atoms at a distance R_j, one obtains (5)–(7), where the disorder factors $Q_j(k)$ and $\phi_j(k)$ are defined by

$$Q_j(k)\exp[i\phi_j(k)] = \langle (R_j/r)^2 \exp[-2(r - R_j)/\lambda]\exp[i2k(r - R_j)]\rangle_j \qquad (15)$$

where

$$\langle f(r)\rangle_j = \int dr\, f(r)P_j(r) \qquad (16)$$

and $P_j(r)$ is a net probability density for the average distribution of atoms in shell j. The net distribution is a convolution of the configuration and thermal distributions. In a crystal, where all atoms within a coordination shell are at the same average radial distance from the absorber, the average over distributions reduces to a thermal average.

The factors r^{-2} and $\exp(-2r/\lambda)$ have weak r dependences and hence may be replaced by their mean values in a lowest order evaluation of (15). The average on the right-hand side of the equation then reduces to the Fourier transform of the distribution function:

$$Q_j(k)\exp[i\phi_j(k)] = \int dr\, P_j(r)\exp[i2k(r - R_j)] \qquad (17)$$

Quite often the function $P_j(r)$ is Gaussian or nearly Gaussian, and in such a case it is convenient to describe the average in terms of cumulants [52–54]:

$$\langle \exp[i2k(r - R_j)]\rangle = \exp\left[\sum_{n=1}^{\infty} \frac{(2ik)^n}{n!} C_{n,j}\right] \qquad (18)$$

In terms of moments about the mean, $\mu_n = \langle (r - R)^n\rangle$, the first few cumulants are [55]:

$$C_1 = 0 \qquad (19)$$

$$C_2 \equiv \sigma^2 = \mu_2 \qquad (20)$$

$$C_3 = \mu_3. \tag{21}$$

$$C_4 = \mu_4 - 3\mu_2^2 \tag{22}$$

The disorder amplitude and phase factors can then be written

$$Q_j(k) = \exp(-2\sigma_j^2 k^2 + \tfrac{2}{3}C_{4,j}k^4 + \cdots) \tag{23}$$

$$\phi_j(k) = -\tfrac{4}{3}C_{3,j}k^3 + \cdots \tag{24}$$

Even-order cumulants contribute to the amplitude, while odd-order cumulants enter the phase. Notice that an asymmetric distribution function (i.e., one having $C_3 \neq 0$) introduces a k-dependent factor to the EXAFS single-shell phase. If not accounted for, such a factor can cause an incorrect interatomic distance determination in the analysis of EXAFS data [56]. Although cumulants provide a general, model-independent way to describe a non-Gaussian distribution, other parameterizations are possible and several have been discussed in the literature [56–58].

In a crystal with harmonic forces between atoms, the distribution function is just a Gaussian:

$$P_j(r) = \frac{1}{\sqrt{2\pi\sigma_j^2}} \exp\left[\frac{-(r - R_j)^2}{(2\sigma_j^2)}\right] \tag{25}$$

where σ_j^2 is the mean-square deviation about the average value R_j. In such a case, the only nonzero cumulant is the second-order one, and the disorder is described by a Debye–Waller-like term:

$$Q_j(k) = \exp(-2k^2\sigma_j^2)$$
$$\phi_j(k) = 0 \tag{26}$$

If structural disorder is also present, it follows from the properties of cumulants [53, 55] that the nth-order cumulant of the net distribution is equal to the sum of the nth-order cumulants of the configuration and thermal distributions. For example, if the structural and thermal distributions are both Gaussian with mean-square deviations Δ_j^2 and $\sigma_j^2(T)$, respectively, then

$$C_{2,j} = \Delta_j^2 + \sigma_j^2(T) \tag{27}$$

Returning to (15), it is possible to take some account of the factors r^{-2} and $\exp(-2r/\lambda)$ by expanding them to first order about their mean values; assuming a Gaussian distribution, one finds that although there is no change to the amplitude, the disorder phase factor becomes [53, 54]

$$\phi_j(k) = -\frac{4\sigma_j^2 k}{R_j}\left(1 + \frac{R_j}{\lambda}\right) \tag{28}$$

where $\bar{\lambda}$ is an average value for the mean free path over a relevant range of k space. Because of its linear k dependence, this phase factor acts like a small distance shift $[\phi_j(k) = 2k\Delta R_{\text{eff}}]$ in the net EXAFS phase. Although it is often quite small, ΔR_{eff} can be as large as a few hundredths of an angstrom.

The general theory of the thermal mean-square relative displacements σ_j^2 was discussed by Beni and Platzman [52]. If $\mathbf{u}_j$ is the displacement of an atom in the jth shell from its equilibrium position and $\hat{\mathbf{R}}_j$ is a unit vector pointing from the central atom toward the atom in the jth shell, σ_j^2 can be written as

$$\sigma_j^2 = \langle [(\mathbf{u}_j - \mathbf{u}_0) \cdot \hat{\mathbf{R}}_j]^2 \rangle \tag{29}$$

where the brackets denote a thermal average. It is related to the atomic mean-square displacements along the bond direction, $u_j^2 = \langle (\mathbf{u}_j \cdot \hat{\mathbf{R}}_j)^2 \rangle$ and $u_0^2 = \langle (\mathbf{u}_0 \cdot \hat{\mathbf{R}}_j)^2 \rangle$, by

$$\sigma_j^2 = u_j^2 + u_0^2 - 2c_j \tag{30}$$

where $c_j = \langle (\mathbf{u}_0 \cdot \hat{\mathbf{R}}_j)(\mathbf{u}_j \cdot \hat{\mathbf{R}}_j) \rangle$ is the displacement correlation function. For nearest-neighbor atoms, the displacement correlation function can be fairly large, so that σ_j^2 can be much smaller than the sum of the mean-square displacements. For more distant neighbors the correlation in the atomic motions is less significant.

Sevillano and co-workers [59] considered several models for calculating mean-square relative displacements in monatomic crystals. In terms of phonon frequencies ω, σ_j^2 can be expressed as

$$\sigma_j^2 = \frac{\hbar}{M} \int d\omega \rho_j(\omega) \frac{\coth(\hbar\omega/2k_B T)}{\omega} \tag{31}$$

where M is the atomic mass, k_B is the Boltzmann constant, and $\rho_j(\omega)$ is the normalized projected density of modes contributing to relative vibrational motion. If a force constant model is available for the system of interest, $\rho_j(\omega)$ can be calculated numerically. A few such calculations have been made and they are generally in good agreement with experimental results [59–61]. Two approximate models were also found [59] to give a reasonable description of the temperature dependence of σ^2. In the Einstein model, it is assumed that only a single phonon mode with frequency ω_E contributes to $\rho_j(\omega)$. This model is quite reasonable for a nearest-neighbor shell because only high-frequency vibrations, whose distribution tends to be sharply peaked in ω space, contribute significantly to relative motion. The projected density of modes is approximated by a Dirac delta function $\delta(\omega - \omega_E^j)$, and (31) reduces to

$$\sigma_j^2 = \frac{\hbar}{M\omega_E^j} \coth\left(\frac{\hbar\omega_E^j}{2k_B T}\right) \tag{32}$$

where the appropriate Einstein frequency depends on the coordination shell index j. While (32) is often used as a simple parameterization for σ^2 with ω_E determined by fitting to experimental data, it has been found [62] for a series of face-centered cubic metals that the Einstein frequency for the nearest-neighbors is equal to the square root of the second moment of the phonon density of states $\langle \omega^2 \rangle^{1/2}$. The other approximation is a correlated Debye model, for which

$$\rho_j(\omega) = \frac{3\omega^2}{\omega_D^3}\left[1 - \frac{\sin(\omega R_j/c)}{\omega R_j/c}\right] \tag{33}$$

where $\omega_D = k_B\Theta_D/\hbar$ is derived from the Debye temperature Θ_D and $c = \omega_D/k_D$, where $k_D = (6\pi^2 n)^{1/3}$ and n is the atomic density. As the Debye model tends to give a poor description of the phonon density of states for high-frequency modes, the correlated Debye approximation generally is more appropriate for describing σ_j^2 for $j > 1$. Nevertheless, several studies [59, 63, 64] have found the correlated Debye model to give reasonable agreement with experimental measurements for the first and higher shells.

One special structure-dependent effect deserves some discussion. When the radial distances for two or more shells of atoms are only slightly different, there is interference between the corresponding EXAFS contributions, resulting in the phenomenon known as *beats* [65]. In such a case, the cumulant expansion is of little use for describing the structural distribution because many cumulants may be significant. For two single-shell EXAFS contributions of the form $A_j \sin(2kR_j + \psi)$, a single amplitude and phase can be obtained using

$$A' \sin(2kR' + \psi') = \text{Im}[A_1 e^{i(2kR_1 + \psi)} + A_2 e^{i(2kR_2 + \psi)}] \tag{34}$$

For example, with two shells of identical atoms, one finds

$$A' = [A_1^2 + A_2^2 + 2A_1 A_2 \cos(2k\Delta R)]^{1/2} \tag{35}$$

$$R' = \tfrac{1}{2}(R_1 + R_2) \tag{36}$$

$$\psi' = \psi + \tan^{-1}\left[\frac{A_1 - A_2}{A_1 + A_2}\tan(k\Delta R)\right] \tag{37}$$

where $\Delta R = R_1 - R_2$.

2.6   L-Edge Spectra

Up to this point it was assumed that the initial state in the X-ray absorption process is a K level. This is the simplest case to treat because with an initial $1s$ state, the final state can have only p symmetry. The EXAFS above an L_1 edge is described by the same formulas since the initial level is a $2s$ state. The situation becomes somewhat more complicated for an L_2 or L_3 edge (denoted collectively by $L_{2,3}$), where a $2p$ level is excited and the final state may have either s or d

symmetry. Mixed final states are also possible where an outgoing d wave gets scattered into an s state and vice versa.

To describe the total $\chi(k)$ for an $L_{2,3}$ edge, it is convenient to define a partial EXAFS interference function $\chi_{l,l'}$ such that

$$\chi_{l,l'}(k) = \sum_j A_j(k) p^j_{l,l'}(\hat{\varepsilon}) \sin[2kR_j + \psi^j_{l,l'}(k)] \tag{38}$$

where $A_j(k)$ is given by (6),

$$\psi^j_{l,l'} = \delta^c_l(k) + \delta^c_{l'}(k) + \theta_j(k) + \phi_j(k) \tag{39}$$

and l and l' are the angular momentum quantum numbers of the outgoing and scattered photoelectron waves. The quantity $p^j_{l,l'}(\hat{\varepsilon})$ is a polarization factor for the l, l' final state and coordination shell j. The function $\delta^c_l(k)$ is the central atom phase shift for a wave with angular momentum l. The EXAFS for an $L_{2,3}$ edge can then be expressed as [66, 67]

$$\chi = (M^2_{21}\chi_{22} + M^2_{01}\chi_{00} + 2M_{01}M_{21}\chi_{20})/(M^2_{21} + \tfrac{1}{2}M^2_{01}) \tag{40}$$

where M_{01} and M_{21} are the radial dipole matrix elements between the core wave function with $l = 1$ and final states with $l = 0$ and $l = 2$, respectively. Because of the multiple contributions, the matrix elements do not cancel out as they do in the K-edge case.

The fact that $L_{2,3}$-edge EXAFS is a sum of three contributions each with a different combination of central atom phase shifts makes analysis appear quite complicated until one appreciates a few fortuitous simplifications. First of all, the s contribution is insignificant relative to the d contribution. Teo and Lee [27] calculated the radial dipole matrix elements for a number of elements and found $M_{01}/M_{21} \approx 0.2$, which is in good agreement with a measurement by Heald and Stern [67]. Second, for an unoriented polycrystalline sample or a material with at least cubic symmetry, the polarization factors p^j_{20} are zero, so that the s–d interference term disappears. One is left with only the $l = 2$ final state term. Of course, for measurements on an oriented anisotropic sample the interference term must be taken into account.

2.7 Polarization Effects

The electric field of an X ray is oriented in a direction $\hat{\varepsilon}$ that is perpendicular to its direction of motion. When an X ray is absorbed, the photoelectron wave tends to be oriented with its greatest amplitude traveling outward in a direction parallel to $\hat{\varepsilon}$. If the environment of the absorbing atoms is anisotropic, then the X-ray absorption coefficient will show a dependence on the orientation of the X-ray polarization. Note that even if the X-ray beam has no net polarization, the electric field is always perpendicular to the direction of motion and anisotropic absorption can still be observed. Use of a linearly polarized X-ray beam, as is obtained from a synchrotron source, provides valuable information about the angular orientation of atomic environments.

Heald and Stern [67] have shown from macroscopic arguments that the X-ray absorption coefficient for dipole transitions must have the transformation properties of a second-rank tensor. It follows that no absorption feature can vary with angle more rapidly than $\cos^2\theta_j$, where $\cos\theta_j = \hat{\varepsilon} \cdot \hat{\mathbf{R}}_j$. Explicit formulas for the K- and L-edge polarization factors can be obtained from the microscopic theory [21, 66, 68]. For K and L_1 edges, the polarization factor for a given coordination shell j is

$$p_j(\hat{\varepsilon}) = 3\langle\cos^2\theta\rangle_j \tag{41}$$

where

$$\langle\cos^2\theta\rangle_j = \frac{1}{N_j}\sum_i \cos^2\theta_i \tag{42}$$

and the sum is over all sites i in shell j. If the material has cubic or higher symmetry or consists of a randomly oriented, polycrystalline powder, then $\langle\cos^2\theta\rangle_j = \frac{1}{3}$ and $p_j(\hat{\varepsilon}) = 1$. For an $L_{2,3}$ edge there are three different polarization factors given by [67]

$$p_{22}^j = \frac{1}{2}(1 + 3\langle\cos^2\theta\rangle_j) \tag{43}$$

$$p_{00}^j = \frac{1}{2} \tag{44}$$

$$p_{02}^j = \frac{1}{2}(1 - 3\langle\cos^2\theta\rangle_j) \tag{45}$$

When $\langle\cos^2\theta\rangle_j = \frac{1}{3}$, one has $p_{22}^j = 1$, $p_{00}^j = \frac{1}{2}$, and $p_{02}^j = 0$.

2.8 XANES

The structure within a few tens of electronvolts of an absorption edge can contain a great deal of information about the electronic and structural environment of the absorbing species. The near-edge structure can be viewed as a measure of the projected density of states of appropriate symmetry (p symmetry for a K or L_1 edge; d and s symmetry at an L_2 or L_3 edge) at the excitation site. Conversely, one can interpret the fine structure in terms of single- and multiple-scattering contributions as is done in the EXAFS regime. Both approaches are equivalent, since the geometry determines the electronic structure while the scattering contributions depend on the electronic structure of the neighboring atoms. A great deal of progress was made in the past several years in terms of our theoretical understanding of and ability to calculate XANES. Despite the recent theoretical developments, however, many of the important applications of XANES spectroscopy in the analysis of unknown atomic environments have depended on empirically determined correlations between edge positions and features, on the one hand, and site symmetry and bond type on the other.

It is now known that the general scattering picture [21, 22] used to derive the formula for EXAFS is also applicable in the near-edge region. The XANES and EXAFS regions differ in that for XANES (1) curved-wave effects must be taken into account, and (2) electron–atom scattering is strong, so that multiple-scattering can be important for scattering paths involving noncollinear arrangements of atoms. As a result, there is no simple inversion procedure for extracting information from XANES. On the other hand, the multiple-scattering contributions contain information on the angular relationships within clusters of atoms that is not available from EXAFS. To unravel all of the structural information one must rely on theoretical calculations. Much of the theoretical workup to the present has involved modeling of known systems to test the theory. Full multiple-scattering calculations have obtained reasonable agreement with experiment for K edges in transition metals [69, 70] and molecular clusters [71–73].

The inclusion of all possible multiple-scattering paths in XANES calculations is often unnecessary. Good results were obtained for the Cu K edge with a single-scattering calculation [31]. In contrast, a comparison between single- and multiple-scattering calculations of near-edge structure for KrF_2 and $FeCl_2$ shows that multiple-scattering effects are definitely important for those molecules [73]. Bunker and Stern [32] have argued that multiple scattering involving noncollinear arrangements of atoms may be significant only when bond lengths are very short. More work is needed to better understand the relative importance of multiple-scattering effects.

Near-edge structure has also been calculated in terms of the projected density of final states obtained from a band-structure approach. By calculating the density of states with the appropriate symmetry relative to an absorption site, good agreement with experimental measurements was obtained for K-edge XANES of transition metals [74–76]; calculations were also made for L and M edges. The good agreement is somewhat surprising because the band-structure calculations do not include any core-hole effects. The lack of importance of the core hole is most likely caused by screening of the core hole by the conduction band electrons. The situation is quite different for insulators such as KCl for which it has been shown that an accurate description of the K- and L-edge XANES requires a proper treatment of the core hole [77–79].

Discrepancies between theoretical and experimental XANES probably result from several factors [80], and the choice of the potential is one of the most important. Most scattered-wave calculations have used a "muffin tin" potential in which atomic charge densities are assumed to have spherical symmetry and are localized in spheres with a constant potential in between. No allowance is made for anisotropic effects such as bond charges. A second problem involves the possible energy dependence of the local scattering potential near the absorption threshold, and it is generally ignored.

Another deficiency in the single-electron calculations is the neglect of possible multielectron excitations. As mentioned in the discussion of many-body effects, it is possible for some of the excitation energy to be transferred to secondary electrons. Such multielectron excitations could, at least in principle, result in sharp features in the near-edge structure. It appears that such effects are not very important in general, as single-electron, multiple-scattering calculations are usually able to account at least qualitatively for all of the features at a particular absorption edge. Nevertheless, multielectron contributions may be significant in certain cases. From a calculation including configuration interactions, Bair and Goddard [81] identified a small peak at the Cu K edge in $CuCl_2$ as resulting from a $1s \rightarrow 4p$ transition with a concurrent transfer of an electron from the ligands to the metal. In another case, Stern [82] has argued that a structure in the metal K-edge XANES in $MnCl_2$, $FeCl_2$, and $CoCl_2$ is caused by a shake-up transition. However, for $FeCl_2$ the alleged multielectron feature is also found in the single-electron calculation of Chou and co-workers [73].

Because of the complications involved in performing ab initio calculations, experimenters have often relied on empirical interpretations of XANES to extract new information from their measurements. For example, it is well known that the energy of the absorption threshold depends on the atomic valence, bond ionicity or covalency, and the number of bonds (see Section 7.1 for an example). It has been shown that a quantity called the *coordination charge*, which is well correlated with the threshold position, can be defined in terms of valence, ionicity, and coordination number [83]. By comparing the absorption threshold energy of a sample with those of a series of well-characterized materials one can determine the coordination charge of the sample [84].

The positions of peaks in the edge structure depend on, among other things, interatomic distances. In K-shell XANES of gas-phase molecules containing only light elements such as C and O, an empirical correlation was observed between the edge position, the σ-shape resonance, and the intramolecular bond length [85]. The correlation has been justified in terms of multiple-scattering theory [86]. It can be used to determine intramolecular distances in simple gas-phase and chemisorbed molecules to an accuracy of ± 0.05 Å.

Because of the presence of empty d states, some special features are present in the XANES of transition metal atoms. For third-row transition metals, a sharp peak associated with a $1s \rightarrow 3d$ transition is sometimes observed below the K-edge absorption threshold. A $1s \rightarrow 3d$ transition is not allowed by the dipole selection rule; however, if the local geometry lacks inversion symmetry, p states can hybridize with $3d$ states, which makes a transition to a mixed $3d$ state possible. The presence of a strong pre-edge peak can then be used to indicate lack of inversion symmetry at an absorption site [87]. A very small $1s \rightarrow 3d$ peak can still occur at a symmetric site because of the quadrupole interaction, as demonstrated by a polarized X-ray study of the square planar $CuCl_4^{2-}$ complex [88]. A vibronically allowed dipole transition can also make a finite contribution.

For a transition metal or rare earth $L_{2,3}$ edge, the presence of empty d states results in a strong *white line*. (The expression *white line* derives from the early days of X-ray absorption spectroscopy when X-ray absorption spectra were recorded predominantly on film. On a photographic record, a strong absorption feature appeared as a white line.) The area under the white line peak is related to the local density of unoccupied d states at the absorption site [89, 90]. By comparing with measurements on well-characterized systems, one can use $L_{2,3}$-edge studies to determine d-band occupancy in transition metal compounds and alloys.

3 DATA ANALYSIS

Equations (5)–(7) show that a great many unknown factors must be determined before quantitative information can be extracted from EXAFS measurements. This observation can make the task of data analysis appear somewhat daunting. Fortunately, most of the elastic and inelastic scattering factors can be determined empirically from measurements on appropriate reference materials. Provided that the atomic species of the absorbing and backscattering atoms are the same in the reference and unknown, and that the interatomic distances are similar, one can determine the coordination number, mean-square relative displacement, and interatomic distance quite accurately for a coordination shell in the unknown *relative* to one in the reference. An understanding of the various factors entering the EXAFS formula is important for judging the transferability of scattering amplitudes and phase shifts from reference compounds to unknowns. Analysis becomes more difficult when a good chemical standard is not available; in that case, one must rely on theoretical calculations.

In this section various steps in the data-reduction process and techniques for extracting information from EXAFS data are discussed. Obviously, the first major step involves the extraction of the interference function $\chi(k)$ from the experimentally measured absorption spectrum. Once $\chi(k)$ is obtained, the type and extent of further analysis depend on one's goals and prejudices. If one wishes merely to compare qualitatively the environment of an absorption site in an unknown sample with that in a reference material, a simple comparison of $\chi(k)$ data may be sufficient. On the other hand, if the goal is to determine the types of neighbors and their precise distribution about an absorber in a sample of interest, then a much more extensive and sophisticated analysis is required.

The $\chi(k)$ function generally involves significant contributions from several coordination shells. While it is possible to extract structural information by attempting to calculate $\chi(k)$ directly [25], a more commonly used approach is to isolate the EXAFS contribution for a single shell by Fourier filtering the data, as explained in Section 3.2. Fourier-filtered single-shell data can then be analyzed by comparison with similarly treated experimental data for a reference material or with theoretical calculations. If the coordination shells are not sufficiently

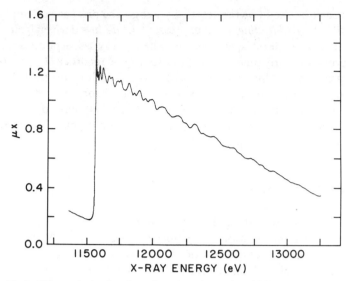

Figure 3.5 Typical X-ray absorption data plotted as absorption thickness μx versus X-ray energy. The plot shows the Pt L_3 edge in Pt metal at 80 K. (An arbitrary constant offset has been applied to the vertical scale.)

well separated in terms of interatomic distance, it may be necessary to Fourier filter several shells together and to apply a nonlinear least-squares fitting procedure to determine structural parameters. These various procedures are described below; for more detailed discussions, see [91] and [92, Lee and co-workers].

To illustrate some of the steps in the analysis, experimental measurements of the Pt L_3-edge EXAFS in Pt metal is used as an example. Figure 3.5 shows the raw absorption data for a sample temperature of 80 K. The absorption edge occurs at an X-ray energy of approximately 11,560 eV, and the EXAFS oscillations extend to at least 1600 eV above the edge.

3.1 Extraction of $\chi(k)$ Data

3.1.1 Removal of Glitches

Although they should be absent if the X-ray detection system is linear and the sample is uniform, sharp glitches caused by fluctuations in X-ray intensity coming from the monochromator are frequently present in the absorption data (see Section 4.3.4). A typical glitch involves one or a few data points and appears δ-functionlike relative to the absorption fine structure. If the data analysis involved only a straightforward least-squares fit to the raw data, the affected points could simply be excluded from the fitting process. However, because the normalized fine structure will be Fourier filtered, it is worthwhile to remove the glitches from the data. As long as the glitches are rapidly varying relative to the

underlying EXAFS, the distorted points can be replaced by a simple interpolation from the contiguous regions. Broad glitches or background fluctuations cannot be removed easily and probably indicate a problem in the
measurement.

3.1.2 Determination of E_0

The wave vector k for the photoelectron is obtained from the X-ray energy
$E = \hbar\omega$ using the relation

$$k = \sqrt{(2m/\hbar^2)(E - E_0)} = \sqrt{0.262(E - E_0)} \qquad (46)$$

where E_0 is the threshold energy for transitions to free photoelectron states and
E and k are measured in electronvolts and reciprocal angstroms, respectively. In
the muffin tin model for the scattering potential of a cluster of atoms, the
photoelectron energy is measured relative to the constant potential level
between atomic potential spheres. This muffin tin zero will vary relative to the
vacuum level depending on the types of atoms involved and the distances
between them. In analyzing experimental measurements, one must associate E_0
with some feature at the absorption edge; a common choice is the first inflection
point at the edge. Unfortunately, there is generally no simple way to associate an
appropriate (theoretical) muffin tin zero with an absorption edge feature for an
edge in an arbitrary material. If the EXAFS data is to be analyzed by
comparison with other experimentally measured data, then the exact choice of
E_0 is not crucial as long as it is made consistently; that is, the results of the
analysis will be relatively insensitive to the choice as long as the difference
between the appropriate muffin tin zero and the chosen E_0 is the same for all
measurements being compared. When the EXAFS is compared with theoretical
data, one must generally allow E_0 to be a free parameter that is varied to get the
best agreement between experiment and theory.

3.1.3 Pre-Edge Background Removal

Besides the absorption from the edge of interest, there is always background
absorption because of other edges in the same element as well as absorption by
other elements in the sample. To facilitate the comparison of fine structure in
measurements on different samples, it is often convenient to correct the data for
the background absorption. The energy dependence of the X-ray absorption
coefficient far from an edge can be parameterized fairly well using the empirical
Victoreen formula [91; 92, Koch and MacGillavry], $\mu(E) = CE^{-3} - DE^{-4}$.
However, the measured pre-edge region, from which the coefficients C and D
must be determined, is often limited, and extrapolation of such a formula over a
large energy range can be unreliable. Therefore, it is often more practical to
approximate the background by fitting a polynomial (a linear or quadratic
function is usually adequate) to the pre-edge region. The resulting curve is then
extrapolated through the region above the edge and subtracted from the data.

3.1.4   Normalization

The proper normalization for the EXAFS interference function is given by (4). Because the background absorption caused by all transitions $\mu_0(k)$ and that caused by the K edge alone $\mu_K(k)$ are generally unknown and must be determined from the data, where they are masked by the EXAFS oscillations, it is more convenient to normalize experimental data according to

$$\chi_{\text{exp}}(k) = \frac{\mu(k)x - \mu_0(k)x}{\Delta\mu(0)x} \tag{47}$$

where $\Delta\mu(0) = \mu_K(0)$ is the jump in absorption at the edge (edge step) and x is the sample thickness. The difference in normalization definitions is not important if the data is to be analyzed by comparison with other similarly normalized data. If, instead, the data is to be compared with theory, the k-dependent part of the normalization can be taken from tabulated X-ray absorption coefficients such as those of McMaster and co-workers [93].

Figure 3.6 shows the Pt L_3-edge absorption data after removal of a linear fit to the pre-edge region and normalization to the edge step. The position of the first inflection point, chosen as the value for E_0, has been subtracted, so that the energy scale now represents photoelectron energy.

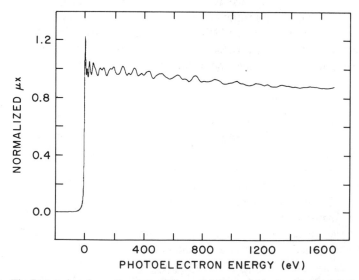

Figure 3.6   The Pt L_3-edge absorption data of Figure 3.5 after subtraction of a straight line fit to the pre-edge region and normalization to the edge step. The edge energy was subtracted from the X-ray energy scale to give the photoelectron energy.

3.1.5   Background Subtraction

The atomic background absorption $\mu_0(k)$ tends to be smoothly varying compared to the EXAFS; however, the absorption of an isolated atom can have some structure [94, 95], and there is no general formula to describe it. It is commonly assumed that the background absorption can be obtained by making a least-squares fit of some appropriate function to the normalized absorption data. As long as the Fourier components of the background curve are of lower frequency than those of all EXAFS oscillations, isolation of the atomic absorption is fairly well defined. One generally approximates the background curve with a polynomial or a segmented cubic spline function. The least-squares fit is typically performed in k space where the EXAFS oscillations have a roughly constant period and are more easily distinguished from structure in the background absorption.

In Figure 3.7 the normalized Pt absorption data are shown on an expanded scale plotted versus the photoelectron wave vector k. The smooth line passing through the EXAFS oscillations is a segmented cubic spline function fit to the data and assumed to describe the atomic background absorption. Subtraction of the background curve yields the EXAFS interference function in Figure 3.8.

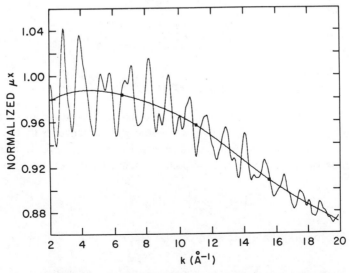

Figure 3.7 The normalized Pt L_3-edge EXAFS as a function of the photoelectron wave vector k. The smooth line passing through the EXAFS oscillations is the background curve (assumed to represent the absorption of an isolated atom) that was fitted to the data.

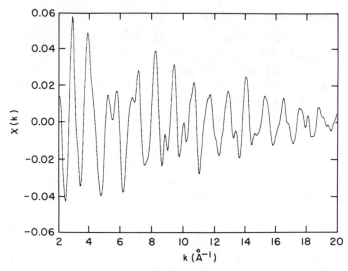

Figure 3.8 The EXAFS interference function $\chi(k)$ obtained for the Pt L_3 edge in Pt metal at 80 K.

3.2   Fourier Filtering

3.2.1   Fourier Transform of $\chi(k)$

The EXAFS formula has the basic form

$$\chi(k) = \sum_j |A_j(k)| \sin[2kR_j + \psi_j(k)]$$
$$= \text{Im} \sum_j A_j(k)e^{i2kR_j}$$

(48)

where

$$A_j(k) = |A_j(k)|e^{i\psi_j(k)}$$

(49)

This form suggests that contributions from different coordination shells can be separated by Fourier transforming $\chi(k)$. While $\chi(k)$ is theoretically defined only for $k > 0$, it is convenient to assume, momentarily, that it is defined for all k, with $A_j(k)$ symmetric about the origin. We then consider the Fourier transform

$$\tilde{\chi}(r) = \frac{1}{2\pi} \int_{-\infty}^{\infty} dk\chi(k)e^{-i2kr}$$

(50)

Substituting for $\chi(k)$ the expression on the second line of (48), one obtains

$$\tilde{\chi}(r) = \frac{1}{2i} \sum_j [\tilde{A}_j(r) * \delta(r - R_j) - \tilde{A}_j^*(r) * \delta(r + R_j)]$$

$$= \frac{1}{2i} \sum_j [\tilde{A}_j(r - R_j) - \tilde{A}_j^*(r + R_j)]$$

(51)

where the $*$ in the first line stands for convolution, and $\tilde{A}_j(r)$ is the Fourier transform of $A_j(k)$. Because $A_j(k)$ tends to peak at small k and decay rapidly to zero as $k \to \infty$, $\tilde{A}_j(r)$ should have a single main peak near the origin, decaying rapidly as r gets large. One can then see that for $r > 0$, $|\tilde{\chi}(r)|$ consists of a series of peaks centered at the various coordination shell distances. Actually, the peaks will be shifted slightly to the low r side of the expected positions because of the k dependence of the phase factor. One can sometimes use a linear approximation for the phase-shift factor, $\psi_j(k) \approx \alpha_j - \beta_j k$, in which case $\tilde{A}(r - R_j)$ is centered at $R_j - \beta_j/2$. There are also complex conjugate peaks $\tilde{A}_j^*(r + R_j)$ located at $-(R_j - \beta_j/2)$, but these contain no independent information and can be ignored.

The Fourier transform function $\tilde{\chi}(r)$ is an effective radial distribution function, with the peaks for the different coordination shells separated according to radial distance. The Fourier transform of each single-shell contribution, $\tilde{A}_j(r - R_j)$, has a relatively narrow central peak but may have broad oscillating tails that depend on the shape of the EXAFS in k space. For further analysis, it is important to isolate individual peaks as much as possible, which means that the wings on the transform peaks, which tend to overlap with neighboring peaks, must be minimized. This can be done by modifying $\chi(k)$ before transforming. One useful type of modification is to multiply $\chi(k)$ by k raised to a power w. The exponent w should be chosen such that $k^w \chi(k)$ has a roughly constant amplitude over the whole data range. Another modification is multiplication by a *window* function, which has a value of one in the middle and tapers off to a small value at the edges of the data range. By shaping the EXAFS amplitude envelope to peak

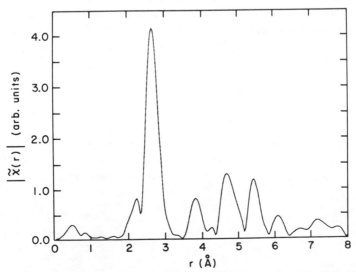

Figure 3.9 Magnitude of the Fourier transform of $\chi(k)$ for the Pt L_3 data at 80 K. The $\chi(k)$ data were multiplied by k^2 and weighted by a Gaussian window centered on the data range and decaying to 10% at the edges. The splitting of the first shell peak at 2.7 Å is caused by structure in the Pt backscattering amplitude.

at midrange and taper off toward the edges one reduces the amplitudes of the wings of $\tilde{A}_j(r - R_j)$, with the consequence that the central peak is broadened. Two common types of window functions are the Hanning function [91, 96], which falls off as $\cos^2 ka$, and a Gaussian centered on the data region [97]. If the backscattering atoms vary significantly in Z, the use of k weighting or a window function can emphasize the contributions of one type of atom over those of another because of the variation in backscattering amplitudes with Z.

As one must determine the Fourier transform with a computer, it is necessary to work with a finite, discrete transform rather than an infinite, continuous one. Furthermore, for speed, a fast Fourier transform (FFT) algorithm is generally used. (For details on discrete transforms and the FFT, see Brigham [96].) Two important points about the finite, discrete Fourier transform are that the range of r space is inversely proportional to the step size in k space, while the step size in r space is inversely proportional to the k range used.

Figure 3.9 shows the magnitude of the Fourier transform, $|\tilde{\chi}(r)|$, of the Pt EXAFS data at 80 K. The $\chi(k)$ data were weighted by k^2 and multiplied by a Gaussian window function decaying to 10% at the window edges before transforming. The crystallographic distances and effective peak positions of the first four coordination shells are listed in Table 3.2. The splitting of the first shell peak is caused by structure in the Pt backscattering amplitude.

3.2.2 The Inverse Transform

While it is possible to extract structural information in r space [98], analysis is somewhat more straightforward if one returns to k space. The Fourier transform serves to spectrally isolate the information from different coordination shells. To obtain single-shell EXAFS $\chi_j(k)$, one can multiply the complex transform by a square window that is zero everywhere except for the region that covers the desired shell. Looking back at (51), if one inverse transforms only the function $\tilde{A}_j(r - R_j)$, the result is a complex quantity $z_j(k)$ given by

$$z_j(k) \approx \frac{1}{2i} A_j(k)e^{i2kR_j} \qquad (52)$$

Table **3.2** Coordination Numbers N_j, Interatomic Distances R_j, and Effective Peak Positions in the Fourier Transform R'_j for the First Four Coordination Shells in Face-Centered Cubic Pt Metal

j	N_j	R_j	R'_j
1	12	2.75	2.66
2	6	3.88	3.82
3	24	4.76	4.69
4	12	5.49	5.43

This formula is only approximate because perfect spectral isolation is impossible. The largest discrepancies occur at the ends of the k range of the data. If the single-shell EXAFS is written as

$$\chi_j(k) = B_j(k) \sin \Phi_j(k) \tag{53}$$

then the amplitude and phase are given by

$$B_j(k) = |z_j(k)|$$

$$\Phi_j(k) = \tan^{-1}\left[\frac{\text{Im } z_j(k)}{\text{Re } z_j(k)}\right] + \frac{\pi}{2} \tag{54}$$

If neighboring shells overlap significantly in the Fourier transform, it is not appropriate to inverse transform the shells separately. Instead, they must be back-transformed as a group. The analysis of interfering shells is more complicated than that for isolated shells, but often it cannot be avoided.

The filtered first-shell data for Pt at 80 K, obtained with an r-space window of 1.7–3.3 Å, are shown in Figure 3.10. The oscillations are much more regular and less complicated than in the full $\chi(k)$ of Figure 3.8. Fluctuations in the amplitude envelope are caused by structure in the Pt backscattering amplitude.

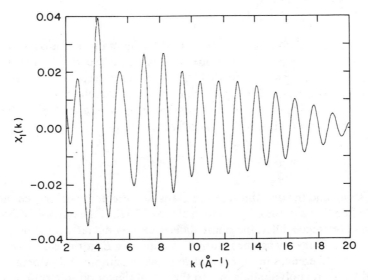

Figure 3.10 Fourier-filtered, first-shell Pt L_3 edge EXAFS data. The r-space window for the inverse transform was 1.7–3.3 Å^{-1}.

3.3   Analysis of Fourier-Filtered Data

3.3.1   The Ratio Method

With the Fourier-filtered, single-shell χ data decomposed into amplitude $B(k)$ and phase $\Phi(k)$ (see Section 3.2.2), one can extract structural information from them using the ratio method [54, 99]. Comparing (53) with (5)–(7), one can see that

$$B(k) = \frac{S_0^2 N}{kR^2} F(k)Q(k)e^{-2R/\lambda} \qquad (55)$$

and

$$\Phi(k) = 2kR + 2\delta_1^c(k) + \theta(k) + \phi(k) \qquad (56)$$

From the amplitude one would like to obtain the number of nearest neighbors and the disorder amplitude factor $Q(k)$, but to do so requires knowledge of the quantities S_0^2, $F(k)$, and $\lambda(k)$. Often these can be determined empirically from a reference compound with the same combination of absorbing and scattering atoms. If the interatomic distances and atomic valences are similar in the sample and reference, then the factors S_0^2, $F(k)$, and $\lambda(k)$ should be very nearly the same for the two single-shell data sets. If one takes the logarithm of the ratio of amplitudes $B_s(k)$ and $B_r(k)$, where s and r denote sample and reference, respectively, then the result is

$$\ln\left[\frac{B_s(k)}{B_r(k)}\right] = \ln\left[\frac{N_s R_r^2}{N_r R_s^2}\right] - \frac{2(R_s - R_r)}{\lambda} - 2(\sigma_s^2 - \sigma_r^2)k^2 \qquad (57)$$

where it is assumed that the pair distribution function is Gaussian and that $\lambda = \lambda_r = \lambda_s$. If the log of the amplitude ratio is plotted versus k^2, a straight line with a slope of $-2\Delta\sigma^2$ and (provided $R_s \approx R_r$) a y intercept of $\ln(N_s/N_r)$ should generally result. Similarly for the phase factor, one can eliminate the unknown scattering phase shifts $\delta_1^c(k)$ and $\theta(k)$ by comparing with a reference phase. The phase difference between sample and reference is

$$\Phi_s(k) - \Phi_r(k) = 2k(R_s - R_r) + \phi_r - \phi_s \qquad (58)$$

If $\Delta\sigma^2$ is small and the distributions are Gaussian, the phase difference should be a straight line through the origin with a slope of $2\Delta R$. If the energy origins of the sample and reference differ, the phase difference may deviate from such a line at low k [91, 92]. Linearity can be restored by adjusting E_0 for one data set relative to the other. Differences in N, σ^2, and R between sample and reference can be obtained by fitting polynomials in k to the log of the amplitude ratio and to the phase difference.

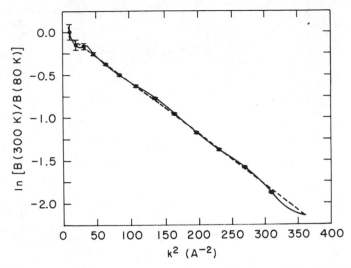

Figure 3.11 First-shell amplitude ratio for Pt metal comparing data at 300 and 80 K. The symbols and solid line represent the natural logarithm of the amplitude ratio plotted versus k^2. The dashed line is a polynomial fit to the data having the form $y = a_0 + a_2 k^2$.

As a simple example of the ratio method, Figure 3.11 shows the ratio of first-shell amplitudes for Pt metal at 300 and 80 K. The symbols and solid line represent the data, while the dashed line is a straight line fit to the data. The fit gives a y intercept of 0.01 ± 0.01 and $\Delta\sigma^2 = (2.99 \pm 0.05) \times 10^{-3} \text{Å}^2$. The linearity of the log ratio and the nearly zero intercept indicate that the unknown amplitude factors cancel out quite accurately in this case. The phase difference is presented in Figure 3.12. The dashed line represents a polynomial of the form $y = ak + bk^3$, which was fit to the data (symbols and solid line). From the fit one finds a distance change $\Delta R = 0.003 \pm 0.001$ Å, consistent with thermal expansion, and a difference in the third cumulant $\Delta C_3 = (4.9 \pm 1.5) \times 10^{-5} \text{Å}^3$, caused by a slight anharmonicity of the interatomic potential at the higher temperature. Although the third cumulant term appears significant here only because of the small distance change and the long data range, neglect of the k^3 term in analysis of the phase difference will in this case result in an apparent contraction of the lattice with increasing temperature.

3.3.2   Curve Fitting

It often happens that for a sample of interest it is not possible to isolate single shells in r space; hence, the ratio method is of little use. Here one must Fourier filter a group of two or more shells and rely on nonlinear least-squares curve fitting to extract information. A $\chi(k)$ function is calculated using reference amplitudes and phases, and the unknown structural parameters are varied until the calculated function is in good agreement with the Fourier-filtered data. The parameter values giving the best fit are assumed to be the correct ones.

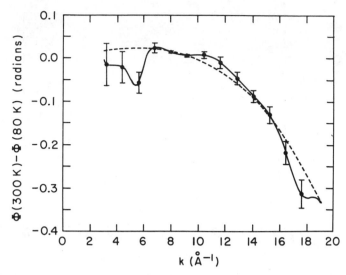

Figure 3.12 Difference between first-shell phases in Pt metal at 300 and 80 K. The symbols and solid line represent the data. The dashed line is a polynomial fit to the data having the form $y = a_1 k + a_3 k^3$.

The reference amplitudes and phases are often obtained empirically from measurements on reference compounds, as discussed for the ratio method. For each shell contributing to the calculated χ function, one may vary the parameters N, R, σ^2, and possibly E_0. The number of independent data points is generally limited; however, and it is important not to vary more parameters than the number of degrees of freedom available. It may be necessary to constrain the number of fitting parameters through the use of independent knowledge about the system of interest. Another problem that can arise involves strong correlations between the parameter pairs (N, σ^2) and (R, E_0) for a given shell. Although parameter correlations may not be important in fitting ideal noise-free data, one must be careful in interpreting the parameter values obtained in practice.

An alternative to empirical amplitudes and phase shifts is the use of theoretical calculations. While theoretical amplitudes and phases can be calculated for any central and backscattering atom pair, one must generally make corrections for the many-body quantities S_0^2 and λ. One must also be cautious in the application of theoretical phases and amplitudes calculated with the plane-wave approximation, as this approximation is not valid at very low k values. The use of curved-wave calculations should yield more satisfactory results [25].

3.3.3   Accuracy of Results

The accuracy of structural parameters extracted from EXAFS data depends on many factors such as the number of parameters being determined, the amount of disorder, the length of the data range, and the quality of the reference data. In the simplest case where one looks for changes in a single, isolated shell as a function of a parameter such as temperature, one may be able to determine a change in distance to within a few thousandths of an angstrom and to obtain the change in the mean-square relative displacement to within a few percent of σ^2. On the other hand, in a situation where several shells must be fit simultaneously and the data range is restricted, one must be satisfied with much less accurate results. Although it is difficult to quote general values, typical accuracies for the determination of parameters in an unknown system are 1% for interatomic distances, 15% for the coordination numbers, and 20% for σ [100].

4   EXPERIMENTAL DETECTION OF EXAFS

Several techniques for detecting the EXAFS signal were developed. They can be classified into two categories: direct techniques in which the X-ray absorption is actually measured, and indirect techniques in which a signal secondary to the absorption process is measured. Indirect techniques rely on there being a direct proportionality between an absorption event and the secondary process being measured. This is true for the emission of fluorescent photons and Auger electrons. In Section 4.1 these and other techniques are described, along with their areas of application. Usually simple statistical arguments can be used to compare the techniques and to determine the most appropriate one for a given experiment. Examples of such comparisons are given in Section 4.2. As with most experiments there are some common experimental pitfalls, and these are discussed in Section 4.3.

4.1   Detection Methods

Of the many techniques for measuring EXAFS the most obvious and widely used method is to measure the absorption of the sample by monitoring the incoming I_0 and transmitted I flux. The absorption is then given by

$$\mu x = \ln(I_0/I) \qquad (59)$$

This method is simple to apply and utilizes all the photons incident on the sample. It is often possible to apply it in conjunction with other detection methods.

The absorption signal can be detected in two ways. The most common uses a monochromator to select a small energy range ΔE and monitors I and I_0 with suitable detectors. The full spectrum is acquired by sequentially stepping the monochromator through the required energies. The second technique allows the full EXAFS energy range (ca. 1000 eV) to impinge on the sample and uses a crystal to disperse the different energies spatially [101–104]. Examples of the various geometries possible are shown in Figure 3.13. A position-sensitive X-ray detector is used to detect the signal. The resulting energy resolution is determined by the spatial resolution of the detector and the dispersive power of the crystal.

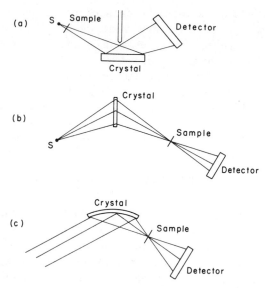

Figure 3.13 Three possible geometries for carrying out dispersive EXAFS experiments using a spatially resolving linear detector: Parts (a) and (b) are most appropriate for laboratory setups, while part (c) is commonly used with synchrotron radiation.

In principle the statistical errors of the two methods are the same since the total number of photons collected per unit energy is the same. In the step mode the total output of the X-ray source at a particular energy is counted for a small portion of the scan time, while in the energy-dispersive mode, because of the need to use Bragg reflectors to disperse the energies, an equivalently small fraction of the output is counted for the total scan time. In practice, the time overhead incurred in stepping the monochromator can be significant for concentrated samples. If the available intensity is large enough, we can save significant time by using the energy-dispersive technique, and its greatest applications are in the study of transient events or short-lived samples. Indeed using laser-driven X-ray sources we can obtain time resolutions approaching nanoseconds [101]. To date the sequential mode yields the most accurate data and is most widely used.

Indirect techniques measure processes that are proportional to the absorption. Two of these, X-ray fluorescence and emission of Auger electrons, are illustrated in Figure 3.14. Monitoring these processes is useful when the EXAFS signal is only a fraction of the total absorption. Isolating the EXAFS from the background in transmission then requires the subtraction of two nearly equal signals, a process requiring very accurate data. Thus, it is desirable to enhance the signal of interest by using a detector that can selectively measure a secondary process.

Auger emission and X-ray fluorescence are competing processes. Their relative strengths depend on the atomic number of the absorber. In light

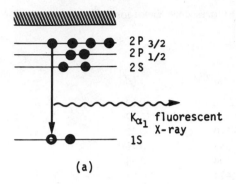

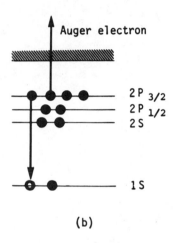

Figure 3.14 Schematic illustration of the decay of the core hole left behind by the excited photoelectron: (*a*) fluorescence decay and (*b*) Auger decay.

elements Auger emission is more probable, while for heavy elements fluorescence is more likely. For the same element fluorescence is more likely for K-shell holes than for L-shell holes. Table 3.3 gives examples of the fluorescence yields for various elements.

As seen in Figure 3.14, the fluorescence radiation results from the filling of the core hole generated by the absorption of an X-ray photon. For the K shell this is dominated by the production of K_α radiation. The energy of this radiation is characteristic of the absorbing element and is less than the original exciting radiation. The background consists of elastically and Compton-scattered radiation, both of which are at higher energies than the fluorescent line. Thus, a suitable energy-dispersive detector can discriminate the background from the signal. It will be shown that the fluorescence technique gives better data than simple transmission when the absorption from the element of interest is less than a few percent of the total absorption in the sample.

Application of the fluorescence technique is not as straightforward as for absorption since, in general, corrections must be made to the measured signal

Table 3.3   *K*- and *L*-Shell Fluorescence Yields for Various Elements[a]

Element		
K Edges	*Energy (keV)*	*Yield*
$_{18}$Ar	3.202	0.12
$_{20}$Ca	4.038	0.17
$_{25}$Mn	6.540	0.32
$_{30}$Zn	9.659	0.49
$_{32}$Ge	11.104	0.54
$_{35}$Br	13.474	0.63
$_{42}$Mo	19.999	0.76
$_{48}$Cd	26.711	0.84
$_{64}$Gd	50.240	0.93
$_{80}$Hg	83.103	0.96
$_{92}$U	115.603	0.97
L₃ Edges		
$_{32}$Ge	1.218	0.015
$_{48}$Cd	3.537	0.059
$_{64}$Gd	7.243	0.16
$_{80}$Hg	12.284	0.33
$_{92}$U	21.756	0.49

[a]Adapted from M. H. Chen and co-workers, *Phys. Rev. A,* **21**, 436 (1980) [105] and M. O. Krause, *Phys. Rev. A,* **22**, 1958 (1980) [106].

$\mu'(E)$. For a thick sample it can be shown

$$\mu'(E) = \frac{I_f}{I_0} \sim \frac{\mu(E)\sin\theta}{[\mu_T(E)/\sin\theta] + [\mu_T(E_f)/\sin\phi]} \qquad (60)$$

where $\mu_T(E)$ is the total absorption coefficient at energy E, E_f is the fluorescence energy, θ is the entrance angle of the incidence X rays, and ϕ is the exit angle of the flourescent X rays. This expression should be integrated over the angles ϕ subtended by the detector. For dilute samples $\mu_T(E)$ is nearly constant, and the correction to $\mu'(E)$ to obtain $\mu(E)$ is small and can be made using tabulated absorption coefficients. Other possible corrections are discussed in Section 4.3.

For *K* edges below about 2000 eV in energy, the probability for fluorescence becomes small, and detection of electrons emitted by the sample may be preferable. This is particularly the case for studies of surface regions since the escape probability for electrons decreases strongly for increasing depth into the sample. Therefore, electron detection may also be desirable for higher energy edges when enhancement of the surface signal is necessary [107].

Detection of Auger electrons [108–110] is analogous to fluorescent detection in that a single line that is characteristic of a particular element is measured. Again, a good discrimination of background is possible, and since the Auger electrons can only escape from within about 30 Å of the surface, a strong surface enhancement is achieved. The Auger yield is related to μ by a relation similar to (60) with $\mu_T(E_f)$ becoming the attenuation coefficient for the Auger electrons μ_A. However, in this case $\mu_A \gg \mu_T(E)$, which means I_{Auger}/I_0 should be proportional to μ to a good approximation. There were some unexplained amplitude distortions in the original experiments of Citrin and co-workers [110], and recent theoretical work indicates that the proportionality between absorption and Auger yield may not be exact [111]. Other possible distortions are discussed in Section 4.3.

Other electron detection channels can be used to obtain EXAFS information. An obvious choice is the detection of the primary photoelectron emitted as the X ray is absorbed. However, the angular distribution of the photoelectrons varies as the photon energy is changed. This introduces an additional variation that causes a different dependence with energy than EXAFS, unless the photoelectron is collected over a full 4π solid angle. Since it is only possible to collect at most a 2π solid angle, such a photoelectron monitoring is unreliable [108, 109], although the photoelectron signal has been used to obtain EXAFS-like signals [112]. A better technique is to measure the total yield of electrons emitted by the surface. In this case the complicated cascade processes involved in producing the secondary electrons provide a form of angular averaging. Such total yield [113, 114] and related partial yield [115] measurements were shown to give fine structure analogous to EXAFS. When the total yield signal was compared with the Auger yield for Al, a small but significant amplitude deviation was found [116]. Thus, the total yield signal may not be exactly the same as EXAFS, but it certainly contains much useful information, especially concerning the near-neighbor distances. Also, uncertainties in absolute amplitudes do not rule out measuring the amplitude dependence on polarization from which important coordination information can be inferred.

Electron detection is not the only method of enhancing surface sensitivity. Since the index of refraction for X rays is slightly less than 1, at glancing incidence angles they will undergo total external reflection. When this occurs the X rays penetrate only 20–30 Å, and the potential of using total external reflection for surface studies was first pointed out by Parratt [117]. A major advantage is the greater penetration ability of X rays, which means a vacuum environment is not required. This opens up new experiments such as surfaces in gaseous or liquid environments [118], which is an important advantage for studies of chemical reactions on surfaces. Similarly the reflection can be made to occur at a solid–solid interface if there is a significant density difference, which allows interface structures to be probed [119].

As with bulk EXAFS the signal can be measured by monitoring either the reflected X rays [120–122] or the fluorescence signal [123]. The signals are larger for the reflected beam, but recent work [123] indicates that fluorescence detection is more useful for surface EXAFS studies. To understand why, it is necessary to consider the reflection process in more detail. The index of refraction for X rays can be written $n = 1 - \delta - i\beta$, where the absorption coefficient is contained in β. However, the reflectivity depends on n as given by the Fresnel equations [117, 120]. Therefore, to extract μ from the measured reflectivity it is necessary to determine δ. It was found by Martens and Rabe [120] that the EXAFS oscillations also show up in δ, which makes a determination of β a difficult task. A proper experiment requires a detailed measurement of the reflectivity as a function of both energy and angle. While the fluorescence signal also depends on β and δ, the EXAFS is a larger fraction of the total signal. Thus, the corrections can be less accurate, and simpler approximations can be used. When very thin surface layers are studied, the reflectivity is dominated by the substrate and the distortions are small.

The fluorescence signal is directly proportional to the probability for absorption. The absorption probability is determined by the strength of the evanescent wave, which penetrates into the substrate [124]. Below the critical angle the evanescent wave penetration depth is typically 30 Å, but its amplitude depends on the incident angle θ. This can be determined by matching the evanescent wave field to the standing wave field set up by the incoming and reflected beams. It is found that for an ideal surface (100% reflectivity) the evanescent wave amplitude varies from 0 at $\theta = 0$ to twice the incident amplitude for $\theta = \theta_c$, where θ_c is the critical angle. The absorption probability depends on the intensity of the evanescent wave, and is, therefore, enhanced by a factor of four near θ_c over the case where no reflection is occurring.

These points are illustrated in Figure 3.15, which shows data taken for approximately monolayer Au films evaporated onto glass and Ag substrates. The reflectivities are plotted in Figure 3.15a. Since the Au films are thin the reflectivity is determined by the substrate, and the more dense Ag substrate has a larger critical angle. The fluorescent and background signals are plotted as a function of angle in Figure 4.15b. As expected the fluorescent signal increases up to the critical angle where it drops by about a factor of 3. The enhancement in this case is not fourfold because the surface is not ideal. Above the critical angle the incoming beam begins to penetrate and the scattered background increases, but the signal-to-background ratio is still respectable. With these counting rates good EXAFS spectra can be obtained in a few seconds per point.

In this example of monolayer films the problem of converting the reflected signal into an absorption is reduced because the reflectivity and δ are dominated by the substrate. However, this means that the change in reflectivity in crossing the Au L_3 edge is small, reducing the signal-to-noise ratio (S/N). A more serious problem, however, is that small changes in the substrate reflectivity with energy dominate the EXAFS, effectively rendering it unusable. It was concluded that because of these problems fluorescence detection is always better than monitoring the reflected signal to determine the EXAFS in the glancing incidence case [123].

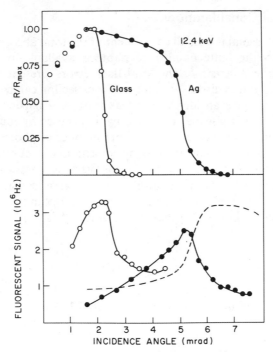

Figure 3.15 (a) X-ray reflectivity for an Au coverage of 2.6×10^{15} atoms/cm^2 on glass and Ag-coated glass. (b) Au L_3 fluorescence signals obtained from the samples in (a) for an incident flux of ca. 5×10^8 photons/s. The dashed line is the scattered background for the Ag substrate.

Some additional methods for measuring EXAFS are under development. The first is to detect the optical photons generated by the absorption of the X rays [125, 126]. The theoretical treatment of the origin of these optical photons is complicated, but such a technique was shown to give EXAFS information in some samples. A possible advantage of the technique is the selective detection of inequivalent sites whose optical activity is different [127].

In solution studies it is sometimes possible to use the solution as a detector by monitoring its photoconductivity [128, 129]. This technique was shown to give EXAFS information and may be useful in some cases. An advantage of the technique is the ability to detect EXAFS in a nontransmissive or black cell, which may make it useful for monitoring solutions in complicated environments.

Finally, surface EXAFS can sometimes be measured by monitoring the ion yield from the sample [130]. The crucial question here is whether the ion yield is proportional to the absorption coefficient, and it appears that the answer depends on the system being studied. Further experimental and theoretical work is required to determine the applicability of ion yield detection to EXAFS studies.

4.2   Statistical Considerations

Since the EXAFS signal is a small component of the total absorption, statistical errors caused by the finite number of photons available are an important consideration. The analysis of the statistical errors present in the various measurement techniques discussed in the previous section can be used to choose between them, and give an idea of the feasibility of a particular experiment. While other sources of noise may overshadow that caused by counting statistics, the ultimate limitation is caused by the finite number of photons available.

For absorption measurements the optimization of counting statistics is discussed by Rose and Shapiro [131] for various cases. In an EXAFS absorption measurement the two quantities to be determined are the fraction f of the incident beam used to determine I_0 and the optimum thickness of the sample. It is easy to show [132] that the optimum f is

$$f = \frac{1}{1 + e^{\mu x/2}} \tag{61}$$

and the resulting S/N in measuring the ratio $R = I_0/I$ is

$$\frac{S}{N} = \frac{I_0^{1/2}}{1 + e^{\mu x/2}} \tag{62}$$

For EXAFS experiments the important quantity is not a measurement of R, but a measurement of changes in R caused by small changes in μ. The optimum μx for measuring $dR/d\mu$ is $\mu x \simeq 2.6$, but reducing distortions caused by thickness effects often require a smaller value. These are discussed in the Section 4.3.1.

For dilute samples the absorption coefficient can be written $\mu = \mu_s + \mu_0$, where the μ_s is the absorption of the dilute species being measured and $\mu_0 \approx \mu$ is the background absorption from all other atoms. Hence, the relevant S/N is that in measuring μ_s rather than μ. The S/N thus becomes

$$\left(\frac{S}{N}\right)_{\mu_s} = \left(\frac{I_0}{2}\right)^{1/2} \frac{\mu_s x}{e^{\mu x/2} + 1} \tag{63}$$

In an ideal fluorescence experiment there is no background, and simplifying equation (60) gives

$$\left(\frac{S}{N}\right)_{\mu_s} = \left(\frac{I_0 \alpha \mu_s}{\mu + \mu_F}\right)^{1/2} \tag{64}$$

where μ_F is the absorption coefficient for fluorescent photons and α is a proportionality constant roughly equal to the product of the fluorescent yield with the collection efficiency of the detector. The two techniques are equal when

$$\mu_s x = \frac{2(e^{\mu x} + 1)^2}{(\mu + \mu_F)x} \alpha \tag{65}$$

Typical values are $\alpha \simeq 0.02$ and $\mu x \simeq \mu_F x \simeq 2.6$, which give $\mu_s x \simeq 0.18$ or $\mu_s/\mu \simeq 0.07$. For smaller values of $\mu_s x$ the fluorescence technique is more efficient.

Equation (65) holds if the detector can discriminate the fluorescent photons from the large component of scattered photons. This requires energy resolution of about 5%, which is achievable only by solid-state detectors. These can be used [133], but they are limited to a total (fluorescence plus background) counting rate of about 2×10^4. Since the fluorescent signal is often much smaller than the background, it is desirable to eliminate the background photons before they reach the detector. Two methods for doing this are shown in Figure 3.16. Either X-ray filters [132, 134] or Bragg reflection can be used. Bragg reflection is very efficient at eliminating the background, but requires a precise alignment of the crystal angle with respect to the incoming fluorescent photons. Thus, to date the best designs only collect approximately 1% of the available photons [135, 136]. The principle of X-ray filters is shown in Figure 3.17. Here the large change in

(a)

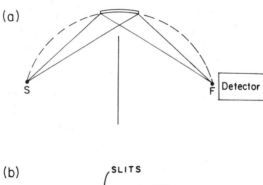

(b)

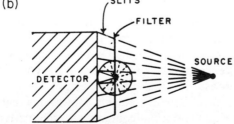

Figure 3.16 Methods for energy discriminating the fluorescence signal: (a) Bragg reflection and (b) filter-slit combination. The slits are designed such that fluorescence from the filter is restricted, while fluorescence from the sample is unattenuated.

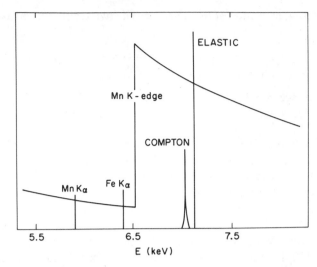

Figure 3.17 Example of the application of a Mn filter to detection of Fe fluorescence. The elastic and Compton background are above the Mn absorption edge and strongly attenuated, while the attenuation of the fluorescence is small.

absorption at an X-ray edge is used to selectively attenuate the background photons. A problem is that the filter reemits many of the absorbed photons as its own fluorescence. To prevent most of these filter fluorescence photons from reaching the detector a set of slits, as shown in Figure 3.16, are used. These reduce the solid angle through which the detector views the filter while permitting essentially all of the fluorescent photons from the sample to pass through.

To compare these detectors a concrete example is useful. A typical concentration in, for example, an Fe metalloprotein is 1 mM. For a total thickness of 2.5 absorption lengths the change in absorption at the Fe K edge is only 2×10^{-3} or about 0.1%. To extract the small EXAFS modulations at the Fe K edge a precision in measuring the Fe component of 10^{-3} is required; therefore, a total precision in measuring μx of about 10^{-6} is required, which is currently impossible.

In fluorescence the task is easier. An unfocused synchrotron beam line might provide 10^{10} photons/s at the sample. Thus, 10^{7} photons/s are absorbed by the Fe component of the sample, and since the Fe fluorescence yield is 0.3 there are about 3×10^{6} fluorescence photons/s. Ideally, a 1-s counting time gives a statistical accuracy in measuring the Fe component of 5×10^{-4}, which is more than adequate for most EXAFS measurements. However, a real detector cannot collect the entire solid angle and must contend with the scattered background. For the present example of low z-scattering atoms the scattering cross section is about 2% of the absorption cross section, resulting in about 2×10^{8} scattered photons or a background-to-signal ratio of 66. This degrades the accuracy to 5×10^{-3}.

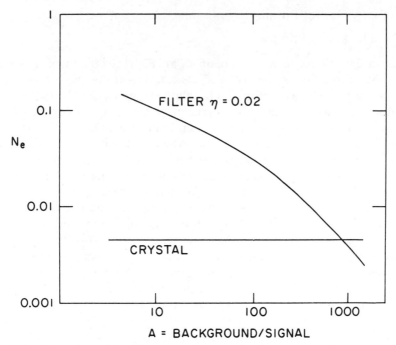

Figure 3.18 Comparison of the effective counting rates for the crystal and filter-slit fluorescence detectors. The crystal calculation assumes a solid angle of $\Omega/4\pi = 0.02$ and a 50% diffraction efficiency, while for the filter the equations in [134] were used to determine the optimum filter thickness and N_e for each value of A.

A crystal detector has essentially perfect background discrimination, but it only collects 1% of the signal or 3×10^4 photons/s. A 1-s integration gives an accuracy for μ_{Fe} of 6×10^{-3}—less than ideal, but approaching feasible levels particularly if longer counting times are used. A typical filter system collects about 30 times more solid angle but lets some background through. Using the equations given in [134] we find that the optimum filter thickness will let through 4×10^5 signal counts and 2.4×10^6 background counts. This gives an accuracy of 3.5×10^{-3}—nearly twice that of the crystal. In terms of counting time the improvement is a factor of 3.

For more dilute samples the background increases and the filter system becomes less effective. Figure 3.18 shows a comparison of the two systems. For the parameters used the two detectors are equal for a background-to-signal ratio of approximately 1000. This would correspond to about a 50-μM sample.

The preceding numbers correspond to a particular case, and, of course, vary for each experiment. However, relevant cross sections are well known [93] so that similar calculations can be made for almost any planned experiment, including electron-detection techniques. Such estimates are particularly important for planning synchrotron experiments where available beam time is often limited.

4.3   Experimental Pitfalls

4.3.1   Thickness Effects

The most difficult parameter to obtain in an EXAFS experiment is accurate amplitude information. A variety of effects can distort the amplitudes severely while leaving the phase information relatively undisturbed, and it is the amplitude information that is often crucial in developing a complete understanding of an unknown system. The amplitude contains information about coordination number; site disorder; and, when polarization studies are made, site symmetry. As the conditions necessary for amplitude transferability become better understood [23], the experimental limitations on the accuracy of EXAFS amplitudes become more important. Fortunately, with proper experimental design inaccuracies in EXAFS amplitudes can be recognized and corrected.

Most of the amplitude distorting factors in an absorption measurement fall under the general heading of *thickness effects* [67, 137–139]. Basically a thickness effect occurs whenever some part of the incoming beam is not attenuated by the sample in the normal fashion. For example, radiation can pass around the sample, through pinholes in the sample, or be at a different energy from the primary radiation. This *leakage* becomes a larger fraction of the total signal as the sample becomes thicker and, thus, the absorption signal appears to depend on sample thickness, which is the origin of the term thickness effect.

Figure 3.19 shows the calculated thickness effects for various levels of leakage based on formulas derived in [138]. Plotted are the errors in both μx and $\chi(k)$ for various leakage levels. Notice that for the optimum μx of 2.6 serious distortions in $\chi(k)$ can occur for leakage levels of approximately 1%. The harmonic content of most monochromators can easily exceed these values unless we take special precautions. Therefore, it is often desirable to use somewhat thinner samples or to employ a harmonic monitor to ensure the harmonics are tuned out.

Another type of leakage originates from the sample. When an X ray is absorbed, it can be reemitted as a fluorescent photon. If this photon reaches either the I or I_0 detector, the measured amplitude is distorted. The distortion is usually most serious for the I detector and depends on the solid angle subtended by the detector [132, 140]. Figure 3.20 shows the magnitude of the effect for various values of $\varepsilon\Omega/4\pi$, where Ω is the solid angle subtended by the detector and ε is the fluorescence efficiency. In most cases the effect of sample fluorescence can be minimized by reducing Ω.

Many solid samples are made from powders. For concentrated materials optimum sample thicknesses are about $10\,\mu$m, which is also the grain size of many powders. This means powdered samples can be very nonuniform resulting in distortions of the EXAFS signal. This problem was studied by Stern and Lu [141], who derived an expression for the distortion for spherical particles. Measurements confirmed the general validity of their result, and it was concluded that the edge step of a single layer of powder should be less than 0.1 to minimize distortions. Such a single layer of powder is usually prepared by

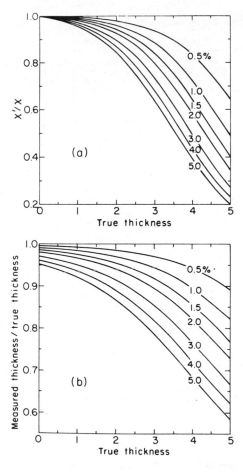

Figure 3.19 Reduction of the measured $\chi(k)$ (a) and absorption coefficient (b) for various levels of leakage.

rubbing the powder onto tape. This technique preferentially chooses the smaller grains in the powder, and the preceding condition can often be satisfied.

From the above considerations it is clear that if care is not taken, the EXAFS amplitudes may be distorted. Generally, however, the necessary precautions are not difficult to apply, and thickness effects can always be checked for by measuring different thicknesses of the same sample. If the results agree, there is a good probability that the amplitudes are correct.

4.3.2   Fluorescence Measurements

For fluorescence EXAFS measurements thickness effects are unimportant both because the sample $\Delta\mu x$ is small and because leakage through the sample has little effect on the fluorescence signal, but other corrections apply. From (60) it is seen that the measured $\mu'(E)$ must be corrected for the self-absorption in the

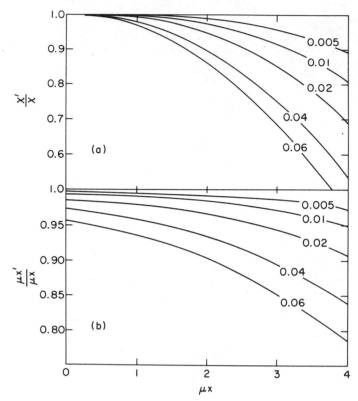

Figure 3.20 Reduction of the measured $\chi(k)$ (a) and absorption coefficient (b) for various values of $\varepsilon\Omega/4\pi$ (see text).

sample to obtain the true $\mu(E)$. Also, if I_0 is measured by a partially transmitting ion chamber, the energy dependence of the chamber efficiency must be considered. As the energy is increased the chamber becomes more transparent, which makes I_0 appear to be smaller. This enhances the apparent signal. Both of these effects are relatively small (5–15%) and can be removed using tabulated absorption coefficients. The corrections are particularly important to make when the data are to be compared to standards taken by absorption measurements.

4.3.3 Electron Detection

As mentioned earlier EXAFS experiments comparing electron-detection results with absorption results often find different amplitudes. These may be caused by effects similar to those found in fluorescence. Auger detection is most analogous to fluorescence detection in that the detected electrons always have the same energy. Since the mean free path for the electrons is very short, the correction in (60), which is needed for fluorescence, is a constant and, thus, not needed. The correction for the energy dependence of the I_0 is still needed, and for electrons it may be more difficult to apply. In this case the I_0 monitor is often a partially transmitting metal grid from which the electron yield is recorded. The gain of such a detector is likely to have an energy dependence requiring a correction if the data is to be compared to absorption measurements. Such a correction has apparently not been applied in most previous studies. Only if the standard and unknown samples are measured using the same apparatus is the correction unnecessary.

When total yield measurements are made there is another possible amplitude distortion from the energy dependence of the yield efficiency. The yield signal has several contributions: direct Auger and photoelectrons, scattered and secondary electrons induced by electrons ejected into the sample, and a small contribution from electrons induced by fluorescent X rays. This can result in the ejection of several electrons for an absorbed photon, which are emitted simultaneously and give a single pulse in a detector. As pointed out by Martens and co-workers [114], if each absorbed photon produces at least one electron that reaches the detector, then the counting rate is directly proportional to the absorption. This is likely to occur in most experiments.

When high signal levels are achieved, the detection of the yield signal is often a current measurement that measures the total number of electrons produced. Thus the constant proportionality of the yield signal to the absorption is not assured. In particular the photoelectron energy and possibly the photoelectron angular distributions are functions of the incoming photon energy. This means the average number of emitted electrons per absorbed photon can also be energy dependent, resulting in distortions of the EXAFS amplitudes. For partial yield measurements the window setting influences the relative sensitivity to the photoelectron- and Auger-derived signals, and the distortion can be modified.

Since these two effects have not been investigated systematically, absolute amplitudes obtained by electron-detection techniques should be used with caution. A better technique is to compare relative amplitudes obtained using the same detectors. In particular, polarization-dependent studies can be most useful in determining the site location of surface species [142, 143].

4.3.4   Other Problems

A common feature in many EXAFS measurements is reproducible sharp noise features, which are often referred to as *glitches*. These are generally caused by variations in the intensity from the crystal monochromator. At a particular angle an additional diffraction channel can subtract intensity from the primary beam. If precautions are taken to linearize the detection system, these glitches can often be reduced [144]. The precautions necessary are essentially the same as those needed to eliminate thickness effects. It is also possible, particularly in *channel cut* monochromators, to have additional reflections from the mono-chromator that can enter the sample [145]. These can affect the experiments in a manner similar to harmonics. However, since they represent different reflection angles on the crystal, they appear at heights different from the fundamental and can be eliminated by appropriate slits.

A related problem can occur with single-crystal samples, where the sample can Bragg reflect the beam at particular angles. This gives the appearance of a sharp increase in absorption. The problem is sometimes eliminated by changing the sample orientation slightly.

It is often found that when data taken at different times are compared, the amplitudes are different. If thickness effects are eliminated the most likely cause is a different energy resolution for the two experiments. This is a particular problem for higher shell data for which the structure is sharper. Corrections can be made [100], but it is often difficult to determine the energy resolution after the fact. The best practice is to measure samples for direct comparison on the same apparatus, preferably at the same time.

5   EXAFS FACILITIES

The EXAFS experiments are carried out at both individual laboratories and large storage ring based synchrotron radiation facilities. The distinct differences in capabilities and operational style between these two cases are summarized in this section. More complete descriptions of synchrotron–radiation instrumentation are found in numerous articles [146–150], reviews [151, 152], and conference proceedings [153–155]. Laboratory facilities are described comprehensively in the proceedings of the 1980 laboratory EXAFS conference [156] and also in articles describing specific laboratory setups [157–160].

5.1   Source Properties

The quality of an X-ray source can be characterized by a spectral brilliance parameter, which is defined as the *flux per unit solid angle per unit area of the source in a defined bandwidth*. Most laboratory facilities are based on X-ray tubes that generate X rays by the bremsstrahlung process, which occurs when high-energy electrons are stopped in a target material [161]. Figure 3.21 shows a schematic bremsstrahlung spectrum. The spectrum is continuous with a high-energy cutoff determined by the electron-accelerating voltage and a low-energy decline caused by self-absorption in the target and absorption in the output windows, if present, of the tube. Superimposed on the continuum are strong emission lines that are characteristic of the anode material. Although such a tube can generate several watts of X-ray power, its brilliance suffers from the large solid angle over which the radiation is emitted. Attempts to increase the output of X-ray tubes are ultimately limited by the physical properties of the anode material and modern tubes approach these limits.

Plasma sources, which are excited either by lasers or intense electrical discharges, achieve very high brightness in the soft X-ray range. Their output power is no longer constrained by material properties, but by the efficiency of

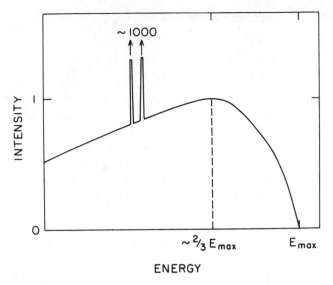

Figure 3.21 Schematic bremsstrahlung spectrum for an X-ray tube run at a voltage E_{max}. The positions of the characteristic lines depend on the anode composition.

coupling to the excitation source. However, the resulting spectra are often contaminated with numerous emission lines that require careful normalization to obtain high-quality EXAFS spectra.

In electron storage rings synchrotron radiation is generated by the acceleration of the electrons as they are caused to move in a circle by bending magnets [152]. The synchrotron output spectrum is shown in Figure 3.22, and it is

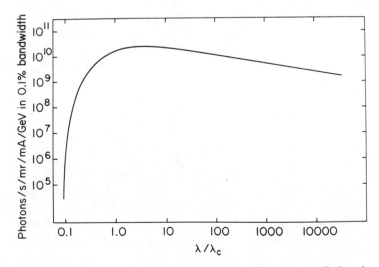

Figure 3.22 X-ray output for a bending magnet source of synchrotron radiation; $\lambda_c = hc/E_c$.

characterized by a critical energy defined as

$$E_c = \frac{3}{2}\hbar\omega_0\gamma^3 = \frac{3\hbar c\gamma^3}{\rho} \tag{66}$$

where ω_0 is the circulation frequency; ρ is the radius of curvature of the electron orbit; and $\gamma = E/mc^2$, where E is the electron energy. A typical storage ring has $E \simeq 3\,\text{GeV}$ and $\omega_0 \simeq 10^6$, which gives $E_c \simeq 5\,\text{keV}$.

At such high electron energies relativistic effects are very important and cause the emitted radiation to be strongly collimated along the direction of travel. The output angle is approximately $1/\gamma$ or about 30 arc-s. As the electron circulates, this narrow cone of radiation sweeps out a circular path that is analogous to the output of a lighthouse. The X-ray power emitted by a storage ring is about 10^5 W, although a single synchrotron radiation experiment can only collect X rays from about 0.1–1% of the total arc. This still amounts to a substantial increase in power over X-ray tubes; and, when combined with the increased brilliance provided by the $1/\gamma$ collimation, it results in an increase in the flux on the sample of about 10^4–10^5.

The polarization and temporal structure of an X-ray source are also important considerations. Synchrotron radiation is highly polarized in the plane of the orbit. Typically the monochromatic radiation from a beam line will be 90–95% polarized. X-ray tube outputs are only slightly polarized, and the monochromatic beam polarization depends primarily on the monochromator operating conditions.

Most modern X-ray tubes are run at a constant potential, which results in a constant output intensity. In a storage ring, however, the circulating electrons are bunched up because of the radio frequency cavities used to replace the power lost to synchrotron radiation. The result is pulses of X rays of approximately 1 ns duration with a frequency of approximately 1–10 MHz. This can modify the saturation properties of counters operating at similar frequencies [162, 163]. The output of a storage ring also decays slowly as the stored electrons are lost, primarily from scattering off residual gas in the ring. For operating pressures of about 10^{-9} torr, lifetimes are several hours.

The bending magnets in an electron storage ring are necessary to cause the electrons to circulate, and are not necessarily optimized for synchrotron radiation output. This has led to the development of specialized sources of synchrotron radiation called *wiggler* and *undulator magnets* [151, 152]. A wiggler magnet employs a periodic magnetic field to produce radiation with no net deflection of the electrons. An experiment can thus be illuminated by the radiation of several bends in the electron trajectory simultaneously. A typical wiggler might contain three complete oscillations of the beam, giving a sixfold increase in intensity. Also, wigglers are often run at higher magnetic fields than bending magnets, which further enhances the intensity and allows higher energy X rays to be obtained.

Undulators are similar to wigglers except that the oscillation angle is very small (ca. $1/\gamma$). This results in interference effects between the radiation emitted by the various poles. The result is that the output radiation is concentrated into narrow energy bands of width approximately $1/N$, where N is the number of bends. Also, the small excursion means that the output is confined horizontally as well as vertically to an angle of approximately $1/\gamma$. The net result is an enormous increase in brilliance over a bending magnet. Proposed designs would offer a 10^4–10^5 gain over bending magnets, which is similar to the bending magnet advantage over bremsstrahlung sources. A disadvantage of undulators is the necessity of higher energy (5–6 GeV) storage rings with better beam characteristics than are currently available. Thus, hard X-ray undulators await the completion of the next generation storage rings. However, when they are completed the intensities on the samples will be so high that radiation and thermal damage are likely to become the limiting factor in carrying out EXAFS experiments on dilute materials.

5.2   Optics

The optics in an EXAFS experiment are used to collect and monochromatize the X rays from the source for delivery to the sample. Since the EXAFS is a weak signal over an extended energy range, the optics must be rapidly tunable while still able to collect the largest possible solid angle. Satisfying these requirements for the quite different characteristics of laboratory and synchrotron sources has resulted in substantially different optical designs for the two cases.

In the laboratory almost all instruments employ a Rowland circle geometry as illustrated in Figure 3.23. The source and focus sit on a circle of radius R, and the monochromating is done by Bragg reflection from a crystal whose lattice planes were bent to a radius $2R$ [164–166]. For the Johannson [165, 166] case (Figure 3.23a) all paths from the source to the focus make the same angle θ on the crystal, and the energy is given by

$$2d \sin \theta = \lambda = 12.4/E \tag{67}$$

To scan the energy it is necessary to vary θ, which means moving two of the components around the Rowland circle. Since it is difficult to move the source the usual choice is to move the crystal by an angle θ and the sample by 2θ. This can be done using independently controlled motions under computer control or by a mechanical linkage scheme. A popular linkage scheme is shown in Figure 3.24 [158]. Using a rotating anode X-ray source such an instrument can provide approximately 10^7 photons/s on the sample. However, the bending of the crystal often degrades the energy resolution in such instruments [167]; and when high resolution near edge studies are desired, some of the flux must be sacrificed.

Since it is difficult to doubly bend a crystal, the output of a laboratory instrument is usually a focal line. Because of the extreme collimation of a

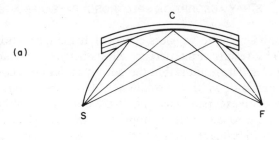

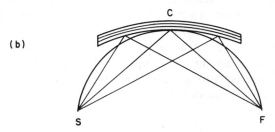

Figure 3.23 Two Rowland circle geometries for focusing a source S to a focus F: (a) Johannson case where the crystal is bent to radius $2R$ and the surface is ground to radius R; (b) Johann case where the crystal is simply bent to a radius $2R$ at some sacrifice in energy resolution.

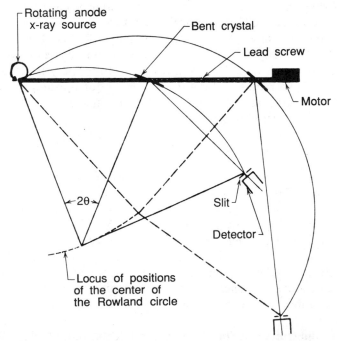

Figure 3.24 Linkage scheme for a laboratory EXAFS facility. As the crystal is translated, the linkage arms cause the crystal to rotate and the detector to move around the Rowland circle. Two positions are indicated by the solid and dashed lines. Reprinted with permission from [158].

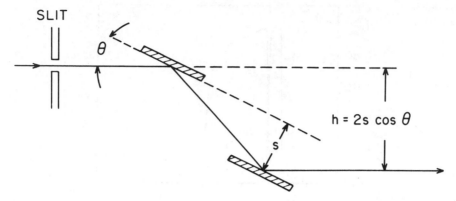

Figure 3.25 Two-crystal monochromator for a synchrotron beam line.

synchrotron source in the vertical direction, a similar focal line can be obtained using flat crystals. Indeed, many EXAFS beam lines at synchrotron facilities simply employ two flat crystals as shown in Figure 3.25. In such a scheme the energy is changed by rotating the pair as a unit, and the beam offset h is given by

$$h = 2s \cos \theta \tag{68}$$

where s is the crystal separation. This requires a small motion of the sample to track the output beam. Such a monochromator also has the nice feature that higher order reflections can be strongly suppressed by detuning the pair to be slightly nonparallel [168]. This principle is shown in Figure 3.26 for a particular case. Suppression of harmonics is very important in minimizing the thickness effects discussed in Section 4.3.1. In the laboratory, harmonics can be eliminated by running the X-ray tube at voltages below the harmonic energy.

A typical synchrotron beam line must be 10–20 m long to accommodate shielding and safety interlocks and to provide enough separation from neighboring experiments. If the horizontal radiation fan is not focused, then for a typical sample size only about $\frac{1}{2}$ mrad of the horizontal fan can be used. Therefore, many beam lines are now being built that provide horizontal focusing. This can be done using total external reflection mirrors or sagittally bent crystals.

In the X-ray energy range the index of refraction is slightly less than one [169, 170]. Therefore, if the incident angles are below a critical angle θ_c, total external reflection will occur in exact analogy to total internal reflection at optical energies where the index is greater than 1. The main difficulty is the very small value of θ_c. For pure materials a good approximation to θ_c is given by

$$\theta_c = \frac{0.029}{E} \left(\frac{Z\rho}{A} \right)^{1/2} \tag{69}$$

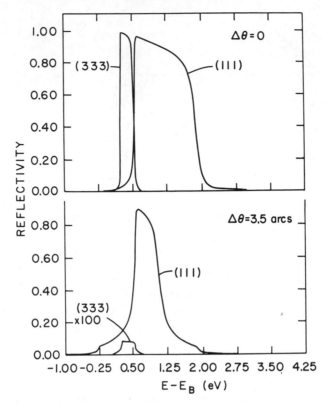

Figure 3.26   Intensity of the fundamental (111) and first harmonic (333) from a Si(111) two-crystal monochromator for two different angular offsets.

where the energy is in kiloelectronvolts, the density ρ is in grams per cubic centimeter, Z is the atomic charge, and A is the atomic weight. Typical values are given in Table 3.4. These extreme glancing angles cause problems for the X-ray optics designer since mirrors quickly become very long when large solid angle collection is attempted [148]. However, the use of glancing angles makes efficient mirrors possible by relaxing the surface finish requirements [171]. In the optical range good low-scatter optics for use at normal incidence require surface finishes to a fraction of the optical wavelength. For X-ray wavelengths of approximately 1 Å this is clearly impossible, but when glancing angles are used the surface must be smooth only to a fraction of $\lambda/\sin\theta$. This relaxes the surface finish requirements to approximately 10 Å, a formidable challenge, but possible for a number of materials [172, 173].

In light of the above challenges, current grazing incident optics can collect about $50/E$ (keV) mrad of horizontal radiation and focus the radiation to an approximately 1-mm^2 spot. Improvement on this performance is likely to occur very slowly. Thus, designers have recently been turning to Bragg reflecting crystals for focusing. The standard Rowland circle designs are impractical

Table 3.4 Critical Angles of Reflection[a] for Selected Elements

	Optical Energy (keV)		
Element	5	10	20
Si	6.2	3.1	1.55
Ni	12	6	3
Ag	12.4	6.2	3.1
Pt	16.8	8.4	4.2

[a]In milliradians.

because the large distances involved at a synchrotron radiation beam line entail large precision motions of the crystal and sample during an energy scan. To avoid this sagittal or nondispersive focusing schemes were developed [174–176]. The basic principle is shown in Figure 3.27. A flat crystal is followed by a crystal bent in the nondispersive direction. If the crystal bend is cylindrical it turns out that in terms of the incident angles the bent crystal behaves just as a flat crystal for $R = F_1 \sin \theta / 2$, where F_1 is the source-to-crystal distance. This results in a crystal-to-focus distance of $F_2 = F_1/3$. Thus, sagittal focusing can be achieved by simply replacing the second crystal in a standard two-flat-crystal monochromator by an appropriately bent crystal.

However, in implementing such a scheme there are two problems that must be overcome. Both are associated with crystal bending. The first is anticlastic bending of the crystal, which is the natural tendency of thin plates to curve along an axis perpendicular to the axis of bending. Such curvature is along the dispersive direction, and it seriously degrades the throughput of the monochromator by causing the Bragg condition to be violated for many of the X rays. The solution is to stiffen the crystal in one direction only by the addition of reinforcing ribs [174]. The crystal then is easy to bend only along the desired axis.

The second problem occurs when energy scans are attempted by changing θ. The ideal radius depends on θ, and changing the radius during an EXAFS scan requires a bender that can respond in less than 1 s. Currently a design [175]

Figure 3.27 Sagittal focusing by replacing a flat crystal in a two-crystal monochromator with a curved crystal.

exists for such a bender, but its performance is not proven. An alternative approach is to keep R fixed during a scan and accept the nonideal throughput condition. For EXAFS-type scans of about 1 keV this is not serious, but the focal length F_2 will vary. Thus, the focus will not remain on the sample unless either the sample or the crystal is translated. This approach is also being attempted currently [175].

Sagittally focusing crystals should be able to efficiently focus 10–20 mrad of synchrotron radiation. They, therefore, offer little advantage over glancing incident optics below 5 keV, and have the disadvantages of the more complex mechanical design of the bender and/or crystal translation. However, crystal focusing can easily operate at energies exceeding 30 keV, and it is in the high-energy region that sagittal focusing schemes should find wide applications. When either focusing scheme is employed on a modern storage ring, the flux available at the sample is about 10^{12} photons/s, some 10^5 times larger than for a laboratory instrument. The question may then arise as to why laboratory insturments continue to be built in light of the overwhelming flux advantage of a storage-ring based instrument. This is discussed next.

5.3   Laboratory Versus Storage Ring Facilities

For concentrated materials good EXAFS data can be obtained with about 10^7 photons per data point. This can be achieved easily with laboratory facilities. In fact, for the higher counting rates at storage rings it is difficult to take full advantage of the increased signal levels for concentrated samples because of detector noise and nonlinearities. To use 10^{10} photons per point properly requires a measurement accuracy of 10^{-5}, which is difficult to achieve in any electronic system. Thus, for routine experiments on concentrated samples the quality of data from laboratory and storage-ring facilities is comparable. For very dilute samples where the total number of photons is the determining factor in the success of an experiment, the storage ring is undoubtedly the only choice. However, for many experiments factors other than the total flux determine the choice.

A laboratory machine can be scheduled according to the needs of the experimenter. By contrast, use of a synchrotron facility usually requires a lengthy proposal and scheduling process. Thus, there are often long delays between the time an idea for an experiment is generated, and the time it can actually be carried out. This problem can be expected to lessen in the future with the growing availability of beam time and as more experimenters become affiliated with particular beam lines; but because travel to a distant facility is required, the samples must still be ready on a given date. Difficult-to-prepare or short-lived samples are not always ready at the particular time a run is scheduled. Also, samples requiring special environments are often easier to handle in a laboratory setting. Finally, for work on proprietary materials, as are often encountered in industrial laboratories, working in the laboratory can avoid many patent and disclosure problems associated with using a national facility.

A laboratory system, however, may require a commitment in its construction, operation, and maintenance larger than can be justified by an occasional EXAFS user. One or two visits a year to a storage-ring facility may be sufficient; and at most facilities the individual experimenter has no responsibilities for construction or maintenance of the beam line. In a university laboratory, on the other hand, constructing and maintaining a laboratory EXAFS facility can provide an important training ground for students. This is not the case at a storage-ring facility, where the major interaction with the equipment is often by computer terminal and where time constraints often make it difficult to allow beginning students to work through their mistakes.

It is ideal to have access to both laboratory and synchrotron facilities because new ideas can be quickly tested and the early part of the learning curve for difficult experiments can be traversed in the laboratory. Optimum use can then be made of synchrotron facilities, where the photons are coming faster and there is less time for mistakes or contemplation about the best measurements to make.

6   COMPARISON WITH OTHER TECHNIQUES

Before considering some examples of applications it is instructive to compare EXAFS and XANES with other techniques. Structural information can be obtained directly as in a diffraction experiment or indirectly by measuring a property that is related to the structure in a more complicated way. Examples of indirect techniques might include nuclear magnetic resonance (NMR), electron paramagnetic resonance (EPR), Raman spectroscopy, and Mössbauer spectroscopy. In all of these there is not a direct transformation that can be applied to the data to obtain the structural information, but all are sensitive to the local environment or bonding. To obtain the structural information it is usually necessary to carry out some form of modeling or comparison with standard materials. The same situation exists for XANES, where at the present stage of knowledge it is usually not possible to obtain quantitative structural information directly. As we have seen, this is because the XANES is also sensitive to the chemical or electronic environment. As with many of the other indirect methods this can be used to advantage in studying other properties, but it complicates the extraction of structural information.

Direct structural techniques such as X-ray and neutron diffraction are possible because the X rays and neutrons only interact weakly with the sample. Thus, the kinematic approximation, which assumes only single scattering events from independent atoms, is often valid. Since the atomic scattering factors are well known, the measured spectra directly reflect the atomic structure and, in principle, can be inverted to obtain the structure. In practice, the long-range order or unit-cell dimensions are very simple to obtain in this fashion. However, to locate the atoms accurately within the unit cell, knowledge of the phase and intensity of the scattered beams is required, while in most experiments only the intensity is recorded. This prevents making a direct inversion to locate all of the atomic positions accurately, and experimenters are forced to model making and/or more complex calculations. This is generally simple for smaller unit cells. However, for very large unit cells, such as for biological protein crystals, the task is very difficult and several clever techniques for extracting the phase information were developed.

When long-range order does not exist, diffraction methods have less utility. The scattering then contains information about the sum of the two particle correlation distances for all of the particles. For complex systems with many types of atoms present it is extremely difficult to separate out the contribution from each type of atom. The separation can be made only if it is possible to make two or more measurements for which the scattering from only one component is changed. For neutrons the scattering from one component can be changed in some cases by making an isotopic substitution. For X rays a promising technique is to tune the X-ray energy above and below an X-ray edge where anomalous scattering effects significantly change the scattering factor of the selected component [177, 178]. However, neither technique can deal with dilute components, which make only a small contribution to the total scattering.

Another type of diffraction technique uses electrons as the interacting particles. Electrons interact strongly with matter, and it is difficult to achieve experimental conditions that satisfy the kinematic approximation. When multiple scattering occurs, even the intensity of the scattered beams is difficult to calculate. Therefore, determination of structure within the unit cell using electrons can be a difficult and uncertain process.

Extended X-ray absorption fine structure is also caused by electron scattering, but it has several characteristics that make it complementary to the preceding techniques. In EXAFS the absorbing atom is both the source and the detector of the scattered electron, and the resulting EXAFS spectrum contains both phase and intensity information. Thus, a direct inversion of the spectrum is possible. The strongly interacting nature of the electrons means that corrections to the phase and amplitudes are necessary to obtain quantitative structural information, but such corrections are often transferable from standards. Even if detailed calculations are required, the local nature of the EXAFS provides important simplifications in minimizing the number of multiple scattering channels that must be considered.

The local nature of EXAFS is its primary strength. It is particularly sensitive to small perturbations within the unit cell and can thus complement the information obtained in diffraction experiments. Since it is possible to focus in on a particular atomic site, changes can be monitored without determining the entire structure. A good example is metalloproteins, where one or a few metal atoms are embedded in many thousand low Z atoms. To solve the complete protein structure is difficult and inevitably leads to inaccuracies in determining the local environment of the metal atom. The EXAFS can determine the local environment without the need to crystallize the protein, and to determine the position of all of the atoms in the unit cell. Also, since EXAFS is a local probe, very dilute materials can be measured. The strength of the EXAFS is unaffected by diluteness, and all that is required is the separation of the absorption component of the species of interest. As shown in Section 4 this can be accomplished for very dilute concentrations. A final important attribute of EXAFS is its wide applicability. It can be routinely performed on elements heavier than Ar, and with somewhat greater difficulty on elements as light as B.

7 APPLICATIONS

Since X-ray absorption measurements are relatively easy to make, there are an enormous number of applications, many more than can be adequately reviewed in a few pages. For a better appreciation of possible applications the reader is referred to the published proceedings of the international EXAFS conferences [179, 180]. Here we briefly highlight some important examples to demonstrate the breadth of applications. First, however, a somewhat more detailed description of work from our own background is given. This is the area of intercalated graphite, which provides a particularly simple demonstration of the amount of information available in X-ray absorption.

7.1 AsF$_5$ and Br$_2$ Graphite Compounds

Graphite can be reacted with several materials to form graphite intercalation compounds. The intercalant molecules, such as AsF$_5$ or Br$_2$, take up ordered positions between the graphite planes, but the order is often insufficient for detailed diffraction studies to determine the intercalant position. There is also the question of whether the molecules retain their identity upon intercalation. Both questions can be studied with X-ray absorption experiments.

For AsF$_5$ intercalation the interesting question is whether a chemical reaction occurs on intercalation. The first X-ray absorption measurements were carried out by Bartlett and co-workers [181] using powdered, highly oriented pyrolytic graphite (HOPG) as the starting material. On the basis of their results they proposed that intercalation is accompanied by the reaction

$$3AsF_5 + 2e^- \rightarrow 2AsF_6^- + AsF_3 \tag{70}$$

This would imply a charge transfer of $\frac{2}{3}e^-$ per AsF$_5$ intercalated. However, measurements by other techniques found smaller values, with an average of about 0.3e$^-$, although the measurements span a wide range of values. This prompted further study using both HOPG and graphite fibers as starting materials [182].

The utility of X-ray absorption measurements in this system is demonstrated in Figure 3.28a where the near-edge spectra for the three constituents involved in the reaction are compared. There are strong white lines in each because of the large density of unfilled 4p states associated with the As—F antibonding levels. As seen from the figure the location and shape of the peaks are different and can

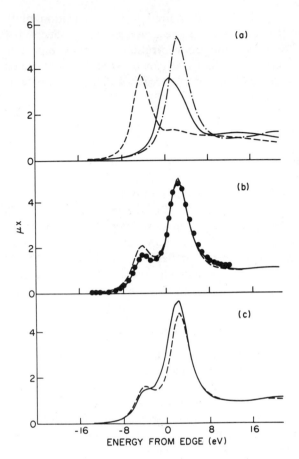

Figure 3.28 Near-edge spectra for As—F compounds: (a) AsF_3 (---), AsF_5 (——), and AsF_6^- (·—·-). (b) AsF_5 intercalated graphite fibers with the X-ray polarization along the fiber axis (---) and perpendicular (——) to the fiber axis. The points are a fit to the solid line using a linear combination of the spectra in (a). (c) Comparison of a spectrum for AsF_5 intercalated HOPG graphite (——) with the solid curve in (b).

be used to identify the species present. These results also show the normal trend of the absorption edge shifting to higher energies as the formal valence is increased. In addition, since the unfilled $4p$ states are associated with the chemical bonds, the strength of the white lines for asymmetric molecules such as AsF_3 and AsF_5 depends on the orientation of the molecule relative to the X-ray polarization direction.

This is demonstrated by the results obtained for AsF_5 intercalated graphite fibers shown in Figure 3.28b. In the fibers the graphite c axis is oriented perpendicular to the fiber axis, and the measurements were taken with the X-ray polarization perpendicular and parallel to the fiber axis. A distinct polarization dependence can be seen in the AsF_3 peak. These molecules are asymmetric, being in the form of a tripod with the As at the apex. The sign of the orientation dependence clearly indicates that the fluorine base of the tripod is aligned parallel to the graphite basal planes.

Although quantitative analysis of edge spectra is often difficult, in this case excellent standards are available and quantitative fits can be carried out. The points in Figure 3.28b are an example of such a fit using a linear combination of the three standards. Excellent agreement is found with the result being that the reaction in (70) is carried essentially to completion, in agreement with Bartlett's results. However, for HOPG, the results are different as shown in Figure 3.28c. A clear difference from the fiber data is seen with the increased absorption between the AsF_3 and AsF_6^- peaks indicating that some AsF_5 remains. Indeed a fit to this data reveals that about 50% of the AsF_5 remains unreacted, although the AsF_3/AsF_6^- ratio remains $\frac{1}{2}$. The HOPG used in this work was not ground up; thus, it appears that the extent of the reaction depends on the surface area of the sample. This could possibly explain the wide variations in change transfers obtained on this system.

For the Br_2–graphite system the analysis concentrated on the EXAFS part of the absorption spectrum. Several samples were measured ranging from those in which the Br_2 molecules are simply adsorbed on the surface to true intercalation compounds [183, 184]. Some examples of the spectra obtained are shown in Figure 3.29 for the adsorbed case. For this study a form of graphite known as graphoil was used that has a large oriented surface area. As for the AsF_5 case a clear polarization dependence is observed for polarizations along and normal to the surface. Also the data show an obvious demarcation into two regions. At low k there are strong oscillations for normal polarization that sharply decrease for the tangential polarization. These are caused by Br—C bonds and die off rapidly because of a large Debye–Waller factor and the characteristic amplitude dependence of low Z backscatterers. The EXAFS caused by Br—Br bonds extends to much higher k and has an opposite polarization dependence. On comparison with the Br_2 vapor results, these data immediately indicate that the Br_2 remains in molecular form and lies flat on the surface. A more detailed analysis indicates that the Br—Br distance increases to 2.31 Å on adsorption as compared to 2.28 Å for the free molecule and that the Br—C distance is about 2.9 Å.

Another interesting aspect of the data is the Br—Br σ^2 results shown in Figure 3.30. These were obtained using the ratio method to compare with Br_2 vapor. Two different coverages are shown. For 0.6 monolayer σ^2 is larger than the vapor but has about the same temperature dependence. For 0.9 monolayer there seems to be an abrupt drop from a high value at room temperature to values similar to the vapor. It is proposed that this is the result of a phase transition from a two-dimensional liquidlike state to an ordered two-dimensional solid. Several of such surface phases were previously observed by low-energy electron diffraction (LEED) [185].

When coverages above one monolayer are attempted, the Br_2 begins to intercalate. This dramatically changes the Br—Br EXAFS as is shown in Figure

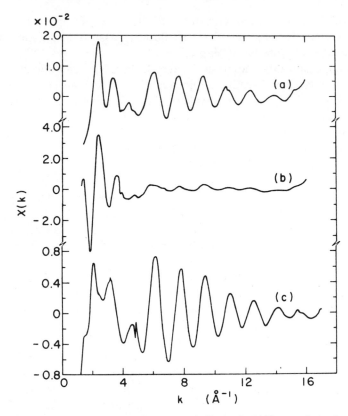

Figure 3.29   EXAFS for a 0.9 monolayer Br_2 on graphoil sample. (*a*) X-ray polarization parallel to the graphite basal planes. (*b*) X-ray polarization perpendicular to the basal planes. (*c*) Br_2 gas as a standard.

3.31. For an equivalent coverage of 1.3 monolayers a strong beating is observed in the EXAFS near $k = 6 Å^{-1}$. However, as the amount of Br_2 is increased the spectrum again becomes similar to the vapor result. This behavior can be explained if the beating in the 1.3 monolayer is examined more closely. This is done in Figure 3.32, where the intercalate data are compared to the vapor results again using the ratio method. Instead of straight lines, obvious beat patterns are obtained. These patterns indicate that there are two types of Br—Br distances. The depth of the amplitude minimum indicates the two components are nearly equal in amplitude, while the position of the minimum gives the distance difference ΔR between the two species. The dashed lines are the results for a two-component fit to the data.

From these results the behavior of the EXAFS spectra is explained. When the threshold for intercalation is crossed, a new expanded Br_2 is formed with a Br—Br distance of 2.53 Å. The coexistence of this form with the adsorbed form gives the observed interference. However, this expanded Br_2 is not intercalated

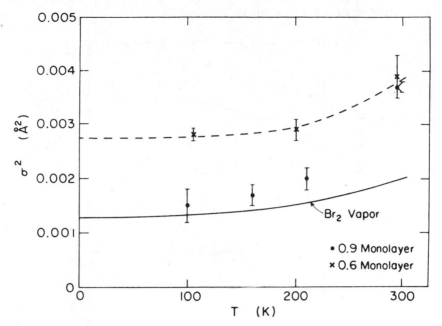

Figure 3.30 Mean-square vibrational amplitudes (σ^2) of the Br—Br distance for adsorbed Br_2 compared to Br_2 vapor. The dashed line was drawn by eye through the 0.6 monolayer points to emphasize the different temperature dependence of the 0.6 and 0.9 monolayer results.

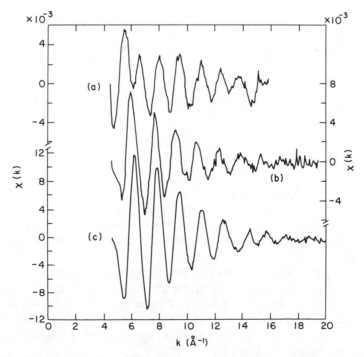

Figure 3.31 EXAFS spectra for (a) 0.27 mol% Br_2 intercalate; (b) 0.75 mol% Br_2; (c) Br_2 vapor.

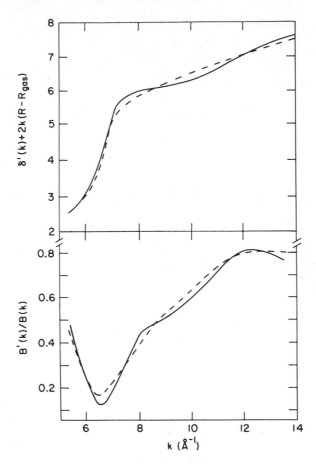

Figure 3.32   Comparison of the experimental (——) phase and amplitude with a two-component fit (———) for a 0.27 mol% Br_2 intercalation compound.

since it is randomly oriented to the graphite planes. As additional Br_2 is added, the molecules go into the true intercalate configuration with a Br—Br distance of 2.34 Å and with their axes aligned parallel to the planes. Since the number of the expanded Br_2 species saturates, the intercalate molecules predominate at higher concentrations, and the interference becomes weak. From these results it is reasonable to speculate that the expanded Br_2 molecules attach to the edges of the planes and act to open the graphite exposing the interior graphite planes for intercalation.

These examples show the wealth of information obtainable from X-ray absorption measurements. They illustrate all of the steps in the analysis, except the final complication, that can occur. This is interference between differing atomic species for which the simple beat expressions no longer apply. Here the data are usually fit in k space using fitting functions derived from either theoretical or experimental models.

7.2   Fluctuating Valence Compounds

Because the transition time for an X-ray absorption event is very short (about 10^{-16} s), the system being probed has little time to adjust. As a result, X-ray absorption measurements can provide information about the initial state of a system. This feature is quite useful for studying mixed-valence, rare earth compounds. The valence of a rare earth ion is determined by the number of filled $4f$ electron levels; and because these states tend to be quite localized and atomiclike, the valence is generally an integer. In mixed-valence compounds, two different integral valences are observed simultaneously, which correspond to either inhomogeneous rare earth sites or valence fluctuations at homogeneous sites. Measurements on fluctuating-valence compounds using Mössbauer spectroscopy, which has a transition time scale of approximately 10^{-9} s, find a single average valence. The valence on a given rare earth ion is believed to fluctuate at a rate comparable to phonon frequencies, and since the ionic radii vary significantly with valence, one may expect to see a strong electron–lattice coupling.

A prototypical fluctuating valence material is SmS, which undergoes an isostructural, first-order transition at a pressure of 6.5 kbar from a state having a Sm valence of $+2$ to a fluctuating valence state with a mean valence between $+2$ and $+3$ and with a commensurate volume collapse. The removal of one $4f$ electron from a rare earth atom causes its L_3-absorption edge to shift by 6–10 eV; and because of the strong white line feature that is usually present, as for the AsF_5 example, it is straightforward to determine the fraction of atoms with each type of valence in a sample by assuming a superposition of absorption edges corresponding to the two integral valences [186]. Measurements [187, 188] of the L_3-edge XANES under pressure clearly show the jump in the mean valence at the transition, as well as the continuous increase in mean valence with pressure in the fluctuating valence phase.

The advantages of determining valences from L_3-edge absorption spectroscopy compared to Mössbauer spectroscopy or lattice parameter measurements were discussed by Röhler [189]. A study of TmSe, which is mixed valent at atmospheric pressure, has shown that valence determinations using L_1 and L_3 measurements agree [190]. L_3-edge absorption spectroscopy is now a standard technique for determining valence, as evidenced by the numerous XANES papers in the Proceedings of the Fourth International Conference on Valence Fluctuations [191]. Two recent studies [192, 193] on Tm compounds have demonstrated the utility of M_5-edge measurements for valence determinations.

Extended X-ray absorption fine structure was used to look for a coupling between valence fluctuations and dynamic distortions of the lattice. Work on $Sm_{0.75}Y_{0.25}S$ [194] and TmSe [195] found no evidence for local distortions in the nearest-neighbor distribution. However, a high-pressure study [196] of SmS found that while there seemed to be a single Sm—S first-shell distance, two Sm—Sm second-shell distances were present at a pressure just above the phase transition, compared to a single distance in the atmospheric pressure phase.

7.3 Disordered Systems

As EXAFS does not require long-range order, it is an ideal technique for studying disordered systems. Some of the general classes of disordered materials that were examined with EXAFS are superionic conductors, amorphous semiconductors, chalcogenide glasses, metallic glasses and amorphous alloys, and liquids. Even in liquids a significant nearest-neighbor contribution may be present, although higher shells are often difficult to see. The dependence of the backscattering amplitude and phase on atomic number allows one to check for chemical ordering in mixed systems. One can also look at the ordering about a dilute component of a disordered system. However, complications in the analysis often occur because of asymmetry of the nearest-neighbor pair distribution function. When asymmetry is present, the lack of low k EXAFS can make it difficult to determine coordination numbers accurately.

In contrast to EXAFS, X-ray scattering measurements do include low k information, so that from the corresponding radial distribution function one can more easily obtain accurate coordination numbers. By using differential anomalous scattering, one can separate the distributions about different elements and obtain results similar to EXAFS [197]. However, EXAFS can be measured out to higher values of momentum transfer, which provides information on the short-range repulsive part of the pair interaction potential [198] and allows chemical differentiation of neighbors, which is not possible with X-ray scattering. Furthermore, with X-ray scattering it is currently practical to study only simple mixtures involving concentrated components. The two techniques are basically complementary to one another.

Eisenberger and Brown [56] were the first to point out that when the nearest-neighbor pair distribution function is non-Gaussian, extra k-dependent terms enter the single-shell phase and amplitude, which, if not accounted for, lead to significant errors in the determination of coordination numbers and interatomic distances. In a study of Zn at high temperatures, including the liquid phase, Crozier and Seary [57] showed that by considering appropriate asymmetric model distributions one can obtain good fits to the EXAFS and extract information on the repulsive part of the interatomic potential.

When the number of nearest-neighbors is known, EXAFS measurements can provide unambiguous information about the shape of the pair distribution function (PDF). Superionic conductors, where at elevated temperatures the cation sublattice softens and diffuses through the more rigid anion sublattice, are good examples of such a situation. Boyce and co-workers [199] studied AgI, CuI, CuBr, and CuCl and find that at and above room temperature the nearest-neighbor PDF is well described by an excluded-volume model in which the cations can sit at any position not occupied by an anion. In amorphous metal alloys the coordination numbers are unknowns, which makes determination of unique PDFs more difficult. Teo and co-workers [200] gave an extensive discussion of the analysis of their EXAFS measurements on amorphous and crystalline MoNi using both symmetric and asymmetric distributions. The EXAFS studies of amorphous metal alloys were reviewed by Cargill [201].

Licheri and Pinna [202] reviewed studies of ionic hydration in aqueous solutions of electrolytes and compared them with X-ray diffraction results. Greaves [203] has discussed the application of EXAFS to silicate glasses. Hayes and Boyce [198] and Gurman [204] reviewed early work on a wide range of disordered systems.

7.4   Catalysts

In supported metal catalysts, the active sites for catalysis are small metal clusters, with typical dimensions of less than 20 Å, which are dispersed on a porous, high surface area support such as alumina, silica, or zeolite. One would like to determine the structure of the metal particles, the nature of their interaction with the support and reactants, and the effects on the clusters of treatments such as reduction or oxidation. Because of the small size of the metal particles and their low concentration in a sample (ca. 1 wt%), EXAFS has obvious advantages over diffraction techniques for studying these systems. Also, since the support atoms are typically low Z and have smaller absorption, it is often possible to study the metal environments using simple transmission EXAFS.

One way to learn about changes in bonding is to study the number of unoccupied d states through L_2- and L_3-edge XANES measurements on $5d$ transition metals. Lytle and co-workers [205] have shown that there is a good correlation between an appropriately defined coordination charge and the difference in white-line areas measured at the L_3 edge for a metal compound and the pure metal. The technique of using L-edge white-line areas to determine the number of vacant d states was made somewhat more quantitative by Mansour and co-workers [206], who extended it to include both L_2 and L_3 edges. This approach was used to study Pt/Al_2O_3 and Pt/SiO_2 catalysts as a function of the support, reduction temperature, and metal concentration [207].

Early EXAFS studies concentrated on determining the nearest-neighbor distribution for monometallic clusters in supported metal catalysts. Studies of Os, Ir, and Pt on silica and alumina found nearest-neighbor distances close to the bulk values, with the number of metal nearest-neighbors reduced by 20–30% and an increase in disorder, consistent with a very small particle size [208]. By comparing an analysis of higher coordination shells and chemisorption measurements with model calculations, one can get an idea of the actual morphology of the metal clusters in a given catalyst; for example, Greegor and Lytle [209] found that the most probable structure of osmium on silica is disklike. With high-quality data and improved analysis techniques we can extract information on metal-support interactions as has been done in the case of Rh on Al_2O_3 [210]. Studies were also done on the structure of bimetallic clusters such as Os—Cu and Rh—Cu on silica [211]. The development of energy-dispersive EXAFS made possible the observation of the time dependence (on the order of minutes) of catalytic treatments, as was demonstrated in a time-dependent study of Pt L_3-edge XANES during the reduction under H_2 of Pt/Al_2O_3 and Pt—Rh/Al_2O_3 catalysts [212].

In a complementary approach, the structure of isolated small metal clusters can be studied. For example, Montano and co-workers [212 (1984)] looked at Ag and Cu [212 (1986)] clusters isolated in solid argon. The Ag clusters showed a significant contraction of the nearest-neighbor distance consistent with the effects of surface stress. For the Cu clusters, on the other hand, a small contraction was observed only for particles less than 15 Å in diameter.

For a detailed review of EXAFS studies on a wide variety of catalysts, see the recent article by Lagarde and Dexpert [213].

7.5 Materials at High Pressure

At high pressures, most crystalline solids exhibit structural transitions to more densely packed phases. X-ray diffraction, especially in the energy-dispersive mode, is the favored technique for determining the structures of high-pressure modifications. However, while the positions of Bragg peaks usually determine the lattice symmetry uniquely, the intensity information is not always sufficient to fix the positions of atoms within the unit cell. The EXAFS can fill in the missing short-range structure, as well as describe the behavior of mean-square relative displacements, while XANES provides information on the local density of states.

The standard apparatus for high-pressure work is the diamond anvil cell, in which a sample held in the center of a metal gasket is squeezed between two opposed diamond anvils. The pressure is typically calibrated by the ruby fluorescence method. If X rays are allowed to pass axially through the diamonds, diffraction by the single-crystal diamonds causes large spikes in the absorption spectrum, making EXAFS work difficult [214]. One practical alternative involves replacing the diamonds by polycrystalline B_4C, which is transparent to X rays but not to ruby fluorescence. The pressure is then determined from EXAFS measurements of the change in nearest-neighbor distance in a calibrant material such as NaBr or RbCl [215, 216]. Another approach is to use a gasket made of beryllium and to shoot the X rays sideways through the gasket, thereby avoiding the diamond anvils [217].

Some of the high-pressure work on the mixed-valence compound SmS is already described. Röhler and co-workers [218, 219] continued their L_3-edge XANES studies on other mixed-valence systems such as Ce and EuO. Changes in near-edge structure were also found to be good indicators of structural phase transitions, as observed at the $B1-B2$ transition in KBr, RbCl, and RbBr and at the multiple transitions in CuBr [216, 220, 221]. Extended X-ray absorption fine structure has been used to study the behavior of zincblende-structure compounds under pressure. In CuBr it was found that both the Cu and Br atoms remain fourfold coordinated at the first transition at 50 kbar before transforming to the sixfold coordinated rock-salt structure at higher pressures [222]. A study of ZnSe has shown that the mean-square relative displacement for nearest neighbors decreases with pressure, as expected, but that for next-nearest neighbors it increases slightly, which indicates an instability of the structure under pressure [216].

7.6   Surface EXAFS

The measurement of surface and interface structure is a growing and expanding field. Traditionally, surface-structure determinations have been dominated by LEED measurements. As discussed, these are excellent for determinations of unit cell parameters, but they can have difficulties in determining details of bonding or in measuring disordered systems. The construction of intense synchrotron sources has also spurred the development of surface X-ray diffraction measurements, but again they cannot be applied to all systems. The diffraction signals are often weak, particularly for low Z adsorbates or disordered systems. Thus, it appears there will always be a demand for the local structural information available from X-ray absorption measurements.

Early surface EXAFS measurements employed the electron-detection techniques described in Section 4. More details can be found in several reviews [223–226]. These techniques are naturally compatible with UHV environments, which are necessary to produce well-characterized surfaces. Electron detection is particularly useful for low-Z elements that have weak fluorescent yields. Indeed much of the early work concentrated on the C, N, and O K edges.

For heavier atoms other techniques become possible. The Br–graphite work described previously is an example where transmission measurements can be made by increasing the surface area of the sample. Fluorescence measurements at glancing angles, as discussed in Section 4, and at nonglancing incidence [227] are increasingly being applied to elements as light as Si. Here the decreased signal is accompanied by a decreased background, and S/N levels remain competitive. An important advantage of nonglancing incidence fluorescence is that amplitude distortions are minimized and accurate coordination numbers can be obtained.

For low-Z atoms there is often a problem with overlapping edges. For example, the C, N, and O K edges are at 284, 410, and 548 eV, respectively. This often limits the available k-space range for EXAFS analysis. Because of this much work was concentrated on the near-edge structure. This is illustrated particularly well by work on the C K edge, where strong correlations between the positions of the σ and π near-edge peaks and the C bond length are found [85, 228]. In addition, the polarization dependence of near-edge structure can be equally useful in determining molecular orientation for low Z molecules [229–230].

7.7   Crystalline Materials

The EXAFS measurements, which provide only local structural information, are not usually sufficient to determine the structure of crystalline materials, and diffraction measurements are usually preferred. However, situations exist where EXAFS can complement the long-range information of a diffraction experiment. These are often situations where some type of randomness exists such that the average structural information from a diffraction experiment cannot be used to provide site-specific information, or where dilute constituents of a crystalline lattice are being studied.

One type of randomness occurs when a lattice constituent is mobile, as for superionic conductors mentioned in Section 7.2. Ferroelectrics are another class of materials where one constituent may have either static or electric field-induced displacements. The randomness here comes from the several possible displacement directions. The EXAFS measurements were made on Nb-doped $KTaO_3$ compounds to study the Nb displacements [231]. Many ceramic materials are mixtures of elements that may be randomly distributed in the lattice. An example is yttria-stabilized zirconia. Here EXAFS measurements were used to study the location of anionic vacancies introduced in the lattice by the yttria additions [232, 233].

Dilute constituents can be difficult to locate in a lattice since their signals may be overwhelmed by bulk contributions. The element specificity of EXAFS measurements make this a much smaller problem, and there have been several diverse applications. Examples include ion-implanted materials [234], trace elements in coal [235], and dilute alloying constituents [236–240]. A problem with EXAFS measurements on these types of materials may occur if the dilute constituents occupy a disordered site such as a grain boundary. Then the EXAFS analysis can be difficult and very model dependent. However, even the knowledge that the constituents are not substitutional is often important.

For many materials a wide range of random solid solutions can be formed. The X-ray lattice parameter is often found to obey Vegard's law, which states that the lattice parameter should vary linearly with composition. However, diffraction measures an average lattice, and when EXAFS measurements were carried out on ternary semiconductors that follow Vegard's law, the near-neighbor distances were found to be remarkably insensitive to composition [241]. This example in particular illustrates the important role the short-range element-specific structural information of EXAFS can have in understanding even simple crystalline systems.

7.8   Biological Applications

For biological materials it is often difficult to obtain structural information. The molecules are large and complex and often available only in small quantities. Elegant X-ray diffraction experiments have solved the structures of several complicated molecules, but the experiments are difficult and can only be applied to materials that can be crystallized. There is also some concern that the crystallized material may not be representative of the structure in vivo.

For materials that contain heavy atoms on which EXAFS measurements can be made, many of the problems can be bypassed. Measurements can be made in solution under a variety of conditions. Since EXAFS focuses on one or a few atomic sites, the structural accuracy for those sites is often higher than in diffraction experiments where the entire structure must be solved. In many cases, with metalloproteins being a prime example, the important properties of the molecule are specific to a few sites, and the EXAFS results can be crucial in understanding chemical changes that may occur.

Because of these advantages, the advent of synchrotron radiation sources has resulted in a great many biological applications. These were reviewed in several articles [242–247]. The usefulness of EXAFS for metalloproteins was first suggested by Sayers and co-workers [248], although they did not actually measure a protein. The first proteins to be measured were rubredoxin [249–250] and hemoglobin [251–254]. The hemoglobin results are interesting in that they demonstrated some of the clear advantages of EXAFS measurements. By focusing on the Fe site, several subtle changes in the Fe environment were observed as the protein was prepared in the oxidized and deoxy forms. These changes are below the resolution of X-ray diffraction measurements, which could only infer changes based on conformational changes in the protein. Although the interpretation of the results by different groups was somewhat controversial, they clearly demonstrated some of the advantages of the EXAFS technique.

More recently, the biological applications have expanded to include a wide variety of proteins, calcium compounds, and anticancer drugs. In addition, work has begun on single-crystal protein studies [255–256]. These make use of the polarization dependence of the near-edge structure and EXAFS to determine details of active site symmetry unavailable from diffraction experiments. In essence, the EXAFS results are used to refine models based on diffraction results.

7.9 Site Symmetry From XANES

As the AsF_5 example demonstrates, XANES features can be used to obtain information about site symmetry. This information can come from two types of experiments. The first, illustrated by the AsF_5 results and polarization studies on single-crystal proteins, is to probe directly the bond orientation using polarized X rays. The chemical bonds in a system often have a large density of empty states or antibonding orbitals associated with them. Since these are generally aligned along the bond directions, monitoring the strength of near-edge features as a function of X-ray polarization direction can determine the bond direction. Of course, a correlation must be established between absorption features and chemical bonds being studied, but often this can be accomplished using model compounds or theoretical studies. The application of this technique also requires that the sample have an intrinsic orientation. This is automatic for surface systems, but requires crystalline samples in other cases. Such samples are not always obtainable for systems on which it is desired to perform X-ray absorption measurements.

The second technique is more generally applicable. It is the correlation of near-edge features either experimentally or theoretically with types of chemical bonding. Early work consisted mainly of empirical correlations between materials and molecular orbital (MO) calculations and is reviewed in [257]. The advent of synchrotron radiation allowed dilute materials to be studied and early synchrotron work concentrated on edge studies of metals in biological materials [258–259]. These generally relied on comparisons with standards since the MO theories do not allow quantitative conclusions to be derived from the data. A good comparison of the MO calculations with experiment is given by Grunes [260] for the $3d$ transition metals, and a poor correlation is found.

As described in Section 2.8, better calculations employ a full multiple scattering approach [69–73]. Again, a primary application is to biological materials such as the hemeproteins [261–263]. In these there is a rigid net of surrounding ligands that gives strong and distinct multiple scattering. While biological systems are not the only possible application, they appear to be a particularly fruitful area because of the lack of other types of structural information and the difficulties in developing good model compounds. For many inorganic compounds the XANES information can often be correlated with well-characterized models. However, as the calculations improve it appears that analysis of the XANES will become an increasingly important source of site-symmetry information.

Acknowledgment

This work was supported by the United States Department of Energy, Division of Materials Science under Contract Numbers DE-AS05-80-ER10742 and DE-AC02-76CH00016.

References

1. W. Friedrich, P. Knipping, and M. Von Laue, *Ber. Bayer. Akad. Wiss.*, 303 (1912).
2. W. L. Bragg, *Proc. Cambridge Philos. Soc.*, **17**, 43 (1912).
3. M. de Broglie, *C. R. Acad. Sci.*, **157**, 924 (1913).
4. H. Fricke, *Phys. Rev.*, **16**, 202 (1920).
5. G. Hertz, *Phys. Z.*, **21**, 630 (1920); *Z. Phys.*, **3**, 19 (1920).
6. W. Kossel, *Z. Phys.*, **1**, 119 (1920); **2**, 470 (1920).
7. B. B. Ray, *Z. Phys.*, **55**, 119 (1929).
8. B. Kievet and G. A. Lindsay, *Phys. Rev.*, **36**, 648 (1930).
9. R. de L. Kronig, *Z. Phys.*, **70**, 317 (1931).
10. J. D. Hanawalt, *Z. Phys.*, **70**, 20 (1931); *Phys. Rev.*, **37**, 715 (1931).
11. H. Peterson, *Z. Phys.*, **76**, 768 (1932); **80**, 258 (1933); **98**, 569 (1936).
12. A. I. Kostarev, *Zh. Eksp. Teor. Fiz.*, **11**, 60 (1941); **19**, 413 (1949).
13. M. Sawada, *Ann. Rep. Sci. Works Fac. Sci. Osaka Univ.*, **1**, 1 (1959).
14. V. V. Shmidt, *Bull. Acad. Sci. USSR Phys. Ser.*, **25**, 998 (1961).
15. L. V. Azaroff and D. M. Pease, in L. V. Azaroff, Ed., *X-ray Spectroscopy*, McGraw-Hill, New York, 1974, Chap. 6.
16. D. E. Sayers, E. A. Stern, and F. W. Lytle, *Phys. Rev. Lett.*, **27**, 1204 (1971).
17. F. W. Lytle, "Determination of Interatomic Distances from X-ray Absorption Fine Structure," in G. R. Mallett, M. J. Fay, and W. M. Mueller, Eds., *Advances in X-ray Analysis*, Vol. 9, Plenum, New York, 1966, p. 398.

18. D. D. Ivanenko and I. Pomeranchuk, *Phys. Rev.*, **65**, 343 (1944).

19. J. Schwinger, *Phys. Rev.*, **70**, 798 (1946); *Phys. Rev.*, **75**, 1912 (1949).

20. F. R. Elder, A. M. Gurewitsch, R. V. Langmuir, and H. L. Pollock, *J. Appl. Phys.*, **18**, 810 (1947).

21. P. A. Lee and J. B. Pendry, *Phys. Rev. B*, **11**, 2795 (1975).

22. C. A. Ashley and S. Doniach, *Phys. Rev. B*, **11**, 1279 (1975).

23. E. A. Stern, B. A. Bunker, and S. M. Heald, *Phys. Rev. B*, **21**, 5521 (1980).

24. W. L. Schaich, *Phys. Rev. B*, **29**, 6513 (1984).

25. S. J. Gurman, N. Binsted, and I. Ross, *J. Phys. C*, **17**, 143 (1984).

26. P. A. Lee and G. Beni, *Phys. Rev. B*, **15**, 2862 (1977).

27. B.-K. Teo and P. A. Lee, *J. Am. Chem. Soc.*, **101**, 2815 (1979).

28. L. I. Schiff, *Quantum Mechanics*, McGraw-Hill, New York, 1968, p. 123.

29. R. F. Pettifer and A. D. Cox, in A. Bianconi, L. Incoccia, and S. Stipcich, Eds., *EXAFS and Near Edge Structure*, Springer-Verlag, Berlin, 1983, p. 67.

30. J. J. Rehr and S.-H. Chou, in A. Bianconi, L. Incoccia, and S. Stipcich, Eds., *EXAFS and Near Edge Structure*, Springer-Verlag, Berlin, 1983, p. 22.

31. J. E. Müller and W. L. Schaich, *Phys. Rev. B*, **27**, 6489 (1983).

32. G. Bunker and E. A. Stern, *Phys. Rev. Lett.*, **52**, 1990 (1984).

33. J. B. Pendry, in A. Bianconi, L. Incoccia, and S. Stipcich, Eds., *EXAFS and Near Edge Structure*, Springer-Verlag, Berlin, 1983, p. 4.

34. V. A. Biebesheimer, E. C. Marques, D. R. Sandstrom, F. W. Lytle, and R. B. Greegor, *J. Chem. Phys.*, **81**, 2599 (1984).

35. B. Lengeler, *Phys. Rev. Lett.*, **53**, 74 (1984).

36. W. L. Schaich, in K. O. Hodgson, B. Hedman, and J. E. Penner-Hahn, Eds., *EXAFS and Near Edge Structure III*, Springer-Verlag, Berlin, 1984, p. 2.

37. J. J. Rehr and E. A. Stern, *Phys. Rev. B*, **14**, 4413 (1976).

38. B.-K. Teo, *J. Am. Chem. Soc.*, **103**, 3990 (1981).

39. J. J. Boland, S. E. Crane, and J. D. Baldeschwieler, *J. Chem. Phys.*, **77**, 142 (1982).

40. J. J. Boland and J. D. Baldeschwieler, *J. Chem. Phys.*, **80**, 3005 (1984).

41. N. Alberding and E. D. Crozier, *Phys. Rev. B*, **27**, 3374 (1983).

42. N. Alberding and E. D. Crozier, in K. O. Hodgson, B. Hedman, and J. E. Penner-Hahn, Eds., *EXAFS and Near Edge Structure III*, Springer-Verlag, Berlin, 1984, p. 30.

43. J. J. Rehr, E. A. Stern, R. L. Martin, and E. R. Davidson, *Phys. Rev. B*, **17**, 560 (1978).

44. U. Fano and J. W. Cooper, *Rev. Mod. Phys.*, **40**, 441 (1968).

45. T. A. Carlson, *Photoelectron and Auger Spectroscopy*, Plenum, New York, 1975.

46. S.-H. Chou, unpublished doctoral dissertation, University of Washington, Seattle, WA, 1982.

47. R. L. Martin and E. R. Davidson, *Phys. Rev. A*, **16**, 1341 (1977).

48. E. A. Stern, S. M. Heald, and B. Bunker, *Phys. Rev. Lett.*, **42**, 1372 (1979).

49. M. P. Seah and W. A. Dench, *Surf. Interface Anal.*, **1**, 2 (1979).

50. M. O. Krause and J. H. Oliver, *J. Phys. Chem. Ref. Data*, **8**, 329 (1979).

51. K. Kim and E. A. Stern, in K. O. Hodgson, B. Hedman, and J. E. Penner-Hahn, Eds., *EXAFS and Near Edge Structure III*, Springer-Verlag, Berlin, 1984, p. 20.

52. G. Beni and P. M. Platzman, *Phys. Rev. B*, **14**, 1514 (1976).

53. J. J. Rehr, *unpublished*.

54. G. Bunker, *Nucl. Instrum. Methods*, **207**, 437 (1983).

55. M. G. Kendall, *The Advanced Theory of Statistics*, Vol. 1, Griffin, London, 1948, pp. 60–64.

56. P. Eisenberger and G. S. Brown, *Solid State Commun.*, **29**, 481 (1979).

57. E. D. Crozier and A. J. Seary, *Can. J. Phys.*, **58**, 1390 (1980).

58. J. B. Boyce, T. M. Hayes, and J. C. Mikkelsen, Jr., *Phys. Rev. B*, **23**, 2867 (1981).

59. E. Sevillano, H. Meuth, and J. J. Rehr, *Phys. Rev. B*, **20**, 4908 (1979).

60. O. H. Nielsen and W. Weber, *J. Phys. C*, **13**, 2449 (1980).

61. J. M. Tranquada and R. Ingalls, *Phys. Rev. B*, **28**, 3520 (1983).

62. G. S. Knapp, H. K. Pan, and J. M. Tranquada, *Phys. Rev. B*, **32**, 2006 (1985).

63. R. B. Greegor and F. W. Lytle, *Phys. Rev. B*, **20**, 4902 (1979).

64. W. Böhmer and P. Rabe, *J. Phys. C*, **12**, 2465 (1979).

65. G. Martens, P. Rabe, N. Schwentner, and A. Werner, *Phys. Rev. Lett.*, **39**, 1411 (1977).

66. P. A. Lee, *Phys. Rev. B*, **13**, 5261 (1976).

67. S. M. Heald and E. A. Stern, *Phys. Rev. B*, **16**, 5549 (1977).

68. E. A. Stern, *Phys. Rev. B*, **10**, 3027 (1974).

69. P. J. Durham, J. B. Pendry, and C. H. Hodges, *Solid State Commun.*, **38**, 159 (1981).

70. G. N. Greaves, P. J. Durham, G. Diakun, and P. Quinn, *Nature (London)*, **294**, 139 (1981).

71. C. R. Natoli, D. K. Misemer, S. Doniach, and F. W. Kutzler, *Phys. Rev. A*, **22**, 1104 (1980).

72. F. W. Kutzler, C. R. Natoli, D. K. Misemer, S. Doniach, and K. O. Hodgson, *J. Chem. Phys.*, **73**, 3274 (1980).

73. S.-H. Chou, F. W. Kutzler, D. E. Ellis, G. K. Shenoy, T. I. Morrison, and P. A. Montano, *Phys. Rev. B*, **31**, 1069 (1985).

74. J. E. Müller, O. Jepsen, O. K. Andersen, and J. W. Wilkins, *Phys. Rev. Lett.*, **40**, 720 (1978).

75. J. E. Müller, O. Jepsen, and J. W. Wilkins, *Solid State Commun.*, **42**, 365 (1982).

76. J. E. Müller and J. W. Wilkins, *Phys. Rev. B*, **29**, 4331 (1984).

77. I. I. Gegusin, V. N. Datsyuk, and R. V. Vedrinskii, *Phys. Status Solidi B*, **109**, 563 (1982).

78. R. V. Vedrinskii, I. I. Gegusin, V. N. Datsyuk, A. A. Novakovich, and V. L. Kraizman, *Phys. Status Solidi B*, **111**, 433 (1982).

79. R. V. Vedrinskii, L. A. Bugaev, I. I. Gegusin, V. L. Kraizman, A. A. Novakovich, S. A. Prosandeev, R. E. Ruus, A. A. Maiste, and M. A. Elango, *Solid State Commun.*, **44**, 1401 (1982).

80. S. Doniach, M. Berding, T. Smith, and K. O. Hodgson, in K. O. Hodgson, B. Hedman, and J. E. Penner-Hahn, Eds., *EXAFS and Near Edge Structure III*, Springer-Verlag, Berlin, 1984, p. 33.

81. R. A. Bair and W. A. Goddard III, *Phys. Ser.*, **31**, 936 (1967).

82. E. A. Stern, *Phys. Rev. Lett.*, **49**, 1353 (1982).

83. I. A. Ovsyannikova, S. S. Batsanov, L. I. Nasonova, L. R. Batsanova, and E. A. Nekrasova, *Bull. Acad. Sci. USSR Phys. Ser.*, **31**, 936 (1967).

84. S. P. Cramer, T. K. Eccles, F. W. Kutzler, K. O. Hodgson, and L. E. Mortenson, *J. Am. Chem. Soc.*, **98**, 1287 (1976).

85. F. Sette, J. Stöhr, and A. P. Hitchcock, *J. Chem. Phys.*, **81**, 4906 (1984).

86. C. R. Natoli, in K. O. Hodgson, B. Hedman, and J. E. Penner-Hahn, Eds., *EXAFS and Near Edge Structure III*, Springer-Verlag, Berlin, 1984, p. 38.

87. F. W. Lytle, *Acta Cryst.*, **22**, 321 (1967).

88. J. E. Hahn, R. A. Scott, K. O. Hodgson, S. Doniach, S. R. Desjardins, and E. I. Solomon, *Chem. Phys. Lett.*, **88**, 595 (1982).

89. M. Brown, R. E. Peierls, and E. A. Stern, *Phys. Rev. B*, **15**, 738 (1977).

90. P. S. P. Wei and F. W. Lytle, *Phys. Rev. B*, **19**, 679 (1979).

91. D. E. Sayers and B. A. Bunker, "Data Analysis," in D. Koningsberger and R. Prins, Eds., *X-Ray Absorption: Principles, Applications, Techniques of EXAFS, SEXAFS, and XANES*, Wiley, New York, 1988.

92. P. A. Lee, P. H. Citrin, P. Eisenberger, and B. M. Kincaid, *Rev. Mod. Phys.*, **53**, 769 (1981). B. Koch and C. H. MacGillavry, "X-Ray Absorption," in K. Lonsdale, Eds., *International Tables for X-ray Crystallography*, Vol. III, Kynoch, Birmingham, England, 1962.

93. W. H. McMaster, N. Nerr Del Grande, J. H. Mallett, and J. H. Hubbell, *Compilation of X-ray Cross Sections*, Lawrence Radiation Laboratory UCRL-50174 Sec. II Rev. 1, 1969.

94. B. W. Holland, J. B. Pendry, R. F. Pettifer, and J. Bordas, *J. Phys. C*, **11**, 633 (1978).

95. C. Bouldin and E. A. Stern, *Phys. Rev. B*, **25**, 3462 (1982).

96. E. O. Brigham, *The Fast Fourier Transform*, Prentice-Hall, Englewood Cliffs, NJ, 1974.

97. G. Martens, P. Rabe, N. Schwentner, and A. Werner, *Phys. Rev. B*, **17**, 1481 (1981).

98. T. M. Hayes, P. N. Sen, and S. H. Hunter, *J. Phys. C*, **9**, 4357 (1976).

99. E. A. Stern, D. E. Sayers, and F. W. Lytle, *Phys. Rev. B*, **11**, 4836 (1975).

100. B. Lengeler and P. Eisenberger, *Phys. Rev. B*, **21**, 4507 (1980).

101. P. J. Mallozzi, R. E. Schwerzel, H. M. Epstein, and B. E. Campbell, *Phys. Rev. A*, **23**, 824 (1981); R. W. Eason, D. K. Bradley, J. D. Killkenny, and G. N. Greaves, *J. Phys. C*, **17**, 5067 (1984).

102. T. Matsushita, "X-ray Absorption Spectrometer with a Dispersive as Well as Focusing Optical System," in E. A. Stern, Ed., *Laboratory EXAFS Facilities—1980*, American Institute of Physics, New York, 1980, p. 109.

103. M. Hida, M. Maeda, N. Kamijo, and H. Terauchi, *Phys. Status Solidi A*, **69**, 297 (1982).

104. A. M. Flank, A. Fontaine, A. Jucha, M. Lemonnier, C. Williams, *J. Phys. Paris Lett.*, **43**, L315 (1982).

105. M. H. Chen, B. Crasemann, and H. Mark, *Phys. Rev. A*, **21**, 436 (1980).

106. M. O. Krause, *Phys. Rev. A*, **22**, 1958 (1980).

107. E. A. Stern, *J. Vac. Sci. Technol.*, **14**, 461 (1977).

108. P. A. Lee, *Phys. Rev. B*, **21**, 4507 (1980).

109. U. Landman and D. L. Adams, *Proc. Natl. Acad. Sci. USA*, **73**, 2550 (1975).

110. P. H. Citrin, P. Eisenberger, and R. C. Hewitt, *Phys. Rev. Lett.*, **41**, 309 (1978).

111. C. Noguera and D. Spanjaard, *Surf. Sci.*, **108**, 381 (1981).

112. G. M. Rothberg, K. M. Choudhary, M. L. den Boer, G. P. Williams, M. H. Hecht, and I. Lindau, *Phys. Rev. Lett.*, **53**, 1183 (1984).

113. W. Gudat and C. Kunz, *Phys. Rev. Lett.*, **29**, 169 (1972).

114. G. P. Martens, P. Rabe, N. Schwentner, and A. Werner, *J. Phys. C*, **11**, 3125 (1978).

115. J. Stöhr, D. Denley, and P. Perfetti, *Phys. Rev. B*, **18**, 4132 (1978).

116. J. Stöhr, C. Noguera, and T. Kendelewicz, *Phys. Rev. B*, **30**, 5571 (1984).

117. L. G. Parratt, *Phys. Rev.*, **95**, 359 (1954).

118. L. Bosio, R. Cortes, and M. Forment, in K. O. Hodgson, B. Hedman, and J. E. Penner-Hahn, Eds., *EXAFS and Near Edge Structure III*, Springer-Verlag, Berlin, 1984, p. 479.

119. S. M. Heald, J. M. Tranquada, D. O. Welch, and H. Chen, "The Characterization of Thin Films and Layered Structures Using X-ray Absorption and Reflection at Grazing Incidence," in J. M. Gibson and L. R. Dawson, Eds., *Layered Structures, Epitaxy and Interfaces*, Materials Research Society, Pittsburgh, PA, 1985, p. 437.

120. G. Martens and P. Rabe, *Phys. Status Solidi*, **58**, 415 (1980).

121. R. Barchewitz, M. Cremonese-Visicato, and G. Onori, *J. Phys. C*, **11**, 4439 (1978).

122. R. Fox and S. J. Gurman, *J. Phys. C*, **13**, L249 (1980).

123. S. M. Heald, E. Keller, and E. A. Stern, *Phys. Lett. A*, **103**, 155 (1984).

124. R. S. Becker, J. A. Golovchenko, and J. R. Patel, *Phys. Rev. Lett.*, **50**, 153 (1983).

125. F. W. Lytle and D. R. Sandstrom, SSRL Report No. 78/09, *5th Annual SSRL User Group Meeting, October, 1978*, Stanford Synchrotron Radiation Laboratory, Stanford, CA, 1978.

126. A. Bianconi, D. Jackson, and K. Monahan, *Phys. Rev. B*, **17**, 2021 (1978).

127. J. Goulon, P. Tola, M. Lemonnier, and J. Dexpert-Ghys, *Chem. Phys.*, **78**, 347 (1983).

128. T. K. Sham and S. M. Heald, *J. Am. Chem Soc.*, **105**, 5142 (1983).

129. T. K. Sham and R. A. Holroyd, *J. Phys. Chem.*, **80**, 1072 (1984).

130. J. Stohr, R. Jaeger, and S. Brennan, *Surf. Sci.*, **118**, 503 (1982).

131. M. E. Rose and M. M. Shapiro, *Phys. Rev.*, **74**, 1953 (1948).

132. S. M. Heald, "Design of an EXAFS Experiment," in D. Koningsberger and R. Prins, Eds., *X-ray Absorption: Principles, Applications, Techniques of EXAFS, SEXAFS and XANES*, Wiley, New York, 1988.

133. J. M. Jaklevic, J. A. Kirby, M. P. Klein, A. S. Robertson, G. S. Brown, and P. Eisenberger, *Solid State Commun.*, **23**, 679 (1977).

134. E. A. Stern and S. M. Heald, *Rev. Sci. Instrum.*, **50**, 1579 (1979).

135. J. B. Hastings, P. Eisenberger, B. Lengeler, and M. L. Perlman, *Phys. Rev. Lett.*, **43**, 1807 (1979).

136. M. Marcus, L. S. Powers, A. R. Storm, B. M. Kincaid, and B. Chance, *Rev. Sci. Instrum.*, **51**, 1023 (1980).

137. L. G. Parratt, C. F. Hempstead, and E. L. Jossem, *Phys. Rev.*, **105**, 1228 (1957).

138. E. A. Stern and K. Kim, *Phys. Rev. B*, **23**, 3781 (1981).

139. J. Goulon, C. Goulon-Ginet, R. Cortes, and J. M. Dubois, *J. Phys. Les Ulis Fr.*, **43**, 539 (1982).

140. S. M. Heald, in A. Bianconi, L. Incoccia, and S. Stipcich, Eds., *EXAFS and Near Edge Structure*, Springer-Verlag, Berlin, 1983, p. 98.

141. E. A. Stern and K.-Q. Lu, *Nucl. Instrum. Methods*, **212**, 475 (1983).

142. P. H. Citrin, P. Eisenberger, and J. E. Rowe, *Phys. Rev. Lett.*, **48**, 802 (1982).

143. J. Stöhr, L. Johansson, I. Lindau, and P. Pianetta, *Phys. Rev. B*, **20**, 664 (1979).

144. E. A. Stern and K.-Q. Lu, *Nucl. Instrum. Methods*, **195**, 415 (1982).

145. V. O. Kostroun, *Nucl. Instrum. Methods*, **172**, 243 (1980).

146. P. Rabe, G. Tolkiehn, and A. Werner, *Nucl. Instrum. Methods*, **171**, 329 (1980).

147. J. B. Hastings, B. M. Kincaid, and P. Eisenberger, *Nucl. Instrum. Methods*, **152**, 167 (1978).

148. S. M. Heald and J. B. Hastings, *Nucl. Instrum. Methods*, **187**, 553 (1981).

149. J. Boulon, M. Lemonnier, R. Cortes, A. Retournard, and D. Raoux, *Nucl. Instrum. Methods*, **208**, 625 (1983).

150. D. E. Sayers, S. M. Heald, M. A. Pick, J. I. Budnick, E. A. Stern, and J. Wong, *Nucl. Instrum. Methods*, **208**, 631 (1983).

151. E. E. Koch, Ed., *Handbook on Synchrotron Radiation*, Vol. 1, North-Holland, Amsterdam, 1983.

152. H. Winick and S. Doniach, Eds., *Synchrotron Radiation Research*, Plenum, New York, 1980.

153. D. M. Mills and B. W. Batterman, Eds., *Synchrotron Radiation Instrumentation*, North-Holland, Amsterdam, 1982.

154. E. E. Koch, Ed., *Proceedings of the International Conference on X-ray and VUV Synchrotron Radiation Instrumentation, Hamburg, FRG, Aug. 9–13, 1982*; *Nucl. Instrum. Methods*, **208** (1983).

155. W. Thomlinson and G. P. Williams, Eds., *Synchrotron Radiation Instrumentation 3*, North-Holland, Amsterdam, 1984.

156. E. A. Stern, Ed., *Laboratory EXAFS Facilities—1980*, American Institute of Physics, New York, 1980.

157. G. S. Knapp, H. Chen, and T. E. Klippert, *Rev. Sci. Instrum.*, **49**, 1658 (1978).

158. G. G. Cohen, D. A. Fischer, J. Colbert, and N. J. Shevchik, *Rev. Sci. Instrum.*, **51**, 273 (1980).

159. S. Khalid, R. Emrich, R. Dujari, J. Schultz, and J. R. Katzer, *Rev. Sci. Instrum.*, **53**, 22 (1982).

160. W. Thulke, R. Haaensel, and P. Rabe, *Rev. Sci. Instrum.*, **54**, 277 (1983).

161. S. T. Stephenson, "The Continuous X-ray Spectrum," in S. Flügge, *Handbuch der Physik*, Vol. 30, Springer-Verlag, Berlin, 1957, p. 338.

162. C. H. Westcott, *Proc. R. Soc. London*, **194**, 508 (1948).

163. U. W. Arndt, *J. Phys. E*, **11**, 671 (1978).

164. H. H. Johann, *Z. Phys.*, **69**, 185 (1931).

165. J. W. M. Dumond and H. A. Kirkpatrick, *Rev. Sci. Instrum.*, **1**, 88 (1930).

166. T. Johansson, *Z. Phys.*, **82**, 507 (1933).

167. K.-Q. Lu and E. A. Stern, "Johann and Johansson Focusing Arrangements: Analytical Analysis," in E. A. Stern, Ed., *Laboratory EXAFS Facilities—1980*, American Institute of Physics, New York, 1980, p. 104.

168. D. Mills and V. Pollock, *Rev. Sci. Instrum.*, **51**, 1664 (1980).

169. A. H. Compton and S. K. Allison, *X-rays in Theory and Experiment*, Van Nostrand, New York, 1935.

170. R. W. James, *The Optical Principles of the Diffraction of X-rays*, Cornell University Press, Ithaca, NY, 1948.

171. A. Franks, *Sci. Prog. Oxford*, **64**, 371 (1977).

172. V. Rehn and R. O. Jones, *Opt. Eng.*, **17**, 504 (1978).

173. R. E. Engdahl, "Chemical Vapor Deposited (CVD) Silicon Carbide Mirror Technology," in M. Howells, Ed., *Reflecting Optics for Synchrotron Radiation*, SPIE Proceedings, **315**, 1981, p. 123.

174. C. J. Sparks, G. E. Ice, J. Wong, and B. W. Batterman, *Nucl. Instrum. Methods*, **195**, 73 (1982).

175. S. M. Heald, *Nucl. Instrum. Methods*, **222**, 160 (1984).

176. S. M. Heald, "EXAFS with Synchrotron Radiation," in D. Koningsberger and R. Prins, Eds., *X-ray Absorption: Principles, Applications, Techniques or EXAFS, SEXAFS, and XANES*, Wiley, New York, 1988.

177. N. J. Shevchik, *Philos. Mag.*, **35**, 805 (1979).

178. P. Fuoss, P. Eisenberger, W. Warburton, and A. Bienenstock, *Phys. Rev. Lett.*, **46**, 1537 (1981).

179. A. Bianconi, L. Incoccia, and S. Stipcich, *EXAFS and Near Edge Structure*, Springer-Verlag, Berlin, 1983.

180. K. O. Hodgson, B. Hedman, and J. E. Penner-Hahn, *EXAFS and Near Edge Structure III*, Springer-Verlag, Berlin, 1984.

181. N. Bartlett, R. N. Bianconi, B. W. McQuillan, A. S. Robertson, and A. C. Thompson, *J. Chem. Soc. Chem. Commun.*, 200 (1978).

182. S. M. Heald, H. A. Goldberg, and I. L. Kalnin, *EXAFS and Near Edge Structure*, Springer-Verlag, Berlin, 1983, p. 141.

183. S. M. Heald and E. A. Stern, *Phys. Rev. B*, **17**, 4069 (1978).

184. S. M. Heald and E. A. Stern, *Synth. Metals*, **1**, 249 (1979/80).

185. J. J. Lander and J. Morrison, *Surf. Sci.*, **6**, 1 (1967).

186. N. N. Efremova, L. D. Finkel'shtein, N. D. Samsonova, and S. A. Nemnonov, *Bull. Acad. Sci. USSR*, **40**, 170 (1976).

187. J. Röhler, G. Krill, J. P. Kappler, and M. F. Ravet, in A. Bianconi, L. Incoccia, and S. Stipcich, Eds., *EXAFS and Near Edge Structure*, Springer-Verlag, Berlin, 1983, p. 213.

188. R. Ingalls, J. M. Tranquada, J. E. Whitmore, R. Ingalls, and E. D. Crozier, in J. S. Schilling and R. N. Shelton, Eds., *Physics of Solids Under Pressure*, North-Holland, Amsterdam, 1981, p. 67.

189. J. Röhler, *J. Magn. Magn. Mater.*, **47–48**, 175 (1985).

190. A. Bianconi, S. Modesti, M. Campagna, K. Fischer, and S. Stizza, *J. Phys. C*, **14**, 4737 (1981).

191. E. Müller-Hartmann, B. Roden, and D. Wohlleben, Eds., "Proceedings of the Fourth International Conference on Valence Fluctuations," *J. Magn. Magn. Mater.*, **47–48** (1985).

192. G. Kaindl, W. D. Brewer, G. Kalkowski, and F. Holtzberg, *Phys. Rev. Lett.*, **51**, 2056 (1983).

193. W. D. Brewer, G. Kalkowski, G. Kaindl, and F. Holtzberg, *Phys. Rev. B*, **32**, 3676 (1985).

194. R. M. Martin, J. B. Boyce, J. W. Allen, and F. Holtzberg, *Phys. Rev. Lett.*, **44**, 1275 (1980).

195. H. Launois, M. Rawiso, E. Holland-Moritz, R. Pott and D. Wohlleben, *Phys. Rev. Lett.*, **44**, 1271 (1980).

196. G. Krill, J. P. Kappler, and J. Röhler, in A. Bianconi, L. Incoccia, and S. Stipcich, Eds., *EXAFS and Near Edge Structure*, Springer-Verlag, Berlin, 1983, p. 213.

197. J. Kortwright, W. Warburton, and A. Bienenstock, in A. Bianconi, L. Incoccia, and S. Stipcich, Eds., *EXAFS and Near Edge Structure*, Springer-Verlag, Berlin, 1983, p. 362.

198. T. M. Hayes and J. B. Boyce, *Solid State Phys.*, **37**, 173 (1982).

199. J. B. Boyce, T. M. Hayes, and J. C. Mikkelsen, Jr., *Phys. Rev. B*, **23**, 2876 (1981).

200. B.-K. Teo, H. S. Chen, R. Wang, and M. R. Antonio, *J. Non-Cryst. Solids*, **58**, 249 (1983).

201. G. S. Cargill III, *J. Non-Cryst. Solids*, **61/62**, 261 (1984).

202. G. Licheri and G. Pinna, in A. Bianconi, L. Incoccia, and S. Stipcich, Eds., *EXAFS and Near Edge Structure*, Springer-Verlag, Berlin, 1983, p.240.

203. G. N. Greaves, *J. Non-Cryst. Solids*, **71**, 203 (1985).

204. S. J. Gurman, *J. Mater. Sci.*, **17**, 1541 (1982).

205. F. W. Lytle, P. S. P. Wei, R. B. Greegor, G. H. Via, and J. H. Sinfelt, *J. Chem. Phys.*, **70**, 4849 (1979).

206. A. N. Mansour, J. W. Cook, Jr., and D. E. Sayers, *J. Phys. Chem.*, **88**, 2330 (1984).

207. A. N. Mansour, J. W. Cook, Jr., D. E. Sayers, R. J. Emrich, and J. R. Katzer, *J. Catal.*, **89**, 462 (1984).

208. F. W. Lytle, G. H. Via, and J. H. Sinfelt, "X-Ray Absorption Spectroscopy: Catalyst Applications," in H. Winick and S. Doniach, Eds., *Synchrotron Radiation Research*, Plenum, New York, 1980, p. 401.

209. R. B. Greegor and F. W. Lytle, *J. Catal.*, **63**, 476 (1980).

210. D. C. Koningsberger, J. B. A. D. van Zon, H. F. J. van't Blik, G. J. Visser, R. Prins, A. N. Mansour, D. E. Sayers, D. R. Short, and J. R. Katzer, *J. Phys. Chem.*, **89**, 4075 (1985).

211. G. H. Via, G. Meitzner, J. H. Sinfelt, R. B. Greegor, and F. W. Lytle, in K. O. Hodgson, B. Hedman, and J. E. Penner-Hahn, Eds., *EXAFS and Near Edge Structure III*, Springer-Verlag, Berlin, 1984, p. 176.

212. D. E. Sayers, D. Bazin, H. Dexpert, A. Jucha, E. Dartyge, A. Fontaine, and P. Lagarde, in K. O. Hodgson, B. Hedman, and J. E. Penner-Hahn, Eds., *EXAFS and Near Edge Structure III*, Springer-Verlag, Berlin, 1984, p. 209. P. A. Montano, W. Schulze, B. Tesche, G. K. Shenoy, and T. I. Morrison, *Phys. Rev. B*, **30**, 672 (1984). P. A. Montano, G. K. Shenoy, E. E. Alp, W. Schulze, and J. Urban, *Phys. Rev. Lett.*, **56**, 2076 (1986).

213. P. Lagarde and H. Dexpert, *Adv. Phys.*, **33**, 567 (1984).

214. R. Ingalls, E. D. Crozier, J. E. Whitmore, A. J. Seary, and J. M. Tranquada, *J. Appl. Phys.*, **51**, 3158 (1980).

215. R. Ingalls, J. M. Tranquada, J. E. Whitmore, E. D. Crozier, and A. J. Seary, "Extended X-Ray Absorption Fine Structure Studies at High Pressure," in B. K. Teo and D. C. Joy, Eds., *EXAFS Spectroscopy: Techniques and Applications*, Plenum, New York, 1981, p. 127.

216. R. Ingalls, J. M. Tranquada, J. E. Whitmore, and E. D. Crozier, in J. S. Schilling and R. N. Shelton, Eds., *Physics of Solids Under High Pressure*, North-Holland, Amsterdam, 1981, p. 67.

217. J. Röhler, J. P. Kappler, and G. Krill, *Nucl. Instrum. Methods*, **208**, 647 (1983).

218. J. Röhler, in K. O. Hodgson, B. Hedman, and J. E. Penner-Hahn, Eds., *EXAFS and Near Edge Structure III*, Springer-Verlag, Berlin, 1984, p. 379.

219. J. Röhler, K. Keulerz, E. Dartyge, A. Fontaine, A. Jucha, and D. Sayers, in K. O. Hodgson, B. Hedman, and J. E. Penner-Hahn, Eds., *EXAFS and Near Edge Structure III*, Springer-Verlag, Berlin, 1984, p. 385.

220. J. M. Tranquada, R. Ingalls, and E. D. Crozier, in K. O. Hodgson, B. Hedman, and J. E. Penner-Hahn, Eds., *EXAFS and Near Edge Structure III*, Springer-Verlag, Berlin, 1984, p. 374.

221. R. Ingalls, J. M. Tranquada, J. E. Whitmore, and E. D. Crozier, in A. Bianconi, L. Incoccia, and S. Stipcich, Eds., *EXAFS and Near Edge Structure*, Springer-Verlag, Berlin, 1983, p. 154.

222. J. M. Tranquada and R. Ingalls, *Phys. Rev. B*, **34**, 4267 (1986).

223. J. Stohr, in R. Prins and D. Koningsberger, Eds., *X-ray Absorption: Principles, Applications, Techniques of EXAFS, SEXAFS and XANES*, Wiley, New York, 1988.

224. D. Norman, *J. Phys. C*, **19**, 3273 (1986).

225. J. Haase, *Appl. Phys. A*, **38**, 181 (1985).

226. J. Stohr, R. Jaeger, and S. Brennen, *Surf. Sci.*, **117**, 503 (1982).

227. J. Stohr, E. B. Kollin, D. A. Fischer, J. B. Hastings, F. Zaera, and F. Sette, *Phys. Rev. Lett.*, **55**, 1468 (1985).

228. J. Stohr, J. C. Gland, W. Eberhardt, D. Outka, R. J. Maix, F. Sette, R. J. Koestner, and U. Dobler, *Phys. Rev. Lett.*, **51**, 2414 (1983).

229. U. Dobler, K. Babersekke, J. Haase, and A. Puschmann, *Phys. Rev. Lett.*, **52**, 1437 (1984).

230. M. Bader, J. Haase, K. H. Frank, A. Puschmann, and A. Otto, *Phys. Rev. Lett.*, **56**, 1921 (1986).

231. O. Petitpierre, E. A. Stern, and Y. Yacoby, in K. O. Hodgson, B. Hedman, and J. E. Penner-Hahn, Eds., *EXAFS and Near Edge Structure III*, Springer-Verlag, Berlin, 1984, p. 509.

232. C. R. A. Cattlow, A. V. Chadwick, G. N. Greaves, and L. M. Moroney, *J. Am. Ceram. Soc.*, **69**, 272 (1986).

233. A. I. Goldman, E. Canova, Y. H. Kao, W. L. Roth, and R. Wong, in K. O. Hodgson, B. Hedman, and J. E. Penner-Hahn, Eds., *EXAFS and Near Edge Structure III*, Springer-Verlag, Berlin, 1984, p. 442.

234. A. Erbil, W. Weber, G. S. Cargill, and R. F. Boehme, *Phys. Rev. B*, **34**, 1392 (1986).

235. J. Wong, L. L. Spiro, D. H. Maylotte, F. W. Lytle, and R. B. Greegor, and C. Spiro, J. Wong, D. H. Maylotte, S. Lamson, B. Glover, F. W. Lytle, and R. B. Greegor, in K. O. Hodgson, B. Hedman, and J. E. Penner-Hahn, Eds., *EXAFS and Near Edge Structure III*, Springer-Verlag, Berlin, 1984, pp. 362 and 369, respectively.

236. A. Fontaine, P. Lagarde, A. Naudon, D. Raoux, and D. Spanjaard, *Philos. Mag. B*, **4**, 17 (1979).

237. D. Raoux, A. Fontaine, P. Lagarde, and A. Sadoc, *Phys. Rev. B*, **24**, 5547 (1981).

238. J. M. Minault, A. Fontaine, P. Lagarde, D. Raoux, A. Sadoc, and D. Spanjaard, *J. Phys. F*, **11**, 1311 (1981).

239. S. M. Heald and J. M. Tranquada, "The Characterization of Cryogenic Materials by X-ray Absorption Methods," in R. P. Reed and A. F. Clark, Eds., *Advances in Cryogenic Engineering-Materials*, Vol. 32, Plenum, New York, 1986, p. 471.

240. G. P. Huffman, F. E. Huggins, L. J. Cuddy, R. W. Schoenberger, F. W. Lytle, and R. B. Greegor, in K. O. Hodgson, B. Hedman, and J. E. Penner-Hahn, Eds., *EXAFS and Near Edge Structure III*, Springer-Verlag, Berlin, 1984, p. 371.

241. J. C. Mikkelson and J. B. Boyce, *Phys. Rev. B*, **28**, 7130 (1983).

242. S. P. Cramer and K. O. Hodgson, *Prog. Inorg. Chem.*, **25**, 1 (1979).

243. S. P. Cramer, in H. B. Stuhrman, Ed., *Uses of Synchrotron Radiation in Biology*, Academic, London, 1979, p. 291.

244. R. G. Shulman, P. Eisenberger, and B. M. Kincaid, *Annu. Rev. Biophys. Bioeng.*, **7**, 559 (1978).

245. S. Doniach, P. Eisenberger, and K. O. Hodgson, "X-ray Absorption Spectroscopyt of Biological Molecules," in H. Winick and S. Doniach, Eds., *Synchrotron Radiation Research*, Plenum, New York, 1980, p. 425.

246. L. Powers, *Biochim. Biophys. Acta*, **683**, 1 (1982).

247. R. A. Scott, in D. L. Roussean, Ed., *Structural and Resonance Techniques in Biological Research*, Vol. 2, Academic, New York, 1984, Chap. 4.

248. D. E. Sayers, F. W. Lytle, M. Weissbluth, and P. Pianetta, *J. Chem. Phys.*, **62**, 2514 (1975).

249. R. G. Shulman, P. Eisenberger, W. E. Blumberg, and N. A. Stombaugh, *Proc. Natl. Acad. Sci. USA*, **72**, 4003 (1975).

250. D. E. Sayers, E. A. Stern, and J. R. Herriot, *J. Chem. Phys.*, **64**, 427 (1976).

251. B. M. Kincaid, P. Eisenberger, K. O. Hodgson, and S. Doniach, *Proc. Natl. Acad. Sci. USA*, **72**, 2340 (1975).

252. P. Eisenberger, R. G. Shulman, G. S. Brown, and S. Ogawa, *Proc. Natl. Acad. Sci. USA*, **73**, 491 (1976).

253. P. Eisenberger, R. G. Shulman, B. M. Kincaid, G. S. Brown, and S. Ogawa, *Nature (London)*, **274**, 30 (1978).

254. M. F. Perutz, S. S. Hasnain, P. J. Duke, J. L. Sessler, and J. E. Hahn, *Nature (London)*, **295**, 535 (1982).

255. R. A. Scott, J. E. Penner-Hahn, K. O. Hodgson, H. Beinert, and C. D. Stout in K. O. Hodgson, B. Hedman, and J. E. Penner-Hahn, Eds., *EXAFS and Near Edge Structure III*, Springer-Verlag, Berlin, 1984, p. 105.

256. J. E. Hahn, R. A. Scott, K. O. Hodgson, S. Doniach, S. R. Dejardins, and E. I. Solomon, *Chem. Phys. Lett.*, **88**, 595 (1982).

257. U. C. Srivasta and H. L. Nigam, *Coord. Chem. Rev.*, **9**, 275 (1972).

258. R. G. Shulman, Y. Yafet, P. Eisenberger, and W. E. Blumberg, *Proc. Natl. Acad. Sci. USA*, **73**, 1384 (1976).

259. S. P. Cramer, K. O. Hodgson, W. O. Gillum, and L. E. Mortenson, *J. Am. Chem. Soc.*, **100**, 3398 (1978).

260. L. A. Grunes, *Phys. Rev. B*, **27**, 2111 (1983).

261. P. J. Durham, A. Bianconi, A. Congiu-Castellano, A. Giovannelli, S. S. Hasnain, L. Incoccia, S. Morante, and J. B. Pendry, *EMBO J.*, **2**, 1441 (1983).

262. A. Bianconi, A. Congiu-Castellano, M. Dell'Ariccia, A. Giovannelli, E. Burattini, M. Castagnola, and P. J. Durham, *Biochem. Biophys. Acta*, **831**, 120 (1985).

263. A. Bianconi, A. Congiu-Castellano, P. J. Durham, S. S. Hasnain, and S. Phillips, *Nature (London)*, **318**, 685 (1985).

Chapter 4

MÖSSBAUER SPECTROSCOPY

Frank J. Berry

1   INTRODUCTION

The discovery in 1958 by Rudolf Mössbauer [1] of recoil-free nuclear resonance fluorescence, together with a theoretical explanation of the effect that now bears his name, gave birth to the technique that is now known as *Mössbauer spectroscopy*. Since the late 1950s Mössbauer spectroscopy has developed from an elegant experiment in nuclear physics to a technique that has made important contributions to such diverse areas of science as nuclear and solid-state physics, structural and analytical chemistry, surface science, molecular biology and biophysics, materials science, archaeology, and the earth sciences.

As an experimental method of scientific investigation, it successfully complements other techniques such as visible, infrared (IR), and nuclear magnetic resonance (NMR) spectroscopy, and the various diffraction methods that can be used to study the way in which atoms exist in solids. However, some advantages of Mössbauer spectroscopy give it a special power in several important situations and applications. This chapter seeks to outline the basic theory of Mössbauer spectroscopy and to give some indication of its scope and limitation, to describe the type of instrumentation that is involved, and to illustrate the value of the technique by highlighting several areas of current application.

2   THEORETICAL PRINCIPLES OF MÖSSBAUER SPECTROSCOPY

Several textbooks [2–7] have dealt in detail with the theory of Mössbauer spectroscopy, and the reader requiring such detail is referred to these books. A more recent book [8] outlines the principles of Mössbauer spectroscopy and gives an account of the type of information that can be obtained from the interpretation of the spectra.

2.1   The Mössbauer Effect

Atomic resonance fluorescence was predicted and discovered in the early part of this century. The process can be envisaged in terms of the decay of an atom in an excited electronic state to the ground state by the emission of a photon that can then be absorbed by another atom during electronic excitation. The subsequent deexcitation and reemission of the photon in random directions gives rise to scattering or resonance fluorescence.

Nuclear resonance absorption can be considered in an analogous fashion. The decay of many radioactive nuclei in an excited state occurs by the emission of gamma (γ) rays, which, as recognized early in the 1920s, may excite other stable nuclei of the same isotope and give rise to nuclear resonant absorption and fluorescence. The initial attempts to detect these resonant processes were prevented by the effects of nuclear recoil that accompany both the emission and the absorption of γ rays by a free nucleus. The physical occurrence of nuclear recoil results in the γ ray being emitted with an energy slightly less than that corresponding to the difference between the two nuclear energy levels of the transition involved. This also precludes the resonant absorption of the γ ray by another nucleus since the γ ray fails to have enough energy to compensate for the recoil of the absorbing nucleus. The situation is described schematically in Figure 4.1, which shows how the proportion of absorption is determined by the overlap between the exciting and excited energy distributions. In one case the γ ray has lost energy E_R as a result of the recoil of a nucleus. For the reverse process, where the γ ray is absorbed by a nucleus, a further increment of energy E_R is required since the γ ray needs to provide both the nuclear excitation energy and the recoil energy of the absorbing atom. The amount of resonance overlap is illustrated by the shaded area in Figure 4.1 and is seen to be very small.

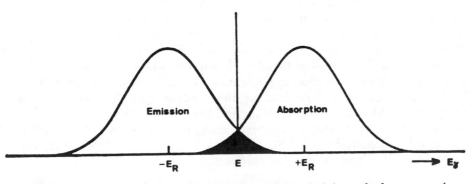

Figure 4.1   The small resonance overlap, as indicated by the shaded area, for free atom nuclear γ resonance.

In 1957 Mössbauer [1] discovered that a nucleus in a solid can sometimes emit and absorb γ rays without recoil because, when in a solid matrix, the nucleus is no longer isolated but is fixed within the lattice. When, under these circumstances, the recoil energy is less than the lowest quantized lattice vibrational energy, the γ ray can be emitted without loss of energy due to the recoil of the nucleus. Since the probability of such recoil-free events occurring depends on the energy of the γ ray, the Mössbauer effect is restricted to those isotopes with low-lying excited states. The probability is also dependent on the temperature and vibrational properties of the solid in which the Mössbauer nucleus is embedded.

In recoil-free events the energy distribution, or line width, of the γ ray depends on the lifetime of the excited nuclear state. Indeed, the ease with which the Mössbauer effect can be observed for a particular isotope is strongly dependent on the γ ray line width and is related to the Heisenberg uncertainty principle. The ground state of the nucleus has an infinite lifetime and therefore has no uncertainty in its energy. The uncertainty in the lifetime of the excited state is defined by its mean life τ and the uncertainty in its energy is described by the width of the statistical energy distribution at half-height Γ. These uncertainties are related by the expression

$$\Gamma\tau \geqslant \hbar$$

where $\hbar = h/2\pi$ is Planck's constant. For a typical nuclear excited-state half-life of $t_{1/2} \sim 10^{-7}$ s, Γ will be 4.562×10^{-9} eV. If the energy of the excited state is 45.62 keV, the emitted γ ray will have an intrinsic resolution of 1 part in 10^{13}. Given that the maximum resolution achieved in atomic line spectra is about 1 in 10^8, the line width of the Mössbauer γ ray gives the technique a very high resolution.

Therefore, it will be recognized that not all isotopes of every element are amenable to exploitation by the practical application of the Mössbauer effect. In general, the Mössbauer effect is optimized for low-energy γ rays associated with nuclei in which $t_{1/2}$ is usually between 10^{-9} and 10^{-7} s and that are strongly bound in a crystal lattice at low temperature.

A tabulation of those elements with isotopes suitable for use in Mössbauer spectroscopy and an indication of the ease by which the Mössbauer effect can be observed are given in Figure 4.2. Although the Mössbauer effect has been detected in over 100 isotopes, the practical application of the technique is restricted to a much smaller number of isotopes where the conditions for observing the Mössbauer effect and obtaining useful information are particularly favorable. Of these ^{57}Fe has the most advantageous combination of properties for Mössbauer spectroscopy, and the great majority of studies have involved the use of this isotope. However, many other isotopes as illustrated in Figure 4.2, especially ^{119}Sn, ^{121}Sb, ^{125}Te, ^{127}I, ^{129}I, ^{151}Eu, ^{182}W, ^{193}Ir, and ^{197}Au, have also found substantial application.

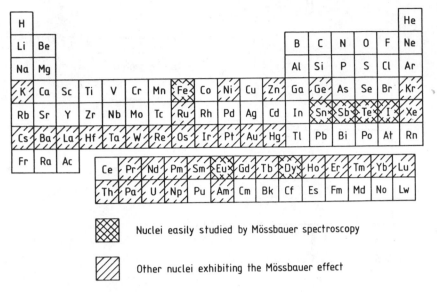

Nuclei easily studied by Mössbauer spectroscopy

Other nuclei exhibiting the Mössbauer effect

Figure 4.2   Elements containing isotopes suitable for Mössbauer spectroscopy.

2.2   Mössbauer Spectroscopy

The energy levels of an atomic nucleus are modified by its electronic environment. We use Mössbauer spectroscopy to examine these perturbations in the nuclear energy levels and to interpret the results in terms of the local environment of the nucleus.

To probe the energy states of a particular nucleus by the Mössbauer effect it is necessary to have a source consisting of the same nucleus in an excited state. For example, the source used in ^{57}Fe Mössbauer spectroscopy is frequently radioactive ^{57}Co, which (Figure 4.3) undergoes a spontaneous electron-capture transition to give metastable ^{57}Fe, which then decays to the ground state by releasing a cascade of γ rays. The 14.4-keV γ radiation is used in ^{57}Fe Mössbauer spectroscopy; and the investigation of iron-containing materials involves the exposure of the sample, which is called the *absorber*, to the radiation and the absorption of the 14.4-keV γ rays by the ^{57}Fe nuclei.

Since the energy states of nuclei depend on the interaction between the nuclei and their electronic environments, it is very seldom that the energy transitions between the ground and excited states in the source and absorber nuclei coincide when the source and absorber are in the form of different compounds. Hence, the investigation of the energy levels of the Mössbauer nuclei in the absorber requires the modification of the energy of the γ rays that are emitted by the source so they can be absorbed by the absorber. This is described schematically in Figure 4.4, which shows that this modification is usually achieved by moving the source relative to the stationary absorber so that an

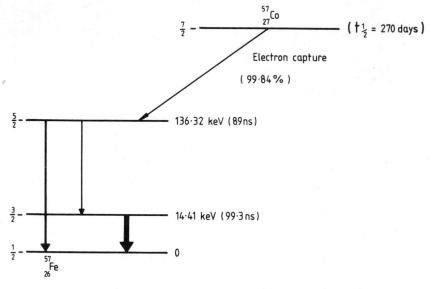

Figure 4.3 The decay of ^{57}Co to stable ^{57}Fe. The 14.4-keV γ ray is the photon normally used in ^{57}Fe Mössbauer spectroscopy.

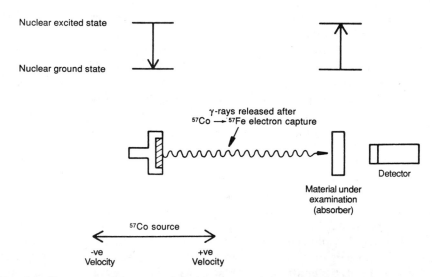

Figure 4.4 The energy of the γ ray emitted by the source is modified by oscillating the source backward and forward with respect to the stationary absorber. Resonant absorption occurs when the energy of the γ ray just matches the nuclear transition energy of the ^{57}Fe nuclei in the absorber. Reproduced, with permission, from F. J. Berry, *Phys. Bull.*, **34**, 517 (1983).

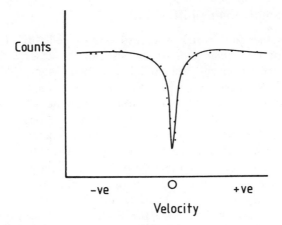

Figure 4.5 A simple representation of a Mössbauer spectrum that is composed of a plot of γ-ray counts against the velocity of the source, which is measured in millimeters per second.

energy shift is imparted to the emitted γ ray as a result of the first-order relativistic Doppler effect. Hence, in a conventional transmission mode Mössbauer spectroscopy experiment, the γ rays emitted by the source pass through the absorber and then enter a suitable detector. Resonant absorption occurs when the energy of the γ ray just matches the nuclear transition energy of the Mössbauer nucleus in the absorber. Therefore the Mössbauer spectrum (Figure 4.5) consists of a plot of γ-ray counts, or relative absorption, against the velocity of the source with respect to the absorber. The velocity of the source is usually measured in millimeters per second. In the very simplest of cases where both the source and the absorber contain the same Mössbauer isotope in the same environment, the spectrum consists of a single absorption line at zero velocity.

2.3   Mössbauer Parameters

A Mössbauer spectrum is characterized by the number, shape, position, and relative intensities of the absorption lines that appear in the spectrum. These features result from the various hyperfine interactions between the nucleus and its surrounding electrons, their time dependence, and any motion of the Mössbauer nuclei.

The three main hyperfine interactions that can be observed by Mössbauer spectroscopy are reflected in the Mössbauer parameters, which are known as the *chemical isomer shift* δ, *the quadrupole splitting* Δ, and the *magnetic splitting*. These Mössbauer parameters and their dependence on temperature, assessments of line width data, and the influence of time and motional aspects of the hyperfine interactions provide the information that can be derived from the Mössbauer spectrum.

2.3.1   Chemical Isomer Shift

The chemical isomer shift, which is also sometimes called the *isomer shift*, the *chemical shift*, or the *center shift*, is the shift of the absorption lines in the Mössbauer spectrum from zero velocity. The chemical isomer shift is a result of the *electric monopole interaction* between the nuclear charge distribution and the electronic charge density over the finite nuclear volume. The chemical isomer shift arises because the volume of the nucleus may be larger or, as in ^{57}Fe, smaller in the excited state than in the ground state. Therefore the consequent change in the nuclear charge density during the γ-ray transition alters the coulombic interaction between the positive nuclear charge and the electron density at the nucleus and causes a change in the energy separation between the nuclear energy levels. When the absorber nuclei are in a different electronic environment from those in the source, the separation between the energy levels in the absorber nuclei will differ from those in the source; and it is this difference that gives rise to a shift in the position of the resonance line. The situation is described schematically in Figure 4.6. In a system where the electric monopole interaction is the only hyperfine interaction affecting the nuclear energy levels, the nuclear ground and excited states are unsplit and are represented by the horizontal lines. The transition energy between the nuclear ground and the first excited states in a bare nucleus is designated by E_n. The electronic environment of the nucleus in the source can lift the nuclear energy levels such that the transition energy for the source is E_s. When the absorber nucleus is in a different chemical environment, the displacement of the energy levels differs from those in the source such that the transition energy E_a also differs. Therefore, the application of a Doppler velocity to the source to attain resonance results in a shift of the position of the resonance line in the spectrum from zero velocity (cf. Figure 4.5). The applied velocity is known as the chemical isomer shift δ and is expressed in millimeters per second. From Figure 4.6 we can see that the chemical isomer shift represents the difference in energy separation between the ground and excited nuclear states in the source and absorber nuclei. Hence the chemical isomer shift is not an absolute quantity, and to compare chemical isomer shift data obtained from different absorbers, the chemical isomer shifts are expressed relative to a standard absorber that is specific to the particular isotope in use (see Section 3.3.3). It is also important to note that thermal vibrations of the nuclei shift the γ-ray energy as a result of the relativistic second-order Doppler effect. Since these thermal vibrations are temperature dependent, it is necessary to consider the temperature of both the source and the absorber when comparing chemical isomer shift data. Therefore the chemical isomer shift is an important parameter in its own right; as such, it sets the position of the center of gravity of the whole Mössbauer spectrum when hyperfine interactions additional to the electric monopole interaction are present. A comprehensive textbook [9] on the theoretical aspects of the electric monopole interaction and the interpretation of chemical isomer shift data is commended to readers requiring further details.

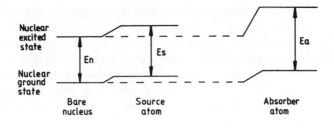

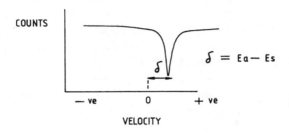

Figure 4.6   Schematic representation of the chemical isomer shift in Mössbauer spectroscopy.

It will now be appreciated that the chemical isomer shift, as with the other hyperfine parameters of the Mössbauer spectrum, depends on both the nuclear and electronic properties of the system. However, independent quantitative information on both these properties cannot be derived from the Mössbauer spectrum alone. Since the nuclear parameters are often constant with the electronic properties being of greatest interest to the experimentalist, the chemical isomer shift data are most frequently used to examine the electronic properties of different systems. By making certain assumptions [3] it can be shown that the chemical isomer shift is given by the expression

$$\delta = \text{constant} \times \frac{\Delta R}{R}(|\psi_s(0)|_a^2 - |\psi_s(0)|_s^2) \qquad (1)$$

where ΔR is the change in the nuclear radius during its transition from the excited to the ground state and R is the radius of the ground state. The value $\Delta R/R$ is characteristic of each Mössbauer transition and may be of either sign such that a positive $\Delta R/R$ indicates that the nucleus shrinks during deexcitation. The terms $|\psi_s(0)|_a^2$ and $|\psi_s(0)|_s^2$ refer to the s-electron densities at the nuclei in the absorber and source, or reference, respectively, and assume importance because the s-electron wave function has a finite value inside the nuclear radius and is directly responsible for the observed change in electrostatic energy. If $\Delta R/R$ is positive, a positive value of the chemical isomer shift δ implies that the s-electron density at the nuclei in the absorber is greater than the s-electron density at the nuclei in the source or reference material. It is important to recognize that

$|\psi_s(0)|^2$ is the s-electron density at the nucleus and not the formal s-electron occupation. Furthermore, $|\psi_s(0)|^2$ includes contributions from all the occupied s-electron orbitals in the atom and is sensitive to changes that occur in the outer valence shells. Although the value of $|\psi(0)|^2$ for p, d, and f electrons is zero, these orbitals do have significant interaction with the nucleus by interpenetration and shielding of the s electrons. For example, a $3d^54s^1$ configuration has a larger value of $|\psi_s(0)|^2$ than $3d^64s^1$ because in the latter case the extra d-electron shields the $4s$ electron from the nucleus.

The chemical isomer shift δ is therefore an important means by which atomic oxidation states, which have sometimes been difficult to determine in the past, can now be directly investigated. For example, in ^{57}Fe where $\Delta R/R$ is negative, higher s-electron densities at the iron nuclei are reflected in decreased, that is, more negative, chemical isomer shifts. Given that the removal of d electrons effectively increases the s-electron density at the iron nuclei, the iron(II) species, which has a d^6 configuration, has a more positive chemical isomer shift than iron(III), which has a d^5 configuration. For ^{119}Sn $\Delta R/R$ is positive, and higher s-electron densities at the Sn nuclei are reflected in a more positive chemical isomer shift; consequently, tin(II) compounds give a ^{119}Sn resonant absorption between $+2.9$ and $+4.5\,\mathrm{mm \cdot s^{-1}}$ relative to barium stannate, while tin(IV) species give chemical isomer shifts in the range from -0.5 to $+2.0\,\mathrm{mm \cdot s^{-1}}$.

It is also to be noted that for Fe and Sn atoms with the same oxidation state and electronic configuration and with identical ligands, the chemical isomer shift is dependent on the number and symmetry of the coordinating ligands [10, 11]. For this reason tetrahedral iron compounds have a lower chemical isomer shift than octahedral compounds with the same ligands. It has also been shown that the chemical isomer shift can be affected in a small way by changes in ligands a number of atoms away from the iron atom [12]. The dependence of the chemical isomer shift on covalency effects and its use in assessing the electron-withdrawing power of electronegative substituents and the bonding properties of ligands have been the subjects of a recent review [13]. It is important to realize, however, that although ^{57}Fe and ^{119}Sn Mössbauer chemical isomer shift data are well suited to this type of interpretation, the analysis of chemical isomer shifts recorded from other isotopes may not be so well disposed to such precise and detailed treatment. For example, the ^{125}Te Mössbauer transition is characterized by a small value of $\Delta R/R$ such that the differences between chemical isomer shifts are small and, given the broad nature of the line width, must be interpreted with caution.

Although the power of Mössbauer spectroscopy for the examination of chemical bonding is considered in slightly more detail later, it is appropriate to mention here that in systems where the bonding is not highly covalent the chemical isomer shift may be an additive function of the ligands [14]. In such situations *partial isomer shifts*, or partial center shifts, can be assigned to each ligand such that the value of δ for the compound under consideration is given by the sum of the partial isomer shifts of all the component ligands (see Section 4.1.3). Scales of partial chemical isomer shifts have been successfully constructed for ligands attached to a variety of transition metal nuclei. The chemical application of partial isomer shifts is given further attention in Section 4.1.3.

2.3.2 Quadrupole Splitting

The principles outlined during the discussion of the electric monopole interaction, which gives rise to the chemical isomer shift, assumed that the nuclear charge distribution is spherical. However, nuclei in energy states with a nuclear angular momentum quantum number $I > \frac{1}{2}$ have nonspherical charge distributions that are characterized by a nuclear quadrupole moment Q. The nuclear charge distribution can be elongated along the intrinsic axis of symmetry, labeled the z axis, and the nuclear quadrupole moment is positive; or it can be compressed along this axis, and Q is negative. The nuclear charge density can interact with asymmetric extranuclear electric fields that result from nonsymmetric arrays of electronic charge, ligands, or ions and that are characterized by a tensor quantity called the *electric field gradient* (EFG). The interaction between the nuclear charge density and the EFG is called the *electric quadrupole interaction* and has been considered in detail in a recent textbook [15]. The axes of the electric field gradient are labeled such that $V_{zz} > V_{xx} > V_{yy}$, and the EFG is normally expressed in terms of the principal component V_{zz}, which is usually written as eq, and an asymmetry parameter η, which is the divergence of the EFG from axial symmetry:

$$\eta = \frac{V_{xx} - V_{yy}}{V_{zz}} \tag{2}$$

The electric quadrupole interaction can be envisaged in terms of the nuclear quadrupole moment aligning itself either with or across the electric field gradient. The resulting coupling of eQ with eq is expressed as the *quadrupole coupling constant* e^2qQ and gives rise to a splitting of the nuclear energy levels. Hence, in ^{57}Fe, the excited state has $I = \frac{3}{2}$ and in the presence of an asymmetric electric field splits into two substates characterized by $M_I = \pm\frac{1}{2}$ and $M_I = \pm\frac{3}{2}$ as is shown in Figure 4.7. The transitions from the degenerate ground state

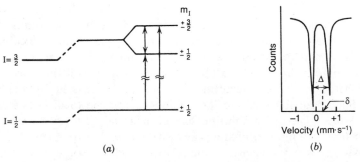

(a)   (b)

Figure 4.7 The Mössbauer quadrupole splitting in ^{57}Fe: (a) the excited state ($I = \frac{3}{2}$) splits into two and (b) the resulting Mössbauer spectrum.

($I = \frac{1}{2}$) to the excited state produce a two-line spectrum in which the separation between the two lines, measured in millimeters per second, is a measure of the quadrupole splitting Δ. The centroid of the two peaks represents the chemical isomer shift δ.

Hence the quadrupole splitting obtained from the Mössbauer spectrum involves both a nuclear quantity, the quadrupole moment (which is normally fixed for a given nuclide), and an electronic quantity, the electric-field gradient. For Mössbauer transitions between nuclear states with spins $\frac{1}{2}$ and $\frac{3}{2}$, for example, ^{57}Fe, ^{119}Sn, ^{125}Te, ^{193}Ir, and ^{197}Au, the quadrupole splitting can be expressed in terms of the quadrupole coupling constant for the $I = \frac{3}{2}$ state.

$$\Delta = \frac{1}{2}e^2qQ(1 + \frac{\eta^2}{3})^{1/2} \qquad (3)$$

When the spins are greater, the spectra are more complex and the data are presented as the quadrupole coupling constant for the ground state, or for the excited state when the ground state has $Q = 0$.

The electric field gradient contains contributions from several different components. A major contribution is that resulting from the electronic environment about the nucleus and is called the *valence term* (q_{val}). This can be envisaged as arising from the valence electrons of the Mössbauer atom and it originates from asymmetry in the electronic structure, which may derive from the unfilled or partly filled electronic shells occupied by the valence electrons. Since the s-electron density is spherically symmetric about the nuclear volume, it is the asymmetric p- and d- electron populations that primarily contribute to the valence term. The sign and magnitude of e^2qQ can be used empirically to measure the asymmetric occupation of the atomic valence orbitals. Another contribution to the electric field gradient originates from surrounding charged entities and is called the *lattice contribution* (q_{lat}). This arises from the asymmetry in the arrangement of atoms around the Mössbauer nuclei and is most important in ionic systems. Other contributions to the electric field gradient include the effects of molecular orbitals and any polarization of the core electrons of the Mössbauer atom.

Therefore, the quadrupole splitting reflects the symmetry of the bonding environment and the local structure in the vicinity of the Mössbauer atom. For example, in high-spin iron(III) compounds the iron nucleus is accommodated in a nearly spherically symmetric d^5 electronic configuration, and such compounds usually have small quadrupole splittings (<1 mm·s^{-1}). However, high-spin iron(II) compounds frequently exhibit large quadrupole splittings arising from the asymmetry in the d^6 electronic environment produced by the sixth d electron. Such contributions are then influenced by the local lattice environment, and the quadrupole splitting has been found [12] to be sensitive to changes in ligands a number of atoms away from iron. The quadrupole splittings can be particularly informative when considered in conjunction with the chemical isomer shift data and can be used in the assessment of the electron populations of oribitals, isomerization phenomena, ligand arrays, semiconductor properties, and the defect structure of solids.

There have been several attempts to formulate semiquantitative treatments of the electric field gradient, and these have been well summarized [13]. In the point-charge model [16–18] the ligands are considered to act as point charges, and each is assigned a parameter that represents the contribution of that ligand to the EFG and is called a *partial quadrupole splitting* parameter (L) (see Section 4.1.4). When the hybridization of the Mössbauer atom experiencing the point charges is considered [19], the value of (L) depends not only on the identity of the ligand, but also on the character of the hybrid orbital by which the ligand is bound to the Mössbauer atom.

Another method for the semiquantitative treatment of the electric field gradient was developed by Townes and Dailey [20] and is well suited for the interpretation of the Mössbauer parameters obtained from xenon-, iodine-, and tellurium-containing compounds. The method assumes that the electric field gradient arises entirely within the valence shell of the atom concerned, that the bonding involves only the p orbitals or has a constant s character, and that π-bonding interactions can be ignored. From the Mössbauer quadrupole splitting and quadrupole coupling constants we can derive the p orbital imbalance (designated U_p), which gives information about the electron population of the p orbitals. The method also allows the calculation of a quantity h_p, which can be used in isomer shift calculations and which describes the number of holes in the p shell.

2.3.3 Magnetic Splitting

When a nucleus is placed in a magnetic field H, a *magnetic dipole interaction* occurs between any nuclear magnetic moment and the magnetic field. Hence, when a nuclear state with an angular momentum quantum number $I > 0$ is placed in a magnetic field, the magnetic dipole interaction removes the degeneracy of the nuclear state and its energy levels split into $2I + 1$ substates. The situation pertaining to ^{57}Fe, where the ground state with $I = \frac{1}{2}$ splits into two substates and the excited state with $I = \frac{3}{2}$ splits into four substates, is shown in Figure 4.8. The selection rules $\Delta_{m_I} = 0, \pm 1$ give rise to a symmetric six-line

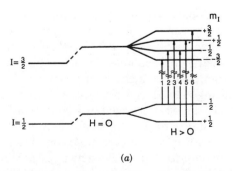

(a)

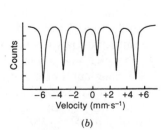

(b)

Figure 4.8 The effect of magnetic splitting on the nuclear energy levels of ^{57}Fe: (a) splitting of the nuclear energy levels in a magnetic field and (b) the resulting Mössbauer spectrum.

Mössbauer spectrum in which the chemical isomer shift is given by the center of gravity of the six peaks. A thorough account of the magnetic dipole interaction and the interpretation of magnetically split Mössbauer spectra are given in a recent textbook [21].

Since the splitting of the spectral lines is directly proportional to the magnetic field experienced by the nucleus, the magnetically split Mössbauer spectrum offers a direct and effective measurement of the magnitude and orientation of the magnetic field. The transition probabilities between the nuclear substates are reflected in the intensities of the lines in the Mössbauer spectrum, and the intensity data can be used to obtain information on both the relative orientation of the magnetic field at the nucleus and the direction of propagation of the γ-ray beam.

The total magnetic field experienced by a nucleus is a vector sum of the magnetic hyperfine field and any external applied magnetic field. The magnetic hyperfine field at the nucleus arises from the spin of any unpaired electrons on the Mössbauer atom and hence depends on the oxidation state and spin state of the atom. It is important to appreciate that only the unpaired electrons contribute to the magnetic hyperfine field; and the situation is therefore less complex than with the chemical isomer shift or quadrupole splitting data where contributions may emanate from all the electrons, both within the atom and also in the molecule and lattice. Magnetic hyperfine fields are observed in the Mössbauer spectra of magnetically ordered systems and can also be seen in the spectra of paramagnetic species with long spin–lattice or spin–spin relaxation times. Relaxation processes, whereby the electron spins can change direction with a characteristic time are frequently reflected by distinctive changes in the Mössbauer spectra (see Section 2.3.4).

The ability to apply an external magnetic field is an important experimental factor when systems showing magnetic hyperfine splitting are under investigation. The total magnetic field at the nucleus then arises from a vector sum of the two contributions, and changes in the splitting of the spectral lines following the application of an external magnetic field are frequently used to interpret the spectra [21]. Hence, the magnetic hyperfine interaction is unique among the three Mössbauer hyperfine interactions since it alone is amenable to modification by an external variable.

The effects of magnetic and quadrupole hyperfine interactions on the Mössbauer spectrum may be significantly more complex when they occur together and the observed spectrum is then highly dependent on their relative magnitudes and orientations. This occurs, for example, in magnetic iron-containing materials where the iron atom occupies a site with noncubic point symmetry such that a quadrupole splitting as well as a magnetic splitting can arise. Although in simple cases the spectra can be interpreted by a consideration of the shift of the levels of the excited state and the difference between the spacings of the pairs of outer peaks, the situation can become more complex and approaches to the problem have been described in some detail [22]. Such matters are important since the application of a magnetic field to a system with no unpaired electrons, and which therefore has no magnetic hyperfine field, produces a magnetic splitting in the Mössbauer spectrum in addition to any quadrupole splitting and the data can then be used to provide information on the sign and magnitude of the electric field gradient at the nucleus.

In general, the data obtained from magnetically split Mössbauer spectra can be used to investigate the magnetic ordering and structure of magnetically ordered systems, the nature of the magnetic interactions, the size of the magnetic moment on particular atoms, and details about the electronic structure of the atom that relates to the magnetic hyperfine field at the Mössbauer nucleus.

2.3.4   Time-Dependent Effects

The hyperfine interactions observed in Mössbauer spectra occur on character-istic time scales, and the spectrum observed from a specific system depends on whether the nuclear environment or the position of the nucleus is changing relative to these times. These time-dependent effects can influence both the spectral line shapes and the observed Mössbauer parameters.

Time-dependent changes in the nuclear environment are often called *re-laxation processes* and may relate to time-dependent structural changes or to changes involving the electronic configuration. The structural changes may be on a macroscopic scale, such as occur during diffusion and melting processes, or they may be more localized structural changes, such as rotation within a molecule. Time-dependent electronic phenomena can be found in systems; for example, with valence fluctuations or materials suffering from the aftereffects of nuclear transformations. Although time-dependent structural and electronic changes can affect all the Mössbauer hyperfine parameters, such changes involving the magnetic hyperfine interaction are most frequently used in Mössbauer studies. These processes are usually referred to as *magnetic re-laxation* phenomena and arise from a time dependence of the magnitude and direction of the magnetic hyperfine field experienced by the nucleus. The influence exerted by relaxation processes and other time-dependent effects on the Mössbauer spectrum is a function of the relative time scales associated with these effects and the time scales of the nuclear transitions and hyperfine interactions. Hence, like the hyperfine interactions previously considered, the time-dependent effects result from the interplay of both nuclear and extranuclear factors.

Given that the Mössbauer effect is closely associated with the motional properties of the emitting or absorbing nucleus the technique is, in principle, a very suitable way to examine either nuclear dynamics or the dynamics of a system where the Mössbauer nucleus acts as a probe. Such motion of the Mössbauer nucleus can influence the Mössbauer spectrum in two ways. First, any motion related to the vibrational properties of the system will influence the number of recoil-free events, called the *recoil-free fraction*, and hence the absorption intensity of the spectrum. Such changes in recoil-free fractions are usually followed as a function of temperature to obtain information on the vibrational properties of the system. Second, the motion of the Mössbauer nucleus in the source or the absorber can also be revealed in the Mössbauer spectral line widths since the motion can be considered as an additional Doppler motion, which may partially obliterate the resonant absorption. Since the line widths also depend on other factors, it is again useful to follow them as a function of temperature, or other relevant variable, to separate the various effects. The dynamic properties of Mössbauer nuclei, which can be monitored by Mössbauer spectroscopy, can arise from the lattice dynamics of the solid in which the nuclei are situated, or they can result from the motion of a localized part of the system, such as molecular motion or the motion of the whole system within its environment. However, it must be appreciated that since the effect on the spectrum depends only on the motion of the nucleus, as opposed to the origin of the motion, it is often difficult to distinguish between the sources of any such motion.

3   INSTRUMENTATION

The Mössbauer spectrum is a record of the resonant interactions occurring in the specimen as a function of the Doppler velocity. In transmission-mode Mössbauer spectroscopy the occurrence of such interactions can be detected by the absorption of γ rays from the beam. Backscatter Mössbauer spectroscopy involves the detection of the decay products from the excited nuclei in the absorber such as γ rays, X rays, and Auger or conversion electrons. A Mössbauer spectrometer is an instrument used for obtaining such spectra, and developments in the construction of such equipment and the computer interpretation of data have been the subjects of recent reviews [23, 24].

3.1   Mössbauer Spectrometers

The Mössbauer spectrometer essentially consists of three sections that (1) drive the source, (2) detect the γ radiation, and (3) handle the data. A typical modern microprocessor-controlled Mössbauer spectrometer is illustrated in Figure 4.9.

3.1.1   The Drive Mechanism

The modification of the energy of the γ rays emitted by the source so that they can be absorbed by an absorber (Section 2.2) is achieved by introducing a relative motion between the source and the absorber thereby changing the γ-ray energy by the Doppler effect. Therefore, we can record the spectrum "point by point" by selecting values of the velocity according to the *constant velocity* method. Alternatively, and also more commonly, a cyclic motion of the source is arranged so that the velocity sweeps through the range of velocities of interest such that if equal times are spent in equal velocity increments, the motion has a constant acceleration, which is called the *constant acceleration* mode.

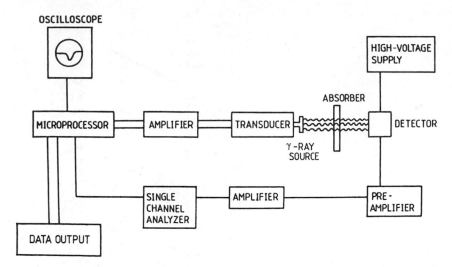

Figure 4.9 Schematic representation of a microprocessor-controlled Mössbauer spectrometer. In ^{57}Fe Mössbauer spectroscopy the detector is set to monitor the 14.4-keV γ rays by the single-channel analyzer. The microprocessor receives and stores the impulses from the detector and the accumulating spectrum can be monitored on the oscilloscope screen. When the spectrum is of satisfactory quality, it is read out for subsequent analysis by a computer.

The drive mechanism should be capable of imparting velocities of up to $\pm 50\,\mathrm{mm\cdot s^{-1}}$ to the source, although, in practice, velocities of about $\pm 10\,\mathrm{mm\cdot s^{-1}}$ are generally sufficient to encompass the spectral absorptions. These velocities are usually generated by a moving coil drive coupled to a velocity transducer, which gives a voltage proportional to the velocity. The combination is connected so that the velocity of the source follows an input waveform. Several types of waveform have been found to be suitable and some are shown in Figure 4.10. Symmetrical waveforms, such as those depicted in Figure 4.10*c* and *e*, which result in the spectrum being collected in both the linear ramp sections, are preferred. More detailed information concerning the operation of the drive mechanism and the different waveforms used to drive the transducer can be found in [23]. Hence, a range of velocities are scanned during the Mössbauer experiment, and the absorption of the Doppler-modified γ-ray energies is sensed by a detector that is placed behind the source. The linear geometric arrangement of the source, absorber, and detector is illustrated in Figure 4.9.

3.1.2   The Detector and Electronics

The detection of the Mössbauer γ rays needs to be as efficient as possible, and it is desirable to discriminate against the remainder of the γ radiation and other radiations such as X rays. The equipment required for such a purpose includes a detector, preamplifier, amplifier, and a discriminator or single-channel analyzer.

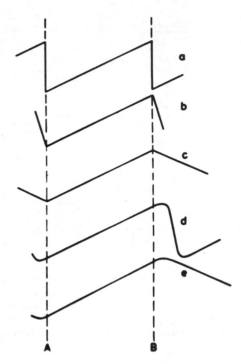

Figure 4.10 Velocity waveforms that can be used in Mössbauer spectrometers. Reproduced, with permission, from T. E. Cranshaw, "Mössbauer Spectrometers and Calibration," in G. J. Long, Ed., *Mössbauer Spectroscopy Applied to Inorganic Chemistry*, Vol. 1, Plenum, New York, 1984, p. 27.

For isotopes such as ^{57}Fe or ^{119}Sn, and other isotopes in which the Mössbauer γ-ray energy is less than about 40 keV, relatively good resolution of the Mössbauer γ radiation can be obtained with a proportional counter. Such proportional counters are normally filled with xenon, krypton, or argon, together with methane or nitrogen as a quenching gas, and operate under high voltage. The resolution of such counters, which are probably the most common form of detectors used in Mössbauer spectroscopy especially for the 14.4-keV γ ray used in ^{57}Fe Mössbauer spectroscopy, is good but at the expense of a low efficiency.

The scintillation crystal detector is frequently used for γ rays with energies in the range from 50 to 100 keV or for work with low counting rates at lower energies. A typical example is the thin NaI/Tl crystal, which has a high efficiency for low-energy γ rays. Hence the scintillator is suitable for the detection of soft γ rays in situations where the radiation background is low and where there are no other X rays or γ rays with energies near to that of the Mössbauer γ ray. The resolution of scintillators deteriorates with decreasing energy, but these counters have a prime advantage in that their efficiency is high.

Germanium or lithium-drifted germanium detectors give much better resolution, especially at higher energies, although the efficiencies fall off quite rapidly above about 60 keV depending on the size of the crystal. The main advantages of these types of detector are found in situations where high-energy Mössbauer γ radiation is of interest and where the Mössbauer γ ray strongly

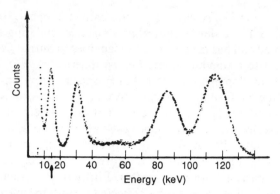

Figure 4.11 Pulse-height spectrum recorded from a ^{57}Co source by a gas-filled proportional counter. Note that the 14.4-keV Mössbauer γ ray, which is arrowed, accounts for only a small fraction of the total radiation detected. Reproduced, with permission, from U. Gonser, Ed., *Topics in Applied Physics, Mössbauer Spectroscopy*, Vol. 5, Springer-Verlag, Berlin, 1975.

overlaps other radiations. Other disadvantages with these types of detectors include their high cost and the need to maintain them at liquid nitrogen temperatures.

Since most Mössbauer sources are not monochromatic but emit γ and X radiation of higher and lower energy than the Mössbauer γ ray, the output of the main amplifier consists of a train of pulses with each pulse being an electrical analogue of the radiation entering the detector. The plot of γ-ray intensity against γ-ray energy is called a *pulse-height spectrum* (Figure 4.11). It is important in all Mössbauer work that the pulse-height spectrum is recorded before any other measurements are made since, by comparing this pulse-height spectrum with that reported in the literature, the Mössbauer spectroscopist is able to check both the equipment and the source.

We can select pulses with the γ-ray energy corresponding to the required Mössbauer γ ray by using the single-channel analyzer (Figure 4.9). The pulses from the output of the single-channel analyzer are selected to correspond to those of the energy of the Mössbauer γ ray, and these become the pulses that are actually counted and passed on to the data-handling system.

3.1.3 Data Acquisition

We can accumulate the Mössbauer spectrum by using a multichannel analyzer in the multiscaler mode. The signals, selected by the single-channel analyzer, are fed to the succession of recording channels in the multichannel analyzer in phase with the movement of the source, the input sweeping the recording channels once during each cycle of motion of the source. Each channel is held open in turn for a short time; and, since the multichannel analyzer and velocity drive mechanism are synchronized, the velocity changes $-v$ to $+v$, which may be associated with increasing channel numbers, correspond to the source always moving at the same velocity when a given channel is open. The multichannel analyzer normally has a device that allows the accumulating spectrum to be visually monitored and has the appropriate facilities to output the data in a form suitable for input to a standard computer peripheral.

More recent developments in Mössbauer spectroscopy have involved the replacement of the multichannel analyzer by a microprocessor. Microprocessor-based systems in which the microprocessor not only collects the data, but also operates the drive, have other advantages such as offering a wide range of output facilities, for example, floppy disks or cassette tapes, for data storage and computation. Under these circumstances stored data can be read back into the memory for comparison with new spectra. Furthermore, when a symmetrical waveform is used, so that a spectrum is collected in both the linear ramp sections of the velocity waveform, the necessary operations of folding and adding the two spectra can be performed before displaying the data on the monitor. This facilitates an assessment of the quality of the spectrum before the data accumulation process is terminated and the spectrum read out. In some large systems it is also possible to perform a least-squares fitting of the data.

3.2   Computation of Data

The interpretation of Mössbauer data involves the determination of the parameters of the curve that best fit the data points comprising the spectrum so the positions, areas, and widths of the peaks in the Mössbauer spectra, together with their standard deviations, can be obtained. The computational methods by which the Mössbauer data can be treated have been well documented in the past and were recently reviewed [24]. For complex spectra it is often essential that some parameters be held constant or a number of parameters be constrained equally throughout all or part of the fitting procedure. Without such constraints it is sometimes not possible to make a complex spectrum converge. Given that some Mössbauer spectra can be very complex with many overlapping lines that visually appear to be almost Lorentzian, it is sometimes not immediately obvious how many peaks are present in a given spectrum. Hence a good fit to data is not unambiguous proof that the fit is correct; indeed, in a complex spectrum it is quite feasible to fit a spectrum that has no physical significance. Therefore, it is clear that a goodness-of-fit parameter such as X^2 must be interpreted with some degree of caution, and, in the final analysis, the interpretation of the Mössbauer data should be considered in terms of its compatibility with other scientific evidence.

3.3   Other Experimental Features

3.3.1   Sources

Sources for Mössbauer spectroscopy consist of a radioactive material that is chosen to give, through radioactive decay, an isomeric nuclear excited state of the element under study (see Section 2.2). Each element to be studied by

Mössbauer spectroscopy must have its appropriate source, and it is seldom that a single source will serve more than a single nuclide. There are, therefore, several practical problems concerning sources that limit the usefulness of Mössbauer spectroscopy.

First, the precursor radioactive nuclide must have a reasonably long half-life. Although half-lives of several months are usually desirable, nuclides with shorter half-lives of, for example, even a day must not be dismissed because they can often be used under appropriate experimental arrangements to accumulate spectra and are sometimes the only means by which certain exotic isotopes can be examined by Mössbauer spectroscopy. It does remain important, however, that the precursor radioactive nuclide can be prepared in high yield. The nuclear decay scheme of ^{57}Co, which gives the metastable ^{57}Fe that is used in ^{57}Fe Mössbauer spectroscopy, has been illustrated in Figure 4.3, and similar decay schemes and relevant data are given for most Mössbauer active nuclides in the *Mössbauer Effect Data Index Covering the 1976 Literature* [25]. Since the ^{57}Co precursor nuclide for ^{57}Fe Mössbauer spectroscopy is readily available and has a suitable half-life, many commercially prepared ^{57}Fe Mössbauer sources have been developed with considerable refinement. Similarly, sources for other Mössbauer isotopes, for example, ^{119}Sn, ^{121}Sb, and ^{125}Te, can be purchased with little difficulty. However, the situation is more complex with some other Mössbauer isotopes. For example, ^{61}Ni Mössbauer spectroscopy requires a ^{61}Co precursor nuclide, which has a half-life of only 99 min. Furthermore, the ^{61}Co source must be prepared either by the $^{62}Ni(\gamma, p)^{61}Co$ reaction using about 100-MeV radiation or by the $^{64}Ni(p, \alpha)^{61}Co$ reaction. Hence, ^{61}Ni Mössbauer spectroscopy can only be performed when the facilities are close to either an electron accelerator or a proton beam from a cyclotron.

Having obtained a suitable precursor nuclide it is necessary to incorporate the species in, or on, a solid matrix so some fraction, called the *recoil-free fraction*, of the emitted γ rays will be emitted without nuclear recoil. The matrix must act as a temperature-independent cubic nonmagnetic environment and be capable of withstanding the radiation damage produced by the decay of the active nuclide. A very suitable source for ^{57}Fe Mössbauer spectroscopy can be prepared by electroplating ^{57}Co onto rhodium and annealing the solid at high temperature in vacuo. Similarly, the incorporation of ^{119m}Sn into $BaSnO_3$ produces a very satisfactory source for ^{119}Sn Mössbauer spectroscopy.

A major requirement for a Mössbauer source is that it should produce γ rays with a narrow energy distribution. As mentioned previously (Section 2.1), this is often called the line width and is controlled by the Heisenberg uncertainty principle such that a sufficiently long mean lifetime for the Mössbauer isomeric excited state is required to produce a γ ray with a narrow energy distribution. Once again, the situation for ^{57}Fe is very favorable since the 97.81-ns lifetime of the 14.4-keV excited state gives a minimum Heisenberg line width of $0.194 \ mm \cdot s^{-1}$. Since chemical isomer shift and quadrupole splitting differences are typically from about 0.2 to $3 \ mm \cdot s^{-1}$, the narrow line width permits accurate resolution and measurement of such shifts and splittings. For ^{119}Sn, where the natural line width is $0.642\,mm \cdot s^{-1}$ and shifts and splittings are approximately the same as in iron, the resolution of the lines is not quite so good but is still sufficient to provide useful and interpretable information. The importance of the line width of the γ radiation means that the chemical effects of the nuclear reactions that precede the emission of the Mössbauer γ ray must be carefully considered when preparing a Mössbauer source. For example, a protracted neutron irradiation of the intended source matrix at high flux may induce significant radiation damage and consequent formation of lattice defects. Since all the excited atoms should be in identical chemical environments, such defects may result in unwanted line-broadening effects. Under these circumstances the source may need to be annealed to restore the regularity of the crystal lattice, or the radioisotope may need to be chemically extracted and incorporated into a new matrix.

3.3.2   Absorbers

Since Mössbauer spectroscopy depends on nuclear recoil-free events, it is restricted to the investigation of materials in the solid state. Although liquids and gases are not amenable to investigation, it is possible to freeze both liquids and solutions to give solids that can be examined. Hence, samples for Mössbauer spectroscopy are often presented as a single crystal, a crystalline powder, an amorphous powder, a glass, or a frozen liquid. The samples are usually encapsulated in a nylon or Perspex cell or between two layers of adhesive tape in a lead holder.

The thickness of the sample is a very important aspect of preparing a specimen for Mössbauer examination. A thick sample is not as transparent to the γ radiation as a thin sample; consequently, the spectrum can only be accumulated over a long period of time. Furthermore, thick samples containing an excessive amount of Mössbauer nuclei give rise to incorrect intensity ratios in the spectral peaks and cause the occurrence of broad lines that may conceal important and subtle fine structure and preclude the accurate interpretation of the spectra. The preparation of excessively thin samples also has undesirable consequences since it gives rise to insufficient absorption.

The ideal thickness of the absorber is therefore a compromise between placing a sufficient amount of the Mössbauer active element in the absorber to maximize the signal intensity and at the same time using a small sample mass to minimize the nonresonant scattering that produces noise at the detector. The acquisition of good quality Fe and Sn Mössbauer spectra is normally achieved from samples containing about 10 mg of iron or tin per square centimeter of the sample holder.

On some occasions the material under examination contains a low concentration of the Mössbauer active element. Under such circumstances it may be possible to improve the S/N ratio, and hence the ability to obtain useful spectra, by using isotopically enriched samples. For example, Fe Mössbauer spectroscopy uses the ^{57}Fe nuclide, which has a natural abundance in iron of only 2.14%. Hence, samples with low iron concentrations are best prepared using the pure ^{57}Fe isotope.

3.3.3   Spectrometer Calibration and Reference Standards

The Mössbauer spectrometer velocity scale is most effectively calibrated by recording the spectrum of a standard absorber for which the Mössbauer parameters are accurately known, and it is now virtually universal practice to use α-iron foil, the spectrum of which is shown in Figure 4.12, for this purpose. This absorber has certain advantages as a standard over others; for example, sodium nitroprusside, which has sometimes been used in the past since the large magnetic field in α-iron produces a six-line spectrum that spans about $10 \, \mathrm{mm \cdot s^{-1}}$ and enables an accurate *calibration* at both small and large velocities. This magnetically split spectrum recorded from natural iron can also be used to determine the linearity of the velocity drive of the spectrometer.

For each different Mössbauer isotope a specific *reference standard* is used to serve as a zero-point reference for all the chemical isomer shift data recorded from that element. Hence, for ^{57}Fe Mössbauer spectroscopy the chemical isomer shifts are usually referred to α-iron or sodium nitroprusside, whereas for ^{119}Sn Mössbauer spectroscopy the chemical isomer shift data are usually quoted with respect to barium stannate, tin(IV) oxide, or tin telluride. The reference standards for other Mössbauer isotopes are well documented [25].

3.3.4   Cryogenic Equipment

Very few Mössbauer resonances are amenable to investigation at room temperature. Iron-57 and ^{119}Sn Mössbauer spectroscopy are exceptions to this requirement since the relatively low energy of, for example, the 14.4-keV ^{57}Fe Mössbauer γ ray produces a rather high, approximately 30%, recoil-free fraction that enables room temperature measurements to be sufficient for most studies.

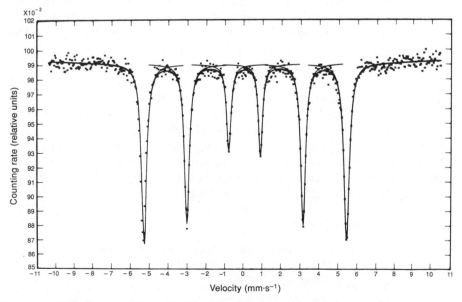

Figure 4.12   Mössbauer spectrum recorded from natural iron foil at 298 K.

However, the higher the γ-ray energy, the lower will generally be the fraction of recoil-free emission or absorption events. The recoil-free fraction can, however, be increased if the source, the absorber, or both are cooled to liquid nitrogen (77 K) or liquid helium (4.2 K) temperatures. Hence, for many Mössbauer isotopes, for example, ^{121}Sb and ^{125}Te, it is necessary to cool at least the absorber and sometimes the source to increase the recoilless fractions. It must also be remembered that some of the hyperfine interactions are temperature dependent; hence, low-temperature Mössbauer experiments are quite common both from the point of view of experimental necessity and for their capacity to give more information on the sample under examination.

The Mössbauer spectroscopy experiments in which both the source and the absorber are held at low temperature still require the source to be vibrated, and it will be recognized that the construction of cryogenic facilities to perform such measurements at temperatures as low as a few millikelvin presents some problems. However, several cryostats for performing measurements at 77 or 4.2 K, such as that illustrated in Figure 4.13, are now commercially available. Cryostats with a variable temperature control are also used in many studies, such as those involving the investigation of phase transitions. For investigations where it is desirable to apply a large external magnetic field to the absorber, superconducting magnets are required, which are capable of producing magnetic flux densities of typically up to 10 T.

3.4   Backscatter Mössbauer Spectroscopy

The discussion so far has been concerned with Mössbauer experiments carried out in the transmission mode. However, it is also possible to observe the Mössbauer resonance by using a backscattering geometry such that the scattered decay products from the excited nuclei in the absorber, that is, those nuclei that have been resonantly excited by the source radiation, are detected. Such decay products are γ rays, X rays, and Auger or conversion electrons. The principal experimental difference in performing these experiments is that a special detector is required that is sensitive to one or other of these radiations.

The main advantage of backscatter spectra is the nonresonant radiation from the source is not recorded; hence, the major contributions to the background are scattered higher energy γ rays and nonresonant Rayleigh scattering. The main disadvantage is the intensity of the scattered radiation is usually weak. The counting of scattered γ rays also involves some particular problems, which are associated with anomalous effects that depend on the scattering angle. The principal application of scattering experiments has been in Mössbauer studies of surface phenomena, which is discussed in Section 5.4.

3.4.1   Backscatter Detectors

The proportions of X-rays, conversion electrons, and Auger electrons resulting from the decay of the $I = \frac{3}{2}$ excited spin state of ^{57}Fe for each γ-ray photon absorbed are given in Table 4.1.

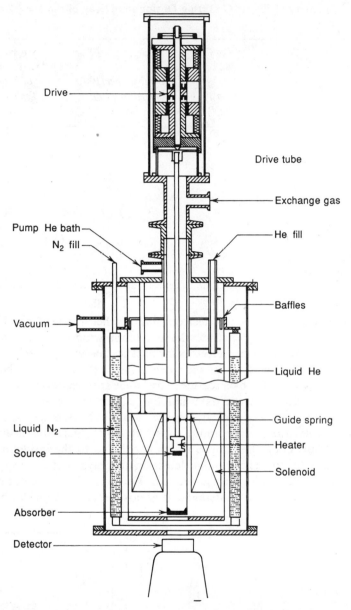

Drive

Drive tube

Exchange gas

Pump He bath

N_2 fill

He fill

Vacuum

Baffles

Liquid He

Liquid N_2

Guide spring

Source

Heater

Solenoid

Absorber

Detector

Figure 4.13 Helium cryostat with vertical beam geometry. Both source and absorber can be cooled. The system is equipped with a superconducting solenoid. If the magnet coil is not in use, the drive tube is extended to bring the source closer to the absorber to obtain a larger solid angle. Reproduced, with permission, from G. K. Shenoy and F. E. Wagner, Eds., *Mössbauer Isomer Shifts*, North-Holland, Amsterdam, 1978, p. 49.

Table 4.1 Summary of Events Occurring During the Decay of the $I = \frac{3}{2}$ Excited Spin State of ^{57}Fe

	Energy (keV)	Number (per 100) of Absorption Events	Approximate Maximum Range of Travel in Typical Solids (nm)
γ rays	14.4	9	
K_α X rays	6.3	27	
K-shell conversion electrons	7.3	81	250
L-shell conversion electrons	13.6	9	900
M-shell conversion electrons	14.3	1	
KLL Auger electrons	5.4	63	
LMM Auger electrons	0.53		

The low energy of conversion electrons inhibits their range of travel in a solid, so they are not able to penetrate the beryllium window of a conventional proportional counter. Hence, detectors for backscatter Mössbauer spectroscopy are frequently constructed so the specimen is inside the detector. The arrangement of the apparatus for backscatter Mössbauer spectroscopy is illustrated in Figure 4.14, the electronic and data-handling equipment are essentially similar to that used in conventional transmission-mode Mössbauer spectroscopy (Figure 4.9). The γ radiation from the source enters the detector through a thin window and falls upon the sample, which is adhered to the inside of the back window of the detector. The selective detection of X rays or electrons is achieved by using different gas fillings in the detector. For example, a detector filled with argon and 10% by volume of methane at atmospheric pressure can be used to

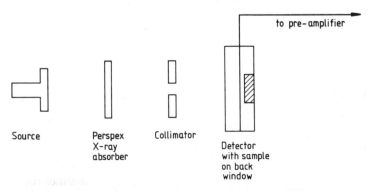

Figure 4.14   Arrangement of source and detector for backscatter Mössbauer spectroscopy.

detect the 6.3-keV k_α X rays resulting from the decay of ^{57}Fe in the sample. A mixture of helium with 10% methane has a low efficiency for X rays and γ rays but is easily able to monitor the 7.3 K-shell conversion electrons emanating from ^{57}Fe. The ionization electrons are collected by the positively charged wire, and a pulse is then produced by the amplifier, which is fed through the spectrometer as described in Sections 3.1 and 3.2.

3.4.2   Backscatter X-Ray Mössbauer Spectroscopy

The use of backscatter X-ray Mössbauer spectroscopy requires that we distinguish the X rays emanating from the specimen from those emitting from the source. Therefore, the source X rays are filtered by placing a Perspex X-ray absorber between the source and the collimator before recording the spectra, as shown in Figure 4.14.

3.4.3   Conversion Electron Mössbauer Spectroscopy

A significant advance in backscatter electron detection has entailed the development of *conversion electron Mössbauer spectroscopy* (CEMS), which is a particularly sensitive method for the examination of solid surfaces [26–28]. The ability of CEMS to investigate solid surfaces is vested in the shallow escape depth of the conversion electrons, which results in about 95% of the 7.3-keV K-shell conversion electrons originating from within 300 nm of the surface, while 66% emanate from within 54 nm.

As mentioned previously, the low energy of these conversion electrons and their ability to penetrate a detector window requires the sample to be inside the detector, which is called a *resonance counter*. A suitable instrument is depicted in Figure 4.15, and I have constructed and used a similar model. The apparatus is essentially an 8-mm-thick flat cylindrical flow counter with Mylar windows. The sample is fixed onto one of these windows, and the pair is set into aluminium rings that are screwed onto a Perspex block containing three parallel 0.1-mm-thick stainless steel anode wires. The inlet and outlet tubes permit the flow of the counting gas composed of 90% helium and 10% methane. The backscattered internally converted electrons are detected as a function of the Doppler modulated incident γ-ray energy, and maxima in electron counts occur at velocities when resonant absorption occurs. The CEM spectrum recorded from natural iron (Figure 4.16) illustrates the type of results obtained by this method of Mössbauer spectroscopy.

A considerable amount of work has been performed on studies of the energy loss of electrons and the possibility of obtaining conversion electron Mössbauer spectra that can be associated with different depths of the solid surfaces. This has involved the energy resolution of the emitted electrons [28] and is sometimes called *depth selective conversion electron Mössbauer spectroscopy* (DCEMS). In addition, the resonance counters are not easily used at low temperature, and to record spectra at low temperatures it is necessary to use channeltrons or other electron detectors in vacuo [29].

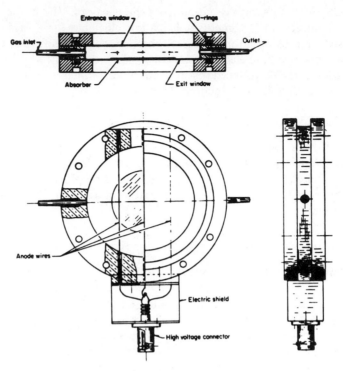

Figure 4.15 A resonance counter. Reproduced, with permission, from *Design and Performance of a Mössbauer Resonance Counter*, Danish Atomic Energy Commission Risø-M-1695.

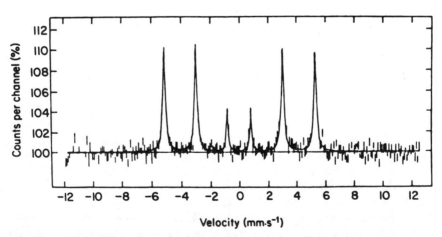

Figure 4.16 Conversion electron Mössbauer spectrum recorded from natural iron. Reproduced, with permission, from J. Fenger, *Nucl. Instr. Methods*, **106**, 203 (1973).

3.5   Mössbauer Emission Spectroscopy

In Mössbauer emission spectroscopy the sample to be studied is prepared with a radioactive isotope and is used as a source. The radiation emitted from the sample is absorbed by a standard absorber that contains the corresponding Mössbauer isotope. For example, cobalt does not have suitable isotopes for Mössbauer spectroscopy, but cobalt-containing materials can be studied by the technique if they are prepared with ^{57}Co and then used as sources in Mössbauer emission spectroscopy. In these experiments information about the cobalt atoms is obtained by studying the ^{57}Fe atoms produced in the decay of ^{57}Co. However, it must be appreciated that in Mössbauer emission spectroscopy the observed oxidation states and spin states of ^{57}Fe may differ from those of the parent ^{57}Co. Furthermore, the decay process can create unstable, but relatively long-living, that is, 10^{-7}s, changes in the chemical surroundings of the Mössbauer nucleus. Although these aftereffects can usually be neglected when considering conducting solids, such aftereffects may be quite important in insulators and can create difficulties in the interpretation of results. Hence, when examining cobalt-containing solids to investigate the structural surroundings of the cobalt atoms by Mössbauer emission spectroscopy, it is important that the immediate surroundings of the resulting ^{57}Fe atoms resemble those of the parent ^{57}Co. Thus, Mössbauer emission spectroscopy is a potentially useful technique in situations where a particular element of specific interest does not contain a suitable Mössbauer isotope. However, it is important to note that such source experiments are often complicated and that the consequences of nuclear transformations that involve both oxidation and reduction are not fully understood.

4   APPLICATIONS OF MÖSSBAUER SPECTROSCOPY

It will be clear from the foregoing that the Mössbauer parameters are capable of giving valuable information about the electronic properties of chemical bonds: the electronic, ligand, and lattice environments of Mössbauer active atoms; magnetic effects in solids; and various forms of dynamic and relaxation processes. Such information is important in many areas of science. The technique is therefore widely used in many disciplines including chemistry, physics, biology, metallurgy, and geology. It is also used by practitioners in many highly specialized fields such as surface science, catalysis, and archaeology. The applications of Mössbauer spectroscopy have recently been reviewed comprehensively [30].

Some aspects of these different applications are outlined in the next sections to give the reader some feeling for the scope of the technique. However, for definitive accounts of the varied applications of Mössbauer spectroscopy and its potential power in so many different areas of science, the reader is referred to the literature and the references therein (see, e.g., [8] and [30]).

4.1   Bonding

Mössbauer spectroscopy is well suited for studies of chemical bonding since two of the principal Mössbauer parameters can be related to the electronic populations of valence shell orbitals. Hence, the capacity of the chemical isomer shift to describe aspects of the electron density associated with an atom and the ability of the quadrupole splitting to give information on the degree of symmetry of the electron distribution about the nucleus renders the Mössbauer spectrum as a sensitive means by which many properties of the Mössbauer atom and its environment can be assessed. For example, oxidation states, valence orbital hybridization, the ionic character or the influence of electronegativity on a bond, bond polarization, coordination number and symmetry, overlap effects, bond multiplicity, and electron back donation are some of the matters that can be elucidated by interpretation of Mössbauer spectra [13].

4.1.1   Oxidation States

The sensitivity of the Mössbauer chemical isomer shift to the oxidation state of the Mössbauer atom was mentioned in Section 2.3.1. This sensitivity arises from the addition or removal of valence electrons. For example, the oxidation of tin(II) to tin(IV) involves the loss of two $5s$ electrons, which is reflected in a large change in the ^{119}Sn Mössbauer chemical isomer shift. Similarly, the oxidation of

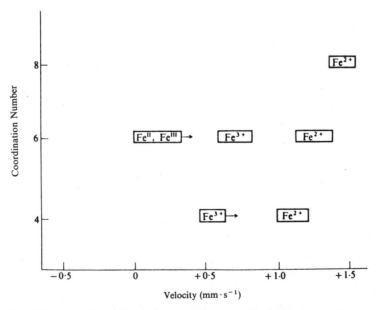

Figure 4.17 ^{57}Fe chemical isomer shifts plotted against coordination number for *ionic* compounds. Arrows indicate that values outside the boxed areas have been observed. Reproduced, with permission, from G. M. Bancroft, *Mössbauer Spectroscopy, An Introduction for Inorganic Chemists and Geochemists*, McGraw-Hill, New York, 1973, p. 157.

Table 4.2 Ranges of Chemical Isomer Shifts (mm·s⁻¹) for
High- and Low-Spin Iron Compounds Relative to Metallic
Iron

Oxidation State	High Spin	Low Spin
Fe(VI)	−0.8 to −0.9	
Fe(IV)	−0.2 to +0.2	+0.1 to +0.2
Fe(III)	+0.1 to +0.5	−0.1 to +0.5
Fe(II)	+0.6 to +1.7	−0.2 to +0.5
Fe(0)		−0.2 to −0.1
Fe(—II)		About −0.2

iron(II) to iron(III) involves the loss of one 3d electron, which reduces the shielding of the outer s electrons. However, within any given oxidation state the chemical isomer shift can vary from compound to compound for a number of reasons, including the effects of covalency, coordination number and symmetry of the Mössbauer atom, and the electronegativity of coordinated ligands. The dependence of the chemical isomer shifts on the coordination number and site symmetry in ^{57}Fe Mössbauer spectroscopy has been the subject of extensive study and is summarized in Figure 4.17.

The sensitivity of chemical isomer shifts recorded from some high- and low-spin iron compounds to effects other than the formal oxidation state of iron is illustrated in Table 4.2. The chemical isomer shifts for the high-spin ionic iron compounds cover rather large ranges, whereas the data for the low-spin compounds, which occur when ligands high in the spectrochemical series producing large crystal fields are involved, fall within more narrow ranges. In general the pairing of spins in iron compounds to give low-spin iron causes a decrease in the chemical isomer shift reflecting an increase in s-electron density at the iron nucleus. This results from the increase in covalency of the metal–ligand bonds, which causes an increase in the 4s-electron density. The accumulation of negative charge on the metal atom also produces an increase in the effective radii of the d orbitals, which deshields the s electrons further. Moreover, any delocalization of d electrons through σ-antibonding, π-bonding, or π-antibonding molecular orbitals also enhances the s-electron density.

4.1.2 Bond Character

The range of chemical isomer shifts within a given oxidation state can also reflect factors relating to bond character. For example, some of the iron(II) halides show [31] a linear relationship between the electronegativity of the anion and the chemical isomer shift (Figure 4.18). The linear increase in δ with increasing electronegativity illustrates the participation of the 4s electrons in the bond, such that, if the electronic configuration of iron is formulated as $3d^64s^x$, the value of x decreases as the electronegativity of the ligand increases. In

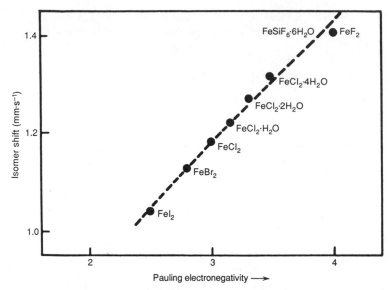

Figure 4.18 Correlation between the chemical isomer shift and the electronegativity of anions in iron(II) halides. Reproduced, with permission, from G. K. Shenoy, "Mössbauer Effect Isomer Shifts," in G. J. Long, Ed., *Mössbauer Spectroscopy Applied to Inorganic Chemistry*, Vol. 1, Plenum, New York, 1984, p. 68.

contrast to the relationship between the electronegativity of the anion and the iron(II) chemical isomer shift just cited, it is relevant to note that the [129]I Mössbauer chemical isomer shift data recorded from alkali metal iodides do not reflect a simple relationship with the electronegativity of the alkali metal atom. In these fairly ionic systems it seems that the [129]I chemical isomer shifts reflect orbital overlap effects [32].

The variation in the chemical isomer shifts of gold halides within a given oxidation state (Figure 4.19) clearly does not follow any simple relationship with the electronegativity of the halide [33]. Figure 4.19 also illustrates the irregular behavior of the chemical isomer shift data for the halides of gold in different oxidation states. To some extent this reflects the different coordination and hence different orbital hybridization of gold in different oxidation states. Thus the gold atoms are involved in linear bonds in the gold(I) compounds, square planar geometry in the gold(III) compounds, and nearly octahedral symmetry in the gold(V) compounds.

4.1.3   Partial Isomer Shifts

As mentioned in Section 2.3.1, systems in which the bonding is not highly covalent show the chemical isomer shift to be an additive function of the ligand [14] so that a *partial isomer shift*, sometimes called a *partial center shift* (pcs), can be assigned to each ligand. In these situations the value of δ for the compound is

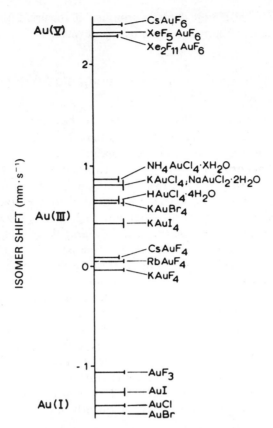

Figure 4.19 Chemical isomer shifts for the halides of gold(I), -(III), and -(V). Reproduced, with permission, from G. K. Shenoy and F. E. Wagner, Eds., *Mössbauer Isomer Shifts*, North-Holland, Amsterdam, 1978, p. 515.

given by the sum of the partial isomer shifts over all the ligands. The treatment requires that the chemical isomer shift be relatively insensitive to small variations in bond angles and bond distances and that the bonding properties of a given ligand be insensitive to the nature of other ligands. Such requirements imply both an absence of significant rehybridization as ligands are changed and a relatively narrow range of covalencies.

Scales of partial isomer shifts can be derived for a number of transition metals, and the reasoning is well illustrated by the scale for low-spin iron(II) where data for a wide variety of complexes are available [14, 34, 35]. Similar scales have been developed for ruthenium(III), ruthenium(II), iridium(II), gold (I), and tin(IV). The partial isomer shifts are indicative of the contribution a ligand makes to the electron density in the valence shell of the metal ion, and the data enable the ligands to be placed in an order of increasing σ-donor power and π-bonding effects [14].

The reader requiring a more detailed appraisal of the usefulness of the partial isomer shift concept is referred to three excellent reviews of different aspects of the subject [5, 7, 13].

4.1.4   Bonding Information From Quadrupole Splitting Data (Partial Quadrupole Splitting)

Given the sensitivity of the quadrupole splitting to the electronic environment of the nucleus (Section 2.3.2) it is not surprising that this Mössbauer parameter is also an important way to examine the nature of chemical bonding. The interpretation of quadrupole splitting data in this context has been excellently described in the recent past [5, 7, 13], and only a few examples involving the semiquantitative treatment of the electric field gradient are given here.

In the point-charge model [16–18] (Section 2.3.2) the ligands are envisaged as point charges that can be assigned a *partial quadrupole splitting* parameter. In the *hybridization treatment* [19] the value of the partial quadrupole splitting parameter is treated as being dependent on both the nature of the ligand and the character of the hybrid orbitals of the Mössbauer atom. Although the hybridization treatment emphasizes the different scales of partial quadrupole splitting that must be used for the different coordinations of metal ions, the two methods frequently lead to analogous expressions and the point-charge model is a convenient means by which many problems can be approached. Partial quadrupole splittings for common ligands coordinated to iron(II), iridium(II), gold(I), ruthenium(II), and tin(IV) have been constructed, and the observed good correlation between the different scales [13] has provided evidence that the bonding characteristics of a ligand are frequently independent of the metal acceptor.

The application of the Townes and Dailey theory [20] (Section 2.3.2) to the interpretation of chemical isomer shift and quadrupole coupling data has been successfully used in studies of several covalent compounds. For example, in xenon compounds the electric field gradient is mainly determined by the number of holes h_p in the $5p$ orbitals [36]. Following the theory of Townes and Dailey the quadrupole coupling caused by a $5p$-electron hole can be expressed in terms of a p-electron imbalance U_p and, for ^{129}Xe, has been calculated as $27.3 \text{ mm} \cdot \text{s}^{-1}$ [36]. From the results of Mössbauer studies of some xenon halides, the values of U_p and h_p have been evaluated [36] (Table 4.3). Based on these results and the chemical isomer shift data, a change in the isomer shift per $5p$ hole was calculated as $0.13 \text{ mm} \cdot \text{s}^{-1}$, and, from such calculations, the $5p$-electron population in other xenon compounds has been assessed.

4.2   Structure

The sensitivity of Mössbauer spectroscopy to the electronic structure of atoms has been mentioned earlier; hence, this section outlines the suitability of Mössbauer spectroscopy for the examination of molecular and lattice

Table 4.3   Chemical Isomer Shifts, Quadrupole Coupling Constants, and Values of U_p and h_p for the Xenon Halides Recorded at 4.2 K[a]

Compound	Chemical Isomer Shift (mm·s⁻¹)	Quadrupole Coupling Constant (e^2qQ mm·s⁻¹)	U_p	h_p
XeF_2	0.10 ± 0.12	39.00 ± 0.10	1.43	1.43
XeF_4	0.40 ± 0.04	41.04 ± 0.04	1.50	3.00
$XeCl_2$	0.17 ± 0.09	28.20 ± 0.14	1.03	1.03
$XeCl_4$	0.25 ± 0.08	25.62 ± 0.10	0.94	1.88

[a]Data adapted from G. J. Perlow, in V. I. Goldanskii and R. H. Herber, Eds., *Chemical Applications of Mössbauer Spectroscopy*, Vol. 1, Academic, New York, 1968, p. 378.

structures. For more comprehensive accounts of the application of Mössbauer spectroscopy to the study of the structural properties of solids the reader is referred to recent textbooks [5, 7, 37].

The interpretation of the Mössbauer spectrum recorded from a molecular solid is highly dependent on the particular situation. For example, in some cases two Mössbauer atoms may be involved in the molecular complex and this provides a means by which the Mössbauer data can be more comprehensively assessed. However, in the majority of compounds studied by Mössbauer spectroscopy only one Mössbauer atom will usually be present, and, while acknowledging the advantages of having available such a sensitive probe of the immediate local environment, there are often problems in commenting on the number of disposition of other atoms or groups around a Mössbauer atom. For this reason, Mössbauer studies of a structural nature have been most successfully performed when the Mössbauer data have been complemented by the results from an independent technique such as X ray or neutron diffraction, electron microscopy, or solid-state NMR.

4.2.1   Molecular Structure

Mössbauer spectroscopy has been a particularly valuable technique for the study of the structural properties of organometallic compounds, and useful data have been obtained from compounds containing a variety of Mössbauer atoms. The Mössbauer results have been especially informative when the compounds under examination have contained the Mössbauer atom in more than one site such as occurs in the iron carbonyls and their derivatives. The principles are well illustrated by a consideration of the structural characterization of $Fe_3(CO)_{12}$, which has recently been reviewed [38].

The iron carbonyl $Fe_3(CO)_{12}$ was first prepared in 1907, was found to be diamagnetic, and in 1930 was shown to possess three $Fe(CO)_4$ units per molecule. Several structures were proposed for the molecule over the following 30 years (Figure 4.20). For example, the linear structures (Figure 4.20a–c) were

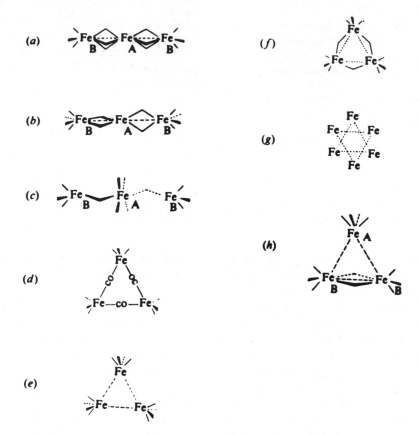

Figure 4.20 Proposed structures of $Fe_3(CO)_{12}$. Reproduced, with permission, from F. J. Berry and D. P. E. Dickson, Eds., *Mössbauer Spectroscopy*, Cambridge University Press, Cambridge, 1986, p. 72.

derived from early IR data that showed $Fe_3(CO)_{12}$ to contain both terminal and bridging carbonyl ligands. Although subsequent studies by X-ray crystallography were not completely successful in determining the structure, the evidence did suggest that the molecule adopted a triangular-type structure (Figure 4.20*d–g*) in which all the iron atoms appeared to occupy similar sites. The ^{57}Fe Mössbauer spectra [39, 40] showed three absorptions (Figure 4.21). The outer two lines were associated with a large quadrupole doublet with a relative area of two and the inner line to a singlet, or a small doublet, with a relative area of one. Therefore, the Mössbauer spectrum showed that $Fe_3(CO)_{12}$ has two different iron sites: one contains only one iron atom, Fe_B, in a high symmetry environment giving the inner singlet, while the other site accounts for two iron atoms, Fe_A, in a lower symmetry site, which gives the large quadrupole splitting. Hence, by simultaneous consideration of the Mössbauer spectra, the IR results, and the triangular structures indicated by the preliminary X-ray

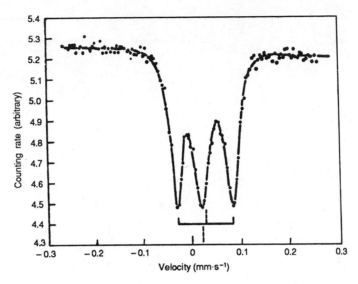

Figure 4.21 ^{57}Fe Mössbauer spectrum recorded from $Fe_3(CO)_{12}$. Reprinted with permission from R. H. Herber, W. R. Kingston, and G. K. Wertheim, *Inorg. Chem.*, **2**, 153 (1963). Copyright 1963 American Chemical Society.

diffraction studies, the structure shown in Figure 4.20*h* was proposed. In this structure the two Fe_A atoms are bridged by two carbonyl ligands and have three terminal ligands each that thereby constitute a low-symmetry site, whereas the Fe_B atom occupies a single high-symmetry site that has four terminal carbonyl ligands and is bonded to the Fe_A atoms. The disordered structure was finally resolved by single-crystal X-ray crystallography [41] (Figure 4.22).

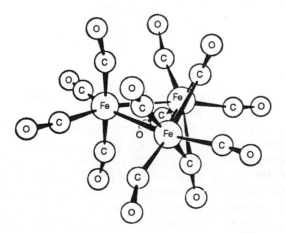

Figure 4.22   The molecular structure of $Fe_3(CO)_{12}$. Reprinted with permission from C. H. Wei and L. H. Dahl, *J. Am. Chem. Soc.*, **88**, 1821 (1966). Copyright 1966 American Chemical Society.

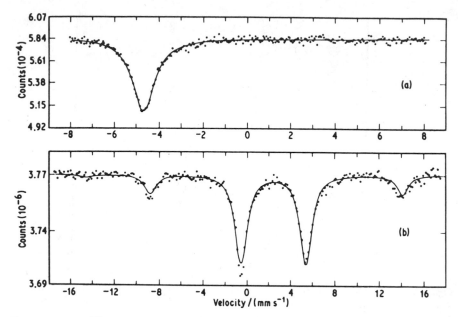

Figure 4.23 The ^{129}I Mössbauer spectra recorded at 90 K from (a) the IF_6^+ cation and (b) the IF_6^- anion. Reproduced, with permission, from S. Bukshpan, J. Soriano, and J. Shamir, *Chem. Phys. Lett.*, **4**, 241 (1969).

The power of Mössbauer spectroscopy for the elucidation of the molecular structure of a very different type of compound is well illustrated by the study of the IF_6^+ cation and the IF_6^- anion [42]. The ^{129}I Mössbauer spectrum of $[IF_6]^+[AsF]^-$ is composed of a single line (Figure 4.23a), which reflects the regular octahedral symmetry of the IF_6^+ cation. However, the spectrum recorded from $CsIF_6$, which contains the IF_6^- anion, shows a large quadrupole splitting (Figure 4.23b) indicating the stereochemical activity of the lone pair of electrons in a pentagonal bipyramidal structure (Figure 4.24).

4.2.2   Isomerism

A number of coordination compounds exhibit cis–trans geometrical isomerism. This is frequently found in octahedral compounds of the type MA_2B_4 where the A groups are either cis or trans; that is, they are either adjacent to or diametrically opposite each other. It has been shown, using a point-charge calculation (Section 4.1.4), that the quadrupole splitting of atom M should be twice as great in the trans isomer and of opposite sign to the cis isomer. The calculations have been verified experimentally during, for example, studies [43–45] of some iron complexes (Table 4.4). Similar results have also been found in six-coordinate organotin compounds by ^{119}Sn Mössbauer spectroscopy.

Mössbauer spectroscopy has also been used to examine linkage isomerism in complexes where simple inorganic ligands such as CN^- and NCS^- are capable

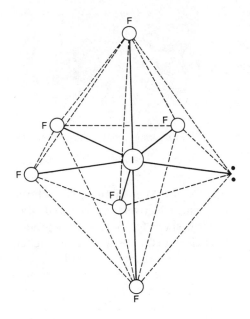

Figure 4.24 A possible structure of the IF$_6^-$ anion. Reproduced, with permission, from S. Bukshpan, J. Soriano, and J. Shamir, *Chem. Phys. Lett.*, **4**, 241 (1969).

of coordinating to a metal cation from either end of the molecule. The application of Mössbauer spectroscopy to these types of problems has been excellently summarized in the recent past [7].

4.2.3 Molecular Rotation in Solids

The use of Mössbauer spectroscopy to elucidate the structural and dynamic properties of a system is well illustrated by recent Mössbauer studies of molecular rotation in solids. These investigations have shown that the molecules of a solid compound can, in certain environments, rotate such that the electric field gradient is no longer static and independent of time. For example, Mössbauer spectroscopy has shown [46] that ferrocene molecules (Figure 4.25a) in the thiourea clathrate of formula $Fe(C_6H_5)_2 : [CS(NH_2)_2]_3$ are encapsulated within channels of the thiourea structure at 77 K with their molecular axes

Table 4.4 ^{57}Fe Mössbauer Quadrupole Splittings for Some Iron Complexes

Compound	$\Delta\,(mm \cdot s^{-1})$	References
$[Fe(CN)(CNEt)_5](ClO_4)$	0.17	[43, 44]
$cis\text{-}Fe(CN)_2(CNEt)_4$	0.29	[43, 44]
$trans\text{-}Fe(CN)_2(CNEt)_4$	0.56	[43, 44]
$[FeCl(p\text{-}MeO \cdot C_6H_4 \cdot NC)_5](ClO_4)$	0.70	[45]
$cis\text{-}FeCl_2(p\text{-}MeO \cdot C_6H_4 \cdot NC)_4$	-0.83	[45]
$trans\text{-}FeCl_2(p\text{-}MeO \cdot C_6H_4 \cdot NC)_4$	$+1.59$	[45]

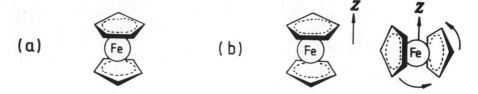

Figure 4.25 (*a*) Ferrocene molecule and (*b*) stationary and rotating ferrocene molecules within the thiourea channels. Reproduced, with permission, from B. W. Fitzsimmons, *Eur. Spectros. News*, **24**, 51 (1979). Copyright 1979 John Wiley & Sons, Ltd.

parallel and perpendicular to the channel axis. Under such circumstances both types of molecule occupy fixed positions and, being effectively immobilized, give quadrupole-split ^{57}Fe Mössbauer spectra as observed for pure ferrocene and characteristic of iron in an asymmetric environment. However, at higher temperatures the perpendicular ferrocene molecules can rotate rapidly with a symmetry axis orientated at 90° to that of the stationary parallel molecules (Figure 4.25*b*) such that a single-line Mössbauer spectrum is observed. The study demonstrates that ferrocene molecules accommodated within the channels of thiourea may rotate in a fashion that is precluded in pure crystalline ferrocene. More recent work has indicated that such processes are not restricted to clathrates but that the holes in ionic lattices are also sometimes large enough to allow unexpected rotational activity.

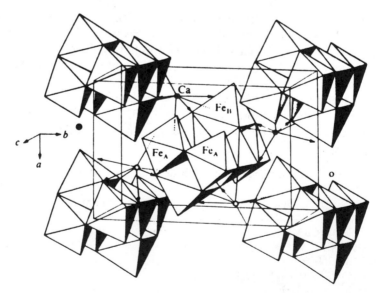

Figure 4.26 The structure of ilvaite. Reproduced, with permission, from T. Yamanaka and Y. Takenchi, *Phys. Chem. Miner.*, **4**, 149 (1979).

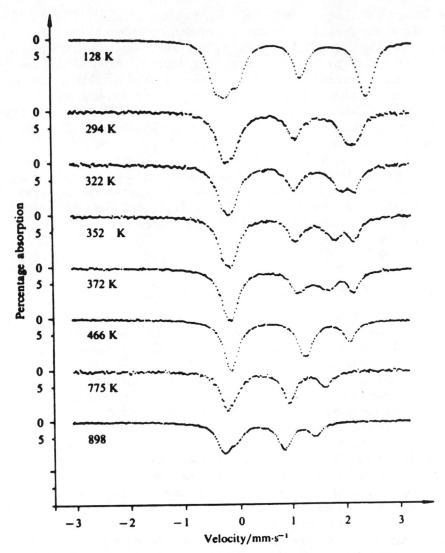

Figure 4.27 ^{57}Fe Mössbauer spectra recorded from ilvaite over a range of temperatures. Reproduced, with permission, from I. V. Heilmann, N. B. Olsen, and J. B. Olsen, *Phys. Scr.*, **15**, 285 (1977).

4.2.4   Mixed-Valence Compounds

Mössbauer spectroscopy is a useful means by which the electronic and structural properties of mixed-valence compounds can be correlated. A good example of such application is provided by the mixed-valence iron silicate mineral known as ilvaite of formula $CaFe(II)_2Fe(III)Si_2O_8(OH)$.

The structure of ilvaite consists [47, 48] (Figure 4.26) of infinite edge-sharing double chains in which one type of iron site, Fe_A, contains equal amounts of

Fe(II) and Fe(III) and where the Fe_A—Fe_A bond distances are 3.03 and 2.83 Å. Iron(II) species, Fe_B, in a different type of site, are bound to these chains at distances of 3.15 and 3.25 Å from the Fe_A sites. The structure therefore provides a variety of pathways for intervalence charge transfer. The Mössbauer spectra (Figure 4.27) recorded [49] from ilvaite between 128 and 898 K showed smooth variations in the region between 1 and $2 \, mm \cdot s^{-1}$ over the temperature range that has been associated [49–51] with an increase in iron(III) character of one of the iron(II) species as it gives up an electron through intervalence charge transfer. The Fe_B iron(II) site remains essentially single valent up to at least 900 K and indicates that the transfer of an electron from Fe_B across the shared face to Fe_A is nonfacile at least in comparison to transfer between the edge-shared Fe_A sites. Indeed the spectra gave evidence of electron exchange along or between the Fe_A double chains, showing that ilvaite exhibits a fixed valence for Fe_B and a mixed-valence interaction for the Fe_A sites.

4.3   Magnetism

4.3.1   Magnetically Split Mössbauer Spectra

The principles of the magnetic dipole interaction that gives rise to magnetically split Mössbauer spectra were outlined in Section 2.3.3. Such magnetically split Mössbauer spectra can be used to investigate a variety of matters including the magnetic ordering and structure of certain types of solids, the nature of the magnetic interactions, the magnitude of magnetic moments in particular atoms, and the electronic structure of atoms that relate to the magnetic hyperfine field at the Mössbauer nuclei. The Mössbauer spectra recorded from magnetically ordered solids frequently show details in the region of the ordering temperature, which can, for example, be associated with phenomena such as the critical slowing down of the spin fluctuation rate, and attempts have been made to develop dynamic models whereby the data can be interpreted quantitatively. The technique can also be used to investigate other materials, for example, those that undergo ferroelectric phase transitions, to monitor the variation in the electric field gradient tensor with temperature and its relationship with the ferromagnetic displacement. The application of Mössbauer spectroscopy to the study of magnetically ordered solids has been described in detail [52].

4.3.2   Examples

IRON OXIDES

An example of a magnetically ordered solid that can be advantageously examined by Mössbauer spectroscopy is the iron oxide Fe_3O_4 called *magnetite* [53]. The oxide is an inverse spinel in which the iron ions occupy two types of interstitial sites between the oxygen atoms, which adopt a cubic close-packed structure. The tetrahedral (A) iron sites are occupied by Fe^{3+} ions, and the octahedral (B) sites are occupied by Fe^{2+} and Fe^{3+} ions. The unit cell contains twice as many B sites as A sites. The iron ions at the B sites are antiferromagnetically coupled to those at the A sites so that the moments on the Fe^{3+} ions cancel and the net ferromagnetic moment arises from the Fe^{2+} ions at the B sites with the easy direction of magnetization being normally along the [III] direction. At the Verwey temperature T_v of 119 K the structure transforms to orthorhombic symmetry with a large increase in electrical resistivity. The low resistivity above T_v has been explained in terms of rapid electron hopping between Fe^{2+} and Fe^{3+} at the octahedral sites. At temperatures below T_v, the electron hopping has been envisaged as being inhibited by an ordering of the Fe^{2+} ions in alternate (100) layers, which gives rise to the orthorhombic symmetry with a higher electrical resistivity.

However, the Mössbauer spectra recorded [53] at temperatures above T_v have supported an alternative explanation for the higher conductivity in terms of a band model for the electrons and has indicated that below T_v the magnetic ordering is not so simple as originally envisaged. The Mössbauer spectrum of Fe_3O_4 at 298 K (Figure 4.28) contains two superimposed six-line patterns corresponding to magnetic fields of 48.8 and 46.0 T in the area ratio 1:2 that can be associated with the A and B sites. The widths of the B-site lines are about 50% broader than those of the A-site lines. The line broadening has been associated

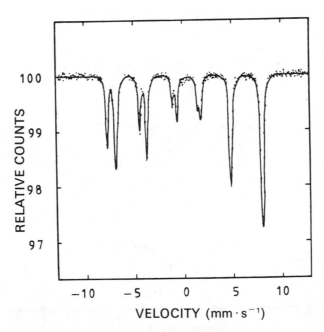

Figure 4.28 ^{57}Fe Mössbauer spectrum recorded from magnetite at 298 K showing two superimposed six-line patterns corresponding to magnetic fields of 48.8 and 46.0 T with area ratios of 1:2 and associated with the A and B sites. Reproduced, with permission, from T. E. Cranshaw and G. Longworth, "Mössbauer Spectroscopy of Magnetic Systems," in G. J. Long, Ed., *Mössbauer Spectroscopy Applied to Inorganic Chemistry*, Vol. 1, Plenum, New York, 1984, p. 171.

with the presence of several fields at the B sites as opposed to relaxation effects involving electron hopping between the Fe^{2+} and Fe^{3+} sites. The deviations in the chemical isomer shift values from those calculated at temperatures exceeding 700 K have been correlated with a maximum resistivity at temperatures above 850 K arising from changes in the band structure. The Mössbauer spectra

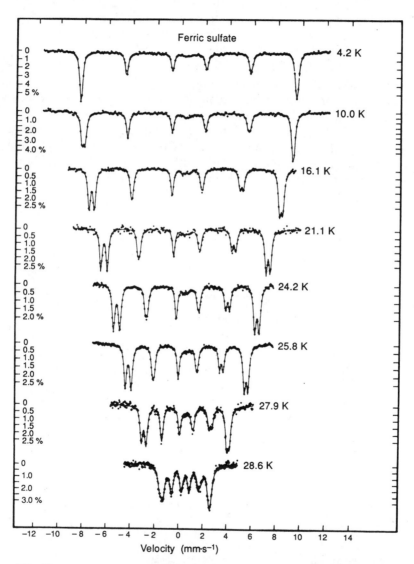

Figure 4.29 ^{57}Fe Mössbauer spectra of iron(III) sulfate recorded at increasing temperatures. Above 4.2 K there are two different field values corresponding to the two iron sites. Reprinted with permission from G. J. Long, G. Longworth, P. Battle, A. K. Cheetham, R. V. Thundathil, and W. Beveridge, *Inorg. Chem.*, **18**, 624 (1979). Copyright 1979 American Chemical Society.

recorded below the Verwey transition have also supported the alternative method. In disagreement with the simple ordering model involving the Fe^{2+} and Fe^{3+} B-site ions on alternative (100) planes and the presence of two hyperfine fields, the Mössbauer spectra together with NMR results have suggested the presence of a larger number of fields. Indeed, the data are consistent with 8 inequivalent Fe_3O_4 molecular units in the unit cell below T_v, which give rise to 8 inequivalent A sites and 16 inequivalent B sites.

IRON(III) SULFATE

Another example of the type of information that can be obtained from Mössbauer studies of magnetic solids is provided by the recent neutron diffraction, magnetic susceptibility, and Mössbauer spectroscopy investigation of anhydrous iron(III) sulfate [54]. In the monoclinic modification the structure is characterized by an infinite network of Fe—O—S linkages where the oxygen atoms coordinate the iron atoms octahedrally and the sulfur atoms tetrahedrally with the octahedra sharing corners with the tetrahedra. Neutron diffraction data have indicated that the material is a two-sublattice antiferromagnet in which each sublattice contains crystallographically distinct iron atoms. Analysis of the Mössbauer spectra showed the Néel point to be 28.7 K and, at temperatures above about 4.2 K, revealed the presence of two different field values corresponding to the two iron sites (Figure 4.29). The two fields showed different variations with temperature and exhibited behavior implying a difference in the exchange coupling, which may reflect differing magnetic superexchange pathways between the iron atoms.

LOW-DIMENSIONAL MAGNETISM

Recent developments in Mössbauer studies of magnetic materials have included investigations of iron-containing compounds with low-dimensional properties [52, 55]. These studies have shown that Mössbauer measurements of the hyperfine fields offer a powerful means by which the large fluctuations in quasi-one-dimensional antiferromagnets that reduce the sublattice magnetization and the ordering temperature can be investigated. Studies of this type may well become important in future investigations of noncrystalline chain materials.

5 APPLICATIONS OF SOCIAL AND TECHNOLOGICAL RELEVANCE

The examination by Mössbauer spectroscopy of pure homogeneous solids to give information on fundamental matters such as bonding, structure, and magnetic properties has provided many valuable insights into the technologically relevant properties of such materials. However, many materials of commercial importance are complex solids in which their technologically interesting properties may result from a heterogeneity of stoichiometries or phases. Hence, Mössbauer spectroscopy is established as an important technique in *materials science* and other applied areas of scientific research where the solids under examination are far from being pure, homogeneous, single-phase materials.

Mössbauer spectroscopy is particularly useful in these types of investigations since the resonance absorption is a purely nuclear process in the material of interest and therefore independent of additional phases or distant parts of the host that can sometimes interfere when other resonance techniques are used. Mössbauer spectroscopy can be applied to a wide range of solids including glasses, disordered alloys, and finely divided samples; even dissolved molecules and complexes become amenable to examination when the solutions are frozen. The technologically relevant applications of Mössbauer spectroscopy have been detailed in several textbooks in recent years [30, 56–58] and only a few selected areas are briefly reviewed in this chapter to give some idea of the different types of materials amenable to investigation and the diversity of areas in which Mössbauer spectroscopy can make a useful contribution.

5.1   Metals and Alloys

The study of metals, alloys, and intermetallic compounds has been a particularly fruitful area for the application of Mössbauer spectroscopy and has received attention in several textbooks [7]. Although several techniques exist for the examination of metallic properties such as electrical conductivity and magnetic susceptibility that originate from the collective-electron band structure of the bulk solid, Mössbauer spectroscopy offers a distinctly different means of investigation since it examines the local environment of individual atoms in the solid. This facilitates, for example, investigations of near-neighbor interactions and the effects of changing compositions on the electronic and magnetic interactions of particular atoms within the alloy. Although the properties of many metals and alloys have been investigated by Mössbauer spectroscopy using isotopes of a variety of metallic elements, the ^{57}Fe isotope continues to be the most widely used means of examination.

Disordered alloys where two or more elements occupy the same crystallo-graphic sites with a random probability have been found to be especially amenable to investigation by Mössbauer spectroscopy. Such materials frequent-ly have a wide range of compositions that affect the structural properties of the alloys and that are reflected in changes in the chemical isomer shift and magnetic flux density at the nucleus under investigation.

Intermetallic compounds differ in that each element shows a strong pre-ference for a particular lattice site, and the resulting regular structure usually contains a nearly stoichiometric composition. For example, binary intermetallic compounds usually have restricted ranges of compositions such as AB_4, AB_3, AB_2, or AB and frequently represent intermediate situations between the ionic lattice with localized electrons and the true random alloy. The uniformity of the atomic environments results in any quadrupole split spectra being amenable to interpretation in terms of the local site symmetry. Magnetically ordered intermetallic compounds and alloys have been found to be very suitable systems for examination by Mössbauer spectroscopy.

Mössbauer spectroscopy also has other applications in the study of alloys. For example, it can be used to measure the thickness of layers formed by the treatment of iron in a molten salt bath containing, for example, cyanides and other compounds [59]. The salts are broken down by interaction with the iron and the carbon and nitrogen and form a hard layer of iron nitrides and carbides. Although the thickness of the layer is known to depend on the time of immersion of the component and the temperature of the treatment, it is often difficult to know with confidence the thickness of the hardened surface layers. The use of Mössbauer spectroscopy to examine the thickness of the hardened layers is based on the knowledge that as one goes down the nitrided layer, the concentration of nitrogen or carbon falls off, and between the outer nitrided layer and the iron base there is a layer of carbon or nitrogen austerite (Figure 4.30). The Mössbauer spectra recorded by detecting backscattered X rays from the surface of iron treated by immersion in a commercial salt bath for times between 2 and 20 min are shown in Figure 4.31. In the top spectrum a trace of martensite or ferrite from the underlying steel is visible. The peak near zero velocity can be associated with the layer of nitrogen austenite between the nitrided layer and the underlying steel. The remainder of the spectrum is characteristic of the hexagonal epsilon nitride, ε-$Fe_2N_{0.7}$. As the immersion time is increased (Figure 4.31), the martensite, and subsequently the austenite, components begin to disappear until only the ε-nitride contribution to the spectrum remains. The results show that the signal from the austenite layer is attenuated by the iron nitride above it. Hence the thickness of the nitride layer can be calculated by making a calibration of the intensity of the austenite spectral peak relative to the nitride spectrum.

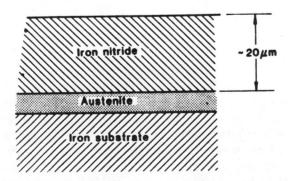

Figure 4.30 Representation of the layers as a nitrided iron component. Reproduced, with permission, from T. E. Cranshaw, "Metal Research of Industrial Significance by Mössbauer Spectroscopy," in G. J. Long and J. G. Stevens, Eds., *Industrial Applications of the Mössbauer Effect*, Plenum, New York, 1986, p. 7.

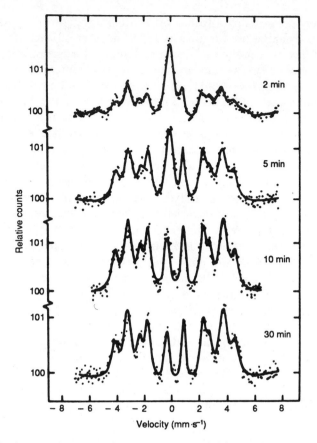

Figure 4.31 Backscattered X-ray Mössbauer spectra recorded from an iron component treated for different times in a commercial nitriding salt bath. Reproduced, with permission, from T. E. Cranshaw, "Metal Research of Industrial Significance by Mössbauer Spectroscopy," in G. J. Long and J. G. Stevens, Eds., *Industrial Applications of the Mössbauer Effect*, Plenum, New York, 1986, p. 7.

5.2 Energy Sources

The application of Mössbauer spectroscopy to problems of contemporary economic and social importance is well illustrated by recent studies of hydrogen storage materials [60] that can be used as a source of fuel in, for example, the operation of automobiles. In this area of interest the many intermetallic compounds of *d*- and *f*-block elements, which reversibly absorb large amounts of hydrogen to form ternary hydrides at easily accessible temperatures and pressures, have developed a special importance; and Mössbauer spectroscopy has been found to be a particularly powerful means by which the chemical nature of hydrogen and its location in the ternary hydrides can be identified. These studies have enabled the formation of the different structural phases and the influence of hydrogen absorption on the electronic and magnetic properties of the host lattice to be monitored. Mössbauer spectroscopy has also been used to examine the stability of hydrogen storage materials against numerous hydrogen absorption–desorption cycles. This is well illustrated by the recent ^{151}Eu Mössbauer spectroscopy study of the degradation of europium-substituted LaNi$_5$ over several absorption–desorption cycles [61]. The europium adopts the trivalent ($4f^6$) state in LaNi$_5$ and, on adsorption of hydrogen, becomes divalent ($4f^7$). These two oxidation states can be readily distinguished by ^{151}Eu Mössbauer spectroscopy (Figure 4.32a and b). The spectra showed that, following the removal of hydrogen from LaNi$_5$, europium

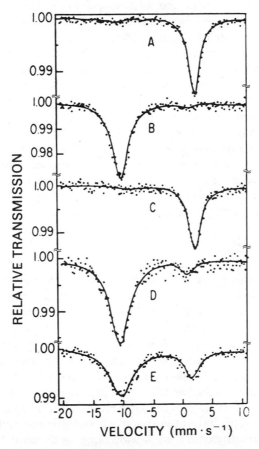

Figure 4.32 ^{151}Eu Mössbauer spectra recorded from $(La_{0.9}Eu_{0.1})Ni_{4.6}Mn_{0.41}$: (A) starting material containing Eu(III), (B) hydrided material containing Eu(II), (C) dehydrided material after 10 absorption–desorption cycles showing conversion to Eu(III), (D) hydrided after 1500 absorption–desorption cycles showing presence of Eu(II) and small amount of Eu(III), (E) dehydrided after 1500 absorption–desorption cycles showing Eu(III) and Eu(II) from permanent europium hydride. Reproduced, with permission, from R. L. Cohen, K. W. West, and J. H. Wernick, *J. Less Common Metals*, **73**, 273 (1980).

adopts the trivalent state (Figure 4.32c), which persists even after 1500 absorption–desorption cycles (Figures 4.32d and e). However, the cycling treatment was also accompanied by a degradation of the total hydrogen storage capacity of $LaNi_5$ and the formation of the rare earth metal hydride (Figure 4.32e). Hence, the Mössbauer studies provided qualitative information on the chemical degradation of the hydrogen storage material.

Mössbauer spectroscopy has also been used to study high-energy density secondary battery materials [62]. For example, ^{57}Fe Mössbauer investigations of the insertion of metal guest species such as lithium, into chalcogenide host lattices, for example, VS_2 or TaS_2, have enabled an enhanced understanding of guest–host interactions and the electronic properties of the host matrix. Such studies also sought to relate the microscopic properties of the hosts to the mechanistic aspects of electrochemical reactions at the cathode and their dependence on the state of charge of the battery.

Although we will later consider the use of Mössbauer spectroscopy for the study of minerals (Section 5.4), it is relevant to mention here a recent development that has involved the examination by Mössbauer spectroscopy of coal and petroleum source rocks to identify the nature of the iron-bearing compounds and mineral matter they contain [63, 64]. In this context the technique has also been used as a means by which the transformation of the iron minerals in coal conversion processes can be examined and as a technique by which the sulfur concentration in coal can be determined. The method has also been developed for the examination of other features such as the stoichiometry of the iron–sulfur oxides that are formed during coal liquefaction processes and that influence the liquid yields and properties.

5.3   Catalysis

Mössbauer spectroscopy is an especially powerful technique for the study of heterogeneous catalysts because it is able to examine authentic solid catalysts in their working gaseous environments. This particular application of Mössbauer spectroscopy has been comprehensively reviewed on two occasions in the recent past [65, 66]. The capacity of Mössbauer spectroscopy for in situ catalyst characterization is a very important feature of its application in this field where information pertaining to solid-state properties and catalytic performance are best recorded under conditions where the gaseous environment, temperature, and pressure can be carefully controlled. Several cells that enable us to record in situ Mössbauer spectra have been described, and it is relevant to note that we can record good quality data from a wide range of elements that are common components of catalytic solids, for example, iron, tin, antimony, tellurium, iridium, ruthenium, tungsten, gold, and europium.

Several metallic catalysts have been successfully studied in situ by Mössbauer spectroscopy. For example, metallic iron, which has been used for many years as a catalyst for the reaction between hydrogen and nitrogen to produce ammonia, has been the subject of several investigations. Iron catalysts have also been used for the hydrogenation of carbon monoxide and hydrogen, which can be derived from coal in a process known as *Fischer–Tropsch synthesis*. This process has been of much recent interest because of problems associated with the supply of petroleum and the need to obtain fuels and chemicals from other routes. Although iron is a very active Fischer–Tropsch catalyst it lacks selectivity; that is, it converts the carbon monoxide and hydrogen to a variety of hydrocarbon and oxygenated products. Hence, much work has been directed toward the development of bimetallic catalysts that are not only active, but are also selective to a restricted range of products. The value of Mössbauer spectroscopy in such investigations is illustrated by a study [67] of silica- and alumina-impregnated iron catalysts modified by the incorporation of ruthenium. The inclusion of ruthenium into the iron gave catalysts that were shown by Mössbauer spectroscopy to be more amenable to reduction by hydrogen at

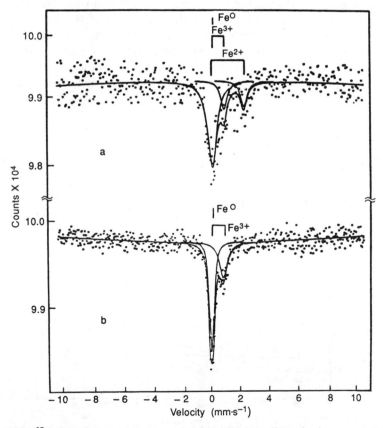

Figure 4.33 ^{57}Fe Mössbauer spectra recorded from 5% iron–5% ruthenium supported on (*a*) alumina and (*b*) silica following treatment at 452°C (6 h) in hydrogen. Reproduced, with permission, from F. J. Berry, L. Lin, C. Wang, R. Tang, S. Zhang, and D. Liang, *J. Chem. Soc. Faraday Trans. 1*, **81**, 2293 (1985).

lower temperature and, in several cases, treatment of the iron–ruthenium catalysts in a reducing atmosphere gave rise to the formation of an iron–ruthenium alloy. Furthermore, the Mössbauer spectra recorded from the silica-supported iron–ruthenium particules showed them (Figure 4.33b) to be more extensively reduced than their alumina-supported counterparts where the metal-support interaction is stronger (Figure 4.33a).

The influence of these solid-state properties on catalytic performance is illustrated in Figure 4.34, which shows the activities of a high-metal-loaded series (HML) of catalysts containing 5 wt% ruthenium with varied iron contents, and a low-metal-loaded series (LML) of catalysts, which contained 1 wt% ruthenium with similar iron/ruthenium ratios. The higher activities of the alumina-supported materials were associated with the Mössbauer results, which showed that the interaction of the metallic phase with alumina is stronger than that with silica and gives rise to the smaller particles in which the reduction of iron is inhibited. The decrease in catalytic activity that accompanies an increase in the iron content of the alumina-supported series was considered in conjunction with the effect of the iron on the selectivity of these catalysts. The hydrocarbon product distribution from LML-alumina-supported catalysts (Figure 4.35b) was, in contrast to the LML silica-supported counterparts (Figure 4.35a), particularly sensitive to the iron content of the catalysts with increases in the iron concentration producing higher yields of the shorter chain hydrocar-

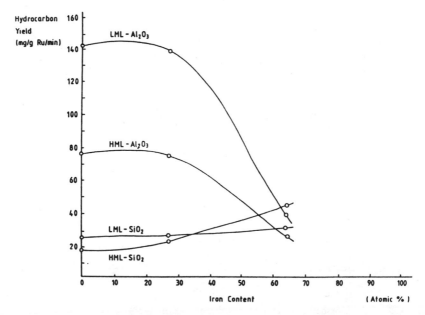

Figure 4.34 Catalytic activity of alumina- and silica-supported iron–ruthenium catalysts as a function of the iron content. Reproduced, with permission, from F. J. Berry, L. Lin, C. Wang, R. Iang, S. Zhang, and D. Liang, *J. Chem. Soc. Faraday Trans. 1*, **81**, 2293 (1985).

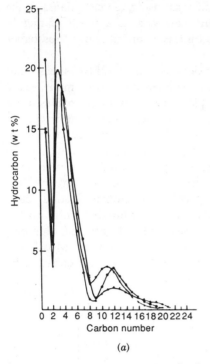
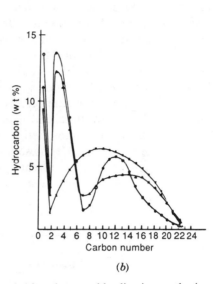

(a)   (b)

Figure 4.35 Hydrocarbon product distribution recorded from low metal loading iron–ruthenium catalysts supported on (a) silica and (b) alumina: Θ, 1% Ru–1% Fe; Δ, 1% Ru–0.2% Fe; × 1% Ru. Reproduced, with permission, from F. J. Berry, L. Lin, C. Wang, R. Iang, S. Zhang, and D. Liang, *J. Chem. Soc. Faraday Trans. 1*, **81**, 2293 (1985).

bons. For example, the alumina-supported 1% Ru–1% Fe catalyst had a selectivity to C_3 hydrocarbons, which approached that of pure ruthenium supported on silica. An indication of the role of iron(II) in the prereduced alumina-supported catalyst, which alters the activity and selectivity, was obtained from the iron(II) chemical isomer shift data, which decreased in magnitude as the ruthenium concentration increased, a trend that corresponds to an increase in the s-electron density at the iron nucleus as a result of a decrease in the d-electron population. Hence, the interaction between iron and ruthenium in the prereduced alumina-supported catalyst was associated with the transfer of d electrons from the iron(II) to the ruthenium. This was considered to inhibit the formation of electron-deficient ruthenium species created by electron donation from ruthenium to the acidic alumina support. The decrease in activity and shift in selectivity toward lower hydrocarbons was therefore associated with a weakening of the ruthenium–alumina interaction. Such a scheme was supported by the lower degree of electron transfer between ruthenium and less acidic silica, which results in the activity and selectivity being less susceptible to influence by the incorporation of iron. Hence, the catalytic performance of the silica-supported ruthenium was shown to be emulated by the alumina-supported iron-rich iron–ruthenium catalysts as a result of metal–metal and metal–support interactions monitored in situ by Mössbauer spectroscopy.

The catalytic properties of mixed metal oxides, consisting of two metals and oxygen, for the oxidation of hydrocarbons have been known for many years; and some recent Mössbauer investigations [68] of tin–antimony oxides are a suitable illustration of the value of the technique for the study of these types of materials. The power of Mössbauer emission spectroscopy has also been demonstrated during studies of cobalt molybdate hydrodesulfurization catalysts [69, 70]. Hydrodesulfurization involves the removal of sulfur from petroleum compounds and is important because it enables the removal of sulfur from fuel oils and feedstocks for the chemical industry. A failure to remove the sulfur leads to the production of atmospheric pollutants such as sulfur dioxide or to the formation of sulfides that poison the catalysts [70]. A sulfided alumina-supported cobalt molybdate catalyst was examined in situ by Mössbauer emission spectroscopy and the spectrum (Figure 4.36b) showed the cobalt to be

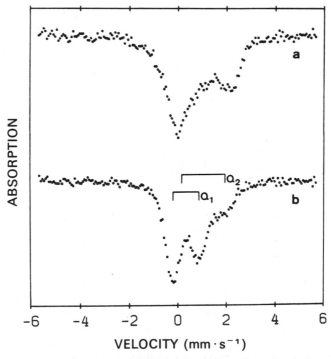

Figure 4.36 Mössbauer emission spectra recorded from an alumina-supported cobalt molybdate catalyst (a) after calcination and (b) after treatment in a gaseous mixture of H_2S/H_2 (2% H_2S). Reproduced, with permission, from H. Topsoe, B. S. Clausen, R. Candia, C. Wirel, and S. Morup, *J. Catal.*, **68**, 433 (1981).

present in two configurations, which gave rise to the two doublets designated as Q_1 and Q_2. The broad-line absorption Q_2, which represented about 30% of the cobalt, was assigned to cobalt located in sites of slightly differing symmetry inside the alumina support and showed that a significant amount of cobalt is unaffected by the sulfiding treatment. The other spectral component Q_1 corresponded to a model compound in which the cobalt had been intercalated in the molybdenum sulfide layers. The results therefore gave for the first time direct evidence for the presence of a cobalt–molybdenum–sulfur phase in real cobalt molybdate catalysts. Subsequent investigations using other techniques have sought to elucidate more fully the location of cobalt in this phase and the working of the cobalt molybdate catalysts.

5.4 Surfaces

A significant advance in the use of backscatter methods of Mössbauer spectroscopy for the investigation of solid surfaces has involved the development of conversion electron Mössbauer spectroscopy (CEMS) (Section 3.4.3) and depth-selective CEMS (DCEMS), which have been the subject of several reviews [26–28].

Many of the exploratory CEMS studies involved the identification of the oxide products that are formed on corroded iron plates and foils. For example, studies of the thermal oxidation of surface-enriched iron foils in dry atmospheres [71] showed that treatment at 450°C gave a nonstoichiometric magnetite surface species composed of either $Fe_{2.9}O_4$ or a mixture of $\alpha\text{-}Fe_2O_3$ and Fe_3O_4, whereas shorter period oxidation at 350°C produced a duplex film composed of Fe_3O_4 and $\alpha\text{-}Fe_2O_3$. Figure 4.37 depicts the variation of the CEM spectra with length of oxidation at 225°C and illustrates the sensitivity of the technique to the small changes in surface composition that can readily accompany minor variations in external conditions. The data were successfully used to estimate the thickness of the oxide film. More recent studies have utilized a combination of CEMS and other surface-sensitive techniques such as Auger electron spectroscopy to investigate more difficult problems. For example, the nature of the passive layer that is formed on steel in a phosphate buffer has been found [72] to consist of $\gamma\text{-}FeOOH$ and the thickness shown to vary from 0.5 to 3.5 nm. The passive layer was also shown to be covered by a 1–6-nm film of ferrophosphate.

Conversion electron Mössbauer spectroscopy has also found important applications in the study of the surface regions of other types of solids. This is illustrated by a recent study [73] of the surface and bulk properties of an extraterrestrial solid using both transmission and conversion electron Mössbauer spectroscopy. It is well known that many meteorites contain two phases: one called kamacite, which is an iron–nickel alloy, and the other taenite, which often occurs as lamellae of about 100 μm thickness lying between the thicker ($\sim$1-mm) plates of kamacite. The transmission mode Mössbauer spectrum recorded from a taenite lamella of 63-μm thickness from the iron meteorite Cape York is shown in Figure 4.38a and can be interpreted in terms of the

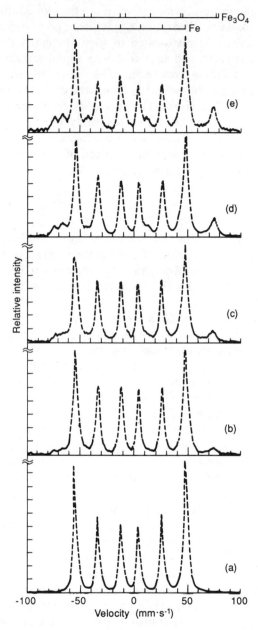

Figure 4.37 Conversion electron Mössbauer spectra recorded from iron oxidized at 225°C for (*a*) 0, (*b*) 5, (*c*) 15, (*d*) 120, and (*e*) 1000 min. Reproduced, with permission, from G. W. Simmons, E. Kellerman, and H. Leidheiser, *Corrosion*, **29**, 277 (1973).

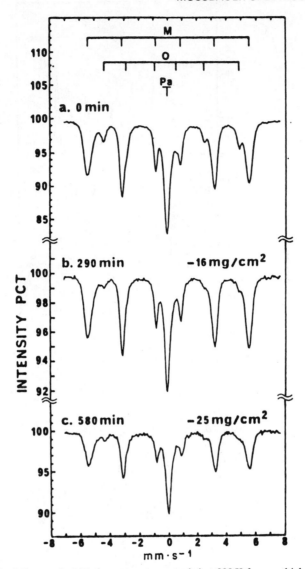

Figure 4.38 Transmission mode Mössbauer spectra recorded at 298 K from a thick lamella from the iron meteorite Cape York after different etching times: M, martensite; O, ordered phase; and Pa, paramagnetic phase. (*a*) Lamella as found, (*b*) after 290 min of etching, and (*c*) after 580 min of etching. Reproduced, with permission, from A. Christiansen, L. Larson, H. Roy-Poulson, N. O. Roy-Poulson, L. Vistisen, and J. M. Knudsen, *Phys. Scr.*, **29**, 94 (1984).

superposition of three spectra: (1) a central line from a paramagnetic atomically disordered iron–nickel alloy containing about 25% nickel (Pa); (2) an asymmetric six-line spectrum showing a magnetic field of 342 kG and characteristic of a ferromagnetic atomically ordered iron–nickel alloy containing approximately 50% nickel (O); and (3) a symmetric six-line spectrum showing a magnetic hyperfine field of 342 kG corresponding to martensite (M).

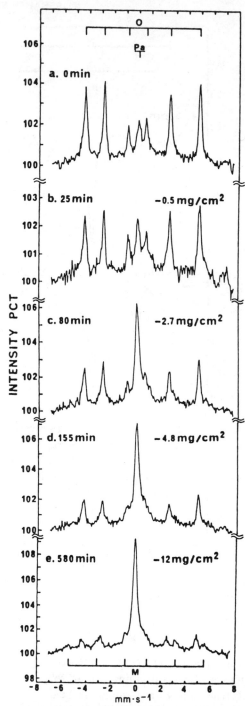

Figure 4.39 Conversion electron Mössbauer spectra recorded at 298 K from a thick taenite lamella from the iron meteorite Cape York after different etching times: O, ordered phase; Pa, paramagnetic phase; M, martensite. (*a*) Lamella as found, (*b*) after 25 min of etching, (*c*) after 80 min of etching, (*d*) after 155 min of etching, and (*e*) after 580 min of etching. Reproduced, with permission, from A. Christiansen, L. Larson, H. Roy-Poulson, N. O. Roy-Poulson, L. Vistisen, and J. M. Knudsen, *Phys. Scr.*, **29**, 94 (1984).

The conversion electron Mössbauer spectrum recorded from the lamella is shown in Figure 4.39a and, by direct comparison with the transmission mode Mössbauer spectrum (Figure 4.38a), demonstrates that a clear difference exists between the surface layer of the lamella and its bulk composition. The transmission mode Mössbauer spectrum is dominated by the martensite spectrum and the central peak from the paramagnetic iron–nickel alloy phase, while the CEM spectrum is dominated by the spectrum from the ordered iron–nickel alloy. The transmission- and conversion electron-Mössbauer spectra recorded after etching the lamella in a solution of 5% hydrochloric acid and 5% nitric acid in water are shown in Figure 4.38b and c and Figure 4.39b–e, respectively. The transmission mode Mössbauer spectra showed that the martensite is mainly present in the inner parts of the lamella, indeed even after 580 min of etching, which corresponds to the removal of 25 mg·cm^{-2} lamella in total, the martensite dominated the spectrum (Figure 4.38c). However, the conversion electron Mössbauer spectra showed that the area of the central peak representing the paramagnetic atomically disordered iron–nickel alloy increased relative to the area of the spectrum corresponding to the ferromagnetic atomically ordered iron–nickel alloy as more and more of the lamella was etched away. The results are consistent with the suggested structure of taenite lamella as illustrated in Figure 4.40. The results show that the combined use of transmission mode Mössbauer spectroscopy, which samples the bulk, and of surface-sensitive conversion electron Mössbauer spectroscopy can give essential information on the distribution of the different phases in a solid as a function of distance from the surface.

5.5   Minerals

The application of ^{57}Fe Mössbauer spectroscopy to the study of mineral materials has been described in detail in a book [5] and in a very recent review [74]. Various applications have been developed in this field; for example, the line intensities can be related to the number of atoms per unit area of the mineral and used to determine Fe(II)/Fe(III) ratios. The technique has also been used for fingerprinting and in petrological analysis; this has been very effectively applied to studies of many materials, including lunar rocks, meteoritic materials, glassy materials, clays and micas, as well as sediments and nodules. Mössbauer

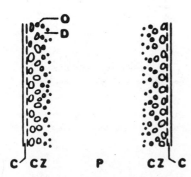

Figure 4.40 Representation of the composition of a thick taenite lamella (thickness ca. 70 μm): C, ordered FeNi alloy; CZ, mixture of ordered FeNi alloy and disordered paramagnetic FeNi alloy; O, domains of ordered FeNi alloy; D, disordered FeNi alloy filling space between the ordered domains; P, martensite. Reproduced, with permission, from A. Christiansen, L. Larson, H. Roy-Poulson, N. O. Roy-Poulson, L. Vistisen, and J. M. Knudsen, *Phys. Scr.*, **29**, 94 (1984).

spectroscopy has also been used to study many fundamental properties of minerals that are of importance in solid-state chemistry in general, such as cationic ordering in silicates, the kinetics and mechanism of mineral solid-state reactions, electron delocalization, and mixed-valence materials. Attempts have also been made to use the technique for dating and for the elucidation of mechanisms in mining and extraction processes.

It is also relevant to note a very recent Mössbauer spectroscopic study [75] of exchangeable iron(II) and iron(III) ions on the clay mineral saponite, which, unlike the smectite hectorite in which substitution occurs in octahedral layers of the aluminosilicate sheet, has a layer charge derived primarily from tetrahedral substitutions. The ^{57}Fe Mössbauer spectrum recorded from iron(II)-exchanged saponite at 77 K (Figure 4.41) showed two components one corresponding to iron(II) and the other to iron(III), which is formed by partial oxidation of iron(II) during the adsorption process. The iron(II) Mössbauer parameters were associated with the presence of $Fe(H_2O)_6^{2+}$. Support for this assignment was provided by the spectrum recorded at 298 K in which the absorption from the solvated species disappeared as its recoil-free fraction decreased. The iron(II) component in the spectrum at 298 K with different Mössbauer parameters to those recorded from the solvated species was associated with the presence of iron(II) in structural sites with octahedral coordination and to arise from either

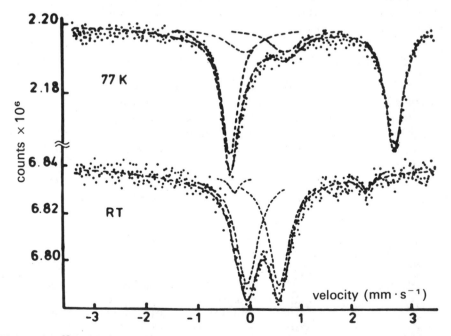

Figure 4.41 ^{57}Fe Mössbauer spectra recorded from iron(II)-exchanged saponite. Reproduced, with permission, from B. A. Goodman, J. A. Helson, and G. Langouche, *Hyperfine Interact.*, **41**, 799 (1988).

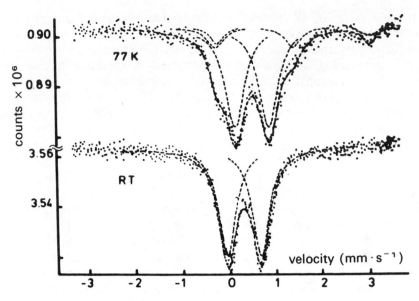

Figure 4.42 ^{57}Fe Mössbauer spectra recorded from iron(III)-exchanged saponite. Reproduced, with permission, from B. A. Goodman, J. A. Helson, and G. Langouche, *Hyperfine Interact.*, **41**, 799 (1988).

the diffusion of a small amount of iron(II) into the lattice or the reduction of some structural iron(III) as a result of the exchange process.

The ^{57}Fe Mössbauer spectrum recorded from the iron(III)-exchanged saponite at 77 K (Figure 4.42) showed the presence of two different iron(III) species and the $Fe(H_2O)_6^{2+}$ ion. The results thereby demonstrated the occurrence of unexpected reduction during the adsorption process that was associated with the presence of trapped free radicals and Mn^{2+} ions and that was contrasted with the behavior previously observed in montmorillonite. The disappearance of one of the iron(III) absorptions when the spectrum was recorded at 298 K was attributed to a solvated dimer of composition $[Fe(OH)]_2^{4+}$. The occurrence of this dimer was associated with adsorption onto clays such as saponite and montmorillonite where there is a contribution from tetrahedral substitution to surface charge. The study is a good illustration of how Mössbauer spectra recorded at different temperatures can be used to identify the nature of more than one iron-containing species present in a solid.

5.6   Archaeology

The nondestructive nature of Mössbauer spectroscopy has resulted in its development as an important technique for the study of archaeological and precious ceramic materials [76]. This is well illustrated by the use of the technique in the study of pottery, which is a major source of information in archaeology. Such studies have provided useful information on matters such as the transformation of clay during firing, the characterization of ancient pottery samples in attempts to identify their origins, studies of surface glazes, and the dating of pottery artifacts. The technique has also been used to study iron oxides in soils at archaeological sites and to interpret the data in terms of the residues of previous human occupation such as buried pits, ditches, or other relics.

Mössbauer spectroscopy is at its most powerful for the examination of archaeological artifacts when used in conjunction with other complementary techniques. This is well illustrated by a study [77] of some Spanish Campanian and Terra Sigillata pottery, which are high-quality coated ceramics produced during the period from about 400 BC to about AD 200. The origin of the distinctive colors of the coatings and the method of fabrication of the ceramics has been the subject of uncertainty for some time. Scanning electron microscopy showed the coatings to be homogeneous and finely textured, whereas the body of the ceramic was composed of an open porous structure with larger particle size. The Terra Sigillata coating of 30–40 μm was thicker than the Campanian coating of 9–10 μm. Energy-dispersive X ray analysis showed the calcium concentration to be significantly higher in the body than in the coat, which was rich in iron and potassium. The results were interpreted in terms of the coatings being prepared from noncalcereus clays such as an illitic clay. The examination by Mössbauer spectroscopy involved the use of the technique in the transmission mode to examine the bulk and in the conversion electron mode to

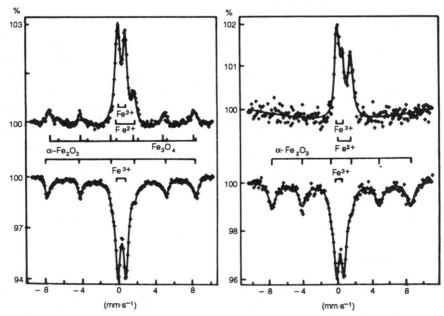

Figure 4.43 ^{57}Fe Mössbauer spectra recorded from coating (upper) and body (lower) of the Campanian ceramic. Left, type A; right, type B. Reproduced, with permission, from J. R. Gancedo, M. Gracia, J. F. Marco, and J. Palacios, *Hyperfine Interact.*, **41**, 791 (1988).

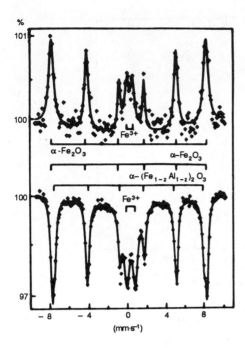

Figure 4.44 Conversion electron Mössbauer spectrum (coating) and transmission Mössbauer spectrum (body) recorded from a Terra Sigillata ceramic from Merida. Reproduced, with permission, from J. R. Gancedo, M. Gracia, J. F. Marco, and J. Palacios, *Hyperfine Interact.*, **41**, 791 (1988).

examine the surface. The results from Campanian pottery originating from two different excavation sites and described as type A and type B are shown in Figure 4.43. The conversion electron Mössbauer spectra demonstrated the presence of Fe^{2+} and some magnetite in the coatings, whereas patterns characteristic of Fe^{3+} in structural clay sites and hematite were observed in the transmission mode spectra recorded from the body of the pottery. The black color of the coatings was associated with the Fe^{2+} and magnetite phases with the Fe^{2+} arising from firing in a reducing atmosphere. The spectra recorded from the Terra Sigillata pottery (Figure 4.44) showed that the coatings and the body of the ceramic contained iron in similar phases, that is, hematite (or aluminium-substituted hematite) and Fe^{3+} in substitutional sites in the clay structure. The absence of Fe^{2+} in the coatings and the quadrupole-splitting data for the Fe^{3+} doublets were associated with the firing of the pottery at a temperature of 900–1000°C in an oxidizing atmosphere. The intense red color of the coating was associated with the presence of hematite in a low calcium-containing surface region.

5.7 Biological Systems

The power of Mössbauer spectroscopy for the study of biological systems derives from the sensitivity of iron to Mössbauer investigation and the fundamental role of this element in many biomolecules. The subject has been reviewed in the recent past [78] in an article that shows how Mössbauer

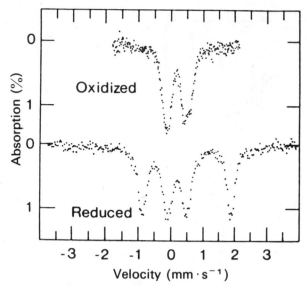

Figure 4.45 Mössbauer spectra recorded from *Scenedesmus ferredoxin* at 195 K. Reproduced, with permission, from C. E. Johnson, *J. Appl. Phys.*, **42**, 1325 (1971).

spectroscopy enables a detailed examination of the local environment of iron in large and complex biological materials, such as proteins and enzymes, and assumes enhanced importance if the iron atoms are involved in active centers at which biologically significant changes occur.

The application is illustrated by a comparison of the spectra of oxidized and reduced plant-type ferredoxins at relatively high temperatures [79]. In the oxidized proteins the spectrum consisted (Figure 4.45) of a quadrupole-split absorption with a small chemical isomer shift and quadrupole splitting similar to that of Fe(III) in rubredoxin. Reduction of the proteins gave spectra showing two doublets, one of which was similar to that in the oxidized proteins and the other, with larger quadrupole splitting, being characteristic of iron(II). Hence the Mössbauer spectra gave clear evidence that the oxidized protein contains two types of iron(III) species, one of which is amenable to reduction to iron(II), and showed that during reduction of the protein the extra reducing electron can be localized on one of the iron atoms thereby leaving the other substantially unchanged.

A recent study [80] of hemosiderins, which are mammalian iron-storage materials related to ferritin, is also worthy of note since it shows how different forms of hemosiderin can be distinguished by Mössbauer spectroscopy. The similar ^{57}Fe Mössbauer spectra recorded at 77 K from hemosiderins in horse spleen, horse liver, reindeer spleen, and reindeer liver (Figure 4.46) were found to be characteristic of other hemosiderins and also ferritin. The spectra recorded from horse spleen hemosiderin between 10 and 70 K (Figure 4.47) showed the occurrence of superparamagnetic behavior with the doublet growing in intensity

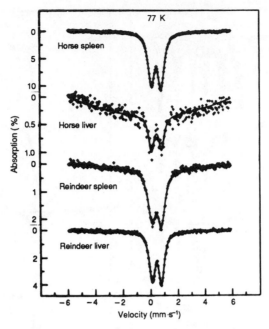

Figure 4.46 [57]Fe Mössbauer spectra recorded from hemosiderins at 77 K. Reproduced, with permission, from D. P. E. Dickson, R. K. Pollard, B. Borch-Johnson, R. J. Ward, and T. J. Peters, *Hyperfine Interact.*, **42**, 889 (1988).

at the expense of the sextet as the temperature increased. The mean superparamagnetic transition temperature was found to be 28 K. The [57]Fe Mössbauer spectra recorded from the other hemosiderins at 20 K are collected in Figure 4.48 and show that differences exist between the hemosiderins originating from horse or reindeer and between different organs. The reindeer spleen hemosiderin was shown to have a transition temperature of 30 K, somewhat higher than the horse spleen hemosiderin, while the horse liver and reindeer liver hemosiderins had rather lower temperatures between 20 and 25 K. All four samples had much lower transition temperatures than the value of 70 K that characterizes hemosiderin extracted from iron-overloaded human spleen. The results thereby demonstrated that different forms of hemosiderin with different iron cores are distinguishable by Mössbauer spectroscopy and are related to the way in which the hemosiderins are produced.

Although many of the Mössbauer studies of biological systems have involved the simpler types of isolated biomolecules, the results have often been used to rationalize the behavior of similar molecules in more complex systems. Furthermore, Mössbauer studies of model compounds resembling centers in biologically significant materials have been found to shed light on the structural properties of active centers in the real systems. Studies such as these, together with the steady accumulation of data from the simpler biomolecules, have led to

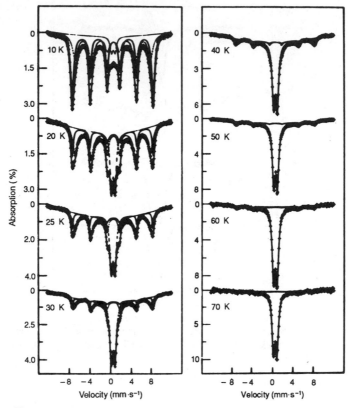

Figure 4.47 ^{57}Fe Mössbauer spectra recorded from horse spleen hemosiderin between 10 and 70 K. Reproduced, with permission, from D. P. E. Dickson, R. K. Pollard, B. Borch-Johnson, R. J. Ward, and T. J. Peters, *Hyperfine Interact.*, **42**, 889 (1988).

an enhanced understanding of the specific catalytic functions of enzymes and other complex molecules in biological systems.

Medicinal applications of Mössbauer spectroscopy have involved studies of tissue samples and, as in some recent investigations of blood disease, have led to the determination of important pathological factors. It is also pertinent to cite the feasibility of in vivo studies, which have, for example, enabled experiments to monitor the uptake and use of a Mössbauer element. Such investigations are illustrative of a further development in the scope and application of Mössbauer spectroscopy.

6   CONCLUSIONS

During the past 25 years Mössbauer spectroscopy has matured from an elegant experiment in nuclear physics to a technique that is being vigorously applied to problems in many areas of science and technology. Although ^{57}Fe and ^{119}Sn

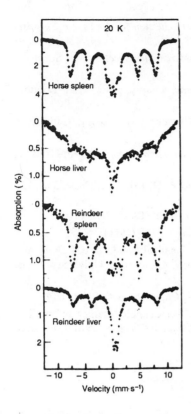

Figure 4.48 ^{57}Fe Mössbauer spectra recorded from hemosiderins at 20 K. Reproduced, with permission, from D. P. E. Dickson, R. K. Pollard, B. Borch-Johnson, R. J. Ward, and T. J. Peters, *Hyperfine Interact.*, **42**, 889 (1988).

offer the most favorable isotopes for Mössbauer spectroscopy, it is quite clear that many other elements are amenable to examination by the technique, and it is apparent that Mössbauer spectroscopy is fulfilling its potential as a way we can examine both fundamental and technologically relevant properties of the solid state. The interdisciplinary nature of the technique and its many applications, only a few of which have been outlined in this chapter, are a reflection of recent developments in scientific concepts and approaches to technological problems, and it is quite reasonable to expect that the future of the technique will be inextricably connected with the development of our science and changes in our technology.

References

1. R. L. Mössbauer, *Z. Phys.*, **151**, 124 (1958).
2. G. K. Wertheim, *Mössbauer Effect Principles and Applications*, Academic, New York, 1964.
3. N. N. Greenwood and T. C. Gibb, *Mössbauer Spectroscopy*, Chapman and Hall, London, 1971.
4. L. May, Ed., *An Introduction to Mössbauer Spectroscopy*, Plenum, New York, 1971.

5. G. M. Bancroft, *Mössbauer Spectroscopy, An Introduction for Inorganic Chemists and Geochemists*, McGraw-Hill, New York, 1973.

6. U. Gonser, Ed., *Topics in Applied Physics, Vol. V, Mössbauer Spectroscopy*, Springer-Verlag, Berlin, 1975.

7. T. C. Gibb, *Principles of Mössbauer Spectroscopy*, Chapman and Hall, London, 1976.

8. F. J. Berry and D. P. E. Dickson, Eds., *Mössbauer Spectroscopy*, Cambridge University Press, Cambridge, 1986.

9. G. K. Shenoy and F. E. Wagner, Eds., *Mössbauer Isomer Shifts*, North-Holland, Amsterdam, 1978.

10. G. M. Bancroft, A. G. Maddock, W. K. Ong, and R. H. Prince, *J. Chem. Soc.*, 723 (1966).

11. T. C. Gibb and N. N. Greenwood, *J. Chem. Soc.*, 6989 (1966).

12. G. Lang and W. Marshall, *Proc. Phys. Soc. London*, **87**, 3 (1966).

13. R. V. Parish, "Mössbauer Spectroscopy and the Chemical Bond," in F. J. Berry and D. P. E. Dickson, Eds., *Mössbauer Spectroscopy*, Cambridge University Press, Cambridge, 1986, p. 17.

14. G. M. Bancroft, M. J. Mays, and B. E. Prater, *J. Chem. Soc. (A)*, 956 (1970).

15. H. Spiering, "The Electric Field Gradient and the Quadrupole Interaction," in G. J. Long, Ed., *Mössbauer Spectroscopy Applied to Inorganic Chemistry*, Vol. 1, Plenum, New York, 1984, p. 77.

16. R. V. Parish, *Prog. Inorg. Chem.*, **15**, 101 (1972).

17. G. M. Bancroft and R. H. Platt, *Adv. Inorg. Chem. Radiochem.*, **15**, 59 (1972).

18. G. M. Bancroft, *Coord. Chem. Rev.*, **11**, 247 (1973).

19. M. G. Clark, A. G. Maddock, and R. H. Platt, *J. Chem. Soc. Dalton Trans.*, 261 (1972).

20. B. P. Dailey and C. H. Townes, *J. Chem. Phys.*, **23**, 118 (1955).

21. M. F. Thomas and C. E. Johnson, "Mössbauer Spectroscopy of Magnetic Solids," in F. J. Berry and D. P. E. Dickson, Eds., *Mössbauer Spectroscopy*, Cambridge University Press, Cambridge, 1986, p. 143.

22. P. G. L. Williams and G. M. Bancroft, in I. J. Gruverman, Ed., *Mössbauer Effect Methodology*, Vol. 7, Plenum, New York, 1971, p. 39.

23. T. E. Cranshaw, "Mössbauer Spectrometers and Calibration," in G. J. Long, Ed., *Mössbauer Spectroscopy Applied to Inorganic Chemistry*, Vol. 1, Plenum, New York, 1984, p. 27.

24. G. Longworth, "Spectral Data Reduction and Refinement," in G. J. Long, Ed., *Mössbauer Spectroscopy Applied to Inorganic Chemistry*, Vol. 1, Plenum, New York, 1984, p. 43.

25. J. G. Stevens and V. E. Stevens, Eds., *Mössbauer Effect Data Index Covering the 1976 Literature*, Plenum, New York, 1978.

26. M. J. Tricker, "Iron-57 Conversion Electron Mössbauer Spectroscopy," in M. W. Roberts and J. M. Thomas, Eds., *Surface and Defect Properties of Solids*, Specialist Periodical Reports, Vol. 6, The Chemical Society, London, 1977, p. 106.

27. F. J. Berry, *Transition Met. Chem.*, **4**, 209 (1979).

28. M. J. Tricker, "Conversion Electron Mössbauer Spectroscopy and Its Recent Development," in J. G. Stevens and G. K. Shenoy, Eds., *Mössbauer Spectroscopy and its Chemical Applications*, Advances in Chemistry Series, Vol. 194, American Chemical Society, Washington, DC, 1981, p. 63.

29. R. Atkinson and T. E. Cranshaw, *Nucl. Instrum. Methods*, **204**, 577 (1983).

30. G. J. Long, Ed., *Mössbauer Spectroscopy Applied to Inorganic Chemistry*, Vols. 1 and 2, Plenum, New York, 1984 and 1987.

31. G. K. Shenoy, "Mössbauer Effect Isomer Shifts," in G. J. Long, Ed., *Mössbauer Spectroscopy Applied to Inorganic Chemistry*, Vol. 1, Plenum, New York, 1984, p. 57.

32. W. H. Flygare and D. W. Hafermeister, *J. Chem. Phys.*, **43**, 789 (1965).

33. H. D. Bartunik and G. Kaindl, "Mössbauer Isomer Shifts in Chemical Systems of Gold," in G. K. Shenoy and F. E. Wagner, Eds., *Mössbauer Isomer Shifts*, North-Holland, Amsterdam, 1978, p. 515.

34. G. M. Bancroft, R. E. B. Garrod, and A. G. Maddock, *J. Chem. Soc. A*, 3165 (1971).

35. G. M. Bancroft and E. T. Libbey, *J. Chem. Soc. Dalton Trans.*, 2103 (1973).

36. G. J. Perlow, in V. I. Goldanskii and R. H. Herber, Eds., *Chemical Applications of Mössbauer Spectroscopy*, Vol. 1, Academic, New York, 1968, p. 378.

37. G. J. Long, "Mössbauer Spectroscopy as a Structural Probe," in F. J. Berry and D. P. E. Dickson, Eds., *Mössbauer Spectroscopy*, Cambridge University Press, Cambridge, 1986, p. 70.

38. R. Desiderato and G. R. Dobson, *J. Chem. Educ.*, **59**, 752 (1982).

39. R. H. Herber, W. R. Kingston, and G. K. Wertheim, *Inorg. Chem.*, **2**, 153 (1963).

40. N. E. Erickson and A. W. Fairhall, *Inorg. Chem.*, **4**, 1320 (1965).

41. C. H. Wei and L. H. Dahl, *J. Am. Chem. Soc.*, **88**, 1821 (1966).

42. S. Bukshpan, J. Soriano, and J. Shamir, *Chem. Phys. Lett.*, **4**, 241 (1969).

43. R. R. Berrett and B. W. Fitzsimmons, *J. Chem. Soc. Chem. Commun.*, 91 (1966).

44. R. R. Berrett and B. W. Fitzsimmons, *J. Chem. Soc. A*, 525 (1967).

45. G. M. Bancroft, R. E. B. Garrod, A. G. Maddock, M. J. Mays, and B. E. Prater, *J. Chem. Soc. Chem. Commun.*, 200 (1970).

46. T. C. Gibb, *J. Phys. C*, **9**, 2627 (1976).

47. N. Haga and Y. Takenchi, *Z. Kristallogr.*, **144**, 161 (1976).

48. T. Yamanaka and Y. Takenchi, *Phys. Chem. Miner.*, **4**, 149 (1979).

49. I. V. Heilmann, N. B. Olsen, and J. B. Olsen, *Phys. Scr.*, **15**, 285 (1977).

50. A. Gerard and F. Grandjean, *Solid State Commun.*, **9**, 1845 (1971).

51. F. Grandjean and A. Gerard, *Solid State Commun.*, **16**, 553 (1975).

52. M. F. Thomas and C. E. Johnson, "Mössbauer Spectroscopy of Magnetic Solids," in F. J. Berry and D. P. E. Dickson, Eds., *Mössbauer Spectroscopy*, Cambridge University Press, Cambridge, 1986, p. 143.

53. T. E. Cranshaw and G. Longworth, "Mössbauer Spectroscopy of Magnetic Systems," in G. J. Long, Ed., *Mössbauer Spectroscopy Applied to Inorganic Chemistry*, Vol. 1, Plenum, New York, 1984, p. 171.

54. G. J. Long, G. Longworth, P. Battle, A. K. Cheetham, R. V. Thundathil, and D. Beveridge, *Inorg. Chem.*, **18**, 624 (1979).

55. C. E. Johnson, "One-Dimensional Magnetism," in G. J. Long, Ed., *Mössbauer Spectroscopy Applied to Inorganic Chemistry*, Vol. 1, Plenum, New York, 1984, p. 619.

56. R. L. Cohen, Ed., *Applications of Mössbauer Spectroscopy*, Vols. I and II, Academic, New York, 1976 and 1980.

57. J. G. Stevens and G. K. Shenoy, Eds., *Mössbauer Spectroscopy and Its Chemical Applications*, Advances in Chemistry Series, Vol. 194, American Chemical Society, Washington, DC, 1981.

58. G. J. Long and J. G. Stevens, Eds., *Industrial Applications of the Mössbauer Effect*, Plenum, New York, 1986.

59. T. E. Cranshaw, "Metal Research of Industrial Significance by Mössbauer Spectroscopy," in G. J. Long and J. G. Stevens, Eds., *Industrial Applications of the Mössbauer Effect*, Plenum, New York, 1986, p. 7.

60. G. K. Shenoy, B. D. Dunlap, P. J. Viccaro, and D. Niarchos, "Hydrogen Storage Materials," in J. G. Stevens and G. K. Shenoy, Eds., *Mössbauer Spectroscopy and Its Chemical Applications*, Advances in Chemistry Series, Vol. 194, American Chemical Society, Washington, DC, 1981, p. 501.

61. R. L. Cohen, K. W. West, and J. H. Wernick, *J. Less Common Metals*, **73**, 273 (1980).

62. M. Eibschutz, "Mössbauer Studies of Battery Materials," in J. G. Stevens and G. K. Shenoy, Eds., *Mössbauer Spectroscopy and Its Chemical Applications*, Advances in Chemistry Series, Vol. 194, American Chemical Society, Washington, DC, 1981, p. 523.

63. P. A. Montano, "Application of Mössbauer Spectroscopy to Coal Characterization and Utilization," in J. G. Stevens and G. K. Shenoy, Eds., *Mössbauer Spectroscopy and Its Chemical Applications*, Advances in Chemistry Series, Vol. 194, American Chemical Society, Washington, DC, 1981, p. 135.

64. R. E. Karl and J. J. Zuckerman, "Qualitative and Quantitative Analysis of Iron-Bearing Minerals in Fossil Fuels and Petroleum Source Rock by Iron-57 Mössbauer Spectroscopy," in J. G. Stevens and G. K. Shenoy, Eds., *Mössbauer Spectroscopy and Its Chemical Applications*, Advances in Chemistry Series, Vol. 194, American Chemical Society, Washington, DC, 1981, p. 221.

65. J. A. Dumesic and H. Topsoe, "Mössbauer Spectroscopy Application to Heterogeneous Catalysis," in D. D. Eley, H. Pines, and P. B. Weisz, Eds., *Advances in Catalysis*, Vol. 26, Academic, New York, 1976, p. 121.

66. F. J. Berry, "Mössbauer Spectroscopy in Heterogeneous Catalysis," in G. J. Long, Ed., *Mössbauer Spectroscopy Applied to Inorganic Chemistry*, Vol. 1, Plenum, New York, 1984, p. 391.

67. F. J. Berry, L. Lin, C. Wang, R. Tang, S. Zhang, and D. Liang, *J. Chem. Soc. Faraday Trans. 1*, **81**, 2293 (1985).

68. F. J. Berry, *J. Catal.*, **73**, 349 (1982).

69. H. Topsoe, B. S. Clausen, R. Candia, C. Wirel, and S. Morup, *Bull. Soc. Chim. Belg.*, **90**, 1189 (1981).

70. H. Topsoe, B. S. Clausen, R. Candia, C. Wirel, and S. Morup, *J. Catal.*, **68**, 433 (1981).

71. G. W. Simmons, E. Kellerman, and H. Leidheiser, *Corrosion*, **29**, 227 (1973).

72. W. Meisel, U. Stumm, C. Thilmann, R. Gancedo, and P. Gutlick, *Hyperfine Interact.*, **41**, 669 (1986).

73. A. Christiansen, L. Larson, H. Roy-Poulson, N. O. Roy-Poulson, L. Vistisen, and J. M. Knudsen, *Phys. Scr.*, **29**, 94 (1984).

74. A. G. Maddock, "Mössbauer Spectroscopy in Mineral Chemistry," in F. J. Berry and D. J. Vaughan, Eds., *Chemical Bonding and Spectroscopy in Mineral Chemistry*, Chapman and Hall, London, 1985, p. 141.

75. B. A. Goodman, J. A. Helson, and G. Langouche, *Hyperfine Interact.*, **41**, 799 (1988).

76. G. Longworth, "Studies of Ceramics and Archaeological Materials," in G. J. Long, Ed., *Mössbauer Spectroscopy Applied to Inorganic Chemistry*, Vol. 1, Plenum, New York, 1984, p. 511.

77. J. R. Gancedo, M. Gracia, J. F. Marco, and J. Palacios, *Hyperfine Interact.*, **41**, 791 (1988).

78. D. P. E. Dickson, "Applications to Biological Systems," in G. J. Long, Ed., *Mössbauer Spectroscopy Applied to Inorganic Chemistry*, Vol. 1, Plenum, New York, 1984, p. 339.

79. C. E. Johnson, *J. Appl. Phys.*, **42**, 1325 (1971).

80. D. P. E. Dickson, R. K. Pollard, B. Borch-Johnson, R. J. Ward, and T. J. Peters, *Hyperfine Interact.*, **42**, 889 (1988).

Chapter **5**

SOLID-STATE NUCLEAR MAGNETIC RESONANCE

P. Mark Henrichs and James M. Hewitt

1   INTRODUCTION

Until the late 1970s solids were studied with nuclear magnetic resonance (NMR) spectroscopy by only a relatively small group of practitioners of *wide-line* NMR. *High-resolution NMR spectroscopy of solids* is now a general purpose tool for all chemists.

What distinguishes a high-resolution spectrum? In most solids too many physical interactions influence the NMR spectrum for the effect of only one interaction to be resolved. A high-resolution spectrum is one that is dominated by only a few interactions, such that uninteresting or undesirable physical interactions do not obscure the useful information. Such a spectrum results naturally from a liquid because molecular tumbling suppresses the effect of many internuclear interactions. To get such a spectrum from solids, one generally uses special techniques to cancel the effects of uninteresting physical interactions. In special cases (e.g., in deuterium spectroscopy) one of the interactions completely dominates all others; hence, one need not suppress other interactions artificially.

This chapter provides a general description of the methods used to generate the high-resolution NMR spectra of solids. The chapter is devoted primarily, however, to a description of those chemical problems involving solids for which NMR provides answers or is likely to provide answers in the future.

Several other reviews on the principles and application of high-resolution NMR of solids are available [1–6]. There have been two comprehensive books on the physical principles of the methods [7, 8] and several general textbooks directed to the working chemist [9–12].

2   PHYSICAL PRINCIPLES

2.1   Hamiltonian

The rules of quantum mechanics require that the behavior of a nuclear spin system be described by a Hamiltonian. For solids the NMR Hamiltonian is

$$\mathcal{H} = \mathcal{H}_Z + \mathcal{H}_{CS} + \mathcal{H}_D + \mathcal{H}_S + \mathcal{H}_Q + \mathcal{H}_{rf} \tag{1}$$

where $\mathcal{H}_Z$, $\mathcal{H}_{CS}$, $\mathcal{H}_D$, $\mathcal{H}_S$, $\mathcal{H}_Q$, and $\mathcal{H}_{rf}$ represent the Zeeman interaction with the magnetic field, the chemical shift, the dipolar coupling between spins, the scalar coupling between spins, the quadrupole coupling (for nuclei with spin greater than $\frac{1}{2}$), and the interaction with external radio frequency fields, respectively.

For a Cartesian coordinate system the Hamiltonian can be expressed as a function of the angular-momentum operators I_z, I_x, and I_y. The Hamiltonian in terms of spherical tensor operators is very convenient for detailed analysis of NMR experiments [7, 8, 13], but it is unfamiliar to many chemists. We defer its discussion to Section 9.

2.2   Zeeman Interaction With the Static Magnetic Field

The Zeeman effect results from the interaction of the nuclear magnetic dipoles with the external magnetic field. This interaction generally dominates all others; it, of course, makes NMR possible. The Zeeman Hamiltonian is

$$\mathscr{H}_Z = -\sum_i \gamma_i B_z I_{zi} = -\sum_i \omega_{0i} I_{zi} \tag{2}$$

where I_{zi} is the angular-momentum operator of the ith nucleus; γ_i is the gyromagnetic ratio of the ith nucleus; B_z is the magnitude of the static applied field, assumed to lie in the z direction; and ω_{0i} is the Larmor frequency of the ith nucleus. Planck's constant is absent from the equation because we express the energy here and throughout in terms of frequency.

2.3   Chemical Shielding

The absorption frequency of a nucleus contained in a chemical compound is shifted from that of a bare nucleus because the magnetic fields created at the nucleus by the molecular electrons augment or diminish the field of the external magnet. In solids the chemical shielding of the nucleus is usually anisotropic; that is, it depends on the orientation of the molecule in the applied magnetic field.

The Hamiltonian for the chemical-shift interaction is

$$\mathscr{H}_{cs} = \sum_i \gamma_i \mathbf{I} \cdot \sigma_i \cdot \mathbf{B} \tag{3}$$

where $\mathbf{I}_i$ is the vector angular-momentum operator for the ith nucleus and has the components $I_x, I_y,$ and I_z; σ_i is a second-rank tensor describing the chemical-shift interaction for the ith nucleus; and $\mathbf{B}$ is the vector magnetic field.

The tensor σ_i is a convenient mathematical device for expression of the orientation dependence, or *anisotropy*, of the chemical shift [7, 8]. In terms of Cartesian coordinates, it is a 3×3 matrix. The elements of the matrix have values that depend on the orientation of the molecule with respect to a *reference axis system*. Rotation of the reference system while the molecule remains fixed changes the elements of the chemical-shift tensor according to a *unitary transformation*:

$$\sigma_i' = \mathbf{R} \cdot \sigma_i \cdot \mathbf{R}^{-1} \tag{4}$$

In (4) σ_i' is the shielding tensor of the ith nucleus in the rotated coordinate system, σ_i is the shielding tensor in the original system, and $\mathbf{R}$ is the matrix of direction cosines that connects the original and the rotated reference systems.

Proper choice of the rotation matrix in (4) leads to a chemical-shift tensor in which all off-diagonal elements of the chemical-shift tensor are zero. The corresponding reference axis system is the *principal axis system* (PAS). The diagonal elements of the chemical-shift tensor in the PAS are the *principal elements* of the tensor. When the environment of the nucleus is axially symmetric, two of these elements are equal.

The chemical shift of a nucleus is determined to first order by the element σ_{zz} of the chemical-shift tensor in a reference axis system that coincides with the *laboratory frame*. In the laboratory frame the magnetic field is aligned with the z axis. The σ_{zz} element can be determined from the chemical-shift tensor in the principal axis system by the equation:

$$\sigma'_{zz} = \sigma'_{33} = \sum_k \sigma_{kk} R_{3k}^2 = \sigma_{11} R_{31}^2 + \sigma_{22} R_{32}^2 + \sigma_{33} R_{33}^2 \tag{5}$$

where σ'_{zz} is the element of the shielding tensor along the z axis in the laboratory frame, σ_{kk} are elements of the shielding tensor in the principal axis system, and R_{3k}^2 are elements of the matrix of direction cosines that describe the orientation of the PAS with respect to the laboratory frame.

The spectrum of a powder is the superposition of the signals for all possible orientations of the molecule in the laboratory frame [i.e., for all possible values of the direction cosines in (5)]. Powder spectra have the characteristic shape shown in Figure 5.1. The form of the spectrum results because nuclei in a variety of orientations absorb at the same frequency [7]. The frequencies of the edges and the spectral maximum correspond to the values of the principal elements of the chemical-shift tensor.

In a liquid, rapid molecular tumbling averages the absorption frequencies for the nuclei in each molecule to the *isotropic* values. These are just the averages of the absorption frequencies of the principal elements of the shielding tensors. In

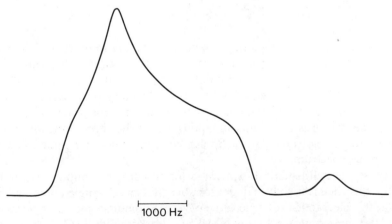

1000 Hz

Figure 5.1 Solid-state ^{13}C NMR spectrum of silver acetate, 9% enriched with ^{13}C in the carbonyl group. The spectrum was acquired with cross polarization for 1 ms on a Bruker Instruments Inc. CXP-100 spectrometer with 20-s delays between scans. The 90° pulse for protons was 4 μs. The small signal to the right of the spectrum results from the ^{13}C in natural abundance in the methyl group.

solids we must subject the sample to motion that simulates the random tumbling in a liquid if we are to get *liquidlike* spectra. This is accomplished by the *magic-angle spinning* to be discussed in Section 4.3.

2.4 Dipolar Coupling

The magnetic field about each nuclear magnet adds to or subtracts from the magnetic field of the external magnet at neighboring nuclei. The nuclei are coupled by these interactions. Spin–spin couplings for solids are very different from the indirect couplings that are important for liquids in that they are transmitted directly through space rather than through the bonding electrons. They are a function of the orientation of the molecule in the magnetic field.

For a given nuclear pair the magnitude of the dipolar coupling is proportional to the inverse cube of the distance between the nuclei and also to $(3 \cos^2\theta - 1)/2$, where θ is the angle between the vector connecting the two nuclei and the direction of the applied magnetic field. For pairs of protons, dipolar couplings can have magnitudes of several kilohertz.

The measured magnitude of the dipolar coupling between a given pair of nuclei can be a valuable source of information about the distance between the nuclei and the orientation of the internuclear vector. In a typical solid, however, any nucleus interacts strongly with many other nuclei, both in and outside the same molecule. Even rare nuclei such as ^{13}C are usually coupled to nearby abundant nuclei such as protons. The effect on the NMR spectrum of any one of the coupling interactions can be observed only in special cases; the dipolar coupling interaction is one of the strongest hindrances to the resolution of detailed features in the NMR spectra of solids.

The dipolar Hamiltonian is

$$\mathcal{H}_D = \sum_{i<k} \mathbf{I}_i \cdot \mathbf{D}_{ik} \cdot \mathbf{I}_k \tag{6}$$

where $\mathbf{D}_{ik}$ is a tensor representing the direct-coupling interaction. The dipolar Hamiltonian has the same general form as does the chemical-shift Hamiltonian. However, the z axis of the PAS for dipolar coupling is always oriented along the internuclear vector. The interaction has axial symmetry with respect to rotation about this axis. For the dipolar coupling the average of the principal elements of the coupling tensor is zero. Thus, in liquids, where the effect of the interaction is averaged to the isotropic value, dipolar couplings have no direct, first-order effect on the spectrum.

Because the mathematical expression for the dipolar coupling is similar to that for the chemical shift, it is reasonable that magic-angle spinning should average the effect to its isotropic value of zero. Sometimes magic-angle spinning alone does give high-resolution NMR spectra of solids [14]. For abundant nuclei such as protons, however, dipolar coupling links many different nuclei, both within the same molecule and in different molecules. The resulting spectral broadening can be 30 or 40 kHz. To give a narrow spectrum, magic-angle spinning would have to exceed this value (Section 4.3). Such high rates of spinning are not currently possible; and magic-angle spinning gives high-resolution proton spectra only for those special samples for which dipolar interactions are relatively weak; for example, in samples where the proton concentration is low.

2.5   Scalar Coupling

Scalar coupling occurs by way of the bonding electrons between nuclei and is (or at least it is generally assumed to be) independent of the orientation of the molecule with respect to the applied magnetic field. As a scalar interaction it may be written

$$\mathscr{H}_J = \sum_{i<k} J_{ik}\mathbf{I}_i \cdot \mathbf{I}_k \tag{7}$$

where J_{ik} is the scalar coupling constant. The scalar coupling is generally small enough to affect the spectrum of a solid only in special cases. It does, of course, give rise to the familiar spin–spin interactions observed in high-resolution spectra of liquids.

2.6   Quadrupolar Coupling

The quadrupolar interaction is the coupling that occurs for nuclei with spin greater than $\frac{1}{2}$ between the nuclear quadrupole and the electric field gradient at the nucleus produced by the bonding electrons. The magnitude of the coupling depends on how the molecule is oriented to the magnetic field and on the size of the electric field gradient.

An important nucleus with spin -1 is deuterium. A calculated spectrum for a powder of a compound containing a single type of deuterium is shown in Figure 5.2. There are three energy levels for deuterium, and for each orientation of the C—D bond there are two nuclear transitions. The spectrum may thus be decomposed into two subspectra. Each subspectrum has the same shape as the powder spectrum of a nucleus influenced by the chemical-shift interaction. However, the local environment about the deuterium nucleus in a C—D bond is almost axially symmetric; two of the quadrupolar principal elements are almost equal. Thus, the maximum intensity in each subspectrum coincides with one of the spectral edges.

The Hamiltonian for the quadrupole interaction is given by an expression similar to those we have already encountered:

$$\mathscr{H}_Q = \sum_{i} \mathbf{I}_i \cdot \mathbf{Q} \cdot \mathbf{I}_i \tag{8}$$

where $\mathbf{Q}_i$ is the quadrupole coupling tensor. The quadrupolar interaction can be quite large; for most quadrupolar nuclei, its magnitude is surpassed only by that of the Zeeman interaction. The average of the principal elements of the

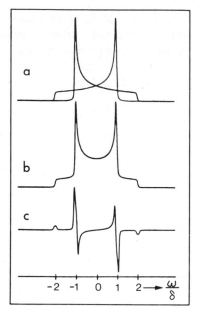

Figure 5.2 Theoretical deuterium NMR powder spectra for deuterium in an axially symmetric environment: (*a*) line shapes for the two NMR transitions; (*b*) complete deuterium spectrum; and (*c*) derivative spectrum, as is obtained in wide-line NMR. Reprinted with permission from H. W. Spiess, "Molecular Dynamics of Solid Polymers As Revealed by Deuteron NMR," *Coll. Polym. Sci.*, **261**, 193 (1983).

quadrupole coupling tensor is zero. Rapid molecular tumbling reduces the first-order effect of the quadrupolar interaction on the spectrum to zero.

Restricted molecular reorientation in deuterated solids sometimes gives rise to deuterium spectra other than those shown in Figure 5.2. Detailed analysis of the observed spectrum provides information about the geometrical restrictions on the motion. Studies at temperatures for which the averaging is only partially effective give information about the rate of motion.

2.7   Radio Frequency Irradiation Hamiltonian

Radio frequency irradiation imposes a time-dependent magnetic field on the nucleus. The effect is represented by the Hamiltonian

$$\mathcal{H}_{rf} = -\cos[\omega_1(t)t + \phi(t)] \sum_i \omega_i I_{xi} \tag{9}$$

where $\omega_1(t)$ is the frequency of the irradiation, ω_i is the frequency response of the ith nucleus to the strength of the irradiating field, and $\phi(t)$ is the phase of the irradiation. In general, the irradiation is applied in pulses. The nature of a specific NMR experiment is determined by the time dependence of $\omega_1(t)$ and $\phi(t)$.

3   IMPORTANT CONCEPTS IN NUCLEAR MAGNETIC RESONANCE OF SOLIDS

3.1   Density Matrix

The density matrix $\rho(t)$ is a mathematical device by which one can describe the state of the nuclear spin system. From it all observable properties of the spin system can be calculated. There are reviews about the density matrix written specifically for NMR spectroscopy [15, 16]. References written specifically for physicists are also available [17, 18]. Here we give only an outline.

The time derivative of the density matrix is given by the Liouville–von Neumann equation:

$$\frac{d\rho}{dt} = i[\rho, \mathscr{H}(t)] \tag{10}$$

where $[\rho, \mathscr{H}(t)]$ is the commutator of the density matrix ρ and the Hamiltonian $\mathscr{H}(t)$. When the Hamiltonian is time independent, (10) has a simple solution:

$$\rho(t) = \exp(-i\mathscr{H}t)\rho(0)\exp(i\mathscr{H}t) \tag{11}$$

The initial condition is specified by $\rho(0)$, which generally can be determined from the fact that, at equilibrium, the density matrix is related to the Hamiltonian:

$$\rho_{eq} = (1/Z)\exp(-\mathscr{H}/kT) \tag{12}$$

where Z is the appropriate partition function, k_B is Boltzmann's constant, and T is the absolute temperature.

An observable property of a nuclear system is determined from the density matrix with (13):

$$\langle \mathbf{O} \rangle = \mathbf{Tr}(\mathbf{O}\rho) \tag{13}$$

Here $\langle \mathbf{O} \rangle$ is the observable property corresponding to the operator $\mathbf{O}$, ρ is the density matrix, and $\mathbf{Tr(O)}$ indicates the trace of the matrix operator $\mathbf{O}$.

Usually the observable in the NMR experiment is one or both of the magnetizations represented by $\langle I_x \rangle$ or $\langle I_y \rangle$. These are the components of magnetization perpendicular to the static magnetic field. When the basis for representation of the density matrix is the set of eigenvectors of the Hamiltonian matrix, the values of these observables are determined by off-diagonal elements of the density matrix. The populations of the various spin states relate to diagonal elements of the density matrix.

When recorded simultaneously, the two components of the transverse magnetization constitute a complex magnetization $M(t)$, which is usually recorded as a function of time following an initial excitation pulse or pulse

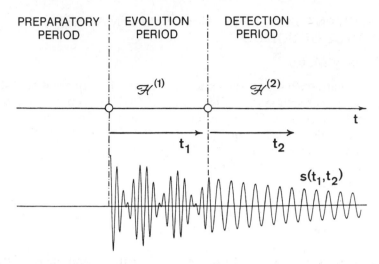

Figure 5.3 Schematic outline of an experiment for the generation of a two-dimensional NMR spectrum. The NMR signal is actually measured only during the period t_2. Reprinted with permission from W. P. Aue, E. Bartholdi, and R. R. Ernst, *J. Chem. Phys.*, **64**, 2229 (1976).

sequence. A frequency spectrum results from a complex Fourier transformation of $M(t)$, which is conveniently accomplished by the minicomputer incorporated into most current NMR spectrometers.

3.2   Two-Dimensional Nuclear Magnetic Resonance Spectra

Sometimes it is possible to perform an NMR experiment so that there are two different time variables, as shown in Figure 5.3. The signal is actually recorded during the period t_2, which is the *observation time* [19]. However, the observed signal depends on how the spin system evolved during an earlier period t_1, called the *evolution* period and on how it was modified during an intermediate interval, called the *mixing period*.

In practice, a series of free-induction-decay signals as a function of t_2 are collected; each corresponds to a particular value of t_1. The free-induction-decay signals are then Fourier transformed to give a set of spectra with a frequency variable v_2. These are transposed into a series of decaying functions with the time variable t_1; each of these corresponds to a particular value of the frequency v_2. These functions are then Fourier transformed to give the complete two-dimensional spectrum as a function of the frequency variables v_1 and v_2.

3.3   Interaction Representations

Generally one term of the NMR Hamiltonian, such as the Zeeman term, is dominant. Often the equations of the density matrix are transformed to eliminate the large term so that attention can be focused on the more subtle effects of the other terms. The new form of the equations is an *interaction representation* [20].

Assume the Hamiltonian in the laboratory frame is

$$\mathcal{H} = \mathcal{H}_0 + \mathcal{H}_1 \tag{14}$$

where $\mathcal{H}_0$ is the time-independent large term of the Hamiltonian and $\mathcal{H}_1$ is the remaining part of the Hamiltonian, which may be time dependent. With the following transformations:

$$\mathcal{H}'_1(t) = \exp(-i\mathcal{H}_0 t)\mathcal{H}_1(t)\exp(i\mathcal{H}_0 t) \tag{15}$$

and

$$\rho'(t) = \exp(-i\mathcal{H}_0 t)\rho(0)\exp(i\mathcal{H}_0 t) \tag{16}$$

the equation of motion of the transformed density matrix becomes

$$\frac{d\rho'}{dt} = i[\rho', \mathcal{H}'_1(t)] \tag{17}$$

The transformations amount to a change of the reference system.

Notice that (17) no longer explicitly refers to $\mathcal{H}_0$. The effect of $\mathcal{H}_0$ has been incorporated into the time dependence of $\mathcal{H}'_1(t)$. In many applications the portions of $\mathcal{H}'_1(t)$ that vary at high frequency with time may be ignored, leading to simplification of the equations.

3.4 Rotating Frame

Mathematically the operations in (15) and (16) are rotations in spin space. In the specialized case where $\mathcal{H}_0$ is the Zeeman Hamiltonian, the transformations are also rotations in real space. Then the interaction representation is called a *rotating frame*.

When $\mathcal{H}_0$ is the Zeeman Hamiltonian, the rotating frame revolves at the Larmor frequency of the nucleus. In the absence of nuclear irradiation the effective magnetic field is zero, and there is no precession of magnetization.

The frame can also be made to rotate at the frequency of applied irradiation. One of the components of the magnetization of the irradiating field is static in

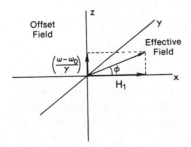

Figure 5.4 Magnetic fields in the rotating frame. The effective field is the vector sum of the irradiation field H_1 and the static field in the z direction that is not canceled by the rotation of the reference frame. The effective field is tilted from the transverse plane by the angle ϕ.

the frame and can induce nuclear transitions. The external field is reduced by a value equivalent to the offset of the irradiation frequency and the Larmor frequency. The effective overall magnetic field in the rotating frame is the vector sum of the irradiation field and the reduced external field (Figure 5.4).

3.5 Spin Temperature

The state of the nuclear system, or a portion of it, can be expressed thermodynamically if a spin temperature can be defined. Such a temperature exists if the off-diagonal elements of the density matrix, which correspond to transverse magnetizations, are zero, and the diagonal elements, which correspond to the populations of the individual levels, have values determined by a Boltzmann distribution of populations. Essentially these requirements mean that a system must be in internal equilibrium. The approach of two nuclear systems defined by spin temperatures to mutual equilibrium can be followed through their spin temperatures.

3.6 Average-Hamiltonian Theory

Many aspects of the NMR of solids (and liquids) are conveniently treated in terms of the average-Hamiltonian theory [7]. This theory stipulates that, if the Hamiltonian relevant to the time development of the nuclear spin system is cyclic, the time development of the system at integral multiples of the cycle time can be described approximately with an *average Hamiltonian* derived from the exact Hamiltonian through a series approximation:

$$\mathcal{H}_{\text{eff}} = \bar{\mathcal{H}}^{(0)} + \bar{\mathcal{H}}^{(1)} + \bar{\mathcal{H}}^{(2)} + \bar{\mathcal{H}}^{(3)} + \cdots \tag{18}$$

where

$$\bar{\mathcal{H}}^{(0)} = 1/t_c \int_0^{t_c} \mathcal{H}(t)dt \qquad \text{and}$$

$$\bar{\mathcal{H}}^{(1)} = -i/(2t_c) \int_0^{t_1} \int_0^{t_2} [\mathcal{H}(t_2), \mathcal{H}(t_1)]dt_1\, dt_2$$

and t_c is the period of the cyclic Hamiltonian $\mathcal{H}(t)$.

The first term of the series is a simple average of the Hamiltonian through one cycle. Higher order terms are increasingly complex. However, in many situations the first term is adequate.

3.7 Homogeneous and Inhomogeneous Effects in Nuclear Magnetic Resonance Spectra

Solid-state spectra are conveniently classed as *homogeneous* or *inhomogeneous*. A homogeneous NMR spectrum was originally defined as one that was affected uniformly by saturating radio frequency irradiation [21]. Portions of an inhomogeneous spectrum can be saturated independently, the result being a spectrum with a "hole" burned in it.

The resonance signals in the spectra of liquids are normally highly resolved. The magnetization components corresponding to each resonance decay almost independently. Thus the spectra of liquids are inhomogeneous.

By contrast, the bands in the spectra of solids are usually broad and only partially resolved. It is important to differentiate homogeneous resonance bands that are inherently broad from inhomogeneous bands that are composites created from many different overlapping narrow signals in the spectra of solids.

All possible molecular orientations are present in a powder. Each crystallite in the sample makes a separate contribution to the total observed spectrum. If the absorption frequency of a nucleus in the crystallite is strongly orientation dependent, the spectrum resulting from the signals from crystallites having all possible orientations is inhomogeneous. The spectra in Figures 5.1 and 5.2 illustrate inhomogeneous spectra resulting from the anisotropic chemical shift and quadrupole interactions, respectively.

For a spectrum to be homogeneous, there must be some mechanism by which the effect of saturating irradiation is transferred to the whole nuclear system, regardless of the irradiation frequency. The phenomenon of *saturation transfer* is occasionally found in liquids, whereby saturation of one resonance in a spectrum with selective irradiation leads to saturation of other signals because the various types of nuclei involved can be interchanged by some chemical process [22]. Such a mechanism suffices to produce a homogeneous spectrum in solids, but more typically the transfer of saturation is accomplished by a mechanism within the nuclear system itself involving mutual spin flip-flops. This mechanism requires that the nuclei be strongly coupled by internuclear interactions and is known as *spin diffusion*.

Nuclear magnetic resonance of solids is different from NMR of liquids because the dipolar coupling that is important for solids is much stronger than is the scalar coupling that is important for liquids. Furthermore, dipolar coupling is transmitted through space and thus links nuclei in different molecules as well as in the same molecule. It provides a very efficient mechanism through which a perturbation applied at any frequency in the NMR spectrum is transmitted throughout the spectrum. For the spectra of abundant nuclei such as protons, homogeneous spectra are the rule rather than the exception.

Maricq and Waugh [23] defined a homogeneous spectrum in terms of magic-angle spinning in a somewhat different manner. In their definition a given interaction leads to an inhomogeneously broadened spectrum if the appropriate Hamiltonian commutes with itself at all times during the spinning process. This is a rigorous definition that avoids some of the ambiguities of the previous definition although it is more difficult to appreciate on an intuitive level.

For solid polymers a distribution of time constants associated with various dynamic processes is very common [24]. Kaplan and Garroway [25] have proposed that such a distribution of rates be considered inhomogeneous if the apparent correlation function is the sum of correlation functions arising from independent sites in the sample. The distribution of rates is homogeneous if the correlation functions for different sites are equivalent but inherently nonexponential.

3.8   Spectral Moments

Spectra composed of broad bands can be characterized by moments defined as

$$M_n = \int_{-\infty}^{\infty} (\omega - \omega_0)^n f(\omega) \, d\omega \tag{19}$$

where M_n is the nth moment and $f(\omega)$ is the spectral intensity at frequency ω. The moments clearly depend on the choice of a reference frequency ω_0. When the spectrum has reflection symmetry about ω_0, all odd moments vanish.

In practice only the second and fourth moments are normally used for describing *broad-line* proton spectra. Higher order moments are highly sensitive to small errors in the measurement of the wings of the spectrum and tend to be unreliable.

3.9   Spin Echos

The generation of spin echos is an integral part of experiments designed to improve the resolution of NMR spectra. The rate at which signals decay as a function of time is inversely proportional to the width of the corresponding resonances in the frequency domain. If improved resolution is to be obtained, the signals must be made to persist over greater periods of time than usual.

Fortunately, magnetic resonance is one of the few physical phenomena for which the time development of certain physical processes are reversible [26]. The NMR signal can be made to stop decaying and return, at least partially, to its initial intensity, usually through application of a pulse after the system is allowed to develop for a period of time τ. Recovery occurs following a second period τ (Figure 5.5).

The simplest type of spin echo, the Hahn echo, is refocused from magnetization acting under the effect of the chemical shielding interaction and magnetic field inhomogeneities. Following an initial excitation pulse, each

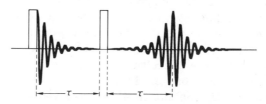

Figure 5.5 Typical pulse sequence used to generate a quadrupolar echo. The maximum echo intensity results when the second pulse is 90° out of phase with the first.

component of the total magnetization precesses in the rotating frame independently of the others. Although the magnetization components initially are all aligned in the rotating frame, they are eventually distributed uniformly in the transverse plane. A 180° pulse reverses the direction of precession of each component but leaves unchanged the precession frequency. After the magnetization has evolved as long after the 180° pulse as before the pulse, its components are again aligned and a signal echo is formed. The time development of magnetization precessing under the influence of the quadrupolar interaction can also be reversed with a pulse, but now a 90° second pulse rather than a 180° pulse results in the largest echo intensity.

The situation is more complicated when signal decay results from a homogeneous process such as dipolar coupling. Now the magnetization components can no longer be treated independently, and the simple concept of the reversal of the process of the precession of the magnetization components in the rotating frame no longer works. Nevertheless, the effects of the dipolar Hamiltonian can be partially reversed with the application of a 90° pulse before the signal has completely decayed. The maximum signal occurs when the second pulse is 90° out of phase with the original pulse. We treat the dipolar echo in a little more detail in Section 4.4.

With magic-angle spinning a type of spin echo is also formed [27]. Such echos differ from the other echos mentioned above because they are not generated by a radio frequency pulse. They will be discussed again in Section 4.3.

4   TECHNIQUES OF HIGH-RESOLUTION NUCLEAR MAGNETIC RESONANCE OF SOLIDS

This section surveys the common methods used to produce high-resolution NMR spectra of solids. *Cross polarization* is a technique by which signals from rare nuclei, such as ^{13}C, are generated through the interaction of the rare nuclei with abundant nearby spins such as protons. *High-power proton decoupling* and *magic-angle spinning* are used selectively to suppress heteronuclear dipolar interactions and the effect of chemical-shift anisotropy in the NMR spectrum. Generation of *solid echoes* allows very broad spectra to be observed with pulse, Fourier-transform NMR spectroscopy. *Multiple-pulse NMR spectroscopy* is a means by which homonuclear dipolar interactions can be suppressed. Finally, *multiple-quantum NMR* is sometimes an alternative technique for getting high-resolution spectra of materials that give very complex single-quantum spectra.

4.1   Cross Polarization

Cross polarization involves a transfer of magnetization from one spin species to another [28]. This transfer is made possible by simultaneous irradiation of both types of nuclei at matched power levels. The irradiation field strengths must satisfy, or nearly satisfy, the equation

$$\gamma^I H_1^I = \gamma^S H_1^S \qquad (20)$$

where H_1^I and H_1^S are the strengths of the irradiating fields at the I and the S nuclei expressed in terms of the magnetic field, and γ^I and γ^S are the gyromagnetic ratios of the two types of nuclei. This is the Hartmann–Hahn condition.

Even when the Hartmann–Hahn condition is satisfied, a net transfer of magnetization between the two systems containing different types of nuclei occurs only if the nuclei are coupled. Furthermore, the two systems of nuclei must have different spin temperatures. To prepare the systems with unequal spin temperatures, one first cools the abundant spin system, which usually consists of protons. This is done by application of a 90° pulse to the protons, followed by continuous irradiation shifted in phase by 90°. The phase shift amounts to a 90° rotation of the effective magnetic field in the rotating frame. The effective field thus lies in the same direction as the proton magnetization after the shift. The arrangement of the effective field and the magnetization is very similar to that in the initial condition, in which the magnetization lay along the direction of the magnetic field in the laboratory frame. However, the effective field in the rotating frame is much smaller than was the original field. Because the magnitude of the magnetization has not changed, the spin temperature must be much lower than the original temperature.

The alignment of the nuclear magnetization along the effective field in the rotating frame is referred to as a *spin lock*. The decay of the magnetization along the effective field to the equilibrium value close to zero is analogous to spin-lattice relaxation in the laboratory frame. Accordingly, it is referred to as *spin–lattice relaxation in the rotating frame*. The time constant for relaxation is given the symbol $T_{1\rho}$.

Once the spin lock is established, irradiation with intensity such that the rare spins rotate around their effective field at exactly the same frequency as the abundant spins rotate around their effective field is added. In the rotating frame the spin temperature of the rare spins is initially high. Polarization thus transfers from the very cold abundant spins to the rare spins, and magnetization builds up along the effective field in the rotating frame for the rare spins.

The transfer of magnetization to the rare spins can be detected in either of two ways. In the indirect method the increase in the spin temperature of the abundant spins is monitored [29–32]. Because the change in temperature is very small for a single transfer, the effect of many different transfers must be superimposed. Various modifications of the indirect method have been devised, and it has been used to detect the anisotropy in the ^{13}C chemical shift [33, 34].

Currently the most common method of using cross polarization is to generate rare-spin magnetization that is detected directly [35–37]. Following cross polarization one simply turns off the rare-spin irradiation and measures the free-induction-decay signal. Experimentally this approach is simpler than the indirect method.

When the magnetization transfer is transferred from high-frequency abundant nuclei (such as protons) to low-frequency rare nuclei (such as ^{13}C), the signal detected in the direct method is stronger than would result from a 90°

pulse alone applied to the rare spins. Some of the inherent sensitivity of the high-frequency nuclei is, in effect, transmitted to the rare nuclei.

During cross polarization of carbon nuclei by proton nuclei, the spin temperature of the abundant protons is perturbed only slightly. After cross polarization the proton irradiation remains on to provide decoupling during observation of carbon signals, and the proton magnetization remains spin locked. The spin temperature of the protons is changed only slightly during observation of the carbon signal because spin–lattice relaxation in the rotating frame is often slow. One may ask whether a second cross polarization to regenerate the carbon signals can be accomplished after the initial observation. Sometimes it is possible to regenerate the transverse magnetization of the carbon nuclei with additional cross-polarization steps [38], but because prolonged irradiation can heat the probe severely, only a few contacts are normally feasible.

Alternatively, the heating problem is avoided if the magnetization of the protons is rotated to the z axis with another radio frequency pulse immediately after signal observation [39]. The magnetization of the protons remains aligned along the z axis with close to the same magnitude for a time approaching the spin–lattice relaxation time T_1. The time between experiments need be only long enough for the probe to cool. If this time is short compared to T_1, several cross polarizations can be done before the proton magnetization returns to equilibrium.

Often the proton irradiation can simply be turned off for a few seconds before repetition of the procedure. If spin–lattice relaxation in the laboratory frame of the abundant spins is rapid, equilibrium is restored during the waiting period. One of the biggest advantages of using cross polarization to generate signals from rare spins is that the abundant spins usually undergo spin–lattice relaxation much faster than do the rare spins [35–37]. Because cross polarization effectively ties the reservoirs of the two spins together, the effective relaxation time of the rare nuclei is shortened.

For ^{13}C NMR of polymers the repetition rate is limited only by the dissipation of heat in the probe itself because proton spin–lattice relaxation times in polymers are typically less than 1 s. Experiments can be repeated as often as once a second. In samples for which the proton spin-lattice relaxation time is long, the addition of paramagnetic relaxation reagents sometimes enhances the relaxation [40, 41].

Problems other than unfavorable relaxation times can arise. The high-power pulses used in the generation of ^{13}C spectra of solids occasionally produce spurious *probe signals* that persist for as long as 100 μs after the end of the pulse and result in baseline artifacts in the spectrum. Most of this problem can be removed by the method of spin-temperature alternation [42]. A reversal of the phase of proton irradiation during cross polarization reverses the sign of carbon magnetization but leaves the spurious signals unaffected. Alternate addition and subtraction of the carbon spectrum cancels the undesired noise. The method is referred to as *alternation of spin temperature* because the alignment of the carbon magnetization with the effective field is reversed in every other experiment. A comparable method to remove artifacts from spectra generated without cross polarization is also possible [43]. The magnetization following a simple excitation pulse is alternately spin locked along the positive and negative y axes of the rotating frame. During the spin-locking period the probe ringing caused by the first pulse dies away. Coaddition of the signals from the two experiments then cancels the probe ringing caused by the spin locking pulse.

Because the carbon magnetization resulting from cross polarization is generated through the dipolar interaction of carbon atoms and protons, it develops faster for those carbon atoms that are strongly coupled to the proton reservoir. A detailed analysis of the dynamics of the process is not easy [44–48], but in the typical experiment, cross polarization monotonically increases to essentially full intensity for all carbon atoms after a millisecond of contact between the carbon atoms and protons [49]. When the coupling to a carbon is dominated by a single dipolar interaction, the transfer of magnetization occurs in an oscillatory manner [50–52]. Such oscillations can be exploited as a source of structural information because they are related to the dipolar coupling between the observed carbon atom and the proton to which it is coupled [53].

There has been recent interest in transfer of polarization from electrons to carbon atoms [53–56]. Transfer of magnetization from electrons to nuclei is generally referred to as *dynamic nuclear polarization*. Because the Larmor frequency of the electron is well above that of any nuclear spin, very large enhancements of the nuclear polarization are theoretically possible. However, the technique is limited because only certain materials contain free radicals.

4.2   High-Power Proton Decoupling

The dipolar coupling of nuclei with low natural abundance to nearby protons is one of the greatest impediments to high-resolution NMR of these nuclei. Strong irradiation in the center of the proton spectrum suppresses the effect of the coupling on the NMR spectrum of the rare nuclei. The procedure appears to be equivalent to the decoupling used to generate high-resolution spectra of liquids, but it differs in several aspects. In liquids the magnitude of heteronuclear scalar couplings between ^{13}C nuclei and attached protons is about 150 Hz. In solids the magnitude of comparable heteronuclear dipolar couplings is about an order of magnitude larger. Amplifiers capable of generating pulse powers of more than 100 W are necessary for effective decoupling. The probes must be built with components capable of carrying this power. Delays of several seconds without the application of proton irradiation are usually used between measurements to allow for heat dissipation.

4.3   Magic-Angle Spinning

Historically magic-angle spinning was the first special technique designed to improve resolution in the NMR spectrum of solids. Its original purpose was to remove the spectral broadening caused by dipolar couplings [57, 58]. To understand the theoretical basis for this method, we must reconsider the Hamiltonian $\mathcal{H}_D$ for the dipolar interaction. For nuclei with negligible chemical shift, the dipolar Hamiltonian may be truncated to a secular portion ($\mathcal{H}_D^{sec}$) that commutes with the much stronger Zeeman term $\mathcal{H}_Z$.

$$\mathcal{H}_D^{sec} = \sum_{i<j} \tfrac{1}{2}\gamma_i\gamma_j Hr_{ij}^{-3}(\mathbf{I}_i\cdot\mathbf{I}_j - 3I_{iz}I_{jz})(3\cos^2\theta_{ij} - 1) \tag{21}$$

The summation is over all pairs of nuclei in the sample. The angle θ_{ij} is the angle made by the vector connecting the two nuclei with the direction of the applied magnetic field.

When the sample is spun about an axis oriented at an angle β to the applied magnetic field, $\cos\theta_{ij}$ can be expressed in terms of β and θ'_{ij}, the angle between the internuclear vector and the spinning axis [57, 58]. The dipolar Hamiltonian then becomes

$$\mathcal{H}_D(t) = \sum_{i<j} \tfrac{1}{2}\gamma_i\gamma_j Hr_{ij}^{-3}(\mathbf{I}_i\cdot\mathbf{I}_j - 3I_{iz}I_{jz})[\tfrac{1}{2}(3\cos^2\beta - 1)(3\cos^2\theta'_{ij} - 1)$$
$$+ \tfrac{3}{2}\sin 2\beta \sin 2\theta'_{ij} \cos(\omega_r t + \phi_{0ij})$$
$$+ \tfrac{3}{2}\sin^2\beta \sin^2\theta'_{ij} \cos 2(\omega_r t + \phi_{0ij})] \tag{22}$$

where ω_r is the angular rotation frequency of the sample and ϕ_{0ij} is a phase factor determined by the angular position of a given internuclear vector at $t = 0$.

When the sample is spinning about an axis oriented at an angle of 54°44′ to the magnetic field (the magic angle), the time-independent first term vanishes. To a first approximation the dipolar broadening of the spectrum disappears.

The remarkable feature of magic-angle spinning is that the rotation of the sample about an axis oriented at a single angle approximates the effect of random rotation in a liquid. The principal difference between the spectra from samples spinning at the magic angle and those from liquids is the presence of sidebands in the spectrum from the solids. These sidebands derive from the time-dependent second and third terms of (22) and are separated from the centerband by integral multiples of the spinning frequency. For spinning at a rate faster than the inherent spectral width, these sidebands are low in intensity and the centerband is narrow.

In the typical proton spectrum, the dipolar broadening might be as much as 40 kHz. This broadening is homogeneous because dipolar coupling ties all of the nuclei tightly together. To be effective, magic-angle spinning must average the effect of each dipolar interaction before communication among the many nuclei can occur. Spinning must be fast compared with the total width of the proton spectrum. The maximum rates of sample spinning are 10–20 kHz, with 4 kHz being more typical. Although magic-angle spinning has been useful for proton NMR spectroscopy in some restricted cases in which the natural line widths are small [59, 60], spinning rates high enough to be effective for the proton spectra of most compounds are not possible. Magic-angle spinning is most commonly used for rare nuclei such as ^{13}C.

It is natural to expect that magic-angle spinning will be more effective in producing narrow spectra in samples that are already partially narrowed by motions in the solid. In fact, in the presence of such motion, magic-angle spinning is usually ineffective. For the spinning to average the dipolar interactions to zero, the interactions must remain static over the rotation period of the spinning. Motion that partially reduces the spectral line width tends to interfere with the effectiveness of the spinning. The exception to this rule occurs when motion that is very fast compared to the rate of spinning does not average the dipolar interactions over all possible angles. For example, magic-angle spinning is useful in reducing the line widths of the NMR spectra of molten polymers, which are constrained in their movements by the restrictions on the chain ends [61, 62].

Magic-angle spinning has become important in recent years because it is a convenient way to remove the effect of chemical-shift anisotropy in spectra from which the dipolar broadenings have been reduced by either high-power decoupling in the heteronuclear case [63] or multiple-pulse decoupling in the homonuclear case [64] (see Section 4.5). Very modest spinning rates are required to average spectra broadened by chemical-shift anisotropy [65, 66]. Nevertheless, it is generally desirable to spin at rates that are much larger than the spectral width in order to move the spinning sidebands out of the spectral region of interest (see Figure 5.6).

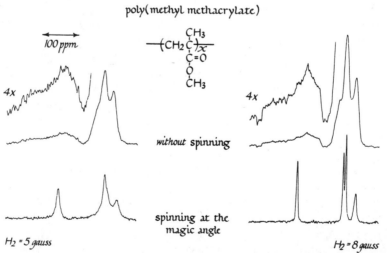

Figure 5.6 Cross-polarization ^{13}C NMR spectra of poly(methyl methacrylate) taken with and without magic-angle spinning at 3 kHz. The importance of high-power proton decoupling is illustrated by the differences between the spectra taken with a decoupling magnetic-field strength of 5 G (left) and the spectra taken with a decoupling magnetic-field strength of 8 G (right). Reprinted with permission from J. Schaefer, E. O. Stejskal, and R. Buchdahl, *Macromolecules*, **10**, 384 (1977). Copyright 1977 American Chemical Society.

At first glance, magic-angle spinning appears to mitigate cross polarization completely. For cross polarization to occur, there must be a static heteronuclear dipolar interaction, and magic-angle spinning suppresses this interaction. In practice, for most materials the dipolar interaction between the abundant nucleus and the rare nucleus is perturbed so strongly by the dipolar interactions among the abundant nuclei that the spinning has a negligible effect on the course of cross polarization. For adamantane, however, polarization transfer takes place most efficiently when the carbon and proton irradiation powers are mismatched by a value equivalent to the spinning frequency [67, 68]. This situation occurs because the weak carbon-hydrogen coupling interaction in this compound is partially averaged by the spinning.

Although the chemical-shift anisotropy can be so large that interesting features of an NMR spectrum are obscured, it is also a potential source of information about the structure and dynamics of solids. Many schemes have been proposed whereby magic-angle spinning can be used to narrow the resonances of the spectrum without completely removing the effects of chemical-shift anisotropy. In one approach the sample is spun at a very slow rate so that the spinning sidebands are closely spaced and large. The outline of the spinning sidebands then approximates the standard chemical-shift anisotropy pattern as shown in Figure 5.7. Mathematical analysis of the sideband intensities yields the chemical-shift anisotropy [65, 66].

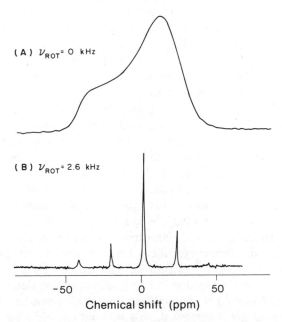

Figure 5.7 Proton-decoupled ^{31}P spectrum taken at 119.05 MHz of phosphatidylcholine: (A) spectrum for a static sample; (B) spectrum for a sample spinning at 2.6 kHz. Reprinted with permission from R. A. Haberkorn, J. Herzfeld, and R. G. Griffin, *J. Am. Chem. Soc.*, **100**, 1296 (1978). Copyright 1978 American Chemical Society.

In another approach the sample is deliberately spun about an axis that is oriented slightly off the magic angle [69]. Bands that resemble the normal chemical-shift patterns but are reduced in width (and reversed for some spinning angles) are the result. Most recently, experiments were performed in which pulses are applied during rotation of the sample to interfere with the effect of magic-angle spinning and to allow the chemical-shift anisotropy to reassert itself [70, 71].

Finally, it has been possible to perform a two-dimensional experiment in which the magnetization evolves under two different sets of conditions during the course of the measurement [72]. In the first part of the experiment, the magnetization evolves normally with magic-angle spinning. A 90° pulse then partially restores the magnetization to the z axis of the rotating frame. The spinning angle is then shifted to some value other than the magic angle. Another pulse then shifts the magnetization away from the z axis so that it evolves under off-axis magic-angle spinning. Double Fourier transformation results in a two-dimensional spectrum in which isotropic shifts are measured along one frequency axis and chemical-shift powder patterns are measured along the other.

In most applications spinning sidebands serve to muddle potentially useful information in the spectrum, and several methods for their suppression have been suggested. One approach is based on the concept of *rotational echos*. Magic-angle spinning is successful because at the end of each spinning cycle the various nuclear magnetizations converge in the rotating frame, even though during the cycle they may take very different pathways. In effect, the reconvergence causes formation of a spin echo. If the magnetization is sampled only at the top of each echo, a pseudo-free-induction-decay curve results, which can be Fourier transformed to a spectrum without spinning sidebands. However, synchronization of data acquisition with the rotation rate can be tedious and this method has not found widespread use.

Dixon [73, 74] designed a pulse sequence that results in a spectrum where the sidebands have a different phase from that of the centerband. The addition of several appropriately chosen spectra of this type gives a spectrum having no sidebands. Variations of the basic procedure have also been proposed [75–77].

Because magic-angle spinning has the potential to remove the effects of either dipolar coupling or chemical-shift anisotropy from the NMR spectra of solids, we can expect that it can also remove the effect of quadrupole coupling. For many nuclei, however, the quadrupole coupling interaction is several megahertz in magnitude. Truncation of the Hamiltonian to a secular term proportional to $(3 \cos^2\theta - 1)$ is no longer justified because the quadrupole interaction approaches the Zeeman interaction in magnitude. Thus the quadrupolar interaction cannot be averaged to zero by spinning. For deuterium, for which the quadrupolar coupling is relatively small, spinning can be effective [78, 79].

Bromine-79 also gives narrow signals with magic-angle spinning, but only with very precise setting of the spinning angle. Because ^{79}Br has a Larmor frequency close to that of ^{13}C, its signal can be used for adjustment of the spinning angle without the need for returning of the probe to the bromine frequency [80, 81].

Recent work shows that for quadrupolar nuclei with half-integral spin substantial narrowing of the central transition can be achieved with magic-angle spinning even though the entire spectrum cannot be narrowed [82]. For these nuclei maximum narrowing is sometimes achieved by spinning at an angle other than the magic angle [83].

Even if the nucleus being observed has spin $-\frac{1}{2}$, the presence of a quadrupolar nucleus can affect the spectrum indirectly. If the quadrupolar interaction is too large to be averaged to zero, the dipolar interactions between the observed spins and the quadrupolar spin also cannot be averaged to zero. Small residual splittings remain in the spectrum [84, 85]. For example, when a carbon atom is located close to ^{14}N, the carbon signal appears as a partially resolved 2 : 1 doublet. Because compounds containing nitrogen are common, this effect occurs often. Simulations of the splittings found in the spectra of carbon atoms attached to nitrogen atoms provides an indirect way to determine the quadrupolar coupling tensor of the nitrogen [86–92].

The isotropic part of the through-bond scalar spin–spin coupling that is commonly observed in liquid samples is not affected by magic-angle spinning. The strong irradiation that is used to remove carbon–proton dipolar coupling from the ^{13}C spectrum (see below) does remove the scalar splittings from the spectrum. Recently, it was demonstrated that if multiple-pulse sequences are used to decouple the protons of adamantane, the scalar coupling of the carbon atoms to protons can be observed (Section 4.5). The magic-angle spinning alone is sufficient to remove the heteronuclear dipolar coupling [93, 94].

There are several practical problems associated with magic-angle spinning. Choice of the material out of which to make the spinner (and sample container) is one. Synthetic polymers offer the features of strength and machinability, which are important. However, most suitable polymers contain carbon atoms that give undesirable signals in the ^{13}C spectrum. In the past, the most common spinner material was poly(oxymethylene), which gives rise to a single signal in the middle of the carbon spectrum that can sometimes be suppressed as a result of favorable proton relaxation properties [95]. Various types of ceramics have recently become the most widely used materials for spinner construction.

Magic-angle spinning at high and low temperatures presents special problems, not the least of which is accurate measurement of the temperature [96–98]. Fully satisfactory compounds suitable as internal thermometers are difficult to find [99]. Commercial probes usually have a thermocouple in the air stream close to the spinning sample. These probes should be calibrated against samples containing an internal standard or a second thermocouple.

4.4   Solid Echos

The spectra of quadrupolar nuclei in powders (in the absence of magic-angle spinning) are extremely broad. In the case of deuterium, for example, the total width is sometimes more than 250 kHz. For other nuclei the spectral range is even greater.

Broad spectra correspond to rapidly decaying free-induction-decay signals. Thus the time domain signal for broad spectra must be measured during a very short interval following an excitation pulse [100]. The probe ringing following a radio frequency that was mentioned in Section 4.1 often obscures a signal that dies away rapidly.

The problem of probe ringing is largely overcome in deuterium NMR through the use of the quadrupole echo. A second pulse 90° out of phase with the first can be applied to generate an echo that reaches its maximum long enough after the second pulse for the probe to recover fully before data acquisition begins.

It is not easy to measure undistorted deuterium spectra even with the echo method. Very intense excitation pulses must be used to cover the large spectral range [101–104]. Composite pulses that compensate for a lack of pulse power have been suggested [105, 106], but they introduce phase distortions into the NMR spectrum. Composite pulses are most useful for inversion of the entire deuterium nuclear spin system in the measurement of nuclear relaxation times [107, 108].

For deuterium NMR all of the tuned circuits, especially that of the probe, must pass a wide band of frequencies with minimal distortion. Finally, the computer must be capable of digitizing data at high speed. Despite the experimental difficulties deuterium NMR is becoming increasingly useful, especially to the biochemist [109, 110].

Echos from protons in solids can be generated superficially in an analogous fashion to echos from quadrupolar nuclei. However, the decay of the free-induction signal resulting from a large system of dipolar coupled spins is inherently different from the decay of the signal from the quadrupolar nucleus because the decay from the dipolar coupling interaction results from a homogeneous interaction. Once the signal from the dipolar coupled spins is completely gone, there is no simple way to recover it. (See, however, the interesting work of Rhim and co-workers [111] for an exception to the rule.) The signal can nevertheless be refocused by a single 90° pulse as long as the pulse is applied before the signal is completely gone.

Observation of the latter half of a dipolar echo circumvents the problems of probe recovery time in the same manner as does observation of a quadrupolar echo. References [112, 113] show that the latter half of the echo is a good approximation of the unseen, full free-induction-decay signal. However, the main significance of dipolar echos is in their relation to the more complex pulse sequences discussed next.

4.5   Multiple-Pulse Nuclear Magnetic Resonance Spectroscopy

It would seem to be a rather simple procedure to prolong the effective duration of a proton free-induction decay indefinitely through a multiple refocusing procedure where each echo is refocused by another pulse as it dies away. Prolongation of the signal in the time domain amounts to a narrowing of the spectrum in the frequency domain. Ostroff and Waugh [114] showed that the lifetime of the NMR signal can indeed be prolonged as a series of spin echos. However, the Ostroff–Waugh procedure is not useful for the generation of high-resolution proton spectra because the effect of the chemical-shift interaction as well as that of the dipolar interaction is refocused. The spectrum derived from such an echo train consists of a single line.

A train of equivalent pulses can be useful for measuring nuclear relaxation even though it does not provide a high-resolution spectrum. The decay rate of the NMR magnetization approaches that for the spin-locking condition as the pulses become closely spaced [115]. The advantage of the pulse train is that the NMR signal can be sampled between each pair of pulses allowing the decay of the echo train to be measured with a single experiment. With continuous irradiation the decay of magnetization can be detected only in a series of experiments in which the irradiation is applied for varying lengths of time and the signal is detected after irradiation is ceased.

The basic breakthrough in the development of multiple-pulse techniques for generation of high-resolution proton NMR spectra of solids was the recognition that trains of pulses can be designed that at least partially preserve the effect of the chemical shift but average the effect of the dipolar interaction. Various schemes using pulses of a variety of phases and durations can be used. The reader is referred to the fundamental work done in the laboratories of John Waugh [116–119] and Peter Mansfield [120–123] and to Haeberlen's book [7] for complete descriptions of the methods. The primary theoretical tool for the analysis of such pulse sequences is average Hamiltonian theory [124], which was introduced in Section 3.6.

With multiple-pulse techniques it is now possible to measure proton chemical-shift anisotropies without dipolar-broadening interference. With a combination of magic-angle spinning to suppress the anisotropy in the chemical shift and multiple-pulse irradiation one can resolve aromatic and aliphatic protons [125–127]. Because the ultimate resolution achievable is modest in the best of cases and the experimental problems are severe, multiple-pulse NMR spectroscopy is done in only a few laboratories.

One of the problems in multiple-pulse NMR is that pulse imperfections play an important role in limiting the ultimate resolution [127, 128]. Thus there has been much effort to devise more and more complex pulse sequences that remove the effects of such imperfections [129]. The design [130] and adjustment [131, 132] of equipment to give short, square-shaped pulses that do not show any change of phase during the rising and falling edges is important [133].

Interestingly, short radio frequency pulses are required because the signal must be observed during periods between the pulses rather than from a fundamental property of the nuclear system [134]. Very effective pulse sequences have been designed for which there is no separation at all between adjacent pulses [135]. Of course, it is not possible to monitor the time behavior of the nuclear system by direct observation during such a sequence, but a single data point can be acquired immediately after the sequence ends. The full free-induction-decay signal can be recovered by repeating the sequence for varying periods of time, even though this is not a practical way to perform NMR spectroscopy.

Multiple-pulse sequences with small delays between pulses are primarily useful for partial decoupling of abundant spins from rare spins. The proton–proton interactions in solids not only broaden the proton spectrum, but also indirectly affect the spectra of nuclei coupled to the protons, making it impossible to observe individual couplings of the second nucleus to the protons. A multiple-pulse sequence applied to the protons decouples the protons from each other but only scales the carbon–proton interaction to a smaller value. Multiple-pulse decoupling thus allow the measurement of individual C—H coupling constants in solids [136, 137].

4.6   Alternatives to Multiple-Pulse Nuclear Magnetic Resonance Spectroscopy

The use of highly deuterated compounds provides a chemical means of getting high-resolution proton spectra. If the level of deuteration is high, there is effectively no interaction among the residual protons in the sample. The dipolar couplings of the deuterium nuclei to the protons must still be removed, but these interactions are small enough that magic-angle spinning alone will remove their effect [138–140].

Magic-angle spinning is unsatisfactory for highly deuterated samples if one wishes to measure proton chemical-shift anisotropies. The magic-angle spinning will destroy the desired information. In this case irradiation to decouple the deuterium nuclei from the protons is required. The deuterium spectrum is too broad to use normal spin decoupling irradiation, however. Irradiation of the double-quantum transitions (Section 4.7) in the center of the spectrum is successful [141].

Another method to get high-resolution proton spectra utilizes the properties of spin-locked magnetizations. When the magnetization is spin locked in the transverse plane of the rotating frame, the magnitude of the dipolar interaction is theoretically reduced to $\frac{1}{2}$ [142]. When the direction of the effective field in the rotating frame is at the magic angle to the magnetic field, the dipolar coupling should disappear entirely [143, 144]. To orient the effective field at this angle, one offsets the radio frequency irradiation from resonance. In practice this approach is not as effective in generating high-resolution spectra as are multiple-pulse sequences.

4.7   Multiple-Quantum Nuclear Magnetic Resonance Spectroscopy

For nuclei with integral spin it is possible to suppress the effect of the quadrupole interaction on the spectrum through the use of multiple-quantum NMR [145–147]. Normally NMR transitions between energy levels differing by more than a unit value of magnetization are forbidden. Nevertheless, elements of the density matrix corresponding to these transitions do exist and can be made nonzero with a series of single-quantum excitations. The resulting states have wave functions with *multiple-quantum coherences*. Because the multiple-quantum coherences cannot be observed directly, their time development must be monitored indirectly after they are converted into single-quantum coherences.

The advantage of multiple-quantum NMR for high-resolution NMR of quadrupolar nuclei is that some of the energy levels of the quadrupolar energy are shifted in parallel by the quadrupolar interaction. Thus the energy of the multiple-quantum transition between such levels is unaffected to first order by the quadrupolar interaction. Detection of the multiple-quantum transition between the first and third levels of deuterium permits the measurement of the deuterium chemical shift in the solid [148]. It has even been possible to use the cross-polarization procedure to generate the double-quantum coherence of interest in such experiments [149].

A technique that is closely related to multiple-quantum NMR is overtone NMR spectroscopy [150]. In this method multiple-quantum coherences are created directly by irradiation at frequencies close to double the Larmor frequency of the nucleus. These transitions are allowed because of slight mixing of the spin states, but nevertheless they are very weak. They have only been used for observation of the ^{14}N nucleus.

5   DETERMINATION OF CHEMICAL STRUCTURE IN SOLIDS

5.1   Experimental Considerations

5.1.1   Measurement of Chemical Shifts With Magic-Angle Spinning

As in liquids, chemical shifts of solids must be measured relative to a signal from a reference compound. For ^{13}C NMR the chemical-shift zero is the signal of tetramethylsilane (TMS). Positive chemical shifts are in the direction of higher frequency. Because TMS is a low-boiling liquid, its use presents practical difficulties. A common approach to measurement of ^{13}C chemical shifts is to use the ^{13}C signal of poly(oxymethylene) as a primary reference and to calculate the shifts relative to TMS by assuming a shift of the poly(oxymethylene) signal from that of TMS of 89.1 parts per million (ppm).

Earl and VanderHart [151] studied various features of ^{13}C chemical shifts in solids. They referenced spectra directly to the signal of TMS contained in a small capillary in the magic-angle spinner but suggest that polyethylene can be a useful secondary standard for many samples. Alternatively, the low-frequency signal of adamantane at 29.5 ppm can be used as a secondary reference.

Accurate measurement of chemical shifts requires not only a good reference standard but also narrow resonances. VanderHart and co-workers [152] investigated the sources of line widths in the ^{13}C spectra of samples spun at the magic angle. A significant source of line broadening for crystalline solids containing aromatic rings can be the anisotropic bulk magnetic susceptibility [153]. Another source of broadening that is especially important for glassy or amorphous materials, most notably polymers, is a distribution of local environments [154]. An important conclusion of VanderHart and co-workers [152] is that ^{13}C line widths in this case scale with the static magnetic field. The overall resolution for many materials, especially synthetic polymers, is apt to be as good at low as at high magnetic field strength. Low-field spectra have an advantage over high-field spectra in that spinning sidebands fall outside the normal range of ^{13}C chemical shifts at low fields, whereas at high fields they can be in the middle of the spectrum. Of course, at higher magnetic field strengths there can be an improvement in sensitivity even if there is no improvement in resolution. For very precise work it should be noted that there is a small second-order dipolar perturbation of chemical shifts of carbon nuclei that is magnetic-field dependent [155].

Chemical shifts of ^{13}C and other nuclides can generally be interpreted in terms of molecular structure by the same rules that are used for solution spectra. In fact, the shifts of metal nuclei in solid complexes can be used as benchmarks for the interpretation of shifts of comparable compounds in solution [156, 157].

Nevertheless, carbon chemical shifts of solids can vary by several parts per million from the corresponding values for the same compounds in solution [158]. Sometimes the shifts of the dissolved compound are affected by specific solvation effects or by hydrogen bonding with the solvent. The shifts of the solid can be influenced by intermolecular interactions reflecting the structure of the crystal lattice. The perturbations of the shifts by the lattice can, of course, provide useful information about the structure of the solid phase (Section 8), but for structure determination they are only an impediment.

5.1.2   Measurement of Scalar Coupling Constants

In liquids the scalar J coupling provides an important source of information about connectivity in a molecule. In solids the spectral splitting caused by indirect coupling of one proton to another is too small to detect, and normally the indirect coupling is removed along with the dipolar coupling by strong proton irradiation. If the protons are decoupled with a multiple-pulse sequence, residual dipolar and scalar couplings of the protons to the carbon atoms remain. The heteronuclear dipolar coupling can be removed by magic-angle spinning to leave a ^{13}C spectrum affected only by chemical shifts and J couplings [93, 94, 159] (Section 4.3). The C—H J couplings in adamantane were observed when the protons were decoupled in this way [159].

5.1.3   Differentiation of Degree of Protonation of Carbon Nuclei

For determination of the molecular structure of an unknown compound, it is important to be able to separate the ^{13}C NMR signals of carbon atoms bearing protons from those of carbon atoms without protons. Opella and Frey [160]

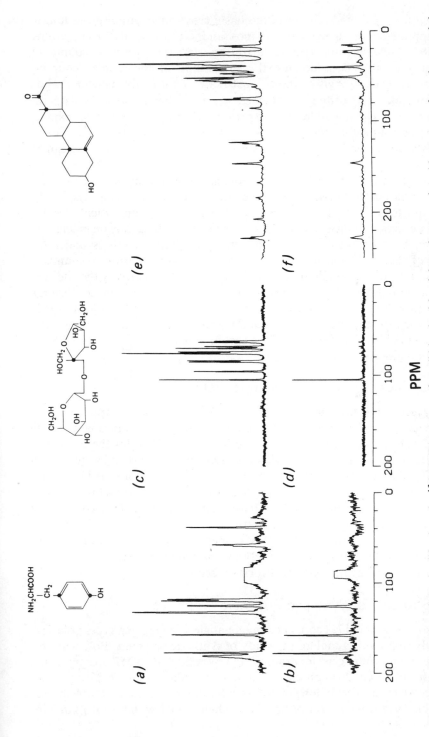

Figure 5.8 Illustration of the simplification of a ^{13}C NMR spectrum of a solid by the *delayed decoupling* procedure: (*a*) spectrum of tyrosine recorded at 37.82 MHz as an average of 1000 scans with 20-s delay between scans. Continuous decoupling was applied after cross polarization; (*b*) spectrum of tyrosine with a delay of 45μs after cross polarization before the decoupler was turned on and signal acquisition was begun. Reprinted with permission from S. J. Opella and M. H. Frey, *J. Am. Chem. Soc.*, **101**, 5854 (1979). Copyright 1979 American Chemical Society.

introduced a simple modification of the basic cross-polarization procedure to allow suppression of signals from mono- and diprotonated carbon atoms. Immediately following the cross-polarization step the proton decoupler is turned off for $40-100\,\mu s$. Because the magnitude of the dipolar coupling interaction depends strongly on the distance between the coupled nuclei, carbon nuclei with one or two directly attached protons interact strongly with the proton reservoir. The signals from these protonated carbon nuclei decay rapidly in the absence of proton irradiation. When the irradiation is turned on after a short delay, those signals are gone. As we see in Figure 5.8, the method can be very effective.

A minor drawback in the *delayed decoupling* method is that signals from the methyl carbon nuclei persist through the period of no decoupling despite the attached protons. The rapid spinning of the methyl groups attenuates the dipolar interaction sufficiently that the signals from these carbon nuclei are long-lived. Another problem is that as the spin system develops under the influence of chemical shifts in the delay period, large phase errors accumulate. The phase problem can be alleviated if the procedure is modified by the addition of simultaneous 180° pulses to the carbon and proton nuclei halfway through the delay [161]. The carbon pulse refocuses the chemical shift (to form a spin echo) but the proton pulse prevents the refocusing of the dipolar coupling. This approach works best if the spacing of the carbon pulses is synchronized with the period of magic-angle spinning [162].

5.2   Examples of Chemical Structure Determination

Naturally occurring materials make attractive targets for the application of NMR spectroscopy, in part because many products of nature are highly insoluble. The energy crisis heightened the interest in NMR for the characterization of coal and oil shales and reviews referencing many articles were written [163, 164]. Other applications are being made to the study of soils [165–167] and even body stones [168]. An unusual use is to determine the oil [169, 170] or protein [171] content of intact seeds so that plant breeding can be accomplished systematically. A detailed coverage of all these applications is beyond the scope of this chapter. To illustrate uses for NMR spectroscopy, we choose instead to concentrate on a few aspects of the study of synthetic polymers and biological materials and some of the extensive work on zeolites.

5.2.1   Structure of Synthetic Polymers

Schaefer and Stejskal [172] combined magic-angle spinning and cross polarization largely to provide a tool for the study of synthetic polymers. Polymers still constitute an important class for examination by solid-state NMR spectroscopy, both for structure clarification and for dynamic studies [173–177].

Polyacetylene consists of long chains of alternating single and double bonds and has unusual semiconductive behavior when doped with various kinds of impurities [178–180]. There are two isomeric forms of polyacetylene. In one, the bonds are in the trans configuration; the other is commonly labeled the cis form although, in fact, only the double bonds are cis and the single bonds are trans.

The cis and trans isomers of polyacetylene give signals separated by about 10 ppm [181]. This has made it possible to follow the kinetics of conversion of the cis into the trans form at elevated temperatures [182]. Interestingly the NMR results do not agree with the kinetic parameters for the same process determined with IR spectroscopy [183].

Another important application of solid-state NMR is to clarify the structure of cross-linked polymers [173, 184–186]. Often these polymers are so insoluble that it is impossible to use solution NMR spectroscopy. However, even for very insoluble polymers the number of units involved in cross-linking may be only a few percent of the total number of units; the signals from the cross-links may be difficult to observe. Fortunately, several typical cross-linking agents, such as formaldehyde, can be obtained enriched in ^{13}C. The spectrum obtained from spectra cross-linked with these materials contain very specific information about the nature of the cross-links. Nuclear magnetic resonance spectroscopy has been used to elucidate the structure of phenolic resins cured with ^{13}C-enriched formaldehyde [174, 187].

5.2.2   Structure of Materials of Biological Origin

The composition of whole kernels of grains can be probed with ^{13}C NMR spectroscopy without destruction of the sample (Schaefer and Stejskal [170] and Haw and Maciel [188]). Much of the guesswork is eliminated from plant breeding when new plants can be grown from seeds that were analyzed.

Nitrogen-^{15}N NMR is also used to determine the protein content of soybeans, but the very low natural abundance of ^{15}N requires isotopic labeling [189]. Viruses that are grown with $^{15}NH_4Cl$ as a nitrogen source result in protein samples that are suitable for ^{15}N analysis. Resolved resonances for 13 of the 14 nitrogen sites in the viral DNA were assigned [190]. Of course, ^{13}C NMR is also important for the analysis of biological samples [191–194].

Biological transformations in highly insoluble materials can now be followed with NMR. For example, the fermentation of lignin was monitored with ^{13}C NMR [195]. Schaefer and co-workers [196] discovered an especially powerful use of cross polarization to clarify the mechanization of biological reactions. The method requires samples to be doubly labeled in ^{13}C and ^{15}N. Cross polarization is accomplished in two steps. In the first step, spin-locked carbon magnetization is created by double irradiation of carbon atoms and protons in the normal manner. In the second step, the carbon magnetization is transferred to nearby ^{15}N nuclei by double irradiation of carbon and nitrogen atoms. The carbon and nitrogen nuclei must be in close contact for cross polarization to be efficient. The linkup of ^{13}C from a particular source with ^{15}N from another source can thus be established unambiguously if cross polarization takes place.

5.2.3   Structure of Zeolites

Zeolites are important as catalysts, absorbing agents, and ion exchangers. They contain silicon and aluminum in various ratios and sodium or other counterions to balance the net negative charge of one for each aluminum atom.

The framework structure of these aluminosilicates is well established (see Figure 5.9). The fundamental unit is a cuboctahedral cage in which silicon and aluminum atoms are linked by oxygen atoms. Each cuboctahedral cage is linked to four other cages through the six-membered rings on the faces. Each cage can thus be represented by a tetrahedron. The linking of the cages produces *supercages* that enclose large cavities.

In a particular material it is difficult to use X-ray diffraction to determine the exact arrangement of the aluminum and silicon atoms. Aluminum and silicon have almost identical X-ray scattering factors, and the structure must be inferred largely from differences in bond lengths. Furthermore, suitable single crystals are often unavailable.

Zeolites are well suited for study with NMR spectroscopy. Silicon NMR with magic-angle spinning is especially fruitful. Normally cross polarization is not used because protons may not be distributed uniformly throughout the material, and it is important to get complete signal intensities from all types of silicon.

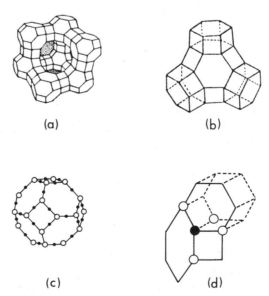

(a)   (b)

(c)   (d)

Figure 5.9 Features of the structure of zeolites: (a) basic structure of faujasite showing the cuboctahedral cage linked through six-membered rings; (b) detail of the tetrahedral cages; (c) cage with tetrahedral atoms given as circles and oxygen atoms as black dots; and (d) schematic representation of the immediate environment of a tetrahedral silicon or aluminum atoms in zeolite. Reprinted with permission from J. K. Klinowski, S. Ramdas, J. M. Thomas, C. A. Fyfe, and J. S. Hartman, *J. Chem. Soc. Faraday Trans. 2,* **78,** 1025 (1982).

However, cross polarization can be a useful way to enhance the signals from silicons in close proximity to adsorbed organic species [197].

Although ^{27}Al is a quadrupolar nucleus, its resonances are sharp enough with magic-angle spinning that separate signals can be resolved for aluminum atoms in tetrahedral and octahedral environments in zeolites, especially with high magnetic field strengths [198–201]. Sodium NMR is a potential source of information about zeolites, but it has not yet proven useful [86].

Within the basic zeolite framework a given silicon may be connected (through oxygen atoms) to one, two, three, or four other silicon atoms. The NMR signals from silicon nuclei in these different sites are well resolved [202] (Figure 5.10). Thus five slightly overlapping ranges of chemical shifts for the different types of silicon atoms may be defined [203, 204]. When the spectra are taken without cross polarization, the areas under the peaks in each range accurately reflect the concentrations of the various types of silicons.

A reliable guideline for many aluminosilicates (but not all [205]) is Loewenstein's rule, which states that no aluminum can be bonded through an oxygen directly to another aluminum in the zeolite framework. With the assumption of this rule, it is possible to calculate the ratio of silicon to aluminum (Si:Al) in zeolites entirely from the silicon spectrum by the following formula:

$$(Si:Al)_{NMR} = \sum_{n=0}^{4} I_{Si(nAl)} \bigg/ \sum_{n=0}^{4} 0.25 n I_{Si(nSi)} \tag{22a}$$

where $I_{Si(nAl)}$ is the normalized intensity of the NMR signal for silicon bonded to n aluminums and $I_{Si(nSi)}$ is the normalized intensity of the NMR signal for silicon

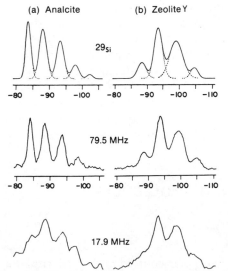

Figure 5.10 Calculated and experimental ^{29}Si NMR spectra of (a) analcite and (b) zeolite Y taken with magic-angle spinning at 79.5 and 17.9 MHz with single-pulse excitation and 5-s delay between scans. The spectra at the top are simulated. Reprinted with permission from C. A. Fyfe, G. C. Gobbi, J. S. Hartman, R. E. Lenkinski, J. H. O'Brien, E. R. Beange, and M. A. R. Smith, *J. Magn. Reson.*, **47**, 168 (1982).

bonded to n other silicons. When the chemical composition is known independently from methods other than NMR, the agreement of the ratio obtained by NMR with the known value provides a way to check the validity of this rule [181]. Usually the agreement is very good.

The silicon NMR spectrum of zeolites contains more information than the ratio of silicon/aluminum atoms, however. Consider for a moment that one has a structure containing no aluminum. The ^{29}Si NMR spectrum consists of a single line corresponding to a silicon surrounded only by other silicons. Such a spectrum was indeed observed for samples in which the aluminum was removed chemically [206, 207]. Aluminum atoms can now be inserted into a framework originally containing only silicon in various ways. This leads to new lines in the NMR spectrum. The relative sizes of the various signals reveals the relative amounts of silicon atoms of different type [208–210].

We close this discussion of the application of solid-state NMR to zeolites with an illustration of some of the potential interpretational problems. Zeolite A has an equal number of silicon and aluminum atoms in its structure. Its ^{29}Si NMR spectrum consists of a single line located -89.6 ppm from the resonance of TMS [211]. The line is quite sharp, as is expected for a very uniform structure. Since the chemical shift of this line corresponds very well to that expected for a silicon attached to only three aluminum atoms, a novel structure that breaks Loewenstein's rule was proposed [212–214].

Other work has focused on a related material, zeolite ZK-4 [215]. This compound and zeolite A have an unusual structure for the framework that incorporates *double four-membered rings*. The ratio of silicon/aluminum in ZK-4 is 1.5 instead of 1.0; therefore, the NMR spectrum consists of five different lines. The resonances can be assigned on the basis of their relative intensities, and the signal for silicon attached to four aluminums falls at -88.4 ppm—very close to the position of the signal in zeolite A. Apparently the range of shifts that is possible for silicon with four attached aluminum atoms is much larger than was previously supposed. The unusual framework of zeolite A and ZK-4 must lead to a displacement of the signal of the silicon with four attached silicon atoms. The chemical shift of the silicon nucleus may depend on both the bond angles and the number of attached aluminum atoms [216]. The story as a whole emphasizes the care with which conclusions about structure based on chemical shifts alone must be made. Of course, this caution applies equally to research on liquids as to solids.

6   DETERMINATION OF DETAILS OF MOLECULAR STRUCTURE

6.1   Internuclear Distances

A collection of randomly oriented pairs of dipolar-coupled nuclei give rise to a spectral pattern known as a *Pake doublet* [217]. It is the same pattern given by a quadrupolar nucleus with spin 1 (Figure 5.2). The spectral splitting in the Pake doublet is inversely proportional to the cube of the distance between the nuclei. Thus the Pake doublet provides a basis for determination of internuclear distances.

6.1.1   Carbon–Carbon Distances

A clean Pake doublet only results from isolated pairs of coupled nuclei. With isotopic labeling one can create pairs of carbon nuclei in an organic compound. However, to minimize intermolecular dipolar coupling the labeled molecules must be isolated from each other. One can do this by locating the enriched molecules in an inert matrix [218] or in the unenriched material [219].

Even in properly prepared labeled samples problems arise in the measurement of dipolar-coupling constants when the magnitude of the chemical-shift anisotropy is comparable to that of the dipolar coupling (Figure 5.11). There are several ways to deal with this problem. If a single crystal can be obtained, it is possible to find both the chemical-shift anisotropy and the dipolar-coupling constant by following the absorption frequencies of the various NMR signals as the crystal is rotated around three different axes [219–222] (Section 6.4). A drawback of using single crystals is that interpreting the results is easy only

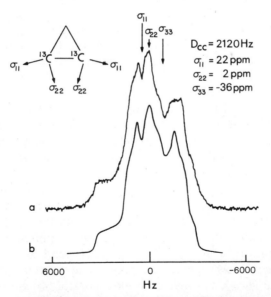

Figure 5.11   (a) Experimental ^{13}C NMR spectrum of dilabeled cyclopropane taken at 20.12 MHz. The presence of both the interaction between the labeled carbon atoms and the chemical-shift anisotropy leads to a complex spectrum. (b) Computer simulation of the observed spectrum. The orientation of the principal elements of the chemical-shift tensor are shown on the chemical structure. Reprinted with permission from K. W. Zilm A. J. Beeler, D. M. Grant, J. Michl, T.-C. Chou, and E. L. Allred, *J. Am. Chem. Soc.*, **103**, 2119 (1980). Copyright 1980 American Chemical Society.

when the crystal structure is already known, in which case the NMR determination of the carbon–carbon bond distances is unimportant.

A method of separating the chemical-shift interaction from the dipolar coupling that avoids the need for single crystals takes advantage of the fact that the chemical shift, in frequency units, scales with the size of the magnetic field, whereas the dipolar coupling is field independent. Spectra taken at low magnetic field approach the ideal Pake pattern more closely than do spectra taken at high magnetic-field strength [223]. Unfortunately, very weak signals result when spectra are simply acquired at extremely low magnetic fields. The sensitivity problem is circumvented by a *rotating-frame* experiment. The spin system is followed as it evolves around the effective magnetic field in the rotating frame rather than about the fixed field in the laboratory frame. Sensitivity is determined by the external field strength, but the evolution of the spin system is controlled by the effective field strength in the rotating frame.

A detailed analysis of the rotating-frame experiment was first made by Barnaal and Lowe [142]. These workers demonstrated that the dipolar interaction term in the Hamiltonian scales according to the expression $(3 \cos^2\theta - 1)/2$, where θ is the angle the effective field makes with the z axis in the laboratory frame. For irradiation on resonance the scaling factor is $-\frac{1}{2}$. Hence, for irradiation on resonance, the nuclear magnetization develops around the direction of the effective field as if the dipolar coupling constant has one-half its normal value and was negative in sign and there was no chemical shift. For most experiments the sign of the dipolar coupling does not affect the appearance of the spectrum. The important fact is that the chemical shift in the rotating frame is negligible compared with the dipolar coupling.

The rotating-frame experiment is not so esoteric as it may seem. It is actually performed every time one excites a nuclear system in a solid with a 90°pulse. During a strong pulse the system develops under the influence of the dipolar as well as the Zeeman interactions, but the fact that the Zeeman term is larger than the dipolar term means that the dipolar interaction can be ignored in most cases. Theoretically, however, the free-induction decay effectively begins halfway through the pulse as a result of the dipolar development during the pulse [142].

Yannoni and Kendrick designed an experiment to make the rotating-frame experiment a practical means to measure internuclear distances [224, 225]. The time development of the magnetization in the rotating frame can be mapped out point-by-point by monitoring the signal at the end of pulses of varying lengths, but this method is time-consuming. A faster approach is to apply a pulse train with observation windows between the pulses. The effectiveness of a pulse train is illustrated in Figure 5.12, where both the laboratory-frame and the rotating-frame spectra are shown. The rotating-frame spectrum is an essentially perfect Pake doublet with no observable chemical-shift effect. The laboratory-frame spectrum would require computer simulation for separation of the combined effects of chemical shift and dipolar coupling.

The Yannoni experiment does present some practical difficulties. The homogeneity of the irradiating field over the entire sample must be good. A

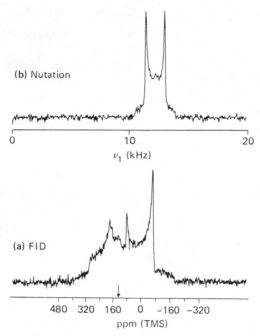

Figure 5.12 Experimental ^{13}C NMR spectra of polycrystalline $^{13}CH_3{}^{13}COOH$ diluted $1:40$ in $^{12}CH_3{}^{12}COOH$: (a) conventional spectrum obtained with proton decoupling and (b) spectrum taken by sampling the time development of the spin system in the rotating frame during a long pulse train. The spectrum in (b) is an almost perfect Pake doublet while the spectrum in (a) is strongly affected by chemical-shift anisotropy and the dipolar interaction. Reprinted with permission from D. W. Horne, R. D. Kendrick, and C. S. Yannoni, *J. Magn. Reson.*, **52**, 299 (1983).

specially designed irradiation coil and small sample sizes may be necessary to ensure radio frequency homogeneity. Because the experimental procedure is similar to that used for multiple-pulse narrowing of a proton spectrum, the same types of pulse imperfections that interfere with multiple-pulse narrowing lead to spectral distortions in the carbon experiment (see Section 4.5).

The rotating-frame experiment, which is also known as *nutation NMR*, has also been useful for simplification of the spectra of quadrupolar nuclei. The quadrupolar interaction is so large for many of these nuclei that a truly high-field experiment is impossible. The alternative of a very low-field experiment works quite well to give a simple spectra that is completely dominated by the quadrupolar interaction [226, 227].

An alternative to the Yannoni method involved actual measurement of an NMR spectrum in a zero magnetic field in the laboratory frame [228]. The low signal intensity that is associated with NMR spectroscopy in low magnetic fields is avoided by initial polarization of the sample in a high magnetic field. The magnetization that has developed in the high magnetic field is then allowed to evolve in zero field through a two-step procedure. Initially the magnetic field is reduced to an intermediate value at a rate that is fast when compared to the spin–spin relaxation time T_2, but is slow when compared to the spin–lattice relaxation time T_1. In the second step the remaining field is turned off rapidly. The spin system is then allowed to evolve in zero magnetic field for a time t_1. The oscillatory behavior during t_1 is monitored by observation of the NMR signal after the system has been returned to the original magnetic field by the same procedure in reverse.

In zero magnetic field there is no chemical shift. Furthermore, every molecular orientation is equivalent to every other because no reference direction is defined by the magnetic field. For nuclei with spin $\frac{1}{2}$ the spectrum consists of three lines [228] separated by a frequency proportional to the dipolar coupling constant. Determination of the internuclear distance can be made with powders to precision comparable to that usually expected for single crystals. The main drawback of this technique is that it requires special equipment for the field cycling. Furthermore spin–lattice relaxation times must be much longer than the time required for the field cycling to prevent loss of most of the magnetization during the field change.

Zero-field NMR has interesting applications outside the measurement of bond distances. Like nutation NMR it also simplifies the spectra of quadrupolar nuclei such as deuterium [229]. Motional averaging of the zero-field spectrum allows very low-amplitude motions to be detected [230]. Complicated pulse sequences similar to those used in high-field NMR can even be devised [231, 232]. However, in zero-field NMR a pulse consists of temporary application of a small static magnetic field instead of radio frequency irradiation.

6.1.2   Carbon–Hydrogen Distances

The well-known difficulty in locating hydrogen atoms precisely with X-ray diffraction makes NMR methods for measurement of carbon–hydrogen bond lengths important. Measurement of carbon–hydrogen dipolar coupling constants presents somewhat different problems than measurement of carbon–carbon coupling constants. Protons in most solids interact among themselves as strongly as they do with any ^{13}C nuclei that happen to be nearby. In most cases resolved carbon–hydrogen dipolar splittings cannot be detected unless the interactions among the protons are suppressed. A less important problem is that a given ^{13}C nucleus generally couples to several nearby protons.

One way to measure the strength of the carbon–hydrogen interaction is to follow the course of the cross-polarization process. If a given carbon is coupled to a relatively isolated proton, cross polarization occurs in an oscillatory fashion, and the frequency of the oscillations are determined by the strength of the dipolar interaction [50]. The rate at which the oscillations decay is largely controlled by proton–proton interactions. To suppress the interproton couplings during cross polarization, one can keep the protons spin locked along an effective field that is oriented at the magic-angle with respect to the static magnetic [233, 234]. Nevertheless, the cross-polarization procedure is probably not the best method for determination of carbon–hydrogen distances. This procedure is difficult to perform, and interpretation of the results requires a complicated mathematical analysis.

The groups of Waugh [137, 235, 236] and Vaughan [136, 237] developed similar methods to measure heteronuclear dipolar coupling constants that involve monitoring the evolution of carbon magnetization in the absence of carbon irradiation while proton–proton interactions are suppressed by a multiple-pulse sequence. The basic procedure of Waugh is outlined in Figure 5.13. Carbon transverse magnetization is created during the initial period by cross polarization. It then evolves during a period t_1 while the protons are multiple-pulse decoupled. The resulting carbon magnetization remaining at the end of this time is observed during a period t_2 while the protons are decoupled with strong, continuous irradiation. The Vaughan method uses an additional 180° pulse applied to the carbon atoms halfway through the period t_1. This pulse cancels the effect of the chemical shift on the carbon atoms during the formation of the spin echo.

In the Waugh experiment the carbon spin system develops under both the chemical shift and the heteronuclear coupling interactions during the period t_1, but only under the chemical shift during the period t_2. In the Vaughan version the carbon system develops only under the dipolar coupling during t_1. Fourier transformation of the results from a Waugh experiment gives a two-dimensional spectrum in which both chemical shifts and dipolar couplings affect the signal displacement along one axis but only chemical shift affects the signal displacement along the other. Double Fourier transformation of the Vaughan experiment gives a spectrum in which chemical shifts can be measured along one axis

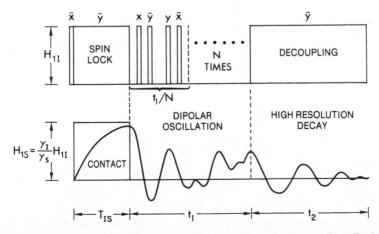

Figure 5.13   Pulse sequence to generate spectra reflecting carbon-dipolar couplings. During period t_1 the magnetizations develop under the influence of carbon–hydrogen interactions. Multiple-pulse decoupling during this period simplifies the results. The signal is observed with normal decoupling during the period t_2. Reprinted with permission from R. K. Hester, J. L. Ackerman, B. L. Neff, and J. S. Waugh, *Phys. Rev. Lett.*, **36**, 1081 (1976).

of the spectrum and only dipolar couplings can be measured along the other. The Vaughan experiment thus gives a little cleaner separation of the interactions, but the distinction is not very important if single crystals are used.

The above methods can be adapted for spinning samples. At first glance it would appear to be unreasonable to spin the sample at the magic angle during a measurement of dipolar coupling constants because the dipolar interaction is suppressed by the spinning. Nevertheless, the coupling information is regained from the spinning experiment through an analysis of the intensities of the spinning sidebands. Griffin and co-workers used this fact to measure the couplings between heteronuclei and protons in spinning samples [161, 238–240]. DiVerdi and Opella [241] measured N—H bond distances in ^{15}N-labeled DNA.

Spinning does introduce an experimental complication. The acquisition of data and the application of pulses must be synchronized with the spinning rate. On the other hand, the effective sensitivity is greater in a spinning than in a nonspinning experiment because intensity is concentrated in the centerband and in the sidebands rather than distributed in a broad powder pattern.

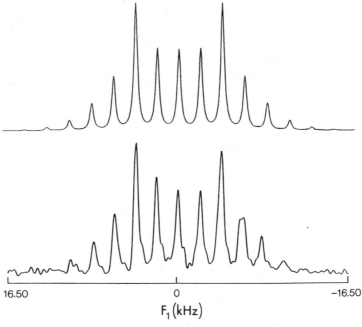

16.50   0   −16.50

F_1 (kHz)

Figure 5.14 Carbon–hydrogen dipolar spectra obtained with magic-angle spinning from a sample of calcium acetate, 30% enriched with ^{13}C. The spectrum was obtained by summing the intensity of a two-dimensional spectrum. The outline of the doublet for the carbon–hydrogen coupling interaction is seen in the peaks of the spinning sidebands. The top spectrum is simulated. Reprinted with permission from M. G. Munowitz, R. G. Griffin, G. Bodenhausen, and T. H. Huang, *J. Am. Chem. Soc.*, **103**, 2529 (1981). Copyright 1981 American Chemical Society.

An example of a projection of a one-dimensional spectrum from a two-dimensional spectrum that shows the effect of carbon–hydrogen coupling is given in Figure 5.14. The individual lines are the spinning sidebands whose outline is the dipolar Pake doublet.

6.1.3   Silicon–Oxygen Distances

One of the principal elements of the silicon chemical-shift tensor corresponds to an axis lying close to the silicate silicon–oxygen bond. The value of this element has been found to correlate with the silicon–oxygen bond distance [242, 243]. Empirical correlations of other NMR parameters with bond lengths are possible in various series of related compounds.

6.1.4   Interpretation of Measured Distances

From the NMR spectrum one measures dipolar coupling constants. These are inversely proportional to the cube of the internuclear distance. However, the calculated distances sometimes disagree with those from other methods beyond the extent that can be explained by systematic errors [244]. The reason for this is well understood [245, 246]. Distances in molecules obtained by any technique represent an average. Unfortunately this average is never of the distances themselves but of the measured quantities. In NMR vibrational motion reduces the observed dipolar couplings to a complicated average involving both orientation and distance parameters. Analysis of the results to get the average of the distance requires detailed knowledge of the molecular vibrations; anharmonic as well as harmonic components of the vibrations may be important. In general the systematic error in carbon–carbon bond lengths produced by neglect of vibrational averaging is probably within 1%. For carbon–hydrogen bonds the error can be several percent.

6.2   Bond Angles

The principal axis system of the quadrupolar coupling tensor for deuterium is normally almost axially symmetric; the axis of symmetry is directed along the C—D bond. Measurement of the relative orientations of the two quadrupolar tensors in D_2O permits a determination of the bond angle [247]. This is done with two-dimensional NMR. Determination of the relative orientation of the principal axes for the chemical shift interaction of two nuclei can be used for bond-angle measurements in rare cases [248]. Griffin and co-workers [237–240] have even correlated the direction in space of the chemical-shift tensor and the dipolar tensor in model compounds.

The shape of the silicon signal in the spectrum of silicate glasses is related to the distribution of silicon–oxygen–silicon bond angles in the sample [249, 250]. The isotropic chemical shift is slightly different in compounds having different bond angles so that the overall signal for the whole sample is inhomogeneously broadened. The dependence of the chemical shift on bond angle is related to the dependence of the shift on bond length mentioned in Section 6.1.3.

6.3   Conformational Analysis

Conformations of a molecule are defined by the dihedral angles within the molecule. Measurement of dihedral angles presents essentially the same problems as measurement of bond angles except that the relative orientations of interaction tensors on two different atoms must be found. Two-, or even three-dimensional NMR spectroscopy sometimes provides this correlation [251].

The number of resolved NMR signals gives information about the symmetry of stable conformations. Molecules are often locked into a particular conformation in the solid even though they may exist as several rapidly interchanging conformations in solution. Thus the presence of multiple NMR signals for solids is sometimes observed when there is a single line in the solution spectrum.

An example is 1,4-dimethoxybenzene, for which three aromatic carbon signals occur [23, 27, 252, 253]. In solution there are only two signals because the methoxy groups rotate rapidly. In the solid phase the molecule is locked into the trans conformation, for which the two methoxy groups point in opposite directions in the plane of the aromatic ring. The resonances from conformationally inequivalent carbon atoms are separated by about 6 ppm.

Synthetic polymers are restricted conformationally by the linkages in the backbone chain. Solid poly(phenylene oxide) has multiple aromatic resonances because rotation of the aromatic rings is restricted [63]. In poly(ethylene terephthalate) rotation around the single bond of the ethylene glycol moiety is inhibited, and there are multiple resonances in the ^{13}C spectrum separated by about 1 ppm [63].

Nuclear magnetic resonance spectroscopy provides an interesting indication of the conformation of crystalline, syndiotactic polypropylene. There are two carbon signals separated by about 9 ppm in the spectrum taken with cross polarization and magic-angle spinning [254]. In the helix structure there are, in fact, two kinds of methylene groups. The chemical shifts may arise from differences in the three-bond interactions experienced by the carbon atoms in the two sites [255].

When conformational interchanges in the solid are fast on the NMR timescale at room temperature, it is sometimes possible to see the spectra of individual conformations when the temperature is reduced. For example, in epoxy resins, which contain a substituted alkoxy benzene group, rotation of the phenyl ring is fast at room temperature but becomes slow on the NMR timescale below about 305 K [256].

Conformational effects can result simply in differences in chemical shifts of solids and liquids [257]. On the basis of the above examples and others [258], we can expect that corresponding carbon atoms in different conformations may differ in chemical shift by as much as 15 ppm.

Sometimes one can get information about molecular conformations in unexpected ways. VanderHart [259] showed that when the ^{13}C spectrum of highly oriented polyethylene is greatly expanded, small peaks resulting from dipolar coupling of the carbon atoms to carbon atoms two and three bonds away can be detected. The spectrum was obtained without magic-angle spinning, of course, because spinning would have averaged the effect of dipolar coupling on the spectra. The high orientation of the polymer means that broadenings from the anisotropic chemical shifts are small. The magnitudes of the observed dipolar couplings are determined by the conformations of the polymer chain.

6.4   Electronic Structure

Chemical shifts in liquids can largely be explained by the electronegativity of the molecular substituents and the anisotropic chemical shielding resulting from ring currents in aromatic rings and double bonds. The complete chemical-shift tensor for nuclei in solids must be interpreted in terms of the three-dimensional distribution of the electrons in the molecular orbitals.

Much of the work to date has simply been in the collection of enough data on model compounds so that trends can be recognized. The books by Spiess [13], Mehring [8], and Haeberlen [7] contain tables of hydrogen, fluorine, and carbon chemical shifts for a variety of compounds. Duncan has written a comprehensive review of carbon chemical-shift parameters [260]. Here we treat the experimental procedures that are used to measure the chemical-shift tensors and the basis for their interpretation very qualitatively.

The simplest experimental procedure that gives the principal elements of the chemical-shift tensor is simulation of an observed chemical-shift powder pattern (such as that in Figure 5.1) [261, 262]. This method does not lead to high precision and, in the absence of obvious molecular symmetry, gives no information as to how the PAS of the chemical-shift tensor is oriented in the molecular structure. If the orientations of the chemical-shift tensors in enough related compounds have been determined, the orientation can be assumed with reasonable certainty, but an exact determination requires the use of single crystals for which an X-ray structure is already available.

Determination of the chemical-shift tensor from a single crystal requires the crystal to be mounted on a goniometer in the NMR probe. The chemical shifts for the nuclei of interest are measured for different angles of rotation of the crystal about an axis whose orientation to the magnetic field is fixed. For a perpendicular axis the observed shifts are given by an expression of the type

$$\sigma = A + B\cos 2\alpha + C\sin 2\alpha \tag{23}$$

where A, B, and C are constants determined by the chemical-shift tensor and the orientation of the PAS with respect to an axis system fixed on the goniometer and α is the angle of rotation about the goniometer axis. Usually rotations are made separately for three different axes. The full chemical-shift tensor and its orientation relative to the crystallographic axis can then be determined by a computer fit of the rotation data. The detailed expressions are given in the books by Mehring [8] and Haeberlen [7]. A two-dimensional NMR experiment that relates the chemical shifts for different nuclei in different molecular orientations is possible [263].

To illustrate typical properties of chemical-shift tensors, we review the results for carbonyl carbon atoms [264]. The carbonyl carbon is shielded by 100 ± 25 ppm relative to liquid carbon disulfide when the molecule is oriented with the plane of the carbonyl group perpendicular to the magnetic field. It is deshielded by 60 ± 40 ppm when the magnetic field is approximately perpendicular to the C=O bond in the carbonyl plane. The chemical shift for the third orientation falls over a much wider range, and the carbonyl carbon is shielded by as much as 80 ppm or deshielded by as much as 70 ppm in extreme cases.

Chemical shifts are classified theoretically as diamagnetic and paramagnetic [265]. For ^{13}C the paramagnetic shifts dominate. The paramagnetic shift results from mixing of the ground-state molecular orbitals and the excited-state orbitals by the magnetic field. The importance of a given ground-state/excited-state pair is inversely proportional to the energy difference between the orbitals. It is also inversely proportional to the cube of the average distance between the nucleus and the electrons in the orbital.

Symmetry controls whether mixing of any given excited state into the ground state can occur for a particular orientation of the molecule in the magnetic field. For example, the chemical shift when the magnetic field is perpendicular to the carbonyl plane is affected only by the σ and σ^* states. The σ-σ^* excitation energy is large, and the overall paramagnetic chemical shift is small. The carbon is highly shielded for this orientation.

On the basis of excitation energies alone, the carbonyl carbon would be predicted to be most deshielded when the magnetic field is directed along the C=O bond. The greatest contribution to the paramagnetic shift for that orientation comes from the n-π^* transition. However, the relative isolation from the nucleus of the n electrons attenuates the effect of the large excitation energy, and the deshielding for this orientation is only moderate, although highly variable. The greatest shielding occurs for the magnetic field perpendicular to the carbonyl bond in the carbonyl plane.

Grant and co-workers [266] have used a similar analysis to explain the chemical-shift tensor for olefins. An important aspect of such studies on solids is that they help to explain the averaged chemical shifts observed in liquids.

7 HYDROGEN BONDING

Berglund and Vaughan [267] compiled data to demonstrate that the isotropic proton chemical shift $\bar{\sigma}$ (the average chemical shift) correlates with the O—O distance R in hydrogen bonds as shown in Figure 5.15. The correlation of chemical shifts and hydrogen-bond length may even be better than it appears in Figure 5.15 because the scatter in points on the graph is comparable to the errors in the measurements of the chemical shifts.

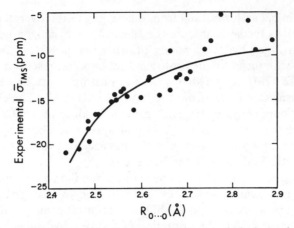

Figure 5.15 Plot of the isotropic proton chemical shift $\bar{\sigma}$ versus the hydrogen-bond distance $R_{O\cdots O}$ for various solid compounds. The solid line is based on a theoretical prediction of the chemical shift. Reprinted with permission from B. Berglund and R. W. Vaughan, *J. Chem. Phys.*, **73**, 2037 (1980).

The correlation is surprising in some respects, however. The chemical-shift anisotropy $\Delta\sigma$ [defined as $\sigma_{33} - (\sigma_{11} + \sigma_{22})/2$] for a typical hydrogen-bonded proton is much greater than the total range of isotropic shifts. A proper correlation should take into account the properties of the entire chemical-shift tensor rather than just the isotropic shift. In practice, the correlation of the chemical-shift anisotropy, $\Delta\sigma$, with the O—O distance R (Figure 5.16) is much worse than is that of the isotropic chemical shift.

Lau and Vaughan [268] proposed that $\Delta\sigma$ is more sensitive to the presence of nearby atoms and ions than is the isotropic shift. The variation in the

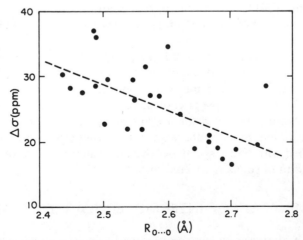

Figure 5.16 Plot of the proton chemical-shift anisotropy $\Delta\sigma$ versus hydrogen-bond distance $R_{O\cdots O}$ for various solid compounds. The dashed line is a least-squares fit through the points. Reprinted with permission from B. Berglund and R. W. Vaughan, *J. Chem. Phys.*, **73**, 2037 (1980).

anisotropic chemical shift resulting from differences in the overall environment around the nucleus tends to obscure the effect of the hydrogen bonding.

One of the strongest hydrogen bonds known is that in KHF_2. Determination of the chemical-shift tensor of the proton in that bond represents a tour de force of multiple-pulse NMR spectroscopy, as the protons must be decoupled not only from themselves but also from the flouride ions [269]. The chemical-shift tensor is nearly axially symmetric with principal values of -39.7, -32.4, and 8.8 ppm relative to TMS. This is the largest spread of chemical-shift–tensor values ever found for a proton. As is usually the case, the proton is most shielded when the magnetic field is along the bond axis.

Less quantitative work has been done with carbon than with proton chemical shifts for hydrogen-bonded structures, but it is known that protonation of a carbonyl oxygen produces an upfield shift in the central principal element of the chemical-shift tensor [270]. The formation of a strong hydrogen bond to other types of oxygen atoms attached to a carbon in the solid phase has been found to shift the isotropic shift of the carbon atoms to lower frequency [271].

The quadrupolar coupling constant is an obvious source of information about deuteriums in hydrogen bonds. The quadrupole-coupling tensor is usually almost axially symmetric, with the symmetry axis pointing along the O—D bond [272]. The magnitude of the quadrupole coupling constant, which reflects the size of the electric field gradient at the deuterium nucleus, correlates with the O—D distance according to the following empirical equation:

$$eQq/h = 310.0 - 3.0[190.6/r^3] \qquad (24)$$

where r is the O—D bond length, and the units are expressed in kilohertz [247].

When the hydrogen bond connects two ^{15}N sites, the heteronuclear dipolar coupling between the nitrogen atoms can be used to test the symmetry of the hydrogen bond [241]. When the nitrogen atoms are nonequivalent, two separate heteronuclear dipolar couplings that describe both N—H bond lengths can be obtained.

Water molecules that are involved in hydrogen bonding can undergo 180° flips. When the flips of deuterated molecules occur rapidly on the scale set by the quadrupolar spectrum, the deuterium is effectively in a site without axial symmetry [273]. If the sample is cooled, the averaging can be slowed so the point that the results expected for a static sample are obtained [274]. Thus dynamic information about hydrogen-bonded structures in the solid state can also be obtained with NMR spectroscopy. The general subject of averaging of the NMR spectra of solids is covered in Section 11.

8   DETERMINATION OF INTERMOLECULAR STRUCTURE

Electrons in a molecule containing an observed nucleus play the most important role in determining the chemical shift. That intermolecular interactions nevertheless play a role is illustrated by the long-chain aliphatic compounds. The resonances from the interior carbon atoms of linear aliphatic compounds of various lengths are shifted by over 0.6 ppm from those of the corresponding carbon atoms in the C_{20} compound because the C_{20} compound has a different crystalline habitat than do the other compounds [143].

Packing effects sometimes cause normally equivalent nuclei within the same molecule to give rise to resolved signals in the solid. In other cases nuclei in inequivalent molecules in the unit cell give rise to resolved NMR signals. We treat these two cases separately in the next two sections.

8.1 Asymmetry of Crystal Sites

The three methyl groups in solid 1,3,5-trimethoxybenzene give rise to resolved carbon resonances [275]. Whereas in solution the time-averaged environments of the methyl groups are equivalent, the local environments at the methyls in the solid are different. The X-ray structure shows that the methoxy groups are twisted by different angles with respect to the aromatic ring. For carbon atoms the lattice probably affects chemical shifts primarily by the way it modifies the local conformation of a molecule.

Ring currents in the aromatic rings of aromatic molecules can significantly alter proton chemical shifts [276]. However, for nuclei other than protons intermolecular shielding from ring currents is much too small to account for observed intermolecular chemical shifts.

Lattice effects on carbon chemical shifts now appear to be limited to 2 or 3 ppm. Because only a few examples have actually been reported [277, 278], this range may expand in the future.

8.2 Crystallographic Nonequivalence of Molecules

When there are no chemically equivalent carbon atoms within the individual molecules, multiple signals for any given type of carbon must result from the presence of molecules in the unit cell that are not related by symmetry. In a simple case, multiple signals were observed in the ^{13}C spectrum of calcium formate, taken with magic-angle spinning [279]. In another example, the methyl signal for 2,4-dinitrotoluene gave two signals separated by almost 3 ppm. There are two noncongruent molecules in the unit cell of 2,4-dinitrotoluene [158].

The resonances of pure tartaric acid are chemically shifted from those of the racemic mixture [280]. The solid-state NMR spectrum provides a novel means of measuring the optical purity of this substance.

8.3 Structure of Single Crystals

The spectra of single crystals reflect the structure of the crystal more strongly than do the spectra of powders spinning at the magic angle. For a given orientation of a single crystal symmetrically equivalent carbon atoms may give rise to resolved NMR resonances, whereas with magic-angle spinning the two nuclei absorb at the same frequency. For example, for an arbitrary orientation of a single crystal of L-histidine hydrochloride monohydrate there are four signals in the ^{15}N NMR spectrum [281]. For histidine there are four molecules in the unit cell.

Single crystals allow determination of the orientation of the PAS of the chemical-shift tensor with respect to the molecule (Section 6.4). The orientation of the shift tensor relative to the molecule may vary somewhat from the orientation expected on the basis of molecular symmetry alone. This again is a crystal-packing effect. For protons ring currents in neighboring molecules may tilt the chemical-shift tensor away from the orientation predicted on the basis of molecular symmetry [276]. The ^{13}C chemical-shift tensor in oxalic acid dihydrate is twisted from the molecular plane by 20° [270]. In this case, the reason is unknown, as other oxalic acid derivatives have normal chemical-shift tensors.

Resonance shifts caused by the quadrupolar interaction and dipolar coupling may also reflect the structure of the crystal. For example, in deuterated potassium bicarbonate, two different quadrupole splittings occur for a given orientation of the single crystal [282]. Likewise, separate dipolar splittings for each of the four molecules in the unit cell of $[1,2-^{13}C_2]$glycine can be observed [222].

9   CHARACTERIZATION OF MORPHOLOGY OF MACROMOLECULAR SAMPLES

Typically a sample of a synthetic polymer contains both highly ordered, or crystalline, regions interspersed with disordered, or amorphous, regions. Solid-state NMR is an important technique for measurement of the relative amounts of the two phases present. Under favorable circumstances an estimate of the domain size is possible.

Block copolymers and polymer blends often contain separated domains differing in chemical composition. Solid-state NMR is also an important method for the study of these materials.

9.1   Proton Nuclear Magnetic Resonance Spectroscopy

Molecules in amorphous regions of polymer samples usually are much more mobile than are those in structured, crystalline regions. As a result, nuclear relaxation processes for protons in the amorphous regions are usually faster than are those for nuclei in the crystalline regions. If the protons in the different regions relax independently and if the relaxation times for the two regions are sufficiently different, the overall relaxation curves can be decomposed into signals for the individual regions. The relative magnitude of the curves at time 0 provides a measure of the abundance of protons in each domain [283, 284].

Sometimes protons in phase-separated regions of amorphous polymers relax together as a result of communication between the two regions by spin diffusion (see Section 3.7). The effectiveness of the spin-diffusion process gives important information about the sizes of the local domains. Because spin diffusion affects different types of relaxation processes in different ways, various types of information about polymer morphology can be obtained through a choice of NMR experiments.

Averaging of relaxation times occurs efficiently if the equilibration process is fast compared to the difference in the rates for the particular relaxation process being studied Spin–lattice relaxation in the rotating frame for protons in synthetic polymers typically requires 1–100 ms. Domain sizes must be on the order of 50 Å if the rate of spin diffusion is to be fast enough to average most spin–lattice relaxation times in the rotating frame. Spin–lattice relaxation rates in the laboratory frame are often more than an order of magnitude slower than spin–lattice relaxation rates in the rotating frame. Consequently, the domain sizes only have to be less than several hundred angstroms in size for efficient averaging to occur [283]. Averaging of the rates of spin–spin relaxation [285] or even the rates of relaxation of states of *dipolar order*, in which the nuclear spins are aligned in the local fields created by neighboring nuclear spins, [286, 287] are sometimes observed. These relaxation measurements are sensitive to still other domain sizes.

Direct measurement of the rate at which magnetization is transferred from one domain to another provides the most detailed information about domain sizes. A direct measurement is possible if the spin systems for the different regions can be prepared with different spin temperatures. The Goldman–Shen method [288, 289] takes advantage of differences in the rate of decay of transverse magnetization created by an initial 90° pulse. Magnetization that is completely decayed after a period of time τ is unaffected by a second 90° pulse; its spin temperature after the pulse is very high. Slowly decaying magnetization is rotated back along the z axis by the second pulse; its spin temperature after the pulse is close to the initial temperature. After the second pulse, the two spin systems move toward a common spin temperature by exchange of magnetization. The extent to which the two systems have equilibrated after a delay following the second pulse can be detected with a third 90° pulse that again generates transverse magnetization that can be detected.

In most materials the signals of protons in different regions of the sample are unresolved. Consequently, it is impossible to manipulate the corresponding systems independently to create different spin temperatures. Furthermore, detection of the rate of equilibration of the spin temperatures is very difficult if the NMR signals are unresolved. Multiple-pulse techniques effectively increase the resolution in the proton spectrum in favorable cases [290] and allow the necessary preparation and detection of the states of the spin systems to be made [291]. A two-dimensional NMR experiment can be done to correlate the initial and final frequencies of the exchanging components of magnetization [292]. The experiments are very demanding, however, and are unlikely to find widespread application.

9.2   Carbon Nuclear Magnetic Resonance Spectroscopy

9.2.1   Carbon–Hydrogen Magnetization Transfer

Transfer of longitudinal magnetization among nuclei with different gyromagnetic ratios is normally prohibited because energy is not conserved. However, energy can be conserved in a transfer when the nuclei interact with their environments as well as with each other. For example, molecular motion can make up the energy difference if it modulates the interactions among the nuclei at frequencies comparable to the energy separation. Cross relaxation between unlike nuclei is well known and is sometimes referred to as the *transient nuclear Overhauser effect*. The occurrence of cross relaxation between two different types of nuclei does require that they be in close contact. Cross relaxation between nuclei in different molecules can be used as a test of the extent that the two types of materials are mixed at the molecular level [293].

The cross-polarization procedure used to generate the carbon spectrum of a solid is a heteronuclear transfer of spin magnetization in which the irradiation makes the process energy conserving. Because cross polarization requires strong dipolar interactions among the nuclei, it can be used in special cases as a probe of polymer compatibility. If a molecule is completely deuterated, ^{13}C signals from it can only be generated from intermolecular carbon–proton dipolar interactions [294–296]. This fact provides the basis of powerful means of exploring the nature of the interactions between protonated and deuterated species.

9.2.2   Carbon–Carbon Magnetization Transfer

Magnetization transfer among the ^{13}C nuclei in an unenriched sample occurs on a timescale of seconds because there is a long average distance between ^{13}C nuclei. Carbon spin–lattice relaxation must be even slower for transfer from carbon to carbon to compete effectively with spin–lattice relaxation. Nevertheless, carbon–carbon magnetization has been observed in nonenriched samples, both with and without magic-angle spinning [297–302].

The rate of spin transfer increases dramatically in isotopically enriched materials. This fact allows carbon–carbon magnetization transfer to be detected in materials that have inherently short carbon spin–lattice relaxation times, such as synthetic polymers [303, 304]. Observation of magnetization transfer between carbon atoms in polymers is an indication that the atoms involved are separated by less than 10 Å. This distance is comparable to molecular sizes. Thus carbon magnetization transfer has the potential for giving specific information about molecular packing [305, 306].

The potential importance of magnetization transfer between rare nuclei to elucidate intermolecular structure in solids has stimulated several studies into the spin dynamics involved [307–309]. For two carbon atoms absorbing at different frequencies, transfer can be energy conserving only with the participation of an external spin reservoir. Because magnetization transfer is allowed to occur with the proton decoupler off, the proton spin system usually absorbs the energy difference between the carbon atoms. The rate of transfer is a complicated function of the frequency difference between the carbon atoms involved and the strength of their interaction with the proton reservoir. The complicated nature of the process makes only semiquantitative determinations of internuclear distances possible.

It has been suggested that magic-angle spinning can be used to remove some of the complications of the spin-dynamics in carbon–carbon magnetization transfer [310, 311]. Careful choice of the spinning frequency overlaps the spinning sidebands of the exchanging carbon atoms. In this case the magic-angle spinning serves to create an energy degeneracy in some of the carbon transitions so that magnetization transfer can be energy conserving. The role of the proton system is much reduced so that quantitative analysis of the transfer rate may be more straightforward.

9.2.3 Proton–Proton Magnetization Transfer Measured With Carbon Spectroscopy

The resolution of the carbon spectrum provides a secondary advantage. With cross polarization one can use carbon NMR to measure separated relaxation times for unresolved proton signals [312, 313].

A typical pulse sequence [314] is illustrated in Figure 5.17. The proton system is prepared in a nonequilibrium state by the initial 180° pulse. Following partial relaxation of the proton system during a time τ, a standard cross-polarization procedure is used to generate the carbon spectrum. Unequal relaxation rates of the amorphous and crystalline portions of a polymer sample

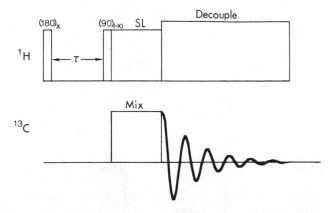

Figure 5.17 Typical pulse sequence used to monitor proton nuclear relaxation through the observed carbon signals. The protons relax after inversion during the time τ. Spin-locked magnetization is then transferred to the carbon nuclei following a 90° pulse to the protons. The intensity of the resulting carbon signal reflects the magnitude of proton magnetization remaining after the relaxation period.

are readily detected by differences in the decay rates of the carbon signals for the individual regions. The reader may readily imagine variations on the same scheme that measure proton relaxation in the rotating frame and spin–spin relaxation.

9.2.4   Illustration of a Use of ^{13}C Nuclear Magnetic Resonance Spectroscopy To Study Polymer Morphology

Although the basic chemical structure of cellulose is known (Figure 5.18), it is difficult to establish details of the structure. Furthermore, there are several common cellulose polymorphs. The native material is cellulose I. It is usually highly crystalline. Regeneration of cellulose from solution usually leads to cellulose II. At high temperatures, cellulose IV results [315].

The ^{13}C spectra of the various polymorphs of cellulose, taken with magic-angle spinning, are distinctly different [316]. In cellulose II the signals for both C-1 and C-4 are partially resolved into two resonances of equal intensity (Figure 5.19). Additional broad tails (for C-6) and even a partially resolved signal (for C-4) occur for cellulose I [317–320].

There appear to be at least two sources of the additional signals in the spectra of cellulose. The splitting of the signal for C-1 is probably a fundamental feature of the structure of crystalline cellulose II. Originally it was proposed that adjacent anhydroglucose units in the cellulose structure are nonequivalent [318]. Further work demonstrated that the splitting also occurs for oligomeric cellulose models containing as few as four anhydroglucose units [315]. In such small structures two lines of equal intensity should not appear if the monomer units alternate in structure. The splittings probably result from the packing of the cellulose molecules.

The broad signals and tails to the signals seen in the spectrum of cellulose I appear to come from amorphous portions of the sample. The intensities of the portion of the signal occurring as a tail of the main resonance decrease as the crystallinity of the sample increases [320]. Earl and VanderHart [319] showed that the upfield tail for C-6 had shorter spin–spin and spin–lattice relaxation times than did the main peak, indicating that the carbon atoms giving rise to it are in a more mobile environment than are those in the bulk of the material. These carbon atoms may simply be in the amorphous regions of the sample or

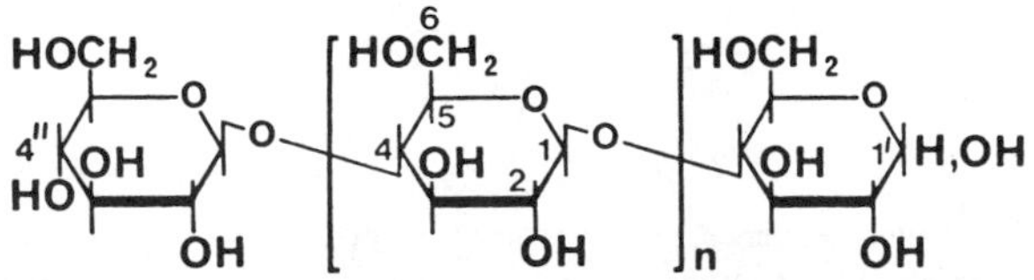

Figure 5.18   Basic structure of cellulose. Reprinted with permission from R. L. Dudley, C. A. Fyfe, P. J. Stephenson, Y. Deslandes, G. K. Hamer, and R. H. Marchessault, *J. Am. Chem. Soc.*, **105**, 1505 (1983). Copyright 1983 American Chemical Society.

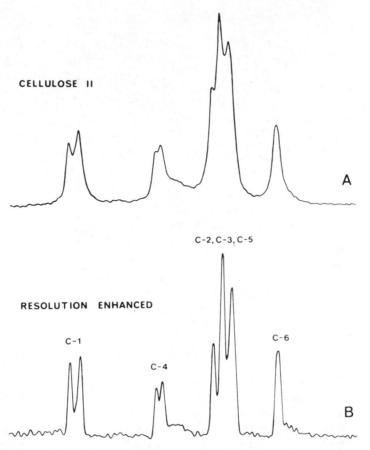

Figure 5.19 Experimental ^{13}C NMR spectra of cellulose II taken at 22.6 MHz with magic-angle spinning. The spectrum at the top (A) was acquired after cross polarization for 1 ms with 1-s delays between scans. The spectrum at the bottom (B) was derived from that at the top by resolution enhancement. Reprinted with permission from R. L. Dudley, C. A. Fyfe, P. J. Stephenson, Y. Deslandes, G. K. Hamer, and R. H. Marchessault, *J. Am. Chem. Soc.*, **105**, 1505 (1983). Copyright 1983 American Chemical Society.

may be influenced by more subtle aspects of the morphology of the cellulose that is not yet clarified [321–323].

9.3 Fluorine Nuclear Magnetic Resonance Spectroscopy

At 259°C the ^{19}F spectrum of poly(tetrafluoroethylene) that has been narrowed with a multiple-pulse sequence can be decomposed into broad and narrow components [324] (Figure 5.20). The crystalline portions of the sample give rise to the pattern expected for nuclei in an axially symmetric environment. The greater degree of motion in the amorphous parts of the sample averages its spectrum into a relatively sharp line. With cooling, the signal for the amorphous

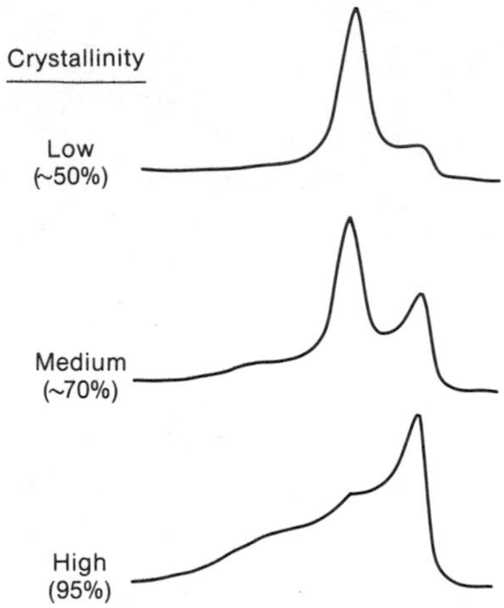

Crystallinity

Low
(~50%)

Figure 5.20 Multiple-pulse-^{19}F spectra of samples of poly(tetrafluoroethylene) differing in the degree of crystallinity. The relatively sharp line in the center of each spectrum derives from the amorphous portion of the sample, whereas the broad resonance derives from the crystalline portion. Reprinted with permission from A. D. English and A. J. Vega, *Macromolecules*, **12**, 353 (1979). Copyright 1979 American Chemical Society.

Medium
(~70%)

High
(95%)

part of the sample results in a partially averaged spectrum. Vega and English [324, 325] used a numerical procedure to decompose the high-temperature spectrum into crystalline and amorphous parts. The degree of crystallinity determined in this fashion agreed well with that obtained by other methods.

9.4   Deuterium Nuclear Magnetic Resonance Spectroscopy

Reorientational motions have pronounced effects on the deuterium spectra of solids. It is not surprising then that the deuterium spectrum for the amorphous regions of a partially crystalline polymer is motionally narrowed while the spectrum of the crystalline parts are broad. In some cases the relaxation times differ sufficiently that complete relaxation of the fast component occurs before there is much change at all in the slow component. It is then possible to separate the spectrum of the slow component from that of the fast component on the basis of the differences in relaxation rates [326, 327].

10   DETERMINATION OF MOLECULAR ORIENTATION

The individual chains in polymer samples often are preferentially oriented in space. The orientation dependence of the NMR interactions is a powerful means for one to measure the degree of molecular orientation in such samples and to relate this orientation to the bulk physical properties that are affected by it.

Proton NMR spectroscopy has been of rather limited value for determination of molecular orientation. Proton spectra of solids are dominated by many-body, dipolar interactions. Typically only second and fourth moments (see Section 3.8) of the proton spectrum can be measured reliably, and these provide only a limited description of molecular orientation.

In contrast to the spectra for protons, the spectra for carbon and deuterium nuclei are usually controlled by single-particle interactions—the chemical-shift anisotropy and quadrupole coupling, respectively. Accordingly, interpretation of the spectral features of these nuclei in terms of molecular orientation is relatively straightforward.

We consider first in this section some general aspects of the orientation dependence of an NMR spectrum. We then discuss some specific examples of the use of dipolar couplings, the chemical-shift anisotropy, and the quadrupolar interaction for determination of molecular order in oriented samples.

10.1   Mathematical Principles

It is necessary to use several different reference coordinate systems to deal with the NMR of oriented materials. Each NMR interaction is most easily defined in terms of its principal axis system (PAS). However, for the chemical and quadrupolar interactions, there is a separate PAS for each nucleus in the molecule, and for dipolar coupling there is a separate PAS for each pair of nuclei. Thus it is also useful to define a *molecular axis system* for the entire molecule. We are interested in the orientation of the molecules in the sample, so we define a *sample axis system* fixed in the bulk sample. Finally, we have the *laboratory frame*, with its z axis along the direction of the magnetic field.

Rotation from one coordinate system into another is completely specified by three Euler angles, for which we use the same definitions as Spiess [13]. The first rotation is positive in a counterclockwise direction about the original z axis, the second rotation is about the resulting y axis (when one looks down the axis toward the origin), and the final rotation is about the z axis resulting after the second rotation.

When the nuclear Hamiltonian is expressed in terms of the operators I_x, I_y, and I_z, the changes in the NMR interaction tensor for rotation of the reference axis system are given by equations similar to (4). More compact expressions result when the Hamiltonian is expressed in terms of spherical tensor operators. The Hamiltonian for each NMR interaction in terms of spherical tensor operators is given by an equation of the form

$$\mathcal{H} = C \sum_{j=0}^{2} \sum_{m=-j}^{j} (-1)^m T_{jm}^k R_{j-m}^k \tag{25}$$

where C is defined for the particular NMR interaction; T_{jm}^k is a *spin tensor*, also defined for the particular interaction; and R_{j-m}^k is a *coupling tensor* [7]. In the principal axis system (see Section 2.3) the coupling tensor is represented by the symbol ρ_{jm}. The complete Hamiltonian is the sum of such terms over all nuclei or pairs of nuclei.

Transformation of spherical tensor operators on rotation from one reference frame to another is made with Wigner rotation matrices. For example, the Hamiltonian in a frame that has an orientation with respect to the PAS described by the set of Euler angles Ω is

$$\mathcal{H} = C \sum_{j=0}^{2} \sum_{m=-j}^{j} \sum_{m'=-j}^{j} (-1)^m T_{jm}^k D_{m'-m}^{(j)}(\Omega) \rho_{jm'}^k \tag{26}$$

where the $D_{m'-m}^{(j)}(\Omega)$ are elements of the Wigner rotation matrices [7, 13].

The Wigner rotational matrices are introduced here because they provide a convenient way to describe orientational order of polymeric samples [328, 329]. Let us define a probability $P(\Omega)$ that the molecular coordinate system of a given molecule is oriented in the sample coordinate system with the set of Euler angles Ω. The probability function $P(\Omega)$ can be expanded in terms of the Wigner rotation matrix elements.

$$P(\Omega) = \sum_l \sum_m \sum_n P_{lmn} D_{mn}^{(j)}(\Omega) \qquad (27)$$

The coefficients P_{lmn} are called *moments of the probability distribution* and are defined by the relation

$$P_{lmn} = 8\pi^2/(2l+1) \int D_{mn}^{(l)}(\Omega) d\Omega \qquad (28)$$

Because even in a highly symmetric sample a large number of moments may have significant values, methods that give information about high-order moments are important. Unfortunately, most techniques give information only for a few of the low-order moments [330]. Methods have been devised with which one can estimate the size of the higher terms based on a statistical assumption [331], but direct measurement is clearly more desirable. Only P_{lmn} with $l = 2, 4, 6$, and 8 can be found from proton second-moment measurements. In theory, however, all of the moments with even values of l can be measured with NMR. The restriction to even-order moments reflects the fact that inversion of a sample leaves the NMR spectrum unchanged.

Spiess and co-workers [332, 333] showed that, in simple cases, there is a direct correlation between certain types of spectral features and the orientational moments. In fact, when the NMR interaction has axial symmetry and there is an axially symmetric distribution of the molecular axes about the sample axis, subspectra can be calculated that correspond to each of the P_{l00}. The subspectra can be combined to match experimental spectra in proportions that are directly proportional to the moments P_{l00}. All other moments are zero because of the axial symmetry. The shape of these subspectra are illustrated in Figure 5.21.

It is difficult to calculate subspectra corresponding to each orientational moment when either the orientational probability function or the NMR coupling interaction lacks axial symmetry. Furthermore, as mentioned previously, the expansion series does not converge rapidly for highly oriented samples. Thus even when subspectra can be determined, the expansion in terms of rotational matrix elements is not always convenient.

Expansion of the orientational probability in terms of functions other than the rotational matrix elements offers an alternative [334, 335]. For example, for samples with axial symmetry the probability function can be expanded in terms of *planar distributions*. For each planar distribution one of the principal axes of the coupling tensor in each molecule lies normal to the plane. The molecules are

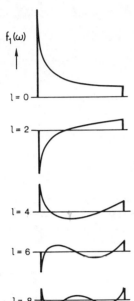

$f_1(\omega)$

$l = 0$

$l = 2$

$l = 4$

$l = 6$

$l = 8$

$\omega_\perp$ 0 $\omega_\parallel$

$\longrightarrow \omega$

Figure 5.21 Subspectra used for generating the NMR spectra of symmetric oriented samples when the NMR interaction also has axial symmetry. Each spectrum corresponds to one of the moments P_{100} as explained in the text. Reprinted with permission from H. W. Spiess, *NMR: Basic Princ. Prog.*, **15**, 55 (1978).

uniformly distributed in the plane. The orientation of each planar distribution is described by the angle that its plane normal makes with the symmetry axis of the sample. The total distribution is the weighted sum over all possible values of this angle. If a Gaussian distribution is assumed, the standard deviation of this angle must be found.

The theoretical spectrum for each of the planar distributions can be expressed analytically. We outline the derivation of Opella and Waugh [336] of an analytical expression for the spectrum of a planar distribution. The same type of pattern results whether the chemical shift is axially symmetric or asymmetric.

The chemical shift is given in terms of polar angles θ and ϕ, which define the orientation of the magnetic field in the principal axis system.

$$\sigma = \sigma_{33}\cos^2\theta + \sigma_{11}\sin^2\theta\cos^2\phi + \sigma_{22}\sin^2\theta\sin^2\phi \tag{29}$$

For the molecules constituting a single planar distribution, the value of θ is fixed. It is then convenient to express the chemical shift in terms of the variable angle ϕ as shown in (30–32).

$$\sigma = \bar{\sigma}(\theta) - \sigma'(\theta)\cos 2\phi \tag{30}$$

$$\bar{\sigma}(\theta) = \sigma_{33}\cos^2\theta + \tfrac{1}{2}(\sigma_{11} + \sigma_{22})\sin^2\theta \tag{31}$$

$$\sigma'(\theta) = \tfrac{1}{2}(\sigma_{22} - \sigma_{11})\sin^2\theta \tag{32}$$

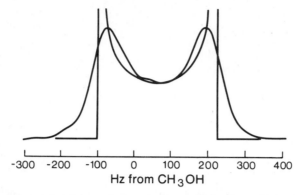

-300   -200   -100   0   100   200   300   400

Hz from CH_3OH

Figure 5.22 Experimental and theoretical spectra for a sample of stretched polyethylene with the draw axis perpendicular to the magnetic field. The experimental spectrum was obtained by summation of a two-dimensional spectrum but reflects only the effect of the chemical-shift anisotropy. Reprinted with permission from S. J. Opella and J. S. Waugh, *J. Chem. Phys.*, **66**, 4919 (1977).

There is a $1:1$ correspondence between the chemical shift and ϕ. The variables can be changed with the methods described by Haeberlen [7], and a little algebra gives

$$l(\sigma) = \sigma'/(\sigma' - \bar{\sigma} + \sigma)^{1/2}(\sigma' + \bar{\sigma} - \sigma)^{1/2}$$

$$l(\sigma) = 0 \quad \text{for} \quad \sigma > \sigma' \quad \text{or} \quad \sigma < \bar{\sigma} - \sigma' \quad (33)$$

The spectral intensity calculated with (33) goes to infinity at $\bar{\sigma} - \sigma'$ and $\sigma' + \bar{\sigma}$. A spectrum for an oriented fiber of polyethylene is shown in Figure 5.22 along with the calculated line shape based on (33).

Expansion of the orientation probability function in terms of other functions is also possible, depending on the symmetry in either the distribution or the coupling interaction. The choice of functions becomes increasingly more difficult as the symmetry decreases.

10.2   Determination of Molecular Orientation With Dipolar Couplings

In his classic work on the dipolar coupling between the protons in the water of hydration in gypsum, Pake [217] showed that the dipolar interaction could be used to get information about the orientation of the internuclear vector connecting the two water protons. The occurrence of two coupled but otherwise isolated nuclei is rare, however. We have already alluded to the limited information that can be obtained from the second and fourth moments of the proton spectra of most materials.

Despite the problems encountered when one is dealing with systems containing many nuclei, proton NMR has given useful results about orientation in various synthetic polymer samples [337–339]. Determination of polymer orientations from the dipolar coupling between other nuclei, such as would occur in compounds doubly labeled with ^{13}C, are not yet reported.

10.3   Determination of Molecular Orientation With Anisotropic Chemical Shifts

Orientation studies based on chemical shift are made with nuclei other than protons. The chemical-shift anisotropy of protons is small, and multiple-pulse techniques to remove the dipolar broadening among the protons from the spectrum are tedious. Furthermore, because most compounds contain several different types of protons, the powder patterns for each type of proton are highly overlapped.

For biological samples ^{31}P NMR is often useful [340]. Phosphorous gives a relatively strong NMR signal and has a large chemical-shift anisotropy. In many biological samples there are only a few different types of phosphorous atoms in the sample of interest so that overlap of the bands is not a problem.

For typical organic compounds carbon NMR is the reasonable tool with which to attack problems of orientation. Carbon NMR is limited because spectral overlap of signals from chemically different carbon atoms is usually difficult to avoid. In selected cases, however, it is possible to enrich a compound with ^{13}C to enhance the signal from one site. The importance of being able to determine orientation in unlabeled polymer samples prepared under standard manufacturing conditions restricts this approach. Fortunately, methods that allow sorting out highly overlapped chemical-shift powder patterns are available (Section 4.3). These techniques make it possible to study orientation in samples with ^{13}C in natural abundance. We outline the general approach.

The principal values of the chemical-shift tensor for the carbon atoms of interest must first be measured. The problem of spectral overlap can sometimes be circumvented by an experiment in which the sample is spun about an axis oriented slightly off the magic angle or through an analysis of the spinning sidebands in a spectrum from a sample slowly spinning at the magic angle. Both approaches were used for measurement of the chemical-shift anisotropy in polyesters [341, 342]. The ^{13}C spectrum of poly(methyl methacrylate) was decomposed by taking advantage of differences in the relaxation properties of the different carbon atoms in the sample to suppress selectively certain signals in the spectrum [343].

The orientation of the chemical-shift tensor relative to the molecular axis system must now be determined. The determination is made most precisely by measurement of the rotation pattern of the chemical shift from a single crystal whose structure can be determined independently with X-ray diffraction (see Section 6.4). Single crystals are not available for most polymers. Thus the orientation of the chemical-shift tensors must be assumed from the known orientation of the PAS in closely related compounds. Fortunately, one of the axes of the principal axis system of the chemical shift often lies close to one of the

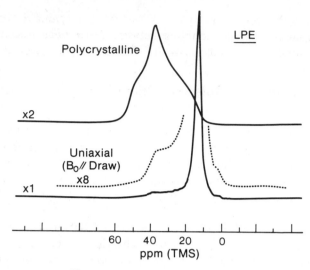

Figure 5.23 Experimental ^{13}C NMR spectra of linear polyethylene. The upper spectrum is characteristic of an unoriented melt-crystallized specimen. The lower spectrum is that of fibers drawn to 15 times their original size. The dashed line is the lower spectrum magnified eightfold. Reprinted with permission from D. L. VanderHart, *Macromolecules*, **12**, 1232 (1979). Copyright 1979 American Chemical Society.

bonds in the molecule so that a reasonable guess as to the orientation of the tensor can be made. The final step is to analyze the spectrum of the oriented material in terms of weighted probabilities that the various orientations are present in the sample.

We illustrate the use of ^{13}C NMR with an example from the work of VanderHart [344] on oriented polyethylene. Figure 5.23 shows a comparison of the cross-polarization spectra of an isotropic sample of polyethylene and a fiber whose draw axis is parallel to the magnetic field. The spectrum of the unoriented material is the typical chemical-shift pattern for a nonaxially symmetric carbon. The spectrum of the oriented material is concentrated at one edge of this pattern. Most of the CH_2 units are oriented with the chain axis parallel to the magnetic field. There is a broad background signal from the less highly oriented parts of the sample that can be seen more clearly in the dotted expansion.

Most of the highly oriented polyethylene is crystalline and relatively immobile. The spectrum of the amorphous region alone can be obtained by a pulse sequence that selects signals from those nuclei whose spin–lattice relaxation times are relatively short. Spectra of this type are shown in Figure 5.24 for a series of samples annealed under various conditions. The amorphous regions of the sample are much less highly oriented than are the crystalline regions.

The information content of a one-dimensional NMR spectrum is limited. For example, the spectrum is insensitive to molecular orientation about the direction of the magnetic field. The combination of data from several experiments in

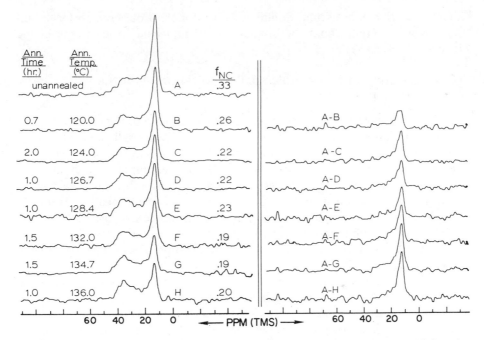

Figure 5.24 Carbon spectra of the noncrystalline portion of polyethylene drawn to 15 times its original size and then annealed under various conditions as shown. At the right are spectra obtained as differences between the spectra of the annealed and unannealed samples. Reprinted with permission from D. L. VanderHart, *Macromolecules*, **12**, 1232 (1979). Copyright 1979 American Chemical Society.

which spectra are recorded for various orientations of the sample in the magnetic field improves the situation. However, for samples having a distribution of molecular orientations, it is usually impossible to identify that those molecules giving rise to signals of one frequency for a given sample orientation correspond to those giving some other frequency for a second molecular orientation. A two-dimensional NMR experiment in which the sample orientation is reoriented dynamically during the course of the time evolution of the magnetization does correlate the frequencies [345]. A three-dimensional NMR experiment is capable in principal of supplying complete orientational information about amorphous samples such as synthetic polymers.

An alternative type of two-dimensional experiment is possible with magic-angle spinning [346, 347]. After a single excitation pulse (or pulses) a particular nucleus gives rise to spinning sidebands whose phase depends on the orientation of the molecule containing the nucleus at the beginning of the experiment. If a series of experiments are performed in which the spinner is allowed to rotate various lengths of time following a reference point, the time development of these phases can be followed in a two-dimensional experiment. The sideband intensities in the spectrum resulting from double Fourier transformation contain information about the distribution of molecular orientations in the sample.

10.4   Determination of Molecular Orientation With the Quadrupolar Interaction

Of the various NMR interactions that are sensitive to molecular orientation, the quadrupolar interaction of the deuterium nucleus is potentially one of the most useful. The quadrupole coupling interaction for deuterium is much larger than either the chemical-shift anisotropy or the dipolar coupling interactions. The sensitivity to orientational differences among samples is thus high. Of course, labeled compounds must be used, but this feature can be turned to advantage. The deuterium can be placed selectively in that portion of the molecule for which the orientation is most likely to be important. Deuterium NMR is used to study orientation and motion in both biological compounds [348] and synthetic polymers.

An interesting use of deuterium NMR was reported for paramagnetic proteins [349, 350]. Because they have a large bulk susceptibility, the microcrystals of these materials can be oriented by the magnetic field of the NMR spectrometer. The orientation shows up as well-resolved resonances in the deuterium spectrum of suitably labeled samples, whereas the unoriented compound gives the usual Pake pattern, which contains only a partially resolved doublet. The quadrupolar interaction for deuterium is so large that apparently any broadening in the spectrum that may result from the paramagnetic center is insignificant. The ability to orient these samples effectively gives the ability to do *single-crystal* studies even though individual crystals large enough to give detectable NMR signals cannot be grown.

Observed and calculated spectra for a sample composed of single crystal mats of deuterated polyethylene are shown in Figure 5.25. That the sample is not isotropic is immediately apparent in the spectral changes as the sample is rotated relative to the magnetic field [333].

The spectra in Figure 5.25 were simulated with a sum of subspectra based on the moment expansion of the orientational distribution function. The coefficients in the expansion were chosen to match a Gaussian distribution of the angle of tilt of the polymer chains from the symmetry axis of the sample. As is shown in the figure, very good fits resulted with the inclusion of only eight terms in the expansion. Best fit resulted with a standard deviation in the angle of tilt of $12°$. A separate calculation showed that for this distribution of orientation angles, moments P_{l00} with $l > 8$ are insignificant.

In deuterated samples with a high degree of molecular mobility the observed quadrupolar splittings are the average of the splittings for all orientations that are sampled by the carbon–deuterium bonds during the course of the measurement. For unstretched rubbery polymers the residual splitting is 0 Hz [351, 352]. In spite of chain entanglements, each carbon–deuterium bond can sample all

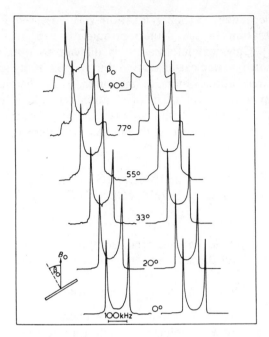

Figure 5.25 Observed and calculated deuterium spectra of single-crystal mats of polyethylene having the plane normal oriented at the angles shown with respect to the magnetic field. The calculated spectra were based on the planar distributions described in the text. Reprinted by permission of the publishers, Butterworth & Co. (Publishers), Ltd. ©, from R. Hentschel, H. Sillescu, and H. W. Spiess, *Polymer*, **22**, 1516 (1981).

possible orientations in space. For onefold extension of the samples residual splittings on the order of 100 Hz occur. This indicates that there is only a very slight molecular ordering as the full deuterium quadrupole splitting can be over 200 kHz. Deuterium NMR is remarkably sensitive to the ordering.

Deuterated polymers are not necessarily required for studies of polymer orientation with deuterium NMR. It is sometimes possible to include a deuterated small molecule in the polymer matrix to supply the NMR signal. Orientation of the small molecule by the polymer matrix is then reflected in the NMR spectrum [353].

10.5   Molecular Orientation and Nuclear Relaxation

Nuclear relaxation results from modulation of one of the orientation-dependent NMR interactions at a rate capable of inducing nuclear transitions [354, 355]. Most commonly the modulation results from molecular reorientation. When the motion is restricted in space, both the magnitude of the interaction and the effect of the motion on the interaction depend on the average orientation of the molecule. Thus the relaxation should be orientation dependent.

Orientation-dependent carbon relaxation was observed in solid benzene [356]. The relaxation of the deuterium nucleus can also be orientation dependent [357]. Chemical information is difficult to obtain from the observation of anisotropic nuclear relaxation, but its possibility should be kept in mind whenever relaxation times are measured from chemical-shift powder patterns in nonspinning samples. The orientation dependence of the deuterium relaxation time can be confused with a distribution of relaxation times resulting from a range of chain environments in amorphous polymers.

11   DYNAMIC PROCESSES AND HIGH-RESOLUTION NUCLEAR MAGNETIC RESONANCE SPECTROSCOPY OF SOLIDS

For convenience we divide dynamic processes into those that average isotropic spectral frequencies, and those that average anisotropic spectral frequencies.

11.1   Processes That Average Isotropic Spectral Frequencies

Conformational interconversions and other chemical processes affect the high-resolution NMR spectrum of a solid in much the same fashion as do comparable processes in the spectrum of a liquid [358]. A very powerful means to characterize the nature of the processes is two-dimensional NMR spectroscopy. A two-dimensional spectrum gives a mapping of the transfer of nuclei by the chemical process [359, 360].

A major advantage of studying dynamic processes in the solid phase is that insoluble materials can be examined, and the studies can be extended to much lower temperatures than would be practical with solutions. Important contributions to the understanding of the structures of organic cations have been made with ^{13}C NMR, both with and without magic-angle spinning [361, 362]. The lowest temperatures used (5 K) would have been impossible for liquid samples [363].

The fluxional behavior of bullvalene was also observed in the solid [364, 365]. The crystal structure strongly influences the molecular rearrangement of bullvalene. Whereas the two fluxional states have equal free energy in the liquid, the local structure in the crystal gives them unequal energies in the solid.

11.2   Processes That Average Anisotropic Spectral Frequencies

Consider an inhomogeneous spectrum composed of orientation-dependent signals from many different molecules. If the molecules interchange orientations during the NMR measurement, the NMR signals are partially scrambled. In the limit of very fast motion, the spectrum is a powder pattern composed of partially overlapping, averaged signals. Simple comparison of the observed pattern with those expected for various types of motional processes can eliminate some of these possibilities from consideration [366, 367]. Analysis of spectra for inter-mediate rates of exchange elucidates both the rate and mechanism of the reorientational process.

An intuitive feeling for the averaging phenomenon can be acquired from a simple example. Assume that two orientations that interchange by rotation about an axis by an angle β can be isolated. Immediately following an excitation pulse there are two types of transverse magnetization; each is associated with one of the two possible orientations. If reorientation is very slow, the two components rapidly separate from each other in the rotating frame and can be detected individually. There are two resolved signals in the frequency spectrum. If reorientation rapidly interchanges the molecular orientations, the mag-netization components precess at the same averaged frequency and are difficult to distinguish. There is a single averaged signal in the frequency spectrum. For a rate of exchange comparable to the frequency difference of the individual magnetization components, interchange leads to rapid decay of transverse magnetization. A broad signal spanning the frequencies of the individual frequencies results.

In a randomly oriented powder, spectral components differing by a variety of frequency values are interchanged by reorientation. The particular selection of frequencies that are interchanged is controlled by the nature of the molecular motion. In principle, the reorientation could be followed at both slow and fast rates because for each rate the motion involves some pair of spectral compo-nents with a comparable frequency separation. In practice it is only the fast motions, with frequencies comparable to the total spectral width, that have a pronounced effect on the overall powder pattern. The spectral changes produced by slower motions are obscured by the overlap of the spectral components.

For solids the analysis of spectral patterns that are partially averaged by motion must be made in light of the fact that many dynamic processes in solids are characterized by a distribution of correlation times. The observed spectral shapes can provide valuable data about inhomogeneous distributions of correlation times in amorphous synthetic polymers (Section 3.7) [368–370].

To make the NMR experiment more sensitive to slow reorientational processes, one needs techniques that effectively isolate the individual spectral components of the inhomogeneous spectrum. Spin–echo methods do essentially this. With the generation of Hahn spin echos generated with 180° pulses, one can measure the decay rates of each portion of a powder pattern controlled by chemical-shift anisotropy (see Section 3.9).

The spectra of individual spectral components can also be isolated by making motional studies on single crystals [371]. Polymer samples drawn into highly oriented fibers also give good results [372]. In either case, interchange among only a few individual orientations can be isolated by the manner in which the sample is oriented.

Spin–echo experiments allow motions occurring at frequencies comparable to either the total spectral width or the width of individual spectral components to be examined. A further set of experiments can be devised that are sensitive to motions occurring on a timescale comparable to the spin–lattice relaxation times. Because spin–lattice relaxation is slow for most solids, these experiments provide a way to study ultraslow motion.

Because the chemical-shift interaction leads to an inhomogeneous spectrum, a narrow hole can be burned into the spectrum with selective irradiation [373]. The rate of disappearance of the hole relates to the rate of interconversion among the various types of nuclei, which differ in the orientation of the molecules in which they are contained. Molecular rotation through small angles results in a slow spread in the width of the hole with time, whereas motions involving large jumps fill in the hole with no overall change in width. This method is limited only by the length of time that the partial saturation of the sample persists, which is determined by the spin–lattice relaxation time.

Three-pulse experiments also facilitate the study of ultraslow motions. We illustrate here a three-pulse experiment for a spectrum that is dominated by the chemical-shift anisotropy. In Section 11.2.5 we describe a comparable experiment for the quadrupolar interaction. The initial pulse simply generates transverse magnetization. For ^{13}C it might be a cross-polarization pulse. The first pulse is followed by an evolution period. The optimum flip angle for the second pulse is 90° when the dominant NMR interaction is the chemical-shift interaction. This pulse serves partially to convert transverse magnetization created by the first pulse back into longitudinal magnetization. Those components of transverse magnetization that are aligned with the direction of the effective field in the rotating frame at the time of the second pulse are unaffected by it. Those that are 90° out of phase with the effective field during the second pulse are completely converted into longitudinal magnetization. Other magnetizations are partially converted. In general the amount of longitudinal magnetization that is formed depends on how the spin system evolved between the first and second pulses.

After the second pulse, longitudinal magnetization is interchanged during a mixing period whose duration is limited only by the spin–lattice relaxation time. After a third pulse, the observed signal reflects the time development of the spin system between the first and the second pulse as well as the result of interchange of spectral components during the time between the second and the third pulse. Analysis of the results as a function of the time between the first and second pulses identifies the spectral components that are interchanged. Observation of the signal after the third pulse as a function of the time between the second and third pulses reveals the rate of the motion.

For carbon atoms the three-pulse experiment can be done with magic-angle spinning as well as without [374–376]. With slow spinning the information about the chemical-shift anisotropy is contained in the spinning sidebands. A two-dimensional NMR spectrum can be created that is sensitive to the interchange of intensity among the spinning sidebands. A restriction of the method is that the pulse sequence must be synchronized rigorously with the spinning. Otherwise off-diagonal sidebands arise that obscure the effect of molecular reorientation.

So far we considered reorientational motion only with respect to inhomogeneous NMR spectra. Motion also affects homogeneous NMR spectra, but the resulting line shapes are more difficult to analyze. *Motional narrowing* in the *wide-line* proton spectra of solids is, in fact, a well-known phenomenon. Because detailed line shape analysis of such spectra is usually difficult, motion is conveniently characterized simply in terms of the spin–spin relaxation time T_2, which is inversely proportional to the homogeneous spectral width. Motion is usually studied by measurement of many different relaxation parameters of which T_2 is only one. We consider the relation between nuclear relaxation and motion in Section 11.7.

11.2.1   Relation Between Reorientational Motion and Line-Narrowing Procedures

Magic-angle spinning was originally used to counter the broadening effect of dipolar couplings on the NMR spectrum [377, 378]. Most modern uses of spinning are for the removal of broadening from the chemical-shift anisotropy. Fortunately, we can discuss the interrelation of motion and magic-angle spinning in a quite general way that applies to all applications.

Because both random and coherent motion (in the form of spinning) result in spectral narrowing, what happens if both occur at comparable frequencies? Naively we may expect that the magic-angle spinning will be more efficient because it is helped along by the partial narrowing already resulting from molecular reorientation. In practice, the reorientation often interferes with the magic-angle spinning [379]. This phenomenon was cited in Section 4.3.

To understand how motion can interfere with the narrowing effect of magic-angle spinning, we must recognize that magic-angle spinning requires the sample to be almost static over the *whole spinning cycle*. If something happens to disrupt the orientation of the molecule before a whole spinning cycle is completed, magic-angle spinning fails completely to narrow the NMR signals. Random changes in the orientation during the spinning cycle result in broadened resonances.

The broadening can be turned to advantage. Waugh and co-workers [380] showed that residual broadening in ^{13}C spectra taken with magic-angle spinning can be used to determine the rates of slow reorientational processes in solids. A detailed mathematical analysis is available [381, 382].

In the limit that many reorientational jumps occur during the course of one spinner rotation, spinning again produces narrow resonances. With fast reorientational motion all of the nuclei experience the same averaged environment during the course of a single revolution. Narrow signals once again result.

Knowledge of the effect of motion on the line-narrowing ability of magic-angle spinning facilitates an understanding of the effect of motion on multiple-pulse experiments. The situation is exactly the same. If reorientation takes place during one of the cycles of the multiple-pulse sequence, the line-narrowing capability of the cycle is diminished. Care must thus be used in the interpretation of experimental results obtained with multiple-pulse NMR of mobile samples. For example, if the shape of a chemical-shift powder pattern in a spectrum acquired with multiple-pulse decoupling is to serve as a source of information about reorientational motion, the spectral shape must be corrected for the interference of the motion with the ability of the multiple-pulse sequence to remove dipolar broadening [383].

Molecular motion can also interfere with sequences used to suppress spinning sidebands from spectra taken with magic-angle spinning. The reappearance of spinning sidebands resulting from such interference provides a unique source of information about molecular motion [384, 385].

Because line-narrowing procedures are comparable to reorientational motion, two such procedures can interfere with each other when their effective timescales are similar [386]. Magic-angle spinning can be used to remove chemical-shift anisotropy broadening from a spectrum at the same time that a multiple-pulse sequence is used to remove homonuclear dipolar couplings, but only if the rotation time for one spinning cycle is much longer than the cycle time for the multiple-pulse sequence.

11.2.2   Mathematical Description of Reorientational Averaging

We treat explicitly here only the effect of motion on a signal that is measured after a single excitation pulse. The basic equations can be extended in a straightforward manner to experiments involving two and three pulses.

To treat motional effects completely, we must expand the density matrix to include the dynamic as well as the nuclear states. The time development of the expanded density matrix is given by the *stochastic Liouville equation* [387–390]. However, it is not necessary for us to consider the full stochastic Liouville equation. If we are primarily interested in line shape analysis, we need only the time development of the transverse components of magnetization associated with the various exchanging orientations. These can be arranged into a vector $\mathbf{g}(t)$. For a given orientation of the motional frame, the time development of this vector is given by

$$\frac{d\mathbf{g}}{dt} = \mathbf{g}(i\omega + \pi) \tag{34}$$

where π is a matrix expressing the transition probabilities per unit time among the motional states.

The solution to (34) is

$$\mathbf{g}(t) = \mathbf{g}(0)\exp(i\omega + \pi)t \tag{35}$$

The observable NMR signal is proportional to the sum of all the magnetization components.

$$\mathbf{g}(t) = \mathbf{g}(0)[\exp(i\omega + \pi)t] \cdot \mathbf{1} \tag{36}$$

The complete free-induction decay is found by integration of the signals for all orientations of the motional frame. These orientations are described by the angles θ and ϕ.

$$G(t) = \iint g(t, \theta, \phi) \sin \theta \, d\theta \, d\phi \tag{37}$$

The spectrum is the Fourier transform of $G(t)$.

11.2.3   Reorientational Averaging of Dipolar Couplings

Reorientational motion changes the apparent second moment of the proton spectrum of a solid. Only general information about the motion can be obtained. In Section 6.1.2 we showed that the splittings from carbon—hydrogen couplings can be resolved in the carbon spectrum if the protons are decoupled from each other with a multiple-pulse sequence. With magic-angle spinning the Pake doublet in the carbon spectrum is outlined by the intensity of the spinning sidebands. Reorientational motion occurring at frequencies comparable to the width of the pattern distorts the intensities of the sidebands. Analysis of the distortions reveals the nature of the motion of the C—H bonds. The technique appears to be a valuable method for determination of the rate and mechanism of motion in synthetic polymers [391].

11.2.4   Reorientational Averaging of Chemical-Shift Anisotropy

Fluorine is suitable for motional studies utilizing the anisotropic chemical-shift anisotropy, but in most compounds it is necessary to decouple the fluorines from any nearby protons as well as from each other. In compounds such as poly(tetrafluoroethylene), which do not contain protons, multiple-pulse sequences suffice to give narrowed spectra [324, 325]. For biological molecules phosphorous NMR gives good results with only high-powered decoupling of the phosphorous nuclei from nearby protons [392]. We already considered the general strategy for use of the chemical-shift anisotropy for motional studies and do not go into further details here.

11.2.5   Reorientational Averaging of Quadrupolar Couplings

Deuterium NMR is a valuable tool for the determination of the orientational and motional properties of biological samples [348]. Spiess and his co-workers have largely been responsible for demonstration of the utility of deuterium NMR for the characterization of synthetic polymers [393].

As we showed in Section 4.4, deuterium spectra must be measured with a spin—echo procedure. However, when the molecules reorient, there can be complications in the formation of the echo. Interchange of spectral components in the time period that the echo is forming can lead to distortion in the spectrum that is derived from the second half of the spin echo [394–396].

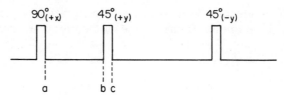

Figure 5.26   Three-pulse sequence used to generate and detect the time development of spin alignment for the study of ultraslow reorientational motion.

The changes in the spectrum induced by motion during the formation of the spin echo can be turned to advantage because the distortions reflect the rates and mechanisms of the reorientational process. Recall that spin–echo techniques provide a way to extend the sensitivity of the NMR experiment to slower motions than can be examined with simple line-shape analysis of a one-pulse experiment alone.

Extension of the sensitivity of the deuterium spectrum to motions as much as five orders of magnitude slower than can be studied with a one-pulse experiment can be made with a three-pulse experiment [397–401]. The overall approach is similar to that we described above for three-pulse experiments involving the chemical-shift anisotropy, but the details are different when the nucleus involved has spin 1 instead of spin $\frac{1}{2}$.

Let us follow the course of such an experiment. The pulse sequence is shown in Figure 5.26. Following a first pulse along the y axis, the two components of deuterium magnetization associated with each orientation are aligned along the x axis (Figure 5.27a). During the time t_1 the magnetization components fan out

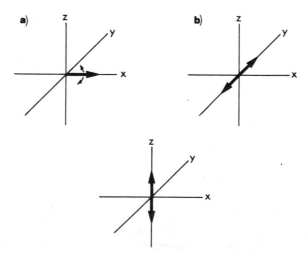

Figure 5.27   Time development of the nuclear magnetizations during the pulse sequence shown in Figure 5.26.

in the rotating frame. Let us assume for the moment that at the time of the second pulse they are aligned in opposite directions along the y axis (Figure 5.27b). A second y pulse at this point would not change the orientation of the magnetization components. An x pulse would leave the magnetization components aligned in opposite directions along the z axis (Figure 5.27c). This state is one of high *quadrupolar order* or *spin alignment*. Maximum alignment is produced by a 45° pulse.

Now let us assume that at the time of the second pulse the magnetization components are aligned along the x axis. In this case an x pulse will not affect the transverse magnetization. On the other hand, a y pulse will partially reform Zeeman order, in which the two magnetization components are aligned in the same direction along the z axis.

Whether or not the second pulse is aligned along the y or the x axis, the amount of Zeeman or quadrupolar order formed depends on the time development of the spin system between the first and the second pulses. If the magnetization components at the time of the second pulse are not aligned exactly along either the x or y axis, less than the optimal amount of longitudinal magnetization or quadrupolar order is formed.

The decay of either quadrupolar order or Zeeman magnetization is slow. Thus the information of how the system evolved during t_1 can be stored for a long time in either state. The choice of which state to use hinges on practical difficulties in controlling experimental artifacts. During the time the magnetization is stored interchange of the magnetization components as a result of molecular motions occurs. The final effect of the interchange is detected after a third pulse regenerates observable transverse magnetization.

The three-pulse experiment can be followed as a function of the time between the second and third pulses to detect the rate of the motional process occurring. It can also be followed as a function of the time between the first and second pulses and the time after the third pulse to give, after double Fourier transformation, a two-dimensional NMR spectrum giving a mapping of the interchange among the spectral frequencies that are interchanged by molecular motion [402–405]. The appearance of the spectrum is quite sensitive to whether the motion involves two- or three-site jumps or is diffusive in nature.

An excellent example of the application of the full range of experiments possible for deuterium NMR appears in a series of studies of deuterated polyethylene by Sillescu and co-workers [326, 406, 407]. The first paper reported work done with traditional *wide-line* NMR [326]. Later measurements were made with pulsed Fourier transform NMR [406], and finally measurements of very slow processes were followed with the foregoing three-pulse experiment [407].

The power of deuterium NMR to give information about the mechanism of polymer reorientation is demonstrated by work with poly(butylene terephthalate) [408]. Spectra at different temperatures for the compound with deuteriums located at the center of the butylene chain are shown in Figures 5.28 and 5.29.

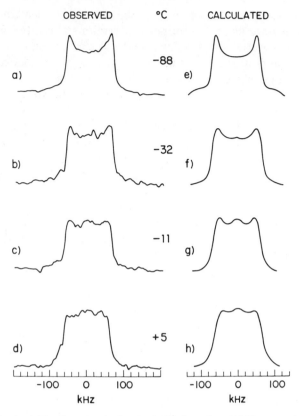

Figure 5.28 Observed (*a*)–(*d*) and calculated (*e*)–(*h*) deuterium NMR spectra of poly(butylene terephthalate) deuterated at the central carbon atom of the butylene group. The temperatures for each spectrum are shown. The calculated spectra were based on the model of a two-site jump between orientations separated by 103°. Reprinted with permission from L. W. Jelinski, J. J. Dumais, and A. K. Engel, *Macromolecules*, **16**, 492 (1983). Copyright 1983 American Chemical Society.

The spectrum at $-88°C$ approximates a rigid-lattice Pake doublet. With increasing temperature the doublet fills in until the spectrum at $21°C$ is almost square. At $85°C$ the spectral averaging has destroyed the axial symmetry that usually results in a Pake doublet for the deuterium spectrum.

The experimental results are fit well by a model in which reorientation occurs as a jump between two orientations of nearly equal energy. Of the various mechanisms possible for local segmental motion in this polymer, the ones most consistent with the NMR results are migration of a gauche bond in the butylene chain fragment and production of a pair of gauche bonds from the all-trans conformation.

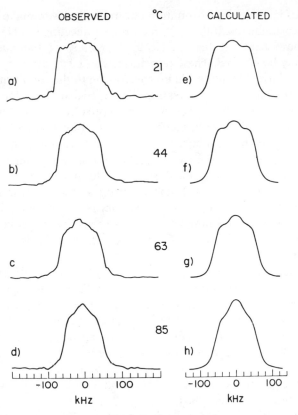

Figure 5.29 Observed (*a*)–(*d*) and calculated (*e*)–(*h*) deuterium spectra of the same sample described in Figure 5.28 for additional temperatures. Reprinted with permission from L. W. Jelinski, J. J. Dumais, and A. K. Engel, *Macromolecules*, **16**, 492 (1983). Copyright 1983 American Chemical Society.

11.2.6   Molecular Reorientation and Nuclear Relaxation

As in our preceding discussion, reorientational processes at frequencies comparable to the overall spectral width alter the shape of the spectrum. Of course, the spectral shape is determined by the time evolution of the free-induction-decay signal. For narrower shapes the decay is slower than for broader shapes. The time for decay of the signal may be approximated with a time constant, the relaxation time T_2^*. If the signal is refocused as a spin echo, its ultimate decay may be followed over a longer time described by a time constant T_2.

Most NMR signals of solids do not decay exponentially. A single relaxation time does not tell so much about the motion as does analysis of the complete free-induction-decay signal (or spectral bandshape). Thus the transverse relaxation times alone provide very limited information.

One way to increase the information about motion is to make other types of relaxation measurements. If the dynamic process affecting the NMR results has frequency components at the transition frequency of the nucleus, nuclear transitions may be induced. These transitions can be detected by measurement of the rate of return of longitudinal magnetization to the equilibrium value after a perturbation. Spin–lattice relaxation times give information about much faster motions than can be studied with transverse relaxation times alone.

Variation of the reference frame for the relaxation process also extends its frequency sensitivity. In the laboratory frame the direction of the magnetic field is along the z axis. Decay of nuclear magnetization along this axis defines the relaxation time T_1, which is most sensitive to frequencies at the Larmor frequency of the nucleus. In a frame rotating at the Larmor frequency the effective magnetic field lies in the xy plane. Decay of magnetization parallel to the magnetic field in the rotating frame is most sensitive to frequencies equal to the effective magnetic field times the gyromagnetic ratio. It is described by the time constant $T_{1\rho}$.

The foregoing relaxation measurements can all be made without the use of the techniques of high-resolution NMR. In *broad-line* proton spectra, however, the internuclear dipolar interactions tend to average the relaxation rates of the various types of protons in the sample through the process of spin diffusion (Section 3.7). Multiple-pulse sequences were specifically designed for the suppression of the dipolar interactions. This suggests that we should follow the course of relaxations in the *toggling frame*, which is used to describe the multiple-pulse experiment [409, 410]. Measurements of this type were made for poly(ethylene terephthalate) [288].

Because there is a low natural abundance of ^{13}C, spin diffusion is not a problem in carbon NMR. In their early work on high-resolution ^{13}C spectra of solid polymers, Schaefer and co-workers [411] showed that independent spin-lattice relaxation in the laboratory frame and the rotating frame can be observed for carbon atoms giving resolved spectral signals. Carbon spin–lattice relaxation times in the rotating frame have attracted special attention because they are controlled by relatively slow polymer motions [63]. For example, the effective frequencies associated with mechanical impact on a polymer fall within the low-to-medium frequency range [63, 412] sampled by carbon $T_{1\rho}$ measurements.

There are several complications in the analysis of ^{13}C relaxation in the rotating frame of polymers. Typically the observed magnetization does not decay exponentially. A single relaxation time does not suffice to describe the relaxation process. On the other hand, the spread in relaxation rates probably gives information about the distribution of local environments possible for the local polymer chains.

A potentially more serious problem is that the carbon spin–lattice relaxation in the rotating frame need not involve motion at all. Typically, relaxation in the rotating frame is sensitive to motions having frequency components of about 50 kHz. In the presence of strong internuclear proton coupling, the carbon–

proton dipolar interactions are modulated by the proton spin diffusion as well as by motion. The rate of proton spin diffusion is roughly proportional to the width of the proton spectrum, which can be as much as 40 kHz. Thus a *spin–spin* contribution resulting from the proton spin diffusion cannot be dismissed.

Fortunately, experiments can be designed to test the mechanism of the carbon relaxation [413]. The consensus now seems to be that for polycarbonates [414], polyesters [415], and epoxy resins [416], and most other polymers the carbon spin–lattice relaxation in the rotating frame is dominated by a *spin–lattice* mechanism. Hence, motion is directly involved in the transfer of energy from the carbon system. In a polymer with a high density of protons, such as polyethylene, spin diffusion among the protons is relatively efficient, and the spin–spin mechanism may play a role in carbon relaxation in the rotating frame [413]. It has been possible to separate spin–spin from spin–lattice relaxation in alanine [417].

Carbon relaxation in the laboratory frame is not complicated by a spin–spin mechanism. Nothing in theory prohibits such a mechanism, but the frequencies associated with proton spin diffusion are much too low to contribute to spin–lattice relaxation in the laboratory frame.

In Section 9.2 we showed the value of carbon NMR in the measurement of proton relaxation times. With magic-angle spinning, the carbon spectrum is more resolved than the proton spectrum, and it is sometimes convenient to transfer proton magnetization to the carbon system to allow indirect detection of the time development of the proton system [418]. Because various relaxation experiments can be performed on the protons, this approach has a wide range of possibilities.

References

1. B. C. Gerstein, *Philos. Trans. R. Soc. London Ser. A*, **299**, 521 (1981).
2. J. R. Lyerla, *Methods Exp. Phys.*, **16A**, 241 (1980).
3. J. R. Lyerla, *Contemp. Top. Polym. Sci.*, **3**, 143 (1979).
4. R. E. Wasylishen and C. A. Fyfe, *Ann. Rep. NMR Spectrosc.*, **12**, 1 (1982).
5. R. G. Griffin, "Solid State Nuclear Magnetic Resonance of Lipid Bilayers," in J. M. Lowenstein, Ed., *Methods in Enzymology*, Vol. 72, Academic, New York, 1981, p. 108.
6. T. M. Duncan and C. Dybowski, *Surf. Sci. Rep.*, **1**, 157 (1981).
7. U. Haeberlen, *High Resolution NMR in Solids*, Academic, New York, 1976.
8. M. Mehring, *High Resolution NMR Spectroscopy in Solids*, Springer-Verlag, New York, 1976.
9. C. A. Fyfe, *Solid State NMR for Chemists*, FC Press, Guelph, 1983.
10. B. C. Gerstein and C. Dybowski, *Transient Techniques in NMR of Solids: An Introduction to Theory and Practice*, Academic, New York, 1985.
11. D. E. Axelson, *Solid State Nuclear Magnetic Resonance of Fossil Fuels*, Multiscience Publications, Ottawa, 1985.

12. R. A. Komoroski, *High Resolution NMR Spectroscopy of Synthetic Polymers in Bulk*, VCH, Deerfield Beach, FL, 1986.

13. H. W. Spiess, *NMR: Basic Princ. Prog.*, **15**, 55 (1978).

14. E. R. Andrew, *Prog. Nucl. Magn. Reson.*, **8**, 1 (1971).

15. R. M. Lynden-Bell, *Prog. Nucl. Magn. Reson. Spectrosc.*, **2**, 163 (1967).

16. J. Jeener, *Adv. Magn. Reson.*, **10**, 1 (1982).

17. U. Fano, "Liouville Representation of Quantum Mechanics with Applications to Relaxation Processes," in E. R. Caianello, Ed. *Lectures on the Many-Body Problem*, Academic, New York, 1964, p. 217.

18. U. Fano, *Rev. Mod. Phys.*, **29**, 74 (1957).

19. W. P. Aue, E. Bartholdi, and R. R. Ernst, *J. Chem. Phys.*, **64**, 2229 (1976).

20. C. P. Slichter, *Principles of Magnetic Resonance*, Springer, New York, 1978.

21. A. M. Portis, *Phys. Rev.*, **91**, 1071 (1953).

22. R. A. Hoffman and S. Forsen, *Prog. Nucl. Magn. Reson. Spectrosc.*, **1**, 15 (1966).

23. M. M. Maricq and J. S. Waugh, *J. Chem. Phys.*, **70**, 3300 (1979).

24. G. Williams, *Adv. Polym. Sci.*, **33**, 59 (1979).

25. J. I. Kaplan and A. N. Garroway, *J. Magn. Reson.*, **49**, 464 (1982).

26. A. Abragam, *The Principles of Nuclear Magnetism*, Oxford, New York, 1961.

27. J. S. Waugh, M. M. Maricq, and R. Cantor, *J. Magn. Reson.*, **29**, 183 (1978).

28. S. R. Hartmann and E. L. Hahn, *Phys. Rev.*, **128**, 2042 (1962).

29. F. M. Lurie and C. P. Slichter, *Phys. Rev.*, **133**, 1109 (1964).

30. M. Schwab and E. L. Hahn, *J. Chem. Phys.*, **52**, 3152 (1970).

31. P. Mansfield and P. K. Grannell, *J. Phys. C*, **4**, L197 (1971).

32. H. E. Bleich and A. G. Redfield, *J. Chem. Phys.*, **55**, 5405 (1971).

33. C. S. Yannoni and H. E. Bleich, *J. Chem. Phys.*, **55**, 5406 (1971).

34. P. K. Grannell, P. Mansfield, and M. A. B. Whitaker, *Phys. Rev.*, **B8**, 4149 (1973).

35. A. Pines, M. G. Gibby, and J. S. Waugh, *J. Chem. Phys.*, **56**, 1776 (1972).

36. A. Pines, M. G. Gibby, and J. S. Waugh, *Chem. Phys. Lett.*, **15**, 373 (1972).

37. A. Pines, M. G. Gibby, and J. S. Waugh, *J. Chem. Phys.*, **59**, 569 (1973).

38. D. E. Demco, S. Kaplan, S. Pausak, and J. S. Waugh, *Chem. Phys. Lett.*, **30**, 77 (1975).

39. J. Tegenfeldt and U. Haeberlen, *J. Magn. Reson.*, **36**, 453 (1979).

40. S. Ganapathy, A. Naito, and C. A. McDowell, *J. Am. Chem. Soc.*, **103**, 6011 (1981).

41. C. E. Brown, *J. Am. Chem. Soc.*, **104**, 5608 (1982).

42. E. O. Stejskal and J. Schaefer, *J. Magn. Reson.*, **18**, 560 (1975).

43. S. Ganapathy, T. M. Eads, V. P. Chacko, and R. G. Bryant, *J. Magn. Reson.*, **62**, 314 (1985).

44. D. E. Demco, J. Tegenfeldt, and J. S. Waugh, *Phys. Rev. B*, **11**, 4133 (1975).

45. A. Pines, and T. W. Shattuck, *J. Chem. Phys.*, **61**, 1255 (1974).

46. D. Suter, and R. R. Ernst, *Phys. Rev. B*, **25**, 6038 (1982).

47. M. Mehring and G. Sinning, *Phys. Rev. B*, **15**, 2519 (1977).

48. M. H. Levitt, D. Suter, and R. R. Ernst, *J. Chem. Phys.*, **84**, 4243 (1986).

49. L. B. Alemany, D. M. Grant, R. Pugmire, T. D. Alger, and K. W. Zilm, *J. Am. Chem. Soc.*, **105**, 2133 and 2142 (1983).

50. L. Mueller, A. Kumar, T. Baumann, and R. R. Ernst, *Phys. Rev. Lett.*, **32**, 1402 (1974).

51. U. Deininghaus and M. Mehring, *Phys. Lett.*, **73A**, 129 (1979).

52. U. Deininghaus and M. Mehring, *Phys. Rev. B*, **24**, 4945 (1981).

53. M. E. Stoll, W. K. Rhim, and R. W. Vaughan, *J. Chem. Phys.*, **64**, 4808 (1976).

54. R. A. Wind, L. Li, H. Lock, and G. E. Maciel, *J. Magn. Reson.*, **79**, 577 (1988).

55. R. A. Wind and C. S. Yannoni, *J. Magn. Reson.*, **72**, 108 (1987).

56. R. A. Wind, M. J. Duijvestijn, C. van der Lugt, A. Manenschijn, and J. Vriend, *Prog. Nucl, Magn. Reson. Spect.*, **17**, 33 (1985).

57. E. R. Andrew, A. Bradbury, and R. G. Eades, *Nature (London)*, **183**, 1802 (1959).

58. I. J. Lowe, *Phys. Rev. Lett.*, **2**, 285 (1959).

59. E. R. Andrew, L. F. Farnell, and T. D. Gledhill, *Phys. Rev. Lett.*, **19**, 6 (1967).

60. E. R. Andrew, M. Firth, A. Jasinski, and P. J. Randall, *Phys. Lett.*, **31A**, 446 (1970).

61. D. Doskocilova and B. Schneider, *Adv. Colloid Interface Sci.*, **9**, 63 (1978).

62. L. L. Sterna and H. C. Smith, *J. Magn. Reson.*, **79**, 528 (1988).

63. J. Schaefer, E. O. Stejskal, and R. Buchdahl, *Macromolecules*, **10**, 384 (1977).

64. B. C. Gerstein, R. G. Pembleton, R. C. Wilson, and L. M. Ryan, *J. Chem. Phys.*, **66**, 361 (1977).

65. R. A. Haberkorn, J. Herzfeld, and R. G. Griffin, *J. Am. Chem. Soc.*, **100**, 1296 (1978).

66. N. J. Clayden, C. M. Dobson, L.-Y. Lian, and D. J. Smith, *J. Magn. Reson.*, **69**, 476 (1986).

67. E. O. Stejskal, J. Schaefer, and J. S. Waugh, *J. Magn. Reson.*, **28**, 105 (1977).

68. M. Sardashti and G. E. Maciel, *J. Magn. Reson.*, **72**, 467 (1987).

69. R. E. Taylor, R. G. Pembleton, L. M. Ryan, and B. C. Gerstein, *J. Chem. Phys.*, **71**, 4541 (1979).

70. Y. Yarim-Agaev, P. N. Tutunjian, and J. S. Waugh, *J. Magn. Reson.*, **47**, 51 (1982).

71. A. Bax, N. M. Szeverenyi, and G. E. Maciel, *J. Magn. Reson.*, **51**, 400 (1983).

72. G. E. Maciel, N. M. Szeverenyi, and M. Sardashti, *J. Magn. Reson.*, **64**, 365 (1985).

73. W. T. Dixon, *J. Magn. Reson.*, **44**, 220 (1981).

74. W. T. Dixon, *J. Chem. Phys.*, **77**, 1800 (1982).

75. W. T. Dixon, J. Schaefer, M. D. Sefcik, E. O. Stejskal, and R. A. McKay, *J. Magn. Reson.*, **49**, 341 (1982).

76. M. A. Heminga, P. A. de Jager, K. P. Datema, and J. Breg, *J. Magn. Reson.*, **50**, 508 (1982).

77. D. P. Raleigh, E. T. Olejniczak, S. Vega, and R. G. Griffin, *J. Magn. Reson.*, **72**, 238 (1987).

78. J. L. Ackerman, R. Eckman, and A. Pines, *Chem. Phys.*, **42**, 423 (1979).

79. R. Eckman, M. Alla, and A. Pines, *J. Magn. Reson.*, **41**, 440 (1980).

80. J. S. Frye and G. E. Maciel, *J. Magn. Reson*, **48**, 125 (1982).

81. C. Ye, B. Sun and G. E. Maciel, *J. Magn. Reson.*, **70**, 241 (1986).

82. D. J. Burton and R. K. Harris, *J. Chem. Soc. Chem. Commun.*, **256** (1982).

83. M. D. Meadows, K. A. Smith, R. A. Kinsey, T. M. Rothgeb, R. P. Skarjune, and E. Oldfield, *Proc. Natl. Acad. Sci. USA*, **79**, 1351 (1982).

84. C. J. Groombridge, R. K. Harris, K. J. Packer, B. J. Say, and S. F. Tanner, *J. Chem. Soc. Chem. Commun.*, **174** (1980).

85. W. W. Fleming, C. A. Fyfe, J. R. Lyerla, H. Vanni, and C. S. Yannoni, *Macromolecules*, **13**, 460 (1980).

86. E. Kundla, A. Samoson, and E. Lippmaa, *Chem. Phys. Lett.*, **83**, 229 (1981).

87. J. G. Hexem, M. H. Frey, and S. J. Opella, *J. Am. Chem. Soc.*, **103**, 224 (1981).

88. J. G. Hexem, M. H. Frey, and S. J. Opella, *J. Chem. Phys.*, **77**, 3847 (1982).

89. N. Zumbulyadis, P. M. Henrichs, and R. H. Young, *J. Chem. Phys.*, **75**, 1603 (1981).

90. E. M. Menger and W. S. Veeman, *J. Magn. Reson.*, **46**, 257 (1982).

91. A. Naito, S. Ganapathy, and C. A. McDowell, *J. Chem. Phys.*, **74**, 5393 (1981); *J. Magn. Reson.*, **48**, 367 (1982).

92. A. Naito, S. Ganapathy, K. Akasaka, and C. A. McDowell, *J. Chem. Phys.*, **74**, 3190 (1981).

93. T. Terao, H. Miura, and A. Saika, *J. Magn. Reson.*, **49**, 365 (1982).

94. K. W. Zilm and D. M. Grant, *J. Magn. Reson.*, **48**, 524 (1982).

95. N. Zumbulyadis, *J. Magn. Reson.*, **49**, 329 (1982).

96. A. D. English, *J. Magn. Reson.*, **57**, 491 (1984).

97. J. F. Haw, *Anal. Chem.*, **60**, 559A (1988).

98. J. F. Haw, G. C. Campbell, and R. C. Crosby, *Anal. Chem.*, **58**, 3172 (1986).

99. G. C. Campbell, R. C. Crosby, and J. F. Haw, *J. Magn. Reson.*, **69**, 191 (1986).

100. A. D. Ronemus, R. L. Vold, and R. R. Vold, *J. Magn. Reson.*, **70**, 416 (1986).

101. M. Bloom, J. H. Davis, and M. Valic, *Can. J. Phys.*, **58**, 1510 (1980).

102. P. M. Henrichs, J. M. Hewitt, and M. Linder, *J. Magn. Reson.*, **60**, 280 (1984).

103. T. M. Barbara, M. S. Greenfield, R. L. Vold, and R. R. Vold, *J. Magn. Reson.*, **69**, 311 (1986).

104. D. J. Siminovitch and R. G. Griffin, *J. Magn. Reson.*, **62**, 99 (1985).

105. D. J. Siminovitch, D. P. Raleigh, E. T. Olejniczak, and R. G. Griffin, *J. Chem. Phys.*, **84**, 2556 (1986).

106. M. H. Levitt, D. Suter, and R. R. Ernst, *J. Chem. Phys.*, **80**, 3064 (1984).

107. N. J. Heaton, R. R. Vold, and R. L. Vold, *J. Magn. Reson.*, **77**, 572 (1988).

108. R. Tycko, E. Schneider, and A. Pines, *J. Chem. Phys.*, **81**, 680.

109. J. H. Davis, *Biochim. Biophys. Acta*, **737**, 117 (1983).

110. R. Hentschel and H. W. Spiess, *J. Magn. Reson.*, **35**, 157 (1979).

111. W. K. Rhim, A. Pines, and J. S. Waugh, *Phys. Rev. B*, **3**, 684 (1971); *J. Magn. Reson.*, **6**, 457 (1972); *Phys. Rev. Lett.*, **25**, 218 (1970).

112. J. G. Powles, and J. H. Strange, *Proc. Phys. Soc.*, **82**, 6 (1963).

113. P. Mansfield, *Phys. Rev.*, **137**, A961 (1965).

114. E. D. Ostroff and J. S. Waugh, *Phys. Rev. Lett.*, **16**, 1097 (1966).

115. J. S. Waugh and C. H. Wang, *Phys. Rev.*, **162**, 209 (1967).

116. J. S. Waugh and L. M. Huber, *J. Chem. Phys.*, **47**, 1862 (1967).

117. J. S. Waugh, L. M. Huber, and U. Haeberlen, *Phys. Rev. Lett.*, **20**, 180 (1968).

118. J. S. Waugh, C. H. Wang, L. M. Huber, and R. L. Vold, *J. Chem. Phys.*, **48**, 662 (1968).

119. U. Haeberlen, J. D. Ellett, and J. S. Waugh, *J. Chem. Phys.*, **55**, 53 (1970).

120. P. Mansfield and D. Ware, *Phys. Lett.*, **22**, 133 (1966).

121. D. Ware and P. Mansfield, *Phys. Lett.*, **25A**, 651 (1967).

122. P. Mansfield and D. Ware, *Phys. Lett.*, **27A**, 159 (1968).

123. P. Mansfield, *J. Phys. C*, **4**, 1444 (1971).

124. U. Haeberlen and J. S. Waugh, *Phys. Rev.*, **175**, 453 (1968).

125. G. Scheler, U. Haubenreiser, and H. Rosenberger, *J. Magn. Reson.*, **44**, 134 (1981).

126. L. M. Ryan, R. E. Taylor, A. J. Paff, and B. C. Gerstein, *J. Chem. Phys.*, **72**, 508 (1980).

127. W. K. Rhim, D. D. Elleman, and R. W. Vaughan, *J. Chem. Phys.*, **59**, 3740 (1973).

128. W. K. Rhim, D. D. Elleman, L. B. Schreiber, and R. W. Vaughan, *J. Chem. Phys.*, **60**, 4595 (1974).

129. D. P. Burum, and W. K. Rhim, *J. Chem. Phys.*, **70**, 3553 (1979); *J. Chem. Phys.*, **71**, 944 (1979).

130. S. Idziak and U. Haeberlen, *J. Magn. Reson.*, **50**, 281 (1982).

131. U. Haubenreiser and B. Schnabel, *J. Magn. Reson.*, **35**, 175 (1979).

132. D. P. Burum, M. Linder, and R. R. Ernst, *J. Magn. Reson.*, **43**, 463 (1981).

133. M. Mehring and J. S. Waugh, *Rev. Sci. Instrum.*, **43**, 649 (1972).

134. M. Mehring, *Z. Naturforsch.*, **27a**, 1634 (1972); *Rev. Sci. Instrum.*, **44**, 64 (1973).

135. D. P. Burum, M. Linder, and R. R. Ernst, *J. Magn. Reson.*, **44**, 173 (1981).

136. M. E. Stoll, A. Vega, and R. W. Vaughan, *J. Chem. Phys.*, **65**, 4093 (1976).

137. E. F. Rybaczewski, B. L. Neff, J. S. Waugh, and J. S. Sherfinski, *J. Chem. Phys.*, **67**, 1231 (1977).

138. A. Pines, D. J. Ruben, S. Vega, and M. Mehring, *Phys. Rev. Lett.*, **36**, 110 (1976).

139. D. Suwelack, M. Mehring, and A. Pines, *Phys. Rev. B*, **19**, 238 (1979).

140. A. Pines, S. Vega, and M. Mehring, *Phys. Rev. B*, **18**, 112 (1978).

141. R. Eckman, *J. Chem. Phys.*, **76**, 2767 (1982).

142. D. Barnaal and I. J. Lowe, *Phys. Rev. Lett.*, **11**, 258 (1963).

143. W. I. Goldburg and M. Lee, *Phys. Rev. Lett.*, **11**, 255 (1963).

144. M. Lee and W. I. Goldburg, *Phys. Rev.*, **140**, A1261 (1965).

145. H. Hatanaka, T. Terao, and T. Hashi, *J. Phys. Soc. Jpn.*, **39**, 835 (1975).

146. H. Hatanaka and T. Hashi, *J. Phys. Soc. Jpn.*, **39**, 1139 (1975).

147. H. Hatanaka and T. Hashi, *Phys. Lett.*, **67A**, 183 (1978).

148. S. Vega, T. W. Shattuck, and A. Pines, *Phys. Rev. Lett.*, **37**, (1976).

149. S. Vega, T. W. Shattuck, and A. Pines, *Phys. Rev. A*, **22**, 638 (1980).

150. R. Tycko and S. J. Opella, *J. Chem. Phys.*, **86**, 1761 (1987).

151. W. L. Earl and D. L. VanderHart, *J. Magn. Reson.*, **48**, 35, (1982).

152. D. L. VanderHart, W. L. Earl, and A. N. Garroway, *J. Magn. Reson.*, **44**, 361 (1981).

153. D. L. VanderHart, *J. Magn. Reson.*, **44**, 117 (1981).

154. J. Schaefer, S. H. Chin, S. I. Weissman, *Macromolecules*, **5**, 798 (1972).

155. D. L. VanderHart, *J. Chem. Phys.*, **84**, 1196 (1986).

156. L. W. Dennis, V. J. Bartuska, and G. E. Maciel, *J. Am. Chem. Soc.*, **104**, 230 (1982).

157. P. G. Mennitt, M. P. Shatlock, V. J. Bartuska, and G. E. Maciel, *J. Phys. Chem.*, **85**, 2087 (1981).

158. G. E. Balimann, C. J. Groombridge, R. K. Harris, K. J. Packer, B. J. Say, and S. F. Tanner, *Philos. Trans. R. Soc. London Ser. A*, **299**, 643 (1981).

159. T. Terao, H. Miura, and A. Saika, *J. Chem. Phys.*, **75**, 1573 (1981).

160. S. Opella and M. H. Frey, *J. Am. Chem. Soc.*, **101**, 5854 (1979).

161. M. G. Munowitz, and R. G. Griffin, *J. Am. Chem. Soc.*, **103**, 2529 (1981).

162. A. Kumar and R. R. Ernst, *J. Magn. Reson.*, **24**, 425 (1976).

163. F. P. Miknis, V. J. Bartuska, and G. E. Maciel, *Am. Lab.*, November, 19 (1979).

164. F. P. Miknis, *Magn. Reson. Rev.*, **7**, 87 (1982).

165. P. G. Hatcher, D. L. VanderHart, and W. L. Earl, *Org. Geochem.*, **2**, 87 (1980).

166. M. A. Wilson, J. Pugmire, K. W. Zilm, K. M. Goh, S. Heng, and D. M. Grant, *Nature (London)*, **294**, 648 (1981).

167. C. M. Preston and J. A. Ripmeester, *Can. J. Spect.*, **27**, 99 (1982).

168. W. R. Woolfenden, D. M. Grant, R. C. Straight, and E. Englert, *Biochem. Biophys. Res. Commun.*, **107**, 684 (1982).

169. V. Rutar and R. Blinc, *Z. Naturforsch.*, **35c**, 12 (1980).

170. J. Schaefer and E. O. Stejskal, *J. Am. Oil Chem. Soc.*, **52**, 366 (1975).

171. V. Rutar, R. Blinc, and L. Ehrenberg, *J. Magn. Reson.*, **40**, 225 (1980).

172. J. Schaefer and E. O. Stejskal, *J. Am. Chem. Soc.*, **98**, 1031 (1976).

173. G. E. Maciel, I.-S. Chuang, and G. E. Meyers, *Macromolecules*, **15**, 1218 (1982).

174. C. A. Fyfe, M. S. McKinnon, A. Rudin, and W. J. Tchir, *J. Polym. Sci. Polym. Lett. Ed.*, **21**, 249 (1983).

175. C. A. Fyfe, A. Rudin, and W. Tchir, *Macromolecules*, **13**, 1322 (1980).

176. M. D. Sefcik, E. O. Stejskal, R. A. McKay, and J. Schaefer, *Macromolecules*, **12**, 423 (1979).

177. G. E. Maciel, N. M. Szeverenyi, T. A. Early, and G. E. Myers, *Macromolecules*, **16**, 59 (1983).

178. S. Kaplan and A. Dilks, *Thin Solid Films*, **84**, 419 (1981).

179. A. Dilks, S. Kaplan, and A. Van Laeken, *J. Polym. Sci. Polym. Chem. Ed.*, **19**, 2987 (1981).

180. S. Kaplan, and A. Dilks, *J. Polym. Sci. Polym. Chem. Ed.*, **21**, 1819 (1983).

181. M. M. Maricq, J. S. Waugh, A. G. MacDiarmid, H. Shirakawa, and A. J. Heeger, *J. Am. Chem. Soc.*, **100**, 7729 (1978).

182. P. Bernier, F. Schue, J. Sledz, M. Rolland, and L. Giral, *Chem. Scr.*, **17**, 151 (1981).

183. H. W. Gibson, J. M. Pochan, and S. Kaplan, *J. Am. Chem. Soc.*, **103**, 4619 (1981).

184. A. C. Wong, A. N. Garroway, and W. M. Ritchey, *Macromolecules*, **14**, 832 (1981).

185. H. A. Resing and W. B. Moniz, *Macromolecules*, **8**, 560 (1975).

186. G. E. Maciel, N. M. Szeverenyi, T. A. Early, and G. E. Myers, *Macromolecules*, **16**, 598 (1983).

187. C. A. Fyfe, M. S. McKinnon, A. Rudin, and W. J. Tchir, *J. Polym. Sci. Polym. Lett. Ed.*, **21**, 249 (1983).

188. J. F. Haw and G. E. Maciel, *Anal. Chem.*, **55**, 1262 (1983).

189. J. Schaefer, E. O. Stejskal, and R. A. McKay, *Biochem. Biophys. Res. Commun.*, **88**, 274 (1979).

190. T. A. Cross, J. A. DiVerdi, and S. J. Opella, *J. Am. Chem. Soc.*, **104**, 1759 (1982).

191. H. R. Kricheldorf and D. Müller, *Macromolecules*, **16**, 615 (1983).

192. M. A. Hemminga, W. S. Veeman, H. W. M. Hilhorst, and T. J. Schaafsma, *Biophys. J.*, **35**, 463 (1981).

193. M. H. Frey and S. J. Opella, *J. Chem. Soc. Chem. Commun.*, 474 (1980).

194. S. J. Opella, M. H. Frey, and T. A. Cross, *J. Am. Chem. Soc.*, **101**, 5856 (1979).

195. J. Schaefer, M. D. Sefcik, E. O. Stejskal, R. A. McKay, and P. L. Hall, *Macromolecules*, **14**, 557 (1981).

196. J. Schaefer, R. A. McKay, and E. O. Stejskal, *J. Magn. Reson.*, **34**, 443 (1979).

197. G. E. Maciel, D. W. Sindorf, and V. J. Bartuska, *J. Chromatogr.*, **205**, 438 (1981).

198. J. Klinowski, J. M. Thomas, C. A. Fyfe, and G. C. Gobbi, *Nature (London)*, **296**, 533 (1982).

199. C. A. Fyfe, G. C. Gobbi, J. S. Hartman, J. Klinowski, and J. M. Thomas, *J. Phys. Chem.*, **86**, 1247 (1982).

200. J. Klinowski, J. M. Thomas, C. A. Fyfe, G. C. Gobbi, and J. S. Hartman, *Inorg. Chem.*, **22**, 63 (1983).

201. D. Freude and H.-J. Behrens, *Cryst. Res. Tech.*, **16**, K36 (1981).

202. C. A. Fyfe, G. C. Gobbi, J. S. Hartman, R. E. Lenkinski, J. H. O'Brien, E. R. Beange, and M. A. R. Smith, *J. Magn. Reson.*, **47**, 168 (1982).

203. M. Ya. Mägi, A. V. Samoson, M. Kh. Tarmak, G. Engelhardt, and E. T. Lippmaa, *Dokl. Akad. Nauk SSSR*, **261**, 1169 (1981), Engl. trans. p. 1159.

204. E. Lippmaa, M. Mägi, A. Samoson, G. Engelhardt, and A. R. Grimmer, *J. Am. Chem. Soc.*, **102**, 4889 (1980).

205. J. M. Thomas, S. Ramdas, G. R. Millward, J. Klinowski, M. Audier, J. Calbet, and C. A. Fyfe, *J. Solid. State Chem.*, **45**, 368 (1982).

206. J. Klinowski, J. M. Thomas, M. Audier, S. Vasudevan, C. A. Fyfe, and J. S. Hartman, *J. Chem. Soc. Chem. Commun.*, 570 (1981).

207. G. Engelhardt, U. Lohse, A. Samoson, M. Mägi, M. Tarmak, and E. Lippmaa, *Zeolites*, **2**, 59 (1982).

208. S. Ramdas, J. M. Thomas, J. Klinowski, C. A. Fyfe, and J. S. Hartman, *Nature (London)*, **292**, 228 (1981).

209. J. Klinowski, S. Ramdas, J. M. Thomas, C. A. Fyfe, and J. S. Hartman, *J. Chem. Soc. Faraday Trans. 2*, **78**, 1025 (1982).

210. G. Engelhardt, U. Lohse, E. Lippmaa, M. Tarmak, and M. Mägi, *Z. Anorg. Allg. Chem.*, **482**, 49 (1981).

211. E. Lippmaa, M. Mägi, A. Samoson, M. Tarmak, and G. Engelhardt, *J. Am. Chem. Soc.*, **103**, 4992 (1981).

212. L. A. Bursill, E. A. Lodge, J. M. Thomas, and A. K. Cheetham, *J. Phys. Chem.*, **85**, 2409 (1981).

213. G. Engelhardt, D. Zeigan, E. Lippmaa, and M. Mägi, *Z. Anorg. Allg. Chem.*, **468**, 35 (1980).

214. J. Klinowski, J. M. Thomas, C. A. Fyfe, and J. S. Hartman, *J. Phys. Chem.*, **85**, 2590 (1981).
215. J. M. Thomas, C. A. Fyfe, S. Ramdas, J. Klinowski, and G. C. Gobbi, *J. Phys. Chem.*, **86**, 3061 (1982).
216. A.-R. Grimmer, F. von Lampe, M. Tarmak, and E. Lippmaa, *Chem. Phys. Lett.*, **97**, 185 (1983).
217. G. E. Pake, *J. Chem. Phys.*, **16**, 327 (1948).
218. K. W. Zilm, A. J. Becher, D. M. Grant, J. Michl, T.-C. Chou, and E. L. Allred, *J. Am. Chem. Soc.*, **103**, 2119 (1981).
219. H. van Willigen, R. G. Griffin, and R. A. Haberkorn, *J. Chem. Phys.*, **67**, 5855 (1977).
220. P. van Hecke, H. W. Spiess, U. Haeberlen, and S. Haussühl, *J. Magn. Reson.*, **22**, 93 (1976).
221. J. C. Pratt and J. A. S. Smith, *J. Chem. Soc. Faraday Trans. 2*, **71**, 596 (1975).
222. R. A. Haberkorn, R. E. Stark, H. Van Willigen, and R. G. Griffin, *J. Am. Chem. Soc.*, **103**, 2534 (1981).
223. L. M. Ishol and T. A. Scott, *J. Magn. Reson.*, **27**, 23 (1977).
224. C. S. Yannoni and R. D. Kendrick, *J. Chem. Phys.*, **74**, 747 (1981).
225. D. Horne, R. D. Kendrick, and C. S. Yannoni, *J. Magn. Reson.*, **52**, 299 (1983).
226. A. Samoson and E. Lippmaa, *J. Magn. Reson.*, **79**, 255 (1988).
227. R. Janssen, G. A. H. Tijink, and W. S. Veeman, *J. Chem. Phys.*, **88**, 15, (1988).
228. D. P. Weitekamp, A. Bielecki, D. Zax, K. Zilm, and A. Pines, *Phys. Rev. Lett.*, **50**, 1807 (1983).
229. J. M. Millar, A. M. Thayer, H. Zimmermann, and A. Pines, *J. Magn. Reson.*, **69**, 243 (1986).
230. J. M. Millar, A. M. Thayer, D. B. Zax, and A. Pines, *J. Am. Chem. Soc.*, **108**, 5113 (1986).
231. J. M. Millar, A. M. Thayer, A. Bielecki, D. B. Zax, and A. Pines, *J. Chem. Phys.*, **83**, 934 (1985).
232. A. M. Thayer and A. Pines, *J. Magn. Reson.*, **70**, 518 (1986).
233. R. K. Hester, J. L. Ackerman, V. R. Cross, and J. S. Waugh, *Phys. Rev. Lett.*, **34**, 993 (1975).
234. R. K. Hester, V. R. Cross, J. L. Ackerman, and J. S. Waugh, *J. Chem. Phys.*, **63**, 3606 (1975).
235. R. K. Hester, J. L. Ackerman, B. L. Neff, and J. S. Waugh, *Phys. Rev. Lett.*, **36**, 1081 (1976).
236. J. S. Waugh, *Proc. Natl. Acad. Sci. USA*, **73**, 1394 (1976).
237. M. E. Stoll, *Philos. Trans. R. Soc. London. Ser. A*, **299**, 565 (1981).
238. M. G. Munowitz and R. G. Griffin, *J. Chem. Phys.*, **76**, 2848 (1982).
239. M. Munowitz, W. P. Aue, and R. G. Griffin, *J. Chem. Phys.*, **77**, 1686 (1982).
240. M. G. Munowitz and R. G. Griffin, *J. Chem. Phys.*, **78**, 613 (1983).
241. J. A. DiVerdi and S. J. Opella, *J. Am. Chem. Soc.*, **104**, 1761 (1982).
242. A.-R. Grimmer, E. Fechner, R. Peter, and G. Molgedey, *Chem. Phys. Lett.*, **77**, 331 (1984).

243. A. R. Grimmer, *Chem. Phys., Lett.*, **119**, 416 (1985).

244. M. G. Munowitz, R. G. Griffin, G. Bodenhausen, and T. H. Huang, *J. Am. Chem. Soc.*, **103**, 2529 (1981).

245. S. Sykora, J. Vogt, H. Bösiger, and P. Diehl, *J. Magn. Reson.*, **36**, 2536 (1979).

246. E. R. Henry and A. Szabo, *J. Chem. Phys.*, **82**, 4752 (1985).

247. G. Soda and T. Chiba, *J. Chem. Phys.*, **50**, 439 (1969).

248. U. Haubenreisser and B. Schnabel, "Investigations of the Chemical Shift Anisotropy of Kieserit ($MgSO_4 \cdot H_2O$) By Means of the Proton Resonance WHH 4-Pulse Experiment," in J. W. Hennel, Ed., *Proceedings First Specialized Colloque Ampere*, Institute of Nuclear Physics, Krakow, 1973, p. 140.

249. U. Selvaray, K. J. Rao, C. N. R. Rao, J. Klinowski, and J. M. Thomas, *Chem. Phys. Lett.*, **114**, 24 (1985).

250. U. Dupree and R. F. Pettifer, *Nature (London)*, **308**, 523 (1984).

251. C. J. Hartzell, T. K. Pratum, and G. Drobny, *J. Chem. Phys.*, **87**, 4324 (1987).

252. M. Maricq and J. S. Waugh, *Chem. Phys. Lett.*, **47**, 327 (1977).

253. E. T. Lippmaa, M. A. Alla, T. J. Pehk, and G. Engelhardt, *J. Am. Chem. Soc.*, **78**, 1929 (1978).

254. A. Bunn, M. E. A. Cudby, R. K. Harris, K. J. Packer, and B. J. Say, *J. Chem. Soc. Chem. Commun.*, 15 (1981).

255. D. M. Grant and E. G. Paul, *J. Am. Chem. Soc.*, **86**, 2984 (1964).

256. A. N. Garroway, W. M. Ritchey, and W. B. Moniz, *Macromolecules*, **15**, 1051 (1982).

257. C. A. McDowell, A. Naito, J. R. Scheffer, and Y. Wong, *Tetrahedron*, **22**, 4779 (1981).

258. A. M. Chippendale, A. Mathias, R. K. Harris, K. J. Packer, and B. J. Say, *J. Chem. Soc. Perkin Trans. 2*, 1031 (1981).

259. D. L. VanderHart, *J. Magn. Reson.*, **24**, 467 (1976).

260. T. M. Duncan, *J. Phys. Chem. Ref. Data*, **16**, 125 (1987).

261. D. W. Alderman, M. S. Solum, and D. M. Grant, *J. Chem. Phys.*, **84**, 3717 (1986).

262. T. G. Oas, G. P. Drobny, and F. W. Dahlquist, *J. Magn. Reson.*, **78**, 408 (1988).

263. C. M. Carter, D. W. Alderman, and D. M. Grant, *J. Magn. Reson.*, **65**, 183 (1985).

264. H. W. Spiess, *Ber. Bunsenges. Phys. Chem.*, **79**, 1009 (1975).

265. N. F. Ramsey, *Phys. Rev.*, **78**, 699 (1950).

266. K. W. Zilm, R. T. Conlin, D. M. Grant, and J. Michl, *J. Am. Chem. Soc.*, **102**, 6672 (1980).

267. B. Berglund and R. W. Vaughan, *J. Chem. Phys.*, **73**, 2037 (1980).

268. K. F. Lau and R. W. Vaughan, *Chem. Phys. Lett.*, **33**, 550 (1975).

269. P. van Hecke, H. W. Spiess, and U. Haeberlen, *J. Magn. Reson.*, **22**, 103 (1976).

270. R. G. Griffin, A. Pines, S. Pausak, and J. S. Waugh, *J. Chem. Phys.*, **63**, 1267 (1975).

271. F. Imashiro, S. Maeda, K. Takegoshi, T. Terao, and A. Saika, *Chem. Phys. Lett.*, **92**, 642 (1982).

272. T. Chiba, *J. Chem. Phys.*, **41**, 1352 (1964).

273. H. van Willigen, R. A. Haberkorn, and R. G. Griffin, *J. Chem. Phys.*, **67**, 917 (1977).

274. C. L. McKnett, C. R. Dybowski, and R. W. Vaughan, *J. Chem. Phys.*, **63**, 4578 (1975).

275. T. R. Steger, E. O. Stejskal, R. A. McKay, B. R. Stults, and J. Schaefer, *Tetrahedron Lett.*, 295 (1979).

276. S. Aravamudhan, U. Haeberlen, H. Irngartinger, and C. Krieger, *Mol. Phys.*, **38**, 241 (1979).

277. N. Zumbulyadis and H. J. Gysling, *Inorg. Chem.*, **21**, 564 (1982).

278. H. C. Dorn, B. E. Hanson, and E. Motell, *J. Organomet. Chem.*, **224**, 181 (1982).

279. R. Richarz and H. Sauter, *J. Magn. Reson.*, **52**, 308 (1983).

280. H. D. W. Hill, A. P. Zens, and J. Jacobus, *J. Am. Chem. Soc.*, **101**, 7090 (1979).

281. G. Harbison, J. Herzfeld, and R. G. Griffin, *J. Am. Chem. Soc.*, **103**, 4752 (1981).

282. A. M. Achlama, *J. Chem. Phys.*, **74**, 3623 (1981).

283. V. J. McBrierty, *Faraday Discuss. Chem. Soc.*, **68**, 78 (1979).

284. B. Crist and A. Peterlin, *J. Polym. Sci. Part A-2*, **7**, 1165 (1969).

285. G. E. Wardell, V. J. McBrierty, and D. C. Douglass, *J. Appl. Phys.*, **45**, 3441 (1974).

286. V. D. Fedotov, A. Ebert, and H. Schneider, *Phys. Stat. Solidi A*, **63**, 209 (1981).

287. K. Ludwigs and D. Geschke, *Ann. Phys.*, **38**, 370 (1981).

288. M. Goldman and L. Shen, *Phys. Rev.*, **144**, 321 (1966).

289. R. A. Assink, *Macromolecules*, **6**, 1233 (1978).

290. A. C. Lind, *J. Chem. Phys.*, **66**, 3482 (1977).

291. T. T. P. Cheung, B. C. Gerstein, L. M. Ryan, R. E. Taylor, and D. R. Dybowski, *J. Chem. Phys.*, **73**, 6059 (1980).

292. P. Caravatti, P. Neuenschwander, and R. R. Ernst, *Macromolecules*, **18**, 119 (1985).

293. D. C. Douglass and V. J. McBrierty, *Macromolecules*, **11**, 766 (1978).

294. J. Schaefer, M. D. Sefcik, E. O. Stejskal, and R. A. McKay, *Macromolecules*, **14**, 188 (1981).

295. J. M. Hewitt, P. M. Henrichs, M. Scozzafava, R. P. Scaringe, M. Linder, and L. J. Sorriero, *Macromolecules*, **17**, 2566 (1984).

296. N. M. Szeverenyi, M. J. Sullivan, and G. E. Maciel, *J. Magn. Reson.*, **47**, 462 (1982).

297. P. Caravatti, J. A. Deli, G. Bodenhausen, and R. R. Ernst, *J. Am. Chem. Soc.*, **104**, 5506 (1982).

298. P. Caravatti, G. Bodenbhausen, and R. R. Ernst, *J. Magn. Reson.*, **55**, 88 (1983).

299. H. T. Edzes and J. P. C. Bernards, *J. Am. Chem. Soc.*, **106**, 1515 (1984).

300. M. H. Frey and S. J. Opella, *J. Am. Chem. Soc.*, **106**, 4942 (1984).

301. K. M. Morden and S. J. Opella, *J. Magn. Reson.*, **70**, 476 (1986).

302. G. C. Gobbi, R. Silvestri, R. P. Russell, J. R. Lyerla, and W. W. Fleming, *J. Polym. Sci. Polym. Lett. Ed.*, **25**, 61 (1987).

303. P. M. Henrichs, J. Tribone, D. J. Massa, and J. M. Hewitt, *Macromolecules*, **2**, 1282 (1988).

304. M. Linder, P. M. Henrichs, J. M. Hewitt, and D. J. Massa, *J. Chem. Phys.*, **82**, 1585 (1985).

305. P. M. Henrichs and M. Linder, *J. Magn. Reson.*, **58**, 458 (1984).

306. D. Suter and R. R. Ernst, *Phys. Rev. B*, **32**, 5608 (1985).

307. P. M. Henrichs, M. Linder, and J. M. Hewitt, *J. Chem. Phys.*, **85**, 7077 (1986).

308. K. Takegoshi and C. A. McDowell, *J. Chem. Phys.*, **86**, 6077 (1987).

309. A. Kubo and C. A. McDowell, *J. Chem. Phys.*, **89**, 63 (1988).

310. D. P. Raleigh, M. H. Levitt, and R. G. Griffin, *Chem. Phys. Lett.*, **146**, 71 (1988).

311. W. E. J. R. Maas and W. S. Veeman, *Chem. Phys. Lett.*, **149**, 170 (1988).

312. E. O. Stejskal, J. Schaefer, M. D. Sefcik, and R. A. McKay, *Macromolecules*, **14**, 275 (1981).

313. B. Schröter and A. Posern, *Makromol. Chem.*, **182**, 675 (1981).

314. N. Zumbulyadis, *J. Magn. Reson.*, **53**, 486 (1983).

315. R. L. Dudley, C. A. Fyfe, P. J. Stephenson, Y. Deslandes, G. K. Hamer, and R. H. Marchessault, *J. Am. Chem. Soc.*, **105**, 1505 (1983).

316. C. A. Fyfe, R. L. Dudley, P. J. Stephenson, Y. Deslandes, G. K. Hamer, and R. H. Marchessault, *J. Macromol. Sci. Rev. Macromol. Chem. Phys.*, **C23**, 187 (1983).

317. R. L. Dudley, C. A. Fyfe, P. J. Stephenson, Y. Deslandes, G. K. Hamer, and R. H. Marchessault, *J. Am. Chem. Soc.*, **105**, 2469 (1983).

318. R. H. Atalla, J. C. Gast, D. W. Sindorf, V. J. Bartuska, and G. E. Maciel, *J. Am. Chem. Soc.*, **102**, 3249 (1980).

319. W. L. Earl and D. L. VanderHart, *J. Am. Chem. Soc.*, **102**, 3251 (1980).

320. F. Horii, A. Hirai, and R. Kitamaru, *Polym. Bull.*, **8**, 163 (1982).

321. G. E. Maciel, W. L. Kolodziejski, M. S. Bertran, and B. E. Dale, *J. Am. Chem. Soc.*, **15**, 686 (1982).

322. J. Kunze, G. Scheler, B. Schröter, and B. Phillipp, *Polym. Bull.*, **10**, 56 (1983).

323. W. L. Earl and D. L. VanderHart, *Macromolecules*, **14**, 570 (1981).

324. A. D. English and A. J. Vega, *Macromolecules*, **12**, 353 (1979).

325. A. J. Vega and A. D. English, *Macromolecules*, **13**, 1635 (1980).

326. D. Hentschel, H. Sillescu, and H. W. Spiess, *Makromol. Chem.*, **180**, 241 (1979).

327. D. Hentschel, H. Sillescu, and H. W. Spiess, *Macromolecules*, **14**, 1607 (1981).

328. R.-J. Roe, *J. Polym. Sci. Part A-2*, **8**, 1187 (1970).

329. R.-J. Roe, *J. Appl. Phys.*, **36**, 2024 (1965).

330. V. J. McBrierty, *J. Chem. Phys.*, **61**, 872 (1974).

331. D. I. Bower, *J. Polym. Sci. Polym. Phys. Ed.*, **19**, 93 (1981).

332. R. Hentschel, J. Schlitter, H. Sillescu, and H. W. Spiess, *J. Chem. Phys.*, **66**, 56 (1978).

333. R. Hentschel, H. Sillescu, and H. W. Spiess, *Polymer*, **22**, 1516 (1981).

334. R. Friesner, J. A. Nairn, and K. Sauer, *J. Chem. Phys.*, **71**, 358 (1979).

335. H. W. Spiess, "NMR in Oriented Polymers," in *Developments in Oriented Polymers*, I. M. Ward, Ed., Applied Science Publishers, London, 1981.

336. S. J. Opella and J. S. Waugh, *J. Chem. Phys.*, **66**, 4919 (1977).

337. T. Shibata and S. Iwayanagi, *Polym. J.*, **10**, 599 (1978).

338. M. Kashiwagi, A. Cunningham, A. J. Manuel, and I. M. Ward, *Polymer*, **14**, 111 (1973),

339. V. J. McBrierty, I. R. McDonald, and I. M. Ward, *J. Phys. D*, **4**, 88 (1971).

340. B. T. Nall, W. P. Rothwell, J. S. Waugh, and A. Ruprecht, *Biochemistry*, **20**, 1881 (1981).

341. L. W. Jelinski, *Macromolecules*, **14**, 1341 (1981).

342. P. D. Murphy, T. Taki, B. C. Gerstein, P. M. Henrichs, and D. J. Massa, *J. Magn. Reson.*, **49**, 94 (1982).

343. H. T. Edzes, *Polymer*, **24**, 1425 (1983).

344. D. L. VanderHart, *Macromolecules*, **12**, 1232 (1979).

345. P. M. Henrichs, *Macromolecules*, **20**, 2099 (1987).

346. B. Blümich, C. Böffel, G. S. Harbison, Y. Yang, and H. W. Spiess, *Ber. Bunsenges. Phys. Chem.*, **91**, 1100 (1987).

347. G. S. Harbison and H. W. Spiess, *Chem. Phys. Lett.*, **124**, 128 (1986).

348. J. Seelig, *Q. Rev. Biophys.*, **10**, 353 (1977).

349. T. M. Rothgeb and E. Oldfield, *J. Biol. Chem.*, **256**, 1432 (1981).

350. E. Oldfield and T. M. Rothgeb, *J. Am. Chem. Soc.*, **102**, 3636 (1980).

351. P. Sotta, B. Deloche, and J. Herz, *Polymer*, **29**, 1171, (1988).

352. B. Deloche, A. Bubault, J. Herz, and A. Lapp, *Europhys. Lett.*, **1**, 629 (1986).

353. B. Deloche and E. T. Samulski, *Macromolecules*, **14**, 575 (1981).

354. D. A. Torchia and A. Szabo, *J. Magn. Reson.*, **49**, 107 (1982).

355. A. Naito, S. Ganapathy, K. Akasaka, and C. A. McDowell, *J. Magn. Reson.*, **54**, 226 (1983).

356. M. G. Gibby, A. Pines, and J. S. Waugh, *Chem. Phys. Lett.*, **16**, 296 (1972).

357. D. J. Siminovitch, M. J. Ruocco, E. T. Olejniczak, S. K. Das Gupta, and R. G. Griffin, *Chem. Phys, Lett.*, **119**, 81 (1985).

358. L. M. Jackman and F. A. Cotton, *Dynamic Nuclear Magnetic Resonance Spectroscopy*, Academic, New York, 1975.

359. N. M. Szeverenyi, A. Bax, and G. E. Maciel, *J. Am. Chem. Soc.*, **105**, 2579 (1983).

360. N. M. Szeverenyi, M. J. Sullivan, and G. E. Maciel, *J. Magn. Reson.*, **47**, 462 (1982).

361. C. S. Yannoni, T. C. Clarke, R. D. Kendrick, V. Macho, R. D. Miller, and P. C. Myhre, *Mol. Cryst. Liq. Cryst.*, **96**, 305 (1983).

362. J. R. Lyerla, *Acc. Chem. Res.*, **15**, 208 (1982).

363. C. S. Yannoni, V. Macho, and P. C. Myhre, *J. Am. Chem. Soc.*, **104**, 7380 (1982).

364. R. D. Miller and C. S. Yannoni, *J. Am. Chem. Soc.*, **102**, 7397 (1980).

365. V. Macho, R. D. Miller, and C. S. Yannoni, *J. Am. Chem. Soc.*, **105**, 3735 (1983).

366. M. S. Greenfield, A. D. Ronemus, R. L. Vold, R. R. Vold, P. D. Ellis, and T. F. Raidy, *J. Magn. Reson.*, **72**, 89 (1987).

367. D. A. Torchia and A. Szabo, *J. Magn. Reson.*, **64**, 135 (1985).

368. E. W. Fischer, G. P. Hellmann, H. W. Spiess, F. J. Hörth, U. Ecarius, and M. Wehrle, *Makromol. Chem.*, *Suppl.*, **12**, 189 (1985).

369. C. Schmidt, K. J. Kuhn, and H. W. Spiess, *Prog. Coll. Polym. Sci.*, **71**, 71 (1985).

370. M. Wehrle, G. P. Hellmann, and H. W. Spiess, *Coll. Polym. Sci.*, **265**, 815 (1987).

371. D. E. Wemmer, D. J. Ruben, and A. Pines, *J. Am. Chem. Soc.*, **103**, 28 (1981).

372. D. L. VanderHart, *Polym. Prepr. Am. Chem. Soc. Div. Polym. Chem.*, **22**, 261 (1981).

373. P. L. Kuhns and M. S. Conradi, *J. Chem. Phys.*, **77**, 1771 (1982).

374. A. P. M. Kentgens, E. de Boer, and W. S. Veeman, *J. Chem. Phys.*, **87**, 6859 (1987).

375. A. F. de Jong, A. P. M. Kentgens, and W. S. Veeman, *Chem. Phys. Lett.*, **109**, 337 (1984).

376. G. S. Harbison, D. P. Raleigh, J. Herzfeld, and R. G. Griffin, *J. Magn. Reson.*, **64**, 284 (1985).

377. E. R. Andrew and R. A. Newing, *Proc. Phys. Soc.*, **72**, 959 (1958).

378. E. R. Andrew, *J. Phys. Chem. Solids*, **18**, 9 (1961).

379. W. P. Rothwell and J. S. Waugh, *J. Chem. Phys.*, **74**, 2721 (1981).

380. D. Suwelack, W. P. Rothwell, and J. S. Waugh, *J. Chem. Phys.*, **73**, 2559 (1980).

381. A. Schmidt, S. O. Smith, D. P. Raleigh, J. E. Roberts, R. G. Griffin, and S. Vega, *J. Chem. Phys.*, **85**, 4248 (1986).

382. A. Schmidt and S. Vega, *J. Chem. Phys.*, **87**, 6895 (1987).

383. A. J. Vega, A. D. English, and W. Mahler, *J. Magn. Reson.*, **37**, 107 (1980).

384. Y. Yang, A. Hagemayer, B. Blümich, and H. W. Spiess, *Chem. Phys. Lett.*, **150**, 1 (1988).

385. Y. Yang, M. Schuster, B. Blümich, and H. W. Spiess, *Chem. Phys. Lett.*, **139**, 239 (1987).

386. U. Haeberlen and J. S. Waugh, *Phys. Rev.*, **185**, 185 (1969).

387. H. Sillescu, *Ber. Bunsenges. Phys. Chem.*, **75**, 283 (1971).

388. S. Alexander, A. Baram, and Z. Luz, *J. Chem. Phys.*, **64**, 4321 (1976).

389. J. H. Freed, G. V. Bruno, and C. F. Polnaszek, *J. Phys. Chem.*, **75**, 3385 (1971).

390. H. W. Spiess, *Chem. Phys.*, **6**, 217 (1974).

391. J. Schaefer, R. A. McKay, E. O. Stejskal, and W. T. Dixon, *J. Magn. Reson.*, **52**, 123 (1983).

392. R. F. Campbell, E. Meirovitch, and J. H. Freed, *J. Phys. Chem.*, **83**, 525 (1979).

393. H. W. Spiess, *Coll. Polym. Sci.*, **261**, 193 (1983).

394. H. W. Spiess and H. Sillescu, *J. Magn. Reson.*, **42**, 381 (1981).

395. T. B. Smith, E. A. Moore, and M. Mortimer, *J. Phys. C.*, **14**, 3965 (1981).

396. D. M. Rice, R. J. Wittebort, R. G. Griffin, E. Meirovitch, E. R. Stimson, Y. C. Meinwald, J. H. Freed, and H. A. Scheraga, *J. Am. Chem. Soc.*, **103**, 7707 (1981).

397. H. W. Spiess, *J. Chem. Phys.*, **72**, 6755 (1980).

398. M. Lausch and H. W. Spiess, *J. Magn. Reson.*, **54**, 466 (1983).

399. E. Rössler, H. Sillescu, and H. W. Spiess, *Polymer*, **26**, 203 (1985).

400. F. Fujara, S. Wefing, and H. W. Spiess, *J. Chem. Phys.*, **84**, 4579 (1986).

401. S. B. Ahmad and K. J. Packer, *Mol. Phys.*, **37**, 47 (1979).

402. C. Schmidt, B. Blümich, S. Wefing, S. Kaufmann, and H. W. Spiess, *Ber. Bunsenges. Phys. Chem.*, **91**, 1141 (1987).

403. C. Schmidt, B. Blümich, and H. W. Spiess, *J. Magn. Reson.*, **79**, 269 (1988).

404. S. Wefing and H. W. Spiess, *J. Chem. Phys.*, **89**, 1219 (1988).

405. S. Wefing and H. W. Spiess, *J. Chem. Phys.*, **89**, 1234 (1988).

406. K. Rosenke, H. Sillescu, and H. W. Spiess, *Polymer*, **21**, 757 (1980).

407. D. Hentschel, H. Sillescu, and H. W. Spiess, *Macromolecules*, **14**, 1606 (1981).

408. L. W. Jelinski, J. J. Dumais, and A. K. Engel, *Macromolecules*, **16**, 492 (1983).

409. A. J. Vega and R. W. Vaughan, *J. Chem. Phys.*, **68**, 1958 (1978).

410. C. Dybowski and R. G. Pembleton, *J. Chem. Phys.*, **70**, 1962 (1979).

411. J. Schaefer, E. O. Stejskal, and R. Buchdahl, *Macromolecules*, **8**, 291 (1975).

412. T. R. Steger, J. Schaefer, E. O. Stejskal, and R. A. McKay, *Macromolecules*, **13**, 1127 (1980).

413. D. L. VanderHart and A. N. Garroway, *J. Chem. Phys.*, **71**, 2773 (1979).

414. J. Schaefer, E. O. Stejskal, and R. Buchdahl, *J. Macromol. Sci. Phys.*, **B13**, 665 (1977).

415. M. D. Sefcik, J. Schaefer, E. O. Stejskal, and R. A. McKay, *Macromolecules*, **13**, 1132 (1980).

416. A. N. Garroway, W. B. Moniz, and H. A. Resing, *Faraday Symp. Chem. Soc.*, **13**, 63 (1978).

417. K. Akasaka, S. Ganapathy, C. A. McDowell, and A. Naito, *J. Chem. Phys.*, **78**, 3567 (1983).

418. J. Schaefer, M. D. Sefcik, E. O. Stejskal, and R. A. McKay, *Macromolecules*, **14**, 280 (1981).

Chapter **6**

APPLICATION OF MAGNETIC RESONANCE TECHNIQUES TO THE STUDY OF DEFECTS IN SOLIDS

J.-Martin Spaeth

1   INTRODUCTION

Point defects such as impurity atoms or ions or intrinsic defects such as
vacancies, interstitials, or antisite defects in binary crystals or aggregates of these
defects very often determine the bulk properties of a solid. For example, this
occurs in ionic crystals, oxides, and, particularly, in semiconductors. Therefore,
it is of prime interest for materials science to have methods that are able to
determine the defect structures and to correlate them with the bulk properties of
materials, like optical or electronic properties. The latter is very important for
microelectronics and optoelectronics in semiconductors or for the production of
tunable or high-power lasers, just to name two fields of actual worldwide
interest. Radiation damage of solids and detection of ionizing radiation are
other fields where the knowledge of the structures of the defects created and their
interactions are absolutely necessary.

There are many methods to *characterize* solids, a term currently used often in
semiconductor research and technology. However, with a closer look, very few
methods can determine defect structures, let alone determine them accurately on
a microscopic or *atomic* scale. Local vibrational mode (LVM) spectroscopy can
often determine the impurity that is involved in a defect and the defect
symmetry, but correlation to other properties is not easily possible. Mössbauer
effect, perturbed angular correlation (PAC), and other methods of the so-called
nuclear solid state physics can frequently offer information about the symmetry
of a defect and about nuclear quadrupole interactions with a very high
sensitivity. The method is restricted, however, to a few suitable nuclei that must
be incorporated into the solid for this investigation, which also limits their use.

Paramagnetic resonance is the most powerful technique; however, it is useful
only when the defects are paramagnetic. Fortunately, this is very often the case,
especially when transition metals or rare earth ions are involved. In radiation
damage, where electron and hole pairs are created, and in semiconductors,
where upon doping the Fermi level can often be suitably shifted, the defects of
interest are or can be made paramagnetic.

Electron spin resonance (ESR) is a widely used method. Unfortunately, for
the investigation of defects in solids it is suitable only in particularly favorable

cases. In solids, generally, the ESR suffers from large line widths in contrast to its powerful use in chemistry. A combination of ESR and nuclear magnetic resonance (NMR), the so-called *electron nuclear double resonance (ENDOR)*, is the most powerful tool, if it works for the specific defect under investigation. In materials science, ENDOR spectra are usually very complicated. This may be the reason why ENDOR spectroscopy in solids is not more popular. However, computer-assisted experiments and analysis of spectra, together with advanced ENDOR methods, made ENDOR spectroscopy quite feasible and of practical use beyond simple model cases, for which this spectroscopy was originally applied following its discovery in 1959 [1].

The number of defects needed for ENDOR in solids is about two orders of magnitude higher than necessary for detection by ESR, which limits the method to rather high defect concentrations of about $10^{15}-10^{16}$cm^{-3} or more in most practical cases. Optical detection of ESR and ENDOR proved to be more sensitive by several orders of magnitude and therefore has received much attention recently, especially in semiconductor physics, where defect concentrations of 10^{15}cm^{-3} and less are of technological importance and where thin layers are important for devices. These samples cannot be studied by conventional ESR and ENDOR because of the small volume and resulting small number of defects. Several other advantageous features of optical detection seem to be causing a revival of this kind of magnetic resonance spectroscopy. Among these features is the possibility of measuring ESR and ENDOR with high spatial resolution, which is very important from technological aspects, for example, when there is a need for electrically homogeneous large wafers for microelectronic and optoelectronic applications. This is a resolution that cannot be obtained for ESR by imaging techniques using magnetic field gradients as is done for NMR imaging.

This chapter reviews briefly the multiple magnetic resonance techniques presently available to determine the defect structures and to correlate defect structures with bulk properties. It outlines what can be accomplished. A precise analysis of the spectra is beyond the scope of this chapter. For convenience the examples used to illustrate the methods are taken largely from the recent work of the group at the University of Paderborn. Therefore, this is not a review in the sense of summarizing all work done in the field. Furthermore, it is assumed that the reader is familiar with the basic concepts and the realization of ESR, which is not discussed in detail. For this reference is especially made to McKinney and Goldberg [2] and textbooks [3–6]. In these textbooks the ESR of point defects in solids is discussed in detail. The ESR of transition metal ion impurities in solids is discussed comprehensively by Abragam and Bleaney [3].

This chapter does not deal with how to measure ENDOR using electron spin–echo methods. These methods have certain advantages, especially at low frequencies, but they also have serious disadvantages, especially in the regime of broad ESR lines and high ENDOR frequencies. The reader is referred to recent articles by Schweiger for these experimental methods [6].

2   STRUCTURE DETERMINATION OF DEFECTS BY MAGNETIC RESONANCE

2.1   Basic Concepts

The ability to determine defect structures by paramagnetic resonance spectroscopy is based on the measurement of the magnetic interaction between the magnetic moment of the unpaired electron(s) or hole(s) of the defect and the magnetic moments of nuclei belonging to the *impurity* atom of the defect (if any) and the atoms of its surrounding lattice. This interaction is usually called *hyperfine (hf) interaction* when it occurs between the unpaired electron (hole) and the central nucleus of the impurity atom, if such a defect is under study. It is called *superhyperfine (shf) interaction* or *ligand hyperfine (lhf) interaction* when it occurs between the magnetic moment of the unpaired electron (hole) and the magnetic moments of the nuclei of the surrounding lattice. In Figure 6.1 a schematic representation of a defect is shown, and the electron spin and the nuclear spins are indicated. The hyperfine interaction is often resolved in ESR; if the impurity nucleus occurs in several magnetic isotopes, then a chemical identification of the impurity is easily possible by taking advantage of the known magnetic moments, the ratio of which determines the ratio of the observed hyperfine splittings (see Section 3), and by measuring the ESR signal intensities, which reflect the natural abundance of the different isotopes. Unfortunately,

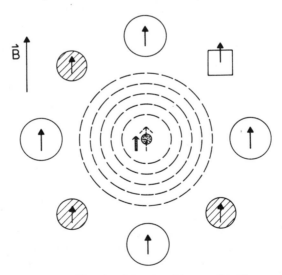

Figure 6.1   Schematic representation of a defect in a binary solid. The unpaired electron spin, electron spin density distribution, and nuclear spins of the lattice nuclei are indicated. The spins, which are aligned in an external static magnetic field **B**, are indicated by arrows. The open square represents a vacancy, which can have an unpaired electron spin due to dangling bonds at the neighboring atoms. The hatched and open circles represent two kinds of atoms or ions, for example, cations and anions in a binary ionic crystal.

there are important and common impurities, like oxygen, where over 99% of the isotopes are nonmagnetic. Here, a specific doping with a magnetic isotope (e.g., ^{17}O) may be necessary.

The g-factors (g-tensors) of the ESR spectra also contain information on the defect structure; they reflect the symmetry of the defects. The deviations of the principal values of the g-tensor from the g-value of the free electron ($g_e = 2.0023$) are due to the mixing of orbital contributions and the spin contribution to the total magnetic moment. This mixing is caused by the spin–orbit interaction. The orbital parts see the electrical crystal field around the defect and reflect therefore its symmetry [4, 5]. For a precise structure determination, apart from symmetry, the g-factor analysis usually does not yield enough reliable information, since excited-state energies, which are mostly unknown, are involved in the g-factor analysis [2, 3–6].

If it were possible to measure the NMR of the nuclei of the surrounding lattice, the local magnetic fields seen by the nuclei would give also the required information on the defect structure, since they are a superposition of the applied static magnetic field and the field equivalent of the superhyperfine or ligand hyperfine interactions. If the local fields of all the nuclei and their symmetry relation to the defect site were known, the structure of the defect, including the presence of vacancies and lattice distortions, can be derived. Unfortunately, however, NMR is not sensitive enough. One needs at least about 10^{19} nuclei like protons to measure NMR, while one deals with defect concentrations of 10^{16} cm^{-3} and less and sample volumes usually well below 1 cm^3.

Generally, in ESR the superhyperfine interaction is not well resolved; if it is at all, it is usually only with the nearest-neighbor nuclei. Therefore, ESR spectroscopy is often suitable for impurity identification by a resolved hyperfine interaction, but not for a precise microscopic structure determination. Recently, in fact, in GaAs several ESR spectra were observed with the same central ^{75}As hyperfine splitting and almost the same line width, which all belong to different defect structures. These different defects have only one common feature: an arsenic antisite; that is, an arsenic atom on a Ga site is involved. However, the remaining structure is different. By measuring only ESR one can be badly misled.

A higher resolution for the superhyperfine interactions and a higher sensitivity than obtained in NMR is achieved by measuring electron nuclear double resonance (ENDOR) [1]. The NMR transition of neighbor nuclei coupled to the unpaired electron causes a change of the electron spin polarization under suitable experimental conditions (partly saturated ESR), which can be detected as ENDOR signals (see Section 4). Thus, the NMR transitions are detected using a quantum transformation to higher quanta, the ESR microwave quanta, which results in a higher sensitivity. If we take one step further, that is, a quantum transformation to optical quanta, the optical detection of ESR and ENDOR is reached, which renders the measurement of superhyperfine interactions even more sensitive. This reason and the fact that there are less NMR lines compared to ESR superhyperfine lines from the same number of neighbor nuclei make the use of modern multiple resonance methods like ENDOR, ODESR (optically detected ESR), and ODENDOR (optically detected ENDOR) very useful when determining defect structures.

By resolving the superhyperfine interaction and the ligand quadrupole interactions one can determine the number and the symmetry relation of neighbor nuclei to the defect center and also the coupling constants such as superhyperfine and quadrupole constants. However, it is not possible to determine from the coupling constants the distance of the identified nuclei from the defect center without having a theoretical wave function for the defect electron (hole) or without assumptions about the radial part of this wave function, for example, whether it falls off monotonously with distance from the defect core. This is very often, but not always, the case. There may be oscillations of the unpaired spin density with distance especially for shallow defects in semiconductors. If a reasonable assumption about this radial part can be made, then a clear defect model can be derived unless there are special symmetry conditions where one cannot decide between possible sites. For example, this occurs for the substitutional and one tetrahedral interstitial site in the diamond structure where the same symmetry types of neighbor nuclei occur, only with a different sequence. If nothing is known about the wave function, one cannot decide on the site from ENDOR alone. For instance, this occurred recently for chalcogen defects in silicon [7].

2.2   Specific Problems in Materials Science

A single defect species in a simple single crystal is a rare case in materials science. Generally, in materials of interest, there are many defect species present simultaneously, very often with overlapping ESR spectra of very different signal intensity. Hence, the measurement of straight ENDOR usually results in such a complicated spectrum that an analysis is impossible. Here the use of a sort of excitation spectrum of ENDOR, *ENDOR-induced ESR* (sometimes called *field-swept ENDOR*), and *electron nuclear triple resonance* (*double ENDOR*) are of great help. With the latter method it is possible to measure separately the ENDOR spectra of one particular defect species (see Section 5.2). The selectivity is also high by using optical detection, since the absorption or emission bands of different defect species seldom overlap completely. Therefore, an optical transition specific to one defect species can be used for its exclusive measurement.

Another problem is connected with the correlation between bulk properties and the defects causing them. Obviously, by optical detection, a correlation can be made to optical properties of the crystals. Another problem is that of correlation to electrical properties that are usually characterized by the energy levels of a particular defect in the energy gap of a semiconductor. Here it is possible to perform photo-ESR, photo-ENDOR, photo-ODESR, and photo-ODENDOR experiments. The idea is to codope the material so that the energy level of the defect in question is not occupied. By irradiating the sample with light of variable energy one can lift electrons from the valence band and populate the energy levels so that a magnetic resonance measurement is possible. In this way a correlation to the energy levels of the defects can be achieved (see Section 6.6.1).

The work discussed in here refers to single crystals only; powders and amorphous materials are beyond the scope of this chapter.

3 ELECTRON SPIN RESONANCE

In a static magnetic field B_0 along the z axis of a laboratory system a free electron has the magnetic moment

$$\mu_z = -g_e \mu_B m_s \tag{1}$$

where g_e is the g-factor of the free electron ($g = 2.0023$), μ_B is the Bohr magneton ($\mu_B = e\hbar/2m_e = 9.274078 \cdot 10^{-24}\,\mathrm{A} \cdot \mathrm{m}^2$), and m_s is the electron spin quantum number, $\pm\frac{1}{2}$. In a magnetic field the potential energy is

$$E = g_e \mu_B B_0 m_s \tag{2}$$

giving rise to the two electron Zeeman levels shown in Figure 6.2a as a function of the static magnetic field B_0. In the basic ESR experiment a magnetic dipole

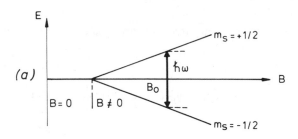

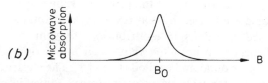

Figure 6.2 (a) Electron Zeeman levels for $S = \frac{1}{2}$ in $B = 0$ and $B \neq 0$. The magnetic dipole transition of the basic ESR experiment is indicated and occurs at $B = B_0$ for the microwave energy $\hbar\omega$. (b) Microwave absorption at $B = B_0$: the ESR line.

transition is induced between the Zeeman levels for $m_s = \pm\frac{1}{2}$ if the resonance condition

$$h\nu_{\text{ESR}} = g_e\mu_B B_0 \tag{3}$$

is fulfilled, where ν_{ESR} is the frequency of the oscillating magnetic field driving the transition [2–5]. The transition is characterized by $\Delta m_s = \pm 1$ as selection rule. Since $\nu_{\text{ESR}}/B_0 = g_e\mu_B/h = 2.802\,\text{MHz}/10^{-4}\text{T}$, the frequency is in the microwave region for the most commonly used magnetic fields. In solids one uses X band around $10\,\text{GHz}$ ($B_0 = 0.35\,\text{T}$), K band around $24\,\text{GHz}$ ($B_0 = 0.86\,\text{T}$), and Q band around $35\,\text{GHz}$ ($B_0 = 1.2\,\text{T}$). For technical reasons one varies the magnetic field and keeps ν_{ESR} constant (Figure 6.2b). Upon resonance a microwave absorption is measured. To be able to use lock-in techniques for sensitivity enhancement one measures the derivative of the absorption as a consequence of magnetic field modulation [5].

In paramagnetic defects the electron is usually bound to an (impurity) atom or occupies a vacancy; therefore, the magnetism of the defect contains both spin and orbital angular momentum contributions. This is taken into account in the formalism of the effective spin and the spin Hamiltonian, which usually both refer to the ground state of the system that is of prime interest for the structure determination. For a justification and further details refer to the textbooks of ESR (e.g., [3–5]).

The spin Hamiltonian for the Zeeman interaction is

$$\mathcal{H} = \mu_B \mathbf{B}_0 \tilde{g} \mathbf{S} \tag{4}$$

where $\mathbf{S}$ is the operator of the effective spin, and $\tilde{g}$ is the g-tensor that reflects the symmetry of the defect [4, 5].

From (4) it follows that

$$E = \mu_B B_0 (g_x^2 l^2 + g_y^2 m^2 + g_z^2 n^2)^{1/2} m_s \quad \text{and} \tag{5}$$

$$E = g\mu_B B_0 m_s \tag{6}$$

where g_x, g_y, and g_z are the principal values of the g-tensor in the principal axis form; l, m, and n are the direction cosines of the tensor orientation with respect to the laboratory frame: $l = \cos(x, x')$, $m = \cos(y, y')$, and $n = \cos(z, z')$, where x, y, z is the laboratory (crystal) axis system and (x', y', z') is that of the principal axes of the g-tensor. The deviations of the principal values of the g-tensor from the free electron value are caused by contributions of orbital angular momentum, which are mixed into the spin contributions by spin–orbit interactions. Their magnitude depends on both the spin–orbit interaction and the energy differences between the ground and excited states [5]. The electrical crystal field experienced by the defect, for example, by an impurity atom at its crystalline site, influences the defect energy through the orbital contributions to the magnetic moment. Therefore the ESR spectra are dependent on the relative orientation of the static magnetic field B_0 to the crystal axes. This is formally expressed by (5) and (6). A measurement of the angular dependence of the ESR spectrum for rotation of the magnetic field B_0 in one or several crystal planes reveals the symmetry of the crystal field experienced by the defect, and thus also reveals the symmetry of its electronic wave function [4, 5]. From ESR investigations often only this is known for a point defect in a single crystal, which then is *characterized* by its symmetry. If the defect has lower point symmetry than the crystalline site and hence several equivalent orientations are possible with respect to crystal symmetry operations, then in the experiment for a particular field orientation B_0 several defect orientations are measured simultaneously. If the g-anisotropy is not very big, the overlaps of such ESR lines can make an analysis difficult. Figure 6.3 shows a defect with C_{2v} symmetry. The defect is the singly ionized thermal donor *NL8* in silicon. The thermal donor has two (110) mirror planes and one twofold axis along [001] [8, 9]. For the orientation chosen in the spectrum of Figure 6.3a several defect orientations overlap. Figure 6.3b shows the angular dependence of the ESR spectra. The center orientations

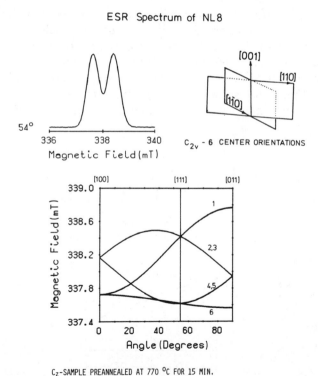

Figure 6.3 Angular dependence of the ESR spectrum of the thermal donors in silicon (so-called *NL8 ESR spectrum*) for rotation of B_0 in a (110) plane between $\langle 100 \rangle$ and $\langle 011 \rangle$. The different lines are caused by the g-factor anisotropy and the different center orientations in the crystal. The thermal donors have C_{2v} symmetry.

are noted. From the angular dependence the defect symmetry was inferred. Electron spin resonance does not reveal more information about the defect structure (see Section 4.4). The defect is formed by oxygen aggregation in silicon [8,9].

When the defect contains several electrons (or holes) with parallel spin, the spin Hamiltonian must have additional terms caused by spin–orbit and crystal field effects. These terms are often called *fine-structure terms*. They depend on both the spin operator and the magnetic field. The term of lowest order in the spin, and the most important, is shown in (7) in addition to the electron Zeeman term [3].

$$\mathcal{H} = \mu_B B_0 \tilde{g} S + S \tilde{F} S \qquad (7)$$

For the simple case of axial symmetry of the defect the energy becomes (for an isotropic g-factor)

$$E = g_e \mu_B B_0 m_s + D/6\{[3\cos^2\theta - 1][3 m_s^2 - S(S+1)]\} \qquad (8)$$

where $D = \frac{3}{2} F_z$. Equation (8) is valid in perturbation theory of first order, that is, for the approximation that the fine-structure term is small compared to the electronic Zeeman term. The angle between the z axis of the fine-structure tensor in its principal axis system and the orientation of B_0 is θ [5]. There is no fine-structure splitting for $S = \frac{1}{2}$. For further details see, for example, [3, 5].

For $S = \frac{3}{2}$ Figure 6.4 shows that there are three allowed transitions for $\Delta m_s = \pm 1$; for $S = \frac{5}{2}$ there are five transitions. As an example Figure 6.5 shows

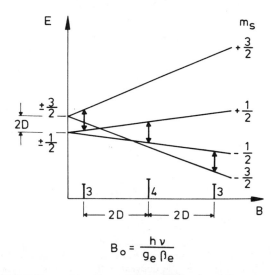

Figure 6.4   The ESR transitions for a defect with $S = \frac{3}{2}$ and fine-structure splitting.

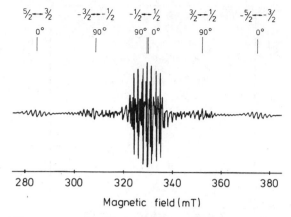

Figure 6.5 The ESR spectrum of Fe^{3+} defects in $RbCdF_3$ in the tetragonal phase. The fine-structure splitting is indicated by the m_s spin quantum numbers. The angles refer to the orientation between the axes of the domains and the external magnetic field, since the crystal contains several domains. Not all fine-structure transitions are marked by the m_s quantum numbers for all domain orientations. $B_0 \| \langle 100 \rangle$, $\nu_{ESR} = 9.26$ GHz, $T = 17$ K. Adapted from P. Studzinski, "Paramagnetische Ionen zur Untersuchung struktureller Phasenübergänge-Eine ENDOR Untersuchung," doctoral dissertation, Universität-GH-Paderborn, 1985.

the ESR spectrum of Fe^{3+} in $RbCdF_3$ in the low-temperature tetragonal phase. A superhyperfine structure with the six nearest ^{19}F neighbors is rather well resolved; Fe^{3+} substitutes for Cd^{2+}. All five fine-structure transitions expected for a $3d^5$ ($S = \frac{5}{2}$) configuration are seen. The transitions $-\frac{3}{2} \rightarrow -\frac{1}{2}$ and $+\frac{3}{2} \rightarrow \frac{1}{2}$ for the tetragonal defect axis parallel to B_0 (0°) are superimposed on those with perpendicular axes (90°), which are marked in Figure 6.5. Not all transitions for perpendicular axes are labeled with their spin quantum numbers in Figure 6.5 [10].

The fine-structure interaction results in a zero-field splitting. Whether a fine-structure splitting is observed depends on the specific configuration of the defect and the size of the splitting relative to the ESR line width. Therefore, it is not always easy to judge the spin state of a defect and thus its charge state from the ESR spectra. Figure 6.6 shows the ESR spectrum for Ni^{3+} in GaP [11, 12]; Ni^{3+} occupies a Ga site and has the configuration $3d^7$. There is no fine-structure splitting being resolved, and the spectrum appears to arise from a defect with $S = \frac{1}{2}$; in fact, it was shown by ENDOR that $S = \frac{3}{2}$ (see Section 4). The superhyperfine interaction with the four nearest ^{31}P neighbors is partly resolved for $B \| \langle 100 \rangle$.

For $S = \frac{1}{2}$ ESR lines are split by the hyperfine interaction. This splitting gives the most information on the microscopic nature of the defect in ESR.

The interaction is basically a magnetic dipole–dipole interaction, and it is described by the hyperfine tensor $\tilde{A}$ in the spin Hamiltonian

$$\mathscr{H} = g\mu_B B_0 S + I\tilde{A}S - g_I \mu_n B_0 I \tag{9}$$

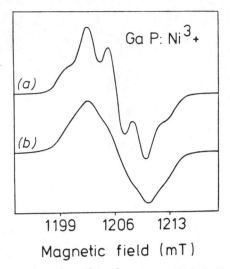

Figure 6.6   The ESR spectrum of Ni^{3+} ($3d^7$, $S = \frac{3}{2}$) on Ga sites in GaP: (a) $B_0 \parallel \langle 100 \rangle$, (b) $B_0 \parallel \langle 110 \rangle$. There is no fine-structure splitting observable. The partly resolved structure for $B_0 \parallel \langle 100 \rangle$ is caused by the superhyperfine interaction with four equivalent nearest ^{31}P neighbors. Reprinted with permission from *Solid State Commun.*, **25**, U. Kaufmann and J. Schneider, "ESR Assessment of $3d^7$ Transition Metal Impurity States in GaP, GaAs and InP," Copyright 1978, Pergamon Journals Ltd.

where μ_n is the nuclear magneton ($\mu_n = m_e/m_p\mu_B$), m_p is the proton mass, and g_I is the nuclear g-factor. For simplicity an isotropic electronic g-factor is assumed. In first-order perturbation theory one obtains the energy

$$E = g\mu_B B_0 m_s + A(\theta)\, m_I m_s \tag{10}$$

and the resonance frequency for $S = \frac{1}{2}$ is

$$h\nu_{ESR} = g_e \mu_B B_{res} + m_I A(\theta) \tag{11}$$

For $I = \frac{1}{2}$ there are two values of the nuclear spin quantum number $m_I = \pm\frac{1}{2}$. Since ν_{ESR} is kept constant, there are two resonance fields, which are separated by $A(\theta)/g_e\mu_B$ (see Figure 6.7).

Figure 6.8 shows how a hyperfine interaction can be used to identify a defect for the example of Te^+ defects in Si [7]. Ninety-two percent of the Te isotopes are diamagnetic and give rise to the central ESR line near 350 mT (measured in X band) corresponding to $g=2$. The two magnetic isotopes ^{125}Te (7% abundant) and ^{123}Te (0.9% abundant) have $I = \frac{1}{2}$ and, according to (11), a doublet splitting. The ESR line intensities follow the isotope abundance, and the relative splittings are in the ratio of the respective nuclear moments [see (17) and (18)]. Therefore the defect is unambiguously identified as being caused by Te impurities. However, no further structure caused by the superhyperfine interactions is resolved. Therefore, the site of Te^+ in the lattice cannot be determined

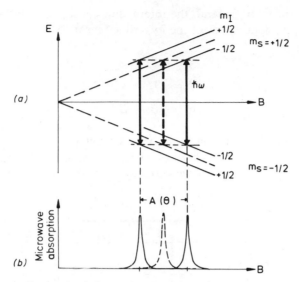

Figure 6.7 Electron Zeeman levels for an electron spin $S = \frac{1}{2}$ hyperfine coupled to a nuclear spin $I = \frac{1}{2}$ and the resulting ESR transitions with hyperfine splitting. The selection rule for the ESR transition is $\Delta m_s = \pm 1$, $\Delta m_I = 0$.

from ESR. The defect Te^+ has an isotropic hyperfine interaction. In general, the hyperfine interaction is anisotropic, and the spectrum is angular dependent as indicated in (11). In first order and for axial symmetry of the defect the resonance fields are given by

$$h\nu_{ESR} = g\mu_B B_{res} + m_I[a + b(3\cos^2\theta - 1)] \tag{12}$$

where a is the isotropic hyperfine constant, b is the anisotropic hyperfine constant, θ is the angle between the magnetic field B_0 and the principal axis z of

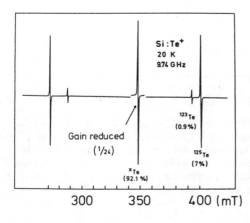

Figure 6.8 An ESR spectrum of Te^+ centers in Si. The g-factor is isotropic. The hyperfine splittings are caused by the two magnetic isotopes ^{125}Te and ^{123}Te with $I = \frac{1}{2}$. Reprinted with permission from H. G. Grimmeiss, E. Janzen, H. Ennen, O. Schirmer, *Phys. Rev. B*, **24**, 4571 (1981).

the hyperfine tensor. In general, the interaction constants a and b and b' are related to the principal values of the hyperfine tensor by

$$\tilde{A} = (a \cdot \tilde{I} + \tilde{B})$$ (13)

$$b = \tfrac{1}{2}B_z$$ (14)

$$b' = \tfrac{1}{2}(B_x - B_y)$$ (15)

where b' describes the deviation from axial symmetry.

Figure 6.9a shows the ESR lines of $Tl^0(1)$ centers in KCl (and some other simultaneously produced defects by X irradiation of the Tl doped KCl crystal at

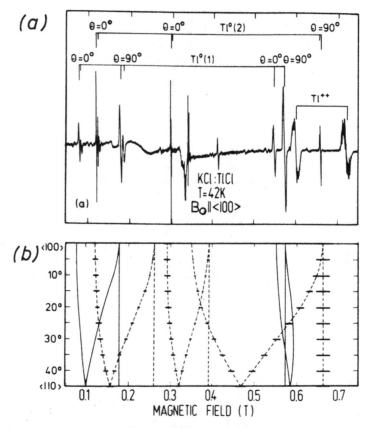

Figure 6.9 (a) ESR spectrum of $Tl^0(1)$ and $Tl^0(2)$ centers in KCL: Tl for $B_0 \| \langle 100 \rangle$ ($T = 4.2\,\text{K}$). $Tl^0(1)$ centers are Tl^0 atoms next to a Cl-vacancy along a $\langle 100 \rangle$ direction, $Tl^0(2)$ centers are Tl^0 atoms located between two Cl^- vacancies along a $\langle 100 \rangle$ direction. (b) The calculated angular variation of the spectra in a (100) plane. Solid lines are caused by $Tl^0(1)$ centers; broken lines are caused by $Tl^0(2)$ centers. The horizontal bars indicate the transition probabilities for $Tl^0(2)$ centers. Reprinted with permission from E. Goovaerts, J. A. Andriessen, S. V. Nistor, and D. Schoemaker, *Phys. Rev. B*, **24**, 29 (1981).

room temperature). Again, since both Tl isotopes ^{203}Tl and ^{205}Tl have $I = \frac{1}{2}$, the spectra consist of doublets. The splitting is, however, angular dependent. The angular dependence in Figure 6.9b for rotation of the magnetic field in a (100)-plane shows this as well as the distribution of the centers in the crystal with respect to their axes. They have axial symmetry about a $\langle 100 \rangle$ direction. The Tl^0(1) centers are Tl^0 atoms next to an anion vacancy, while Tl^0(2) centers have two anion vacancies, one on each side of the atom [13]. The angular dependence (i.e., symmetry) and isotope splittings lead here to structural models. Electron spin resonance cannot reveal, however, whether Tl^0 resides precisely on a cation lattice site or whether it is relaxed toward the vacancy. Also, whether the vacancy is filled with another impurity is unknown since the angular dependence reflects only the tetragonal symmetry of the defect and no further superhyperfine structure is resolved. The distinction between Tl^0(1) and Tl^0(2) also requires a theoretical interpretation of the spin Hamiltonian parameters.

Figure 6.9a shows another feature of conventional ESR spectroscopy. Since the ionizing radiation producing these centers creates many defects simultaneously, it is not easy to assign unambiguously all the ESR lines to particular centers and then to follow their angular dependence. On what follows we show that with optical detection of the ESR each defect can be measured selectively, which greatly facilitates the analysis.

Figure 6.10 shows the ESR spectrum of atomic hydrogen on interstitial sites and on cation and anion vacancy sites in KCl [14]. The central hyperfine splitting with the proton ($I = \frac{1}{2}$) in the three cases is practically the same and nearly that of the free hydrogen atom. Only for the interstitial site is a superhyperfine interaction with nearest neighbors [14] resolved. The substitutional sites cannot be inferred from the ESR spectrum. They can be established only by resolving the superhyperfine interactions with ENDOR experiments [15].

If there is a superhyperfine interaction as in Figure 6.10a, the spin Hamiltonian of (9) must be extended to include all interacting nuclei

$$\mathcal{H} = g\mu_B \mathbf{B}_0 \mathbf{S} + \sum_i (\mathbf{I}_i \tilde{A}_i \mathbf{S} - g_{I,i}\mu_n \mathbf{B}_0 \mathbf{I}_i) \tag{16}$$

The sum runs over all nuclei with which a superhyperfine (hyperfine) interaction is measured. The intensity ratio of the superhyperfine lines is a characteristic feature and can be used to determine the number of interacting nuclei and their spins. This information is usually difficult to obtain from the analysis of ENDOR spectra. Therefore it is often useful or even necessary to simulate the ESR spectrum using the superhyperfine data obtained from the analysis of the ENDOR spectra to obtain the number of interacting nuclei. Especially, if there are several nuclei with the same superhyperfine interaction, which can occur for a specific orientation of the magnetic field, one can define a total nuclear spin $N \cdot I$, which leads to a $(2N \cdot I + 1)$-fold splitting of the ESR spectrum. Figure 6.11 shows the level scheme and the transitions of $N = 1$, 2, and 3 for $I = \frac{1}{2}$. The

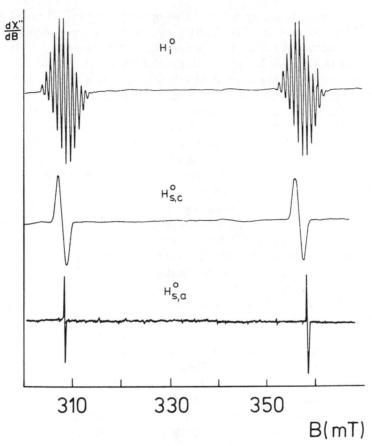

Figure 6.10   The ESR spectra of three atomic hydrogen centers in KCl. $B_0 \parallel \langle 100 \rangle$, $T = 77$ K. The hydrogen atoms occupy interstitial sites (H_i^0 centers), cation vacancy sites ($H_{s,c}^0$ centers), and anion vacancy sites ($H_{s,a}^0$ centers). The splitting caused by the hyperfine interaction with the proton is almost that of the free hydrogen atom. Reprinted with permission from J.-M. Spaeth, "Atomic Hydrogen as a Model Defect in Alkali Halides," in V. M. Tuchkevich and K. K. Shvarts, Eds., *Defects in Insulating Crystals*, Springer, New York, 1981, p. 232.

relative ESR signal height is given by the statistical weights or degeneracy of the levels. For $N = 3$ they are 1:3:3:1 for the four equidistant superhyperfine lines.

Tables 6.1 and 6.2 give the statistical weights of the states M_I for N equivalent nuclei with $I = \frac{1}{2}$ and $\frac{3}{2}$. From Table 6.2 one finds for four equivalent nuclei of $I = \frac{3}{2}$ the weights 1:4:10:20:31:40:44:40:31:20:10:4:1 for the 13 superhyperfine lines. This is observed in Figure 6.10a for the interstitial atomic hydrogen in KCl, which interacts predominantly with four nearest ^{35}Cl nuclei that have $I = \frac{3}{2}$.

In Section 6 the interpretation of the superhyperfine (hyperfine) and quadrupole (superhyperfine) interaction constants is discussed briefly. Here are some introductory remarks that may help to understand the forthcoming sections.

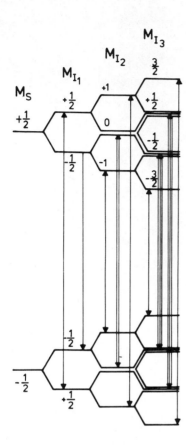

Figure 6.11 Electron Zeeman level scheme for an electron with superhyperfine coupling to three equivalent nuclei with $I = \frac{1}{2}$ and the resulting ESR transitions.

Table 6.1 Statistical Weights of the M_I States for N Equivalent Nuclei With Spin $\frac{1}{2}$

	M_I						
N	0	$\frac{1}{2}$	1	$\frac{3}{2}$	2	$\frac{5}{2}$	3
1		1					
2	2		1				
3		3		1			
4	6		4		1		
5		10		5		1	
6	20		15		6		1

Table 6.2 Statistical Weights of the M_I States for N Equivalent Nuclei With Spin $\frac{3}{2}$

									M_I										
N	0	$\frac{1}{2}$	1	$\frac{3}{2}$	2	$\frac{5}{2}$	3	$\frac{7}{2}$	4	$\frac{9}{2}$	5	$\frac{11}{2}$	6	$\frac{13}{2}$	7	$\frac{15}{2}$	8	$\frac{17}{2}$	9
1		1		1															
2	4		3		2		1												
3		12		10		6		3		1									
4	44		40		31		20		10		4		1						
5		155		135		101		65		35		15		5		1			
6	580		546		456		336		216		120		56		21		6		1

The hyperfine interaction constants are determined by the electronic wave function of the defects and the nuclear moments of the nuclei. In a simple one-particle approximation the isotropic constant a is given by [4]

$$a_l = \tfrac{2}{3}\mu_0 g\mu_B g_I \mu_n |\psi(r_l)|^2 \qquad (17)$$

where $\psi(r)$ is the wave function of the defect and $\psi(r_l)$ is its amplitude at the site r_l of a particular nucleus. The anisotropic tensor elements are given by

$$B_{ik} = \frac{\mu_0}{4\pi} g\mu_B g_I \mu_n \int \left(\tfrac{3}{5}x_i x_k - \frac{1}{r^3}\delta_{ik}\right)|\psi(r)|^2 dV \qquad (18)$$

where r is the radius vector from the nuclear site of concern (origin) and the origin is spared in the integral of (18). Thus, the hyperfine constants are proportional to g_I; therefore, the interaction constants of different isotopes must be in the ratio of their respective g_I factors [4,5]. This was used for their identification.

4 ELECTRON NUCLEAR DOUBLE RESONANCE

4.1 Stationary Electron Nuclear Double Resonance

In solids the superhyperfine structure of defects is usually not or only partly resolved by ESR. This is not because it is too small, but because the ESR spectrum is a superposition of too many superhyperfine lines with splittings that are very similar. In Figure 6.12 this is shown schematically for a $S = \tfrac{1}{2}$ defect and interactions with two shells of equivalent nuclei, each of which contains n_1 and n_2 nuclei with the spins I_1 and I_2, respectively. For both electron Zeeman levels there are $N = (2n_1 \cdot I_1 + 1)(2n_2 \cdot I_2 + 1)$ sublevels; therefore, there are up to N ESR transitions. If there is no quadrupole interaction, there are, however, only four ENDOR transitions, two for each shell of equivalent nuclei, which follows from the ENDOR selection rules as NMR transitions (see Section 4.2). If we consider more interactions, the result is an exponential increase of the number of superimposed ESR lines that are not resolved any more. For solid-state defects the interactions with up to 10 shells of neighbor nuclei are not unusual. The ESR shows only the envelope of all the superhyperfine transitions yielding an *inhomogeneously* broadened ESR line. Figure 6.3 is an example of the width of the ESR lines that is caused by unresolved superhyperfine interactions with several shells of ^{29}Si nuclei. In ENDOR there are only two lines for each superhyperfine-coupled nucleus distinguishable by their superhyperfine inter-actions. Therefore in ENDOR there is an excellent resolution of superhyperfine interactions.

There are two methods to measure ENDOR, without including spin–echo techniques: the dispersion method, which is suited for especially long spin–lattice relaxation times T_1. This method was introduced by Feher [1]. The other

Superhyperfine structure

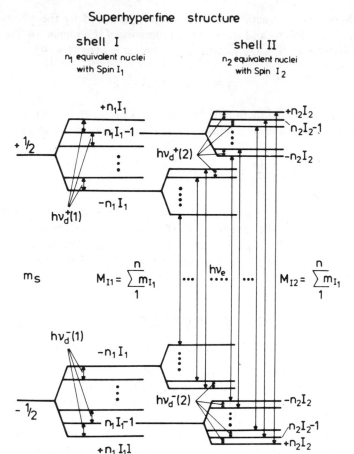

Figure 6.12 Energy-level scheme for an electron with superhyperfine coupling to several shells of equivalent nuclei. The ESR line is an envelope over all superhyperfine lines, which usually are not resolved because of the high number of superimposed lines.

one introduced by Seidel [16] is the stationary ENDOR method, which is more versatile. A quantitative description of both methods is difficult, since for solid-state defects the situation is so complicated by the relaxation paths and couplings that there is not yet a good, general understanding of the ENDOR effect. Since the purpose of this chapter is to show the application of ENDOR to the determination of defect structures, only the more commonly used stationary ENDOR technique is discussed by means of a simple working model to illustrate the experiments done.

Figure 6.13 shows schematically the allowed ESR transitions with resolved superhyperfine structure for $S = \frac{1}{2}$ and for one nucleus with $I = \frac{3}{2}$. In an ENDOR experiment one of the allowed ESR transitions is partially saturated; that is, one chooses the microwave power high enough so that the transition

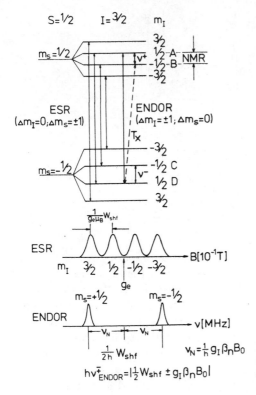

Figure 6.13 Energy-level diagram to explain stationary electron nuclear double resonance (ENDOR).

probability $R_{MW} = \gamma B_1^2$ is of the order or larger than the spin lattice relaxation rate $R_{REL} = 1/T_1$; γ is the gyromagnetic ratio of the electron, and B_1 is the microwave field amplitude. If that is the case, then the spin population of the levels connected by the microwave transitions deviates from the Boltzmann equilibrium distribution. If $R_{MW} \gg R_{REL}$, then these levels become equally populated. This results in a decrease of the observable microwave absorption, since microwave absorption and emission probabilities are equal. The levels unconnected by the microwave transitions are not affected. Therefore, for example, the level population $m_s = +\frac{1}{2}$, $m_I = +\frac{1}{2}$ becomes inverted with respect to the level $m_s = +\frac{1}{2}$, $m_I = -\frac{1}{2}$ (see Figure 6.13). If the two levels are connected by an NMR transition, the level populations can be equalized, which results in a population decrease of the level $m_s = +\frac{1}{2}$, $m_I = +\frac{1}{2}$ leading to desaturation of the (partially) saturated transition $m_s = -\frac{1}{2}$, $m_I = +\frac{1}{2}$ to $m_s = +\frac{1}{2}$, $m_I = +\frac{1}{2}$. This desaturation is monitored. It occurs for two NMR frequencies in the example of Figure 6.13, since the NMR frequencies for $m_s = +\frac{1}{2}$ and $m_s = -\frac{1}{2}$ are different (see Section 4.2). Thus, each nucleus gives rise to two ENDOR lines (for $S = \frac{1}{2}$). A cross relaxation T_x (see Figure 6.13) allows the stationary observation of the desaturation (stationary ENDOR) [16]. If several nuclei with the same or similar interactions are coupled to the unpaired electron, then the ESR pattern becomes complicated and the superhyperfine structure is usually

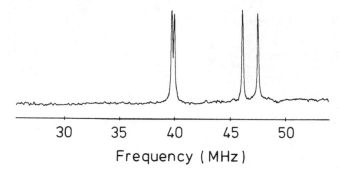

Frequency (MHz)

Figure 6.14   The ENDOR lines due to the nearest ^{19}F neighbors of $F(Cl^-)$ centers in SrFCl; $F(Cl^-)$ centers are electrons trapped at Cl^- vacancies in SrFCl.

not resolved. In ENDOR all nuclei with the same interaction give rise only to two (for $S = \frac{1}{2}$) ENDOR lines, which greatly enhances the resolution. Electron nuclear double resonance lines, as NMR lines, are typically 10–100 kHz wide, which is about three orders of magnitude narrower than homogeneous ESR lines. Thus, in ENDOR one uses the sensitivity enhancement caused by a quantum shift from frequencies of about megahertz to the microwave frequencies of about gigahertz and the increased resolution power caused by the smaller NMR line width and the reduction of the number of lines.

Figure 6.14 shows as an example ENDOR lines (ESR desaturation) of the nearest ^{19}F neighbors of $F(Cl^-)$ centers in SrFCl (for $m_s = -\frac{1}{2}$), where the unpaired electron occupies a Cl^- vacancy. The structure model of BaFCl, which has the same structure, is shown in Figure 6.30 (see Section 5).

4.2   Analysis of Electron Nuclear Double Resonance Spectra

Generally, in addition to the superhyperfine interaction, there is a quadrupole interaction if $I > \frac{1}{2}$ for the interacting nuclei. This interaction is caused by the interaction between the electrical field gradient at the site of a nucleus and its nuclear quadrupole moment [4]. Therefore, the spin Hamiltonian describing the ENDOR spectra is

$$\mathcal{H} = g\mu_B B_0 S + \sum_I (S\tilde{A}_1 I_1 - g_I \beta_I B_0 I_1 + I_1 \tilde{Q}_1 I_1) \tag{19}$$

The sum runs over all nuclei interacting with the unpaired electron. For simplicity it is assumed that g is isotropic (which is generally not the case); $\tilde{Q}$ is the traceless quadrupole interaction tensor with the elements

$$Q_{ik} = \frac{eQ}{2I(2I-1)} \frac{\partial^2 V}{\partial x_i \partial x_k}\bigg|_{r=0} \tag{20}$$

where Q is the quadrupole moment and V is the electrical potential. The spectra are usually analyzed in terms of the quadrupole interaction constants:

$$q = \tfrac{1}{2}Q_z$$
$$q' = \tfrac{1}{2}(Q_x - Q_y) \tag{21}$$

The selection rule for ENDOR transitions as NMR transitions is

$$\Delta m_s = 0$$
$$\Delta m_I = \pm 1 \tag{22}$$

If the superhyperfine and quadrupole interactions are small compared to the electron Zeeman term, then the quantization of the electron spin is not influenced by these interactions and the nuclei are independent of each other. They can be treated separately, and the sum in (19) can be omitted. In first-order perturbation theory, that is, with the conditions

$$|B_{ik}|, |Q_{ik}| \ll \left| a \pm \frac{1}{m_s} g_I \mu_n B_0 \right| \tag{23}$$

$$\nu_{\text{ENDOR}}^{\pm} = \left| \frac{1}{h} m_s W_{\text{shf}} \pm \nu_n \pm \frac{1}{h} m_q W_Q \right| \tag{24}$$

with the following abbreviations:

$$W_{\text{shf}} = a + b(3 \cos^2\theta - 1) + b' \sin^2\theta \cos 2\delta \tag{25}$$
$$W_Q = 3\{q(3 \cos^2\theta' - 1) + q' \sin^2\theta' \cos 2\delta'\} \tag{26}$$

θ, δ and θ', δ' are the polar angles of B_0 in the principal superhyperfine and quadrupole axis system, respectively.

$$\nu_n = \frac{1}{h} g_I \mu_n B_0 \tag{27}$$

where ν_n is the Larmor frequency of a free nucleus in the magnetic field B_0.

$$m_q = m_I + m_I' \tag{28}$$

where m_q is the average between the two nuclear spin quantum numbers, which are connected by the transition [16].

The basic concepts for the analysis of ENDOR spectra is described assuming that (24) holds for the frequency positions of the ENDOR lines. This is generally not the case, since both the anisotropic and quadrupole interactions can become

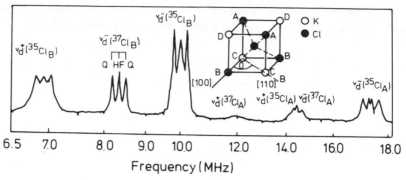

Figure 6.15 The ENDOR spectrum of nearest halogen neighbors of interstitial atomic hydrogen centers in KCl. $B_0 \| \langle 110 \rangle$. Reprinted with permission from J.-M. Spaeth, "Atomic Hydrogen as a Model Defect in Alkali Halides," in V. M. Tuchkevich and K. K. Shvarts, Eds., *Defects in Insulating Crystals*, Springer, New York, 1981, p. 232.

too large with respect to the isotropic term and g-anisotropies, and the fine-structure interaction can influence the spectra. A detailed discussion of the complications is beyond the scope of this chapter. Some particularly important consequences of higher order contributions to the frequency positions will, however, be discussed where appropriate.

Figure 6.15 shows an ENDOR spectrum of interstitial atomic hydrogen in KCl for $B_0 \| \langle 110 \rangle$, the ESR spectrum of which is shown in Figure 6.10a [for $B_0 \| \langle 100 \rangle$]. The ENDOR spectrum contains the interaction with the nearest Cl neighbors [15]. The identification of the chemical nature of nuclei can be achieved in various ways. According to (24), for $S = \frac{1}{2}$ and no quadrupole interactions, each nucleus gives a pair of lines separated by $2v_n$ if $\frac{1}{2} W_{shf} > h v_n$ and by W_{shf} if $h v_n > \frac{1}{2} W_{shf}$. Since v_n can be calculated according to (27), the nuclei can be identified either from the line pairs separated by $2v_n$ or by symmetric line patterns about v_n.

If there are several magnetic isotopes present (as ^{37}Cl and ^{35}Cl in Figure 6.15), then their line positions must be in the ratio of their respective nuclear moments (if $Q = 0$). The line intensity roughly reflects the isotope abundance. Unfortunately, the ENDOR line intensity is very little understood quantitatively in solids because of the many and complicated relaxation paths. Therefore, according to (24) the two lines of each nucleus cannot always be observed, particularly for low frequencies in stationary ENDOR. However, upon shifting the magnetic field through the ESR line the ENDOR line positions are also shifted according to (24) and (27). The ENDOR line shift is caused by the shift of v_n and is thus proportional to g_I, which is characteristic for a particular nucleus. If the ESR line width is too small to detect an ENDOR line shift (which depends on the ENDOR line width), then one can either perform additional experiments with a different ESR band (e.g., K band or Q band) or change the resonant frequency of the cavity by inserting a material of suitable dielectric constant and changing the microwave source frequency accordingly.

If there is a quadrupole splitting, then each *hyperfine* ENDOR line is split into a characteristic multiplet; for example, for $I = \frac{3}{2}$ it is split into a triplet. This is easily recognized in Figure 6.15 as a triplet structure of each ENDOR line. The additional structure indicated in the ENDOR lines is due to second-order superhyperfine structure [14].

To determine the defect structure and the interaction parameters, the dependence of the ENDOR line positions on variation of the magnetic field with respect to the crystal orientation must be measured and analyzed. This is the major problem in an ENDOR analysis and the essential tool for the determination of the defect structure.

Figures 6.16a–c show such an angular dependence for a cubic crystal, such as an alkali halide, calculated according to (24) for the first three neighbor shells of a defect on a lattice site. The patterns are characteristic for (100), (110), and (111) *symmetry* of the neighbor nuclei. For each m_s value such a pattern is observed. From the number of such patterns according to (24) one can infer the electron spin of the defect and thus often its charge state.

Each nucleus has its own principal axis system for the superhyperfine and quadrupole tensors. Often, their orientation in a crystal is determined by symmetry. Otherwise, they must be determined from the analysis of the angular dependence of the ENDOR spectra. If the defect center (impurity) and the respective nucleus are in a mirror plane, then two principal axes must be in this mirror plane. If the connection line between the nucleus and the center is a threefold or higher symmetry axis, then the tensor is axially symmetric with its axis along this symmetry axis.

If the angular patterns are separated in frequency, they are easily recognized and the analysis is fairly straightforward.

Figure 6.17 shows such an angular dependence for substitutional Te^+ in Si (see Figure 6.8 for the ESR spectrum). In practice, one proceeds as follows: Assume a center model and then calculate the expected angular dependence according to the appropriate spin Hamiltonian making use of the symmetry properties of the assumed model. Comparison with the experimental angular dependence then verifies or falsifies the assumption. If the first-order solution of the spin Hamiltonian is not sufficient, the symmetry patterns of the neighbor shells may not be easily recognized. For the tetrahedral symmetry of a substitutional point defect in a diamond lattice or zincblende lattice the superhyperfine interactions of nearest neighbors should give the simple pattern of Figure 6.18a. If $I = \frac{3}{2}$ for the nearest neighbors and a small quadrupole interaction is present, the pattern looks like Figure 6.18b, where the quadrupole triplet splitting of the ENDOR lines is easily recognized. However, for larger values of the quadrupole interaction constant q after diagonalization of the spin Hamiltonian the pattern changes considerably (Figure 6.18c) and its origin is not so easily recognized. This complication through quadrupole interactions often makes the analysis difficult.

A diamond or zincblende structure (silicon or III–V compounds) presents the particular difficulty that the symmetry of the substitutional site and tetrahedral

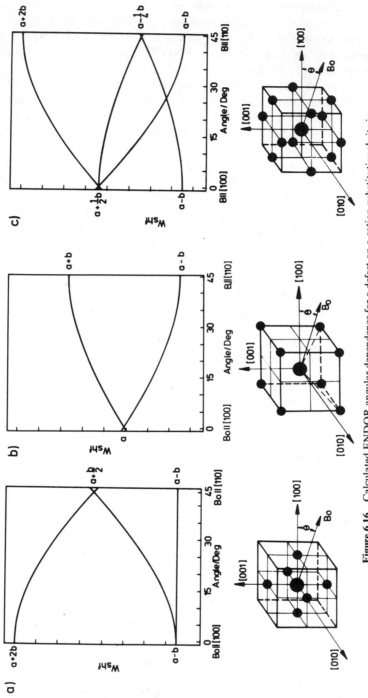

Figure 6.16 Calculated ENDOR angular dependence for a defect on a cation substitutional site in an alkali halide. (*a*) (100) neighbors (anions), (*b*) (111) neighbors (anions), (*c*) (110) neighbors (cations). The magnetic field is rotated in the (001) plane.

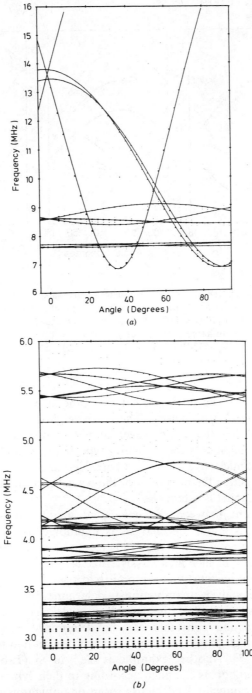

Figure 6.17 Angular dependence of ^{29}Si ENDOR lines of Te$^+$ defects in Si for rotation of the magnetic field in a (110) plane; $0° = \langle 100 \rangle$. The solid lines are the calculated angular dependencies with the parameters of Table 6.4. Reprinted with permission from *Solid State Commun.*, **46**, J. R. Niklas and J.-M. Spaeth, "ENDOR Investigation of Tellurium Donors in Silicon," Copyright 1983, Pergamon Journals, Ltd.

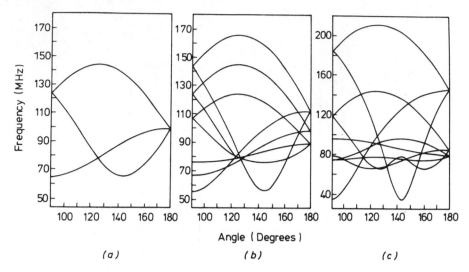

Figure 6.18 Calculated ENDOR angular dependence of an assumed $AsAs_4$ antisite defect in GaAs due to four equivalent nearest ^{75}As neighbors for $m_s = -\frac{1}{2}$ and rotation of the crystal in a (110) plane from $\langle 110 \rangle = 90°$ to $\langle 100 \rangle = 180°$. The simulation is for the following ligand hyperfine and quadrupole constants (tensor axes $\parallel \langle 111 \rangle$ directions): (a) $a/h = 198$ MHZ, $b/h = 53$ MHz, $q/h = 0$; (b) $a/h = 198$ MHz, $b/h = 53$ MHz, $q/h = 3.5$ MHz; (c) $a/h = 198$ MHz, $b/h = 53$ MHz, $q/h = 11.1$ MHz.

interstitial site cannot be distinguished as to the symmetry type of neighbor nuclei with respect to both defects. There are neighbors with (100), (110), and (111) symmetry of their tensors; however, only the sequence is different on both cases. From experiment only the symmetry type of nuclei can be determined, not the distance of the nuclei from the defect core. Here theoretical arguments about the site of the defect or the character of the wave function must be considered before a definite model is established.

If the agreement between the calculated angular dependence and the experimental one is good, the analysis is unambiguously correct. There are many more experimental data than parameters to be extracted from them. As an example of the result of such an analysis Table 6.3 gives the superhyperfine and quadrupole interaction constants of atomic hydrogen on anion sites in KCl ($H^0_{s,a}$ centers) [15]. The analysis unambiguously determines the site for the hydrogen atom from the symmetry of the neighbor nuclei. The superhyperfine and quadrupole interaction constants can be determined down to very small interactions with high precision. The spin Hamiltonian had to be diagonalized numerically [15, 17]. Another example is Te^+ in Si (Table 6.4). Here the sequence of Si neighbors is classified according to their symmetry types. Again, only theoretical arguments lead finally to an unambiguous assignment of the Te^+ to the substitutional site [18, 19].

In a straight ENDOR analysis only the relative signs of the superhyperfine constants a, b, b' can be determined. The signs of the quadrupole interaction

Table 6.3 Superhyperfine and Quadrupole Constants of $H_{s,a}^0$ Centers in KCl (in kHz) $(T = 40 \, \text{K})^a$

Shell[b]	Constant	$H_{s,a}^0$
$^{39}\text{K}^{\text{I}}$	a	253
	b	219
	q	198
$^{35}\text{Cl}^{\text{II}}$	a	57
	b	312
	b'	-3
	q	-88
	q'	-94
$^{35}\text{Cl}^{\text{IV}}$	a	37
	b	54
	q	± 45
$^{39}\text{K}^{\text{V}}$	a	4
	b	11
	b'	ca. 0
	ϕ_{B}	$26.0^0 \pm 0.2^0$
	q	± 39
	q'	± 17
	ϕ_{Q}	$13.5^0 \pm 0.2^0$

[a]Adapted from G. Heder, J. R. Niklas, and J.-M. Spaeth, *Phys. Status Solidi B*, **100**, 567 (1980).
[b]The Roman numeral superscripts denote the shell numbers. They are the sums of the squares of the Miller indices to describe the position of the nuclei with the defect at the center (000).

constants with respect to the superhyperfine interaction constants cannot be determined either. With ENDOR-induced ESR and double-ENDOR more about the signs of the interactions can be extracted from experiment.

In solids a particular situation is met when there are superhyperfine (and quadrupole) interactions with several neighbor nuclei that have exactly the same interactions and tensor orientations with respect to the magnetic field orientation B_0; in this case the nuclei are precisely equivalent. An additional line splitting can occur, which can be very confusing if it is not recognized as such (second-order superhyperfine splitting). The splitting pattern can be quite complicated. If there is only an isotropic superhyperfine interaction (or $b \ll a$), then the splitting between the ENDOR lines is about $a^2/2g_e\mu_B B_0$. For small interaction constants it is hardly visible, but it is clearly visible for large

Table 6.4 Superhyperfine Interactions of the Te$^+$ and S$^+$ Centers With ^{29}Si[a,b]

Type	Si:Te$^+$				Si:S$^+$	
	a/MHz	b/MHz	b'/MHz	θ_{shf}[c]	a/MHz	b/MHz
111	17.7	9.8			32.7	12.0
	11.4	0.47			9.0	0.62
	1.3	0.03			2.94	0.02
					1.35	0.09
100	4.6	<0.005	<0.005		2.04	0.05
	1.8	0.02	<0.005			
110	2.83	0.51	−0.08	53.1	8.94	0.47
	5.26	0.21	0.12	67.1	3.84	0.42
	9.52	−0.09	−0.03	0	4.77	0.03
	1.95	0.1	0.02	−5		
	2.46	0.04	<0.005	50.9		
	0.9	0.04	<0.005	47.6		
	0.67	0.04	−0.01	47.2		

[a]Adapted with permission from *Solid State Communications*, **46**, J. R. Niklas and J.-M. Spaeth, "ENDOR Investigation of Tellurium Donors in Silicon," Copyright © 1983, Pergamon Press PLC.
[b]Experimental error is about ±0.1 MHz.
[c]θ_{shf} is given with respect to a ⟨110⟩ axis.

interactions (for $a/h = 100$ MHz, the splitting is about 500 kHz in X band) [20]. In particular, if b and q are not small with respect to a, rather large splittings and a complicated pattern can occur. A recent example shows this in the ENDOR spectrum for the Ga vacancy in GaP (Figure 6.19). For the particular orientation of $B_0 \| [111]$ one would expect without this superhyperfine structure of second order only four ENDOR lines due to the four nearest ^{31}P neighbors, since for each m_s value B_0 is parallel to the superhyperfine axes of one neighbor, the other three nuclei have the same angle with respect to B_0, thus yielding only two ENDOR lines for each value. In the frequency range of Figure 6.19 there should be only two lines observed because of the four equivalent ^{31}P neighbors for $m_s = \pm\frac{3}{2}$. However, there are nine lines observed with large splittings of the order of several megahertz. Initially, whether the vacancy is distorted was unclear. The distortion can also give rise to splittings. However, double-ENDOR experiments showed that this splitting is caused by the second-order superhyperfine structure. A full diagonalization of the spin Hamiltonian for the case of four equivalent nuclei confirmed that the splitting is caused by the second-order effects [21].

In practice, another fact renders the ENDOR analysis rather difficult. If there are many lines caused by one defect within a narrow frequency range, as for the atomic hydrogen centers on anion sites, then an analysis is only feasible by employing computer-aided methods. Figure 6.20a shows the angular de-

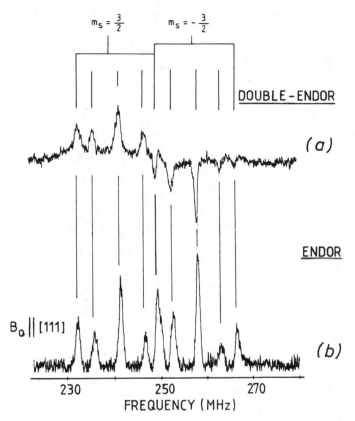

Figure 6.19 (a) Double-ENDOR lines. (b) ENDOR lines of ^{31}P nearest neighbors of the gallium vacancy (V_{Ga}) in GaP. B_0 approximately parallel to [111]. Reprinted with permission from J. Hage, J. R. Niklas, and J.-M. Spaeth, *Mater. Sci. Forum*, **10–12**, 259 (1986).

pendence of these centers. Another example is that of Ni^{3+} in GaP, where for each magnetic field orientation over 600 ENDOR lines were observed [22]. The angular pattern can also become complicated because of the many center orientations of low-symmetry defects and because the angular dependence must be described by a complete diagonalization of the spin Hamiltonian. The result of this for hydrogen centers is shown in Figure 6.20*b* using the superhyperfine and quadrupole constants of Table 6.3.

Hence, with conventional ENDOR spectrometers only comparatively simple problems can be solved. In complicated cases the angular dependence must be measured in small angular steps, and because of a low signal-to-noise (S/N) ratio observed the experiments must be repeated many times. In recent years considerable progress was made by setting up computer-controlled ENDOR spectrometers and by using computers for the data processing and analysis of the spectra.

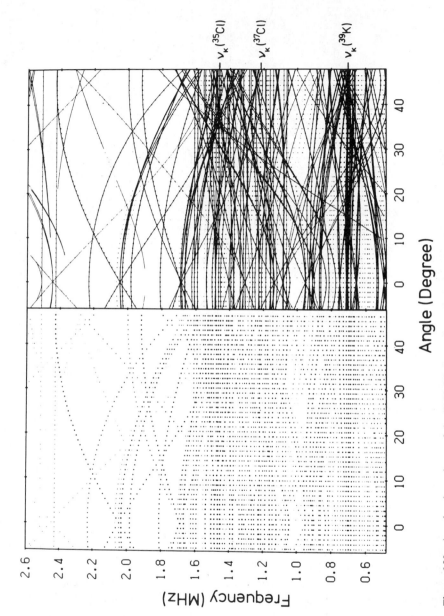

Figure 6.20 Angular dependence of the ENDOR lines of atomic hydrogen centers on anion sites in KCl for rotation of the magnetic field in a (100) plane between ⟨100⟩ (0°) and ⟨110⟩ (45°). Left: experimental results. Right: calculated angular dependence with the superhyperfine and quadrupole constants given in Table 6.3. Reprinted with permission from G. Heder, J. R. Niklas, and J.-M. Spaeth, *Phys. Status Solidi B*, **100**, 567 (1980).

4.3 Experimental Aspects

Observation of a stationary ENDOR effect from solid-state defects depends mainly on two essential parts of the spectrometer: (1) a suitable ENDOR cavity, which allows the generation of a strong enough radio frequency amplitude B_2 without unduly affecting the quality factor Q of the cavity, since the microwave amplitude B_1 must be high enough to partially saturate the ESR signal; (2) a cryosystem, which allows one to vary the sample temperature in a wide range, preferably between 1.5 K and room temperature. If the radio frequency field is produced by a coil, in which radio frequency current flows, one needs a TM-mode cavity [23, 24]. A particularly useful cavity construction is that of Figure 6.21, where the radio frequency current flows through solid rods in a TE_{011} resonator [25, 26]. The sample temperature must be varied in a wide range because to observe the stationary ENDOR effect the electron spin–lattice relaxation rate $1/T_1$, the electron cross-relaxation rate $1/T_x$, and the microwave-transition rate must all be of the same order of magnitude and they must be in certain relations with respect to each other. Since there is a complex relaxation scheme in solids, the optimum experimental conditions cannot be predicted. The best approach is to be able to vary both microwave power and temperature and

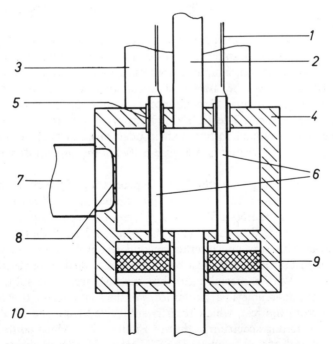

Figure 6.21 The ESR–ENDOR resonator for variable temperature measurements between 3 and 300 K. The cylindrical resonator works in the TE_{011} mode: (1) radio frequency connection, (2) tube to insert sample, (3) tube for helium exhaust gas, (4) resonator wall, (5) Teflon insulating rings, (6) four silver tubes, (7) microwave guides, (8) iris, (9) heat exchange, (10) liquid helium supply.

to find experimentally the optimum conditions. Experience teaches us that ENDOR measurements are often possible only in a very narrow temperature range. The temperature must be neither too high nor too low, and this temperature range varies with the defect system. As a general rule, s-type electron ground states require higher temperatures (typically 30–70 K), and p-type ground states (hole centers, acceptors in semiconductors) require rather low temperatures. The unavailability of suitable cavity–cryostat combinations for solid-state investigations is the most serious deficiency of commercial ENDOR spectrometers.

At low temperatures the nuclear spin–lattice relaxation times are long; hence, one cannot use fast modulation of the radio frequency fields. Therefore, it was suggested that superheterodyne spectrometers be built that were also sensitive for low-modulation frequencies (20–1000 Hz). However, low-noise microwave amplifiers are available and were used to preamplify the signal before the microwave diode, and this proved to be approximately as sensitive as a superheterodyne setup with the advantage of being able to use the more simple and carefree homodyne spectrometers [27].

In practical cases of interest many ENDOR lines appear, and the angular dependence of the ENDOR lines must be measured in small angular steps, which means taking continuous measurements for several weeks. This procedure required computer-aided techniques, for both the experiment and the analysis of the spectra.

In the computer-controlled ENDOR spectrometer the computer controls the following experimental parameters: radio frequency (NMR frequencies), magnetic field, crystal orientation, sample temperature, and cavity matching. Thus, the angular dependence can be measured automatically, also as a function of temperature. A full angular dependence can take up to several weeks of continuous measurement. The ENDOR lines are stored in the computer and with a specially developed software their positions can be determined automatically and a computer plot (see Figure 6.20a) of the angular dependence can be made.

The application of digital methods to the processing of the experimental data is summarized in Figure 6.22 for the example of interstitial Fe^0 centers in Si [28]. Figure 6.22a shows the ENDOR lines of interest between 6 and 8 MHz, which are very weak compared to those from the *distant ENDOR* that are caused by the ^{29}Si nuclei with very low interaction and strong background signal. The latter and the distant ENDOR must be subtracted to deal more clearly with the lines of interest. With a special algorithm the background is subtracted. The algorithm does not assume a particular form of the background. It *eliminates* the sharp peaks from the rest, which then is subtracted from the total spectrum [26, 29]. The resulting spectrum is that of Figure 6.22c, which contains several ENDOR lines and, of course, noise. The S/N ratio of ENDOR spectra is usually not too good. The ENDOR effect in solid-state defects is mostly below 1% of the ESR signal. Low defect concentration and the limited ENDOR effect are the major reasons for the poor S/N ratio. Here the use of digital filtering has proved

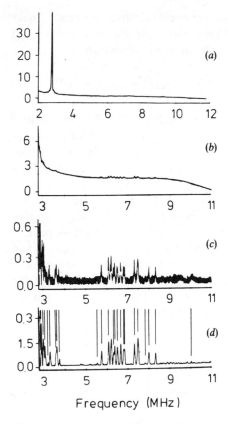

Figure 6.22 Digital data processing for an ENDOR spectrum of interstitial Fe^0 centers in Si: (a) spectrum as measured, (b) subtraction of the strong ENDOR signal at v_n (^{29}Si) that is caused by distant nuclei, (c) subtraction of a smooth background line, and (d) digital filtering and application of the peak search algorithm. Reprinted with permission from S. Greulich-Weber, J. R. Niklas, E. Weber, and J.-M. Spaeth, *Phys. Rev. B*, **30**, 6292 (1984).

to be very advantageous. The major ideas behind this follow. Primarily, one needs to smooth the spectra without disturbing the signals. Conventional resistor capacitor filters give a poor trade-off among noise reduction, signal distortion, and speed of measurement. A simple digital method is a running average algorithm, which replaces a data point by the average of the original data points and its unfiltered left and right neighbor points. This symmetric average over $(2N + 1)$ data points produces the classic noise reduction of any averaging process of uncorrelated data. This idea can be improved by assigning different weights to the neighboring data points.

$$f[K] = \sum_{i=-N}^{N} a[i] \cdot Y[K - i] \tag{29}$$

Here Y and f denote the unfiltered and the filtered data points, respectively. The weights $a[i]$ describe the digital filter used. For a simple running average one has

$$a[i] = 1/(2N + 1) \tag{30}$$

If one requests that the filtering process conserves additive constants, linear slopes, parabolic peaks, area, second and third moments, and minimizes noise under these constraints, it can be shown that the weight function

$$a[i] = 3[(3N^2 + 3N - 1) - 5i^2]/[(2N - 1)(2N + 1)(2N + 3)] \qquad (31)$$

is optimal for any line shape [30]. Such a filter is called a *DISPO filter* (digital smoothing with polynomial coefficient) [31]. Compared to a resistor capacitor filter it typically decreases the signal distortion by a factor of 20. For small signal distortion (1%) and equal scan speed this filter reduces noise by an additional factor of 5 compared to the resistor capacitor filter [32, 33].

Figure 6.22d shows the application of such a filter to the spectrum of Figure 6.22c. The line positions are determined by a special peak-search algorithm, in which the second derivative of the smoothed spectrum is calculated. A peak is resolved where this has a local minimum [26]. This is exact for symmetrical ENDOR lines of Gaussian or Lorentzian shape. The measurement must make sure that the line shape remains symmetrical for this to be applied. Therefore, for this kind of ENDOR spectroscopy, the method of transient ENDOR originally introduced by Feher [1] is not applicable, since in transient ENDOR the line shapes are nonsymmetrical.

When too many ENDOR lines overlap, the application of the peak-search algorithm is not sufficient. An improvement can be reached by applying a deconvolution algorithm, which decomposes the spectra in an iterative process, since the exact shape and width of the single ENDOR lines are not known beforehand.

Figure 6.23a shows a section of a spectrum; Figure 6.23b is obtained after the application of the deconvolution procedure. In Figure 6.24a and b the ENDOR angular dependence of $H_{s,a}^0$ centers in KCl, with and without application of the deconvolution, is compared. The advantages of this procedure are clearly visible in the sections between 1.45 and 1.50 and 1.20 and 1.25 MHz. The angular dependence can be followed more efficiently and it can be analyzed. The analysis yields the H^0 atom at the site of the anion vacancy and the values of the superhyperfine and quadrupole interaction constants given in Table 6.3 as an example of typical ENDOR results and their precision. The spin Hamiltonian was diagonalized numerically [15, 17].

4.4   Defect Reactions and Dynamical Effects Studied by Electron Nuclear Double Resonance

Electron nuclear double resonance is not only used to study the structure of defects *given* in a certain crystal, but also to study defect reactions in solids. The transformation of a particular defect into another species can be followed on an atomic microscopic scale.

For example, the conversion of F centers in KCl into $F_H(F^-)$ centers by bleaching the optical F center absorption band in KCl doped with F^- at a

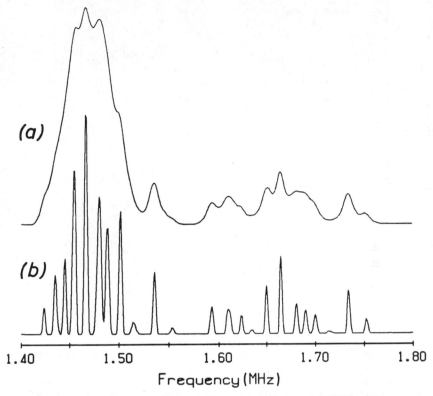

Figure 6.23 Deconvolution of spectra to enhance the resolution of ENDOR: (*a*) spectrum as measured and (*b*) after application of the deconvolution algorithm. Adapted from J. R. Niklas, "Elektronen-Kern-Doppelresonanz-Spektroskopie zur Strukturuntersuchung von Festkörperstörstellen," Habilitationsschrift, Universität-GH-Paderborn, 1983.

temperature where the anion vacancies are mobile is shown in Figure 6.25. The F centers in alkali halides are anion vacancies that have trapped an electron; F_H centers are F centers aggregated with an impurity halogen ion into an F impurity halogen pair center. Thus, $F_H(F^-)$ centers in KCl are Cl^- vacancies with a trapped electron next to F^- impurity halogen. In the lower trace a section of the second shell Cl ENDOR spectrum of the F centers before the conversion is shown, above the same section after conversion of about 50% of the F centers into $F_H(F^-)$ centers. New ENDOR lines appear between the original ones. In ESR the new centers cannot be distinguished from the F centers. Figure 6.26 shows the structure model derived from the analysis of the $F_H(F^-)$ spectrum [34]; F^- occupies a fourth shell position, not a second shell position as may have been assumed. A separation of the ENDOR spectra can be difficult if the "old" spectrum is only partially converted and overlaps strongly the "new" one. There are, however, methods to separate these ENDOR spectra, as will be shown in the Section 5.

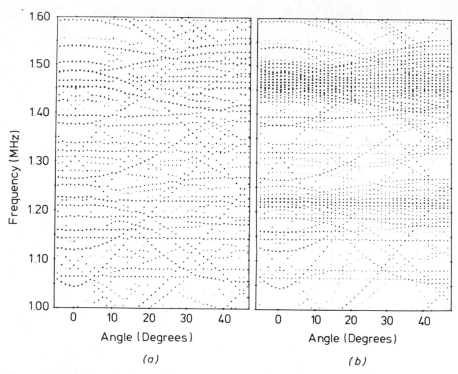

Figure 6.24 (*a*) Section of the ENDOR angular dependence of Figure 6.20 (atomic hydrogen on anion sites in KCl) obtained after digital filtering of the experimental lines and application of the peak-search algorithm. (*b*) The same section after the additional application of the deconvolution algorithm. Adapted from J. R. Niklas, "Elektronen-Kern-Doppelresonanz-Spektroskopie zur Strukturuntersuchung von Festkörperstörstellen," Habilitationsschrift, Universität-GH-Paderborn, 1983.

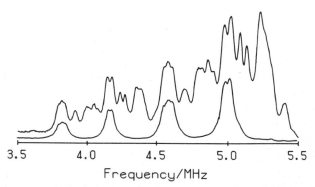

Figure 6.25 Conversion of F centers into $F_H(F^-)$ centers in KCl as measured by ENDOR. Lower trace: section of the ENDOR spectrum of F centers in KCl:F^- prior to conversion. The lines are caused by second-shell ^{35}Cl and ^{37}Cl neighbors. Upper trace: The same section after conversion of about 50% of the F centers into $F_H(F^-)$ centers. The (F^-) model is shown in Figure 6.26. Adapted with permission from H. Söthe, P. Studzinski, and J.-M. Spaeth, *Phys. Status Solidi B*, **130**, 339 (1985).

470

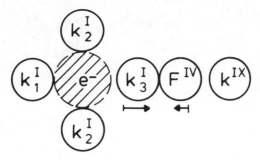

Figure 6.26 Model of the $F_H(F^-)$ centers in KCl as derived from the ENDOR analysis. The superscripts I, IV, and IX are the shell numbers of the neighbor ions. They are the squares of the Miller indices describing the position of the neighbors with respect to the F center at $(0,0,0)$.

Another recent example is the formation of thermal donors in silicon (see Section 3), which are electrically active shallow donors. They are formed when Czochralski-grown, oxygen-rich Si (containing $10^{18}\,\mathrm{cm}^{-3}$ interstitial oxygen) is annealed at about 450°C. By their infrared absorption bands they were identified to consist of about nine defects, which grow one after the other starting from an electrically active smallest one by adding atoms to this aggregate center as the annealing time is increased [35].

Interstitial oxygen in Si is not electrically active. Oxygen was thought to be involved in this structure of thermal donors, but it could not be proven. The ESR spectrum of the singly ionized species was already shown in Figure 6.3. By measuring a sample into which the magnetic ^{17}O ($I = \frac{5}{2}$) was diffused [36], ENDOR experiments not only proved the incorporation of oxygen into the thermal donors, but, interestingly for the present discussion, also the way the thermal donors grow. Very decisive results were obtained.

Figure 6.27 shows a section of the ^{29}Si ENDOR spectrum for several annealing times. Five different thermal donors are distinguishable and labeled (A)–(E). They are superimposed in the ESR line (see Figure 6.3), and in the ENDOR spectrum (lowest trace) five species are also present simultaneously. Interestingly, on further annealing only the relative intensities of the ENDOR lines change (Figure 6.27), not their positions; that is, the inner part of the defect structure is not changed when further atoms are added. Since the absolute ENDOR line intensities are not well understood, it is the change of the relative intensities that reflects the change of the numbers of the different thermal donor species upon annealing. The analysis of the angular dependence shows that the superhyperfine tensor orientations are the same for all species; and they are only distinguished by the magnitude of the interaction constants, which decrease upon thermal donor growth. The fact that no symmetries are changed upon growth shows that the growth occurs under full conservation of symmetry; that is, because of C_{2v} the growth must occur under addition of an even number (at least two) of atoms. Thus, ENDOR can make very precise statements about defect reactions, even in such a complicated case as an aggregate center [36, 37].

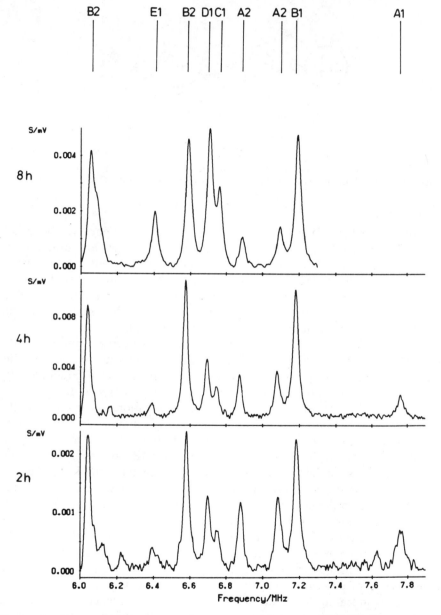

Figure 6.27 Section of the ^{29}Si ENDOR spectrum of thermal donors (NL8) in silicon. Shown is the effect of annealing at 460°C on the ^{29}Si ENDOR spectrum. Upon annealing the relative ENDOR signal intensities change because of the gradual formation of different species of the thermal donors, the ENDOR lines of which are superimposed: (A)–(E) denote the different thermal donors; (1)–(2) are the different ^{29} Si shells. Reprinted with permission from J. Michel, J. R. Niklas, J.-M. Spaeth, and C. Weinert, *Phys. Rev. Lett.*, **57**, 611 (1986).

With ENDOR dynamical properties of defects can also be studied by measuring the temperature dependence of the ENDOR line positions. Since each ENDOR line position can be determined with high precision (to 1–10 kHz, depending on the line width), comparatively small effects can also be seen. In Figure 6.28 the temperature dependence of the isotropic superhyperfine interaction constant of K_3^I and F^{IV} neighbors of $F_H(F^-)$ centers (see Figure 6.26) and of the corresponding K^I and Cl^{IV} neighbors in F centers is plotted between 77 and 300 K [38]. The strong temperature dependence of K_3^I of $F_H(F^-)$ centers reflects that K_3^I, which is displaced by 11% toward F^-, has more room compared to K^I in the F center; and it experiences a large local mode vibration, while F^{IV} is knocked farther away through these vibrations.

Another example is the strong temperature dependence of the superhyperfine interaction observed for atomic hydrogen centers on anion or cation sites [14, 39, 40], which have local vibrational modes with very high amplitudes caused by the light hydrogen mass.

In crystals with structural phase transitions the lattice changes its symmetry at T_c. This can be *seen* by a paramagnetic probe, which therefore can be used to study such transitions. A recent example is the investigation of Mn^{2+} in $RbCdF_3$, which goes from a cubic to a tetragonal phase at 124 K. With ENDOR it was possible to study the order parameter as a function of temperature, which is directly reflected in the splitting of the ENDOR lines with temperature on going through T_c for suitable field orientations. It turned out that the order parameter measured is smaller than the intrinsic value, since the lattice relaxes around the probe, and this *decouples* it from the lattice somewhat [41, 42].

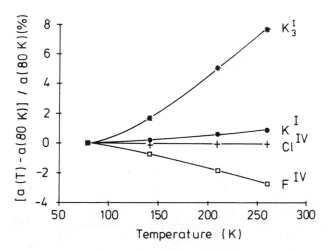

Figure 6.28 Temperature dependence of the isotropic superhyperfine constant a of the K_3^I and F^{IV} neighbors of $F_H(F^-)$ centers in KCl (see Figure 6.26) compared to the corresponding temperature dependence of K^I and Cl^{IV} neighbors of F centers in KCl.

4.5   Photo-Electron Nuclear Double Resonance

Particularly in semiconductor physics a correlation between the determination of a defect structure and other properties, like the energy levels associated with them, is very important to understand and improve the materials. In general, the energy levels are determined by *deep level transient spectroscopy* (DLTS) or related techniques (see, e.g., [43]), which, however, cannot identify the microscopic nature of the defect, while ESR (or) ENDOR alone can only determine the defect structure. A correlation between the energy levels and the defect structure can be achieved by measuring two kinds of crystals: one, where the level of the defects in question is occupied by an unpaired electron and another, where the Fermi energy is lower (e.g., by pinning the Fermi level at shallow impurities) so that the level in question is empty. In the former the ESR–ENDOR analysis is performed. The latter does not show ESR or ENDOR signals unless it is illuminated with light of sufficient energy such that electrons are raised from the valence band to occupy the level. Figure 6.29 shows such a

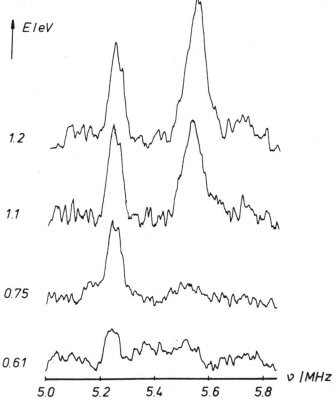

Figure 6.29   Photo-ENDOR of S^+ and $(S—S)^+$ defects in silicon. Adapted from S. Greulich-Weber, "ENDOR-Untersuchungen an Chalkogenen in Silizium," doctoral dissertation, Universität-GH-Paderborn, 1987.

photo-ENDOR experiment in Si doped with S (and codoped with B to lower the Fermi level to be nearer to the valence band). The crystal contained two paramagnetic defects: S^+ and $(S—S)^+$ pairs. The ESR–ENDOR signals of S^+ began to appear with light of $hv > 0.58$ eV, while those of the $(S—S)^+$ pair defects appeared only for $hv > 0.8$ eV. In Figure 6.29, for example, at $hv = 0.75$ eV, only ENDOR lines of the S^+ appear [44]. The two levels identified at $E_v \pm (0.59 \pm 0.05)$eV and $E_c - (0.37 \pm 0.05)$eV in this way agree well with those determined by optical spectroscopy and DLTS [45]. Apart from the correlation between energy levels and defect structures the method can be used to separate the ENDOR spectra of different defect species, whose spectra overlap in the same frequency range. Their analysis can be very difficult without being able to identify and correlate the ENDOR lines to one particular defect species. Similarly, optical ionization of occupied levels can be used to change the ENDOR line intensity of a specific defect and thus identify its lines from other defect lines present simultaneously. For example, this was applied successfully to the investigation of thermal donors in Si [37].

5   ADVANCED ELECTRON NUCLEAR DOUBLE RESONANCE METHODS

5.1   Electron Nuclear Double Resonance-Induced Electron Spin Resonance

Overlapping ESR spectra of different defects cause overlapping ENDOR spectra. It can be very difficult of not impossible to analyze their angular dependence, especially if many ENDOR lines occur in a narrow frequency range. Furthermore, there may be weaker ESR spectra buried under stronger ones having observable ENDOR lines, which can be analyzed, but the ESR spectrum cannot be measured. The two types of F centers possible in BaFCl, for example, are produced simultaneously. Their ESR spectra overlap strongly (see Figure 6.30). Here one can measure a kind of excitation spectrum of a particular ENDOR line that belongs to one defect, which gives an image of the corresponding ESR spectrum that belongs to the same defect (ENDOR-induced ESR spectrum). This can be seen from Figure 6.13. The ENDOR transitions can be measured by setting B_0 to any one of the four superhyperfine ESR transitions for $m_I = \frac{3}{2}, \frac{1}{2}, -\frac{1}{2}$, and $-\frac{3}{2}$; thus the ENDOR line intensity measured should follow the ESR line pattern in the middle of Figure 6.13. However, according to (24), the frequency of an ENDOR line depends on v_n and hence on B_0. Because of this one must "correct" the ENDOR frequency for the variation of B_0 when varying B_0 through the ESR spectrum. This can be done easily with the computer-controlled spectrometer or in other ways when the corresponding nuclear g_I factor is known. In measuring the ENDOR-induced ESR spectrum one monitors the ENDOR line intensity of a particular ENDOR line while varying the magnetic field through the ESR spectrum and correcting the frequency position according to (24). The resulting spectrum is an image of the

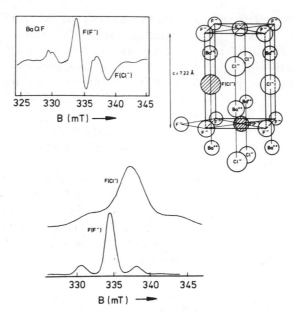

Figure 6.30 Model of two F centers in BaFCl [$F(Cl^-)$ and $F(F^-)$ centers], their superimposed ESR spectra (derivative), and the ENDOR-induced ESR spectra of both centers measured using a ^{19}F ENDOR line of each center. Reprinted with permission from J. R. Niklas, R. U. Bauer, and J.-M. Spaeth, *Phys. Status Solidi B*, **119**, 171 (1983).

(integrated) ESR spectrum of that defect, to which the ENDOR line (nucleus) belongs. In this way the ESR spectrum of different defects can be separated.

Figure 6.30 shows this for the two F centers in BaFCl. The two ENDOR-induced ESR spectra were measured using ^{19}F ENDOR lines of both centers. When $S = \frac{1}{2}$ and no quadrupole interaction is experienced by the nucleus, the ENDOR-induced ESR spectrum corresponds to the true ESR line shape if the cross relaxation does not depend on m_I and is not the dominating electron spin relaxation process. The true line shape was observed in several cases [46].

The line shape of the ENDOR-induced ESR spectrum is not necessarily identical with that of the ESR spectrum. To discuss this let us consider Figure 6.31, where the level scheme for the simplified case $S = \frac{1}{2}, I = \frac{3}{2}, W_{shf}$ and $W_q > 0$ is assumed. If one uses the ENDOR transitions between levels A and B for the ENDOR-induced ESR experiment, only the two ESR transitions between levels B and C and A and D are affected. For instance, the partially saturated ESR transition $D-A$ is desaturated by the NMR transition $A-B$. Similarly the ESR transition $C-B$ is experiencing a population change because of the NMR transition $A-B$. The other two possible ENDOR transitions are not affected. Thus, using any of the three quadrupole ENDOR lines, only for two ESR transitions, the m_I states of which are connected to the ENDOR transition, one fulfills the double resonance condition. They can be seen as an ENDOR-induced ESR spectrum only if the following holds: (1) only the simple type of cross-

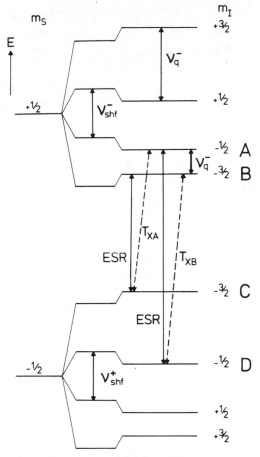

Figure 6.31 Schematic representation of the energy levels for the system $S = \frac{1}{2}$, $I = \frac{3}{2}$ with superhyperfine and quadrupole interactions. The arrows denote ESR–ENDOR and cross-relaxation transitions.

relaxation process $(m_s m_I) \rightarrow (m_s - 1, m_I \pm 1)$ occurs, and (2) no significant relaxation takes place between different m_I states within a given m_s state.

An example of this is shown in Figure 6.32, where the two-line ENDOR-induced ESR spectrum was measured by locking the ENDOR frequencies successively to the three ENDOR lines of the Ba^{2+} quadrupole triplet of $F(Cl^-)$ centers in BaFCl. The assumption seems to hold quite well, although not perfectly, as the small intensity of the two further lines in the ENDOR-induced ESR spectra indicates.

The true line shape of the full ESR spectrum is observed when there is no quadrupole interaction and the relaxation times T_x do not depend on the nuclear spin states.

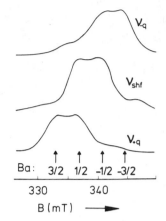

Figure 6.32 An ENDOR-induced ESR spectra of $F(Cl^-)$ centers in BaFCl using the ^{137}Ba quadrupole triplet ENDOR lines v_q^-, v_{shf}, and v_q^+. Reprinted with permission from J. R. Niklas and J.-M. Spaeth, *Phys. Status Solidi B*, **101**, 221 (1980).

Neglecting isotope mixture, one can apply the preceding arguments for one neighbor also to a nucleus, labeled 1, out of many neighbors, the ENDOR line of which is taken for the ENDOR-induced ESR measurement. Nucleus 1 only experiences another effective magnetic field as in the case of a single neighbor because of the superhyperfine interactions of the other nuclei. To clarify this let us assume that nucleus 1 has the largest superhyperfine interaction. Then the ESR spectrum is split because the W_{shf} of each line is inhomogeneously broadened.

$$B = \frac{1}{g_c \mu_n} h\nu - W_{shf}^{(1)} \cdot m_I^{(1)} - \sum_{\substack{i=1 \\ i \neq 1}}^{N} W_{shf}^{(i)} m_I^{(i)} \tag{32}$$

The ESR signal for a given $m_I^{(i)}$ value and particular field B within the $m_I^{(1)}$-superhyperfine line is given by the total number of combinations of the $m_I^{(i)}$ of the other nuclei that lead to this value of B. The line width of the $m_I^{(i)}$-superhyperfine line is determined by the last term in (32).

If nucleus 1 has no quadrupole interaction, the whole ESR spectrum is observed with ENDOR-induced ESR. If nucleus 1 has quadrupole interaction, then only those ESR transitions leading to the two $m_I^{(1)}$ levels that are connected by the ENDOR transition chosen for the experiment are desaturated. Thus, all the lines of the ENDOR-induced ESR spectrum appear at fields B in (32), for which $-I^{(i)} \leqslant m_I^{(i)} \leqslant I^{(i)}, i \neq 1$, and $m_I^{(1)} = \bar{m}_I^{(1)}$ or $m_I^{(1)} = \bar{m}_I^{(1)} + 1$ with $m_I(1)$ having one of the values $-I^{(1)}, -I^{(1)} + 1, \cdots, I^{(1)} - 1$ determined by the ENDOR transition chosen. Therefore, if a quadrupole interaction is present, the ENDOR–ESR spectrum is never a true ESR spectrum because of the necessary selection of nuclear spin states.

Figure 6.33 illustrates the ENDOR-induced ESR experiment when a quadrupole ENDOR line v_{-q}^+ is taken of a nucleus with a very small superhyperfine interaction (I_2). The superhyperfine splitting of the ESR line is determined by

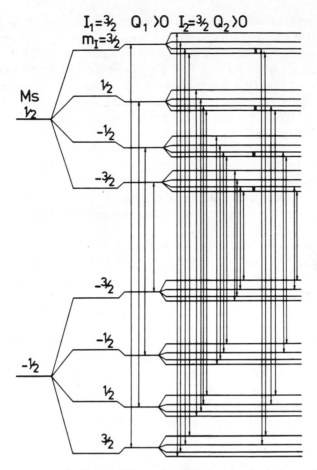

Figure 6.33 Schematic representation of ESR- and ENDOR-induced ESR transitions for a system of one unpaired electron and two nuclei having different superhyperfine and quadrupole interactions. The ENDOR-induced ESR transitions are measured using the nucleus I_2 with the smaller interactions.

another nucleus with a larger interaction (I_1). The ENDOR-induced ESR spectrum represents all superhyperfine lines caused by the larger interaction of I_1 but not the full spectrum, since some lines are missing because of the selection of nuclear spin states of I_2. The line width of the ENDOR-induced ESR lines is smaller, and the whole spectrum is shifted in field. If I_2 has only a very small interaction, then the difference of the ENDOR-induced ESR spectrum and the true ESR spectrum is only small. For details about line width and line position see [46, 47].

The selection of spin states allows one to determine the relative signs of W_q and W_{shf}. Figure 6.31 assumes that both W_{shf} and W_q have the same sign. The observation in Figure 6.32 is explained with the level scheme of Figure 6.31, that is, with the same sign for W_{shf} and W_q.

Certain complications arise when different isotopes are present. Suppose there is only one nucleus with different isotopes, for example, ^{37}Cl and ^{35}Cl, and a superhyperfine-ENDOR line of one isotope is used for ENDOR-induced ESR; then the line shape is not the true ESR line shape. To obtain this, one must add the ENDOR-induced ESR spectra caused by all the corresponding ENDOR lines of the different isotopes to form the total spectrum. The intensity ratio of the different ENDOR-induced ESR spectra should correspond to the abundancy of the different isotopes. If this is not so, then the cross-relaxation rates are different for the isotopes. To obtain the true line shape in this case requires a knowledge of this latter effect to correct the superposition of the isotope ENDOR-induced ESR spectra.

If there are more nuclei of a type with different isotopes, similar arguments apply. One has to add the isotope ENDOR-induced ESR spectra of the nuclei used for the measurement.

More serious complications arise if several strongly coupled nuclei show effects of second-order superhyperfine splitting [20, 21]. One then must work in the total spin representation, and the understanding of the ENDOR-induced ESR spectra is less straightforward. A detailed discussion of this problem is beyond the scope of this chapter. Results we obtained for F^+ centers in Na-β-alumina where two Al nuclei are strongly coupled are described in [48].

For systems with $S > \frac{1}{2}$ there is usually a fine structure interaction splitting of the electron Zeeman levels. In this case with ENDOR-induced ESR one selects spin states in a way analogous to the way one uses to select m_I states because of the quadrupole interaction discussed previously. One measures only some of the possible ESR transition.

This is demonstrated for Fe^{3+} centers in $KMgF_3$, which were produced by X irradiation of $KMgF_3$ doped with Fe^{2+}. In Figure 6.34a the integrated ESR spectrum of the five ESR transitions is reproduced with partly resolved superhyperfine structure with six nearest ^{19}F neighbors. Figure 6.34b shows the ENDOR-induced ESR spectrum measured with a ^{19}F ENDOR line at 42.5 MHz displaying two of the five Fe^{3+} ESR transitions, while Figure 6.34c shows the ENDOR-induced ESR spectrum measured with a ^{19}F ENDOR line at 21.0 MHz. Clearly, another ESR spectrum appears. It was buried under the Fe^{3+} spectrum beyond recognition and turned out to be caused by simultaneously produced F centers [49]. A detailed analysis also allows the determination of the relative signs of quadrupole interaction constants, superhyperfine interaction constants, and fine structure constants [46, 47].

5.2 Double-Electron Nuclear Double Resonance

Although with ENDOR-induced ESR experiments each ENDOR line can be *labeled* to a particular defect when there is the simultaneous presence of several defects, this can be a tedious task, especially if one must follow a complicated angular dependence. Therefore, a method is called for with which the ENDOR spectra of different defects can be measured separately. This can be done by

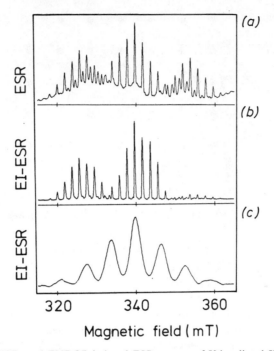

Figure 6.34 The ESR- and ENDOR-induced ESR spectra of X-irradiated $KMgF_3$ doped with Fe^{2+}: (a) Integrated ESR spectrum after X irradiation at room temperature. $B_0 \parallel \langle 100 \rangle$. (b) The ENDOR-induced ESR spectrum for the ^{19}F-ENDOR lines at 42.5 MHz. The spectrum is caused by Fe^{3+}. (c) An ENDOR-induced ESR spectrum for the ^{19}F-ENDOR line at 21.0 MHz. The spectrum is caused by F centers. Reprinted with permission from R. C. DuVarney, J. R. Niklas, and J.-M. Spaeth, *Phys. Status Solidi B*, **97**, 135 (1980).

measuring a triple resonance where two NMR frequencies are applied simultaneously together with the microwaves.

In an ENDOR experiment the radio frequency-induced NMR transitions between the nuclear Zeeman levels of a neighbor nucleus coupled to the unpaired electron by a superhyperfine interaction change somewhat the polarization of the electron spin in the partially saturated situation. This coupling between neighbor nucleus and unpaired electron is indicated schematically by a "spring" in Figure 6.35. If a second NMR transition is induced simultaneously with a second radio frequency at another nucleus coupled to the same unpaired electron, then the induced change of the electron spin polarization is different from what it would be if the first NMR transition does not occur simultaneously. Thus, the polarization change caused by the second NMR transition is dependent on the occurrence of the first NMR transition. The total electron desaturation (ENDOR effect) is a function of the product of the effects of the two NMR transitions. Thus, when modulating the two NMR transitions with different frequencies and using double lock-in techniques, one can, for example, induce one particular ENDOR transition of a particular neighbor nucleus and

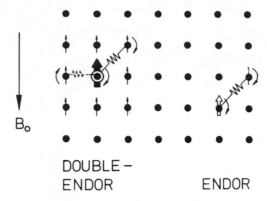

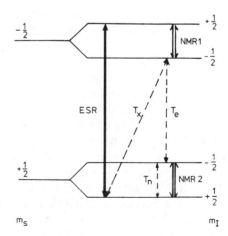

Figure 6.35 Schematic representation of ENDOR and double-ENDOR.

monitor the change of the ENDOR signal height as a function of the simultaneous second ENDOR transition. One observes changes of the NMR line intensity as a function of the second frequency swept radio frequency source. A change is observed when a second ENDOR transition is induced at a nucleus, which is also coupled to the same electron, but not if the nucleus belongs to a different center. The ESR of this different center may also be saturated if the ESR spectra of the different centers overlap. Therefore, this triple resonance experiment can be used to separate the ENDOR spectra of different defects if their ESR spectra overlap. Such a separation may well be impossible otherwise if the centers have many ENDOR lines with complicated angular dependencies. However, the method functions well only when the neighbors have a high abundance of magnetic nuclei. In Si, for example, where only 4.7% nuclei are magnetic (^{29}Si), the probability to have simultaneously two magnetic nuclei as defect neighbors is too low to allow for double-ENDOR experiments.

The simplest experiment, the so-called *special triple resonance* (special double-ENDOR) is shown schematically in Figure 6.36 for the simple case of

Figure 6.36 Level scheme to explain the special triple-resonance experiments (double-ENDOR).

$S = \frac{1}{2}$, $I = \frac{1}{2}$. If stationary ENDOR is measured for the transition *NMR1* between $m_s = -\frac{1}{2}$, $m_I = \frac{1}{2}$ and $-\frac{1}{2}$, then the signal height is determined by the ESR transition probability (i.e., by B_1^2), the NMR1 transition probability [i.e., B_{rf}^2 (NMR1)], and the cross-relaxation time T_x. It is assumed (and a condition for the experiment) that T_x is greater than T_e, the electron spin lattice relaxation time, because of the comparatively long nuclear spin lattice relaxation time T_n. If then a second radio frequency is applied between the levels $m_s = \frac{1}{2}$, $m_I = \frac{1}{2}$ and $-\frac{1}{2}$(NMR2), then T_n is effectively shortened by this transition and therefore T_x is also shortened, which results in an enhancement of the monitored ENDOR signal at the frequency NMR1. In the experiment one irradiates with the fixed ENDOR frequency NMR1, monitors the ENDOR line intensity of the line at NMR1, and sweeps the second radio frequency. When the transition NMR2 is induced the ENDOR line intensity NMR1 increases. The increase is the double-ENDOR signal, and it is detected with a double lock-in technique.

In the stationary double-ENDOR spectrum positive and negative signals are observed [50, 51]. Negative signals occur if the second NMR frequency is induced between nuclear states belonging to the same m_s quantum number. Figure 6.37 shows the double-ENDOR spectrum for the two F centers in BaFCl. In Figure 6.37a the ENDOR lines of both F centers are superimposed; hence, a full analysis was not possible. In Figure 6.37b the fixed ENDOR frequency NMR1 was set to one ENDOR line belonging to $F(Cl^-)$ centers, and NMR2 was swept between 1 and 9 MHz and the double-ENDOR effect was recorded. In Figure 6.37c the analogous experiment was made for an $F(F^-)$ ENDOR line. Both double-ENDOR spectra show only lines caused by the $F(Cl^-)$ or $F(F^-)$ centers alone. Especially around 5 MHz both centers have many ENDOR lines, which otherwise could not have been separated [52].

Figure 6.19a shows as another example the double-ENDOR spectrum measured for the Ga vacancy in GaP for $B_0 \| \langle 111 \rangle$. Comparison of Figure 6.19a with Figure 6.14b demonstrates that all the lines measured in single ENDOR also appear in double-ENDOR. This proves that all ENDOR lines belong indeed to one defect and that the vacancy is not distorted, with the consequence that the ENDOR spectrum may be a superposition of several vacancy configurations. The occurrence of so many ENDOR lines, at first unexpected for a simple tetrahedral surrounding of four equivalent ^{31}P neighbors, is indeed caused by a large and hitherto undescribed effect of second-order superhyperfine structure [21].

Double-electron nuclear double resonance is also very important when analyzing low-symmetry defects. The defects are distributed over several orientations in the crystal. The ESR and ENDOR spectra of these orientations overlap. In a sense each defect orientation is equivalent to a new defect species. With double-ENDOR the spectra of one particular defect orientation can be measured separately, which greatly facilitates the analysis or makes it possible. In a recent investigation of O^- centers in α-Al_2O_3, which had very low symmetry (that is *no* symmetry), a definite assignment of the quadrupole ENDOR lines to their corresponding *hyperfine* ENDOR lines was only possible after one particular center orientation could be measured separately [53].

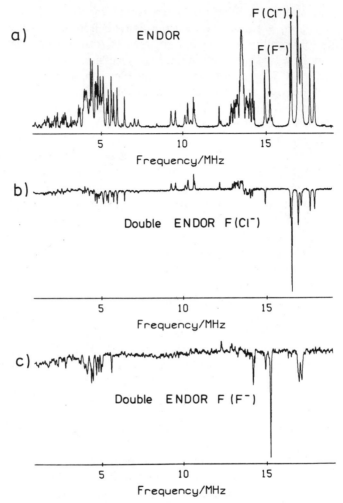

Figure 6.37 (*a*) Part of the ENDOR spectrum of $F(Cl^-)$ and $F(F^-)$ centers simultaneously present in BaFCl. (*b*) Double-ENDOR spectrum obtained for setting one frequency to an ENDOR line of $F(Cl^-)$ centers (see arrow in Figure 6.37*a*) and sweeping the second radio frequency. (*c*) Double-ENDOR spectrum for setting one radio frequency to an $F(F^-)$-ENDOR line (see arrow in Figure 6.37*a*). Reprinted with permission from J. R. Niklas, R. U. Bauer, and J.-M. Spaeth, *Phys. Status Solidi B*, **119**, 171 (1983).

6   OPTICALLY DETECTED ELECTRON SPIN RESONANCE AND ELECTRON NUCLEAR DOUBLE RESONANCE

6.1   Introductory Remarks

There are several ways to detect ESR optically. Selection of a technique depends on the system to be studied and on the kind of question to be answered. Compared to *conventional* ESR the optical detection, if possible, has several advantages about the identification of defects: (1) higher sensitivity, (2) higher selectivity, (3) direct correlation between structural and bulk properties of a crystal, (4) possibility to investigate optical excited states, (5) possibility to investigate ESR and ENDOR with high spatial resolution.

Sensitivity enhancement can be up to 10^4-10^5, compared to the conventional ESR, depending on the method used and the physical parameters of the system. The high selectivity results from the use of single absorption or emission bands for the detection. The bands of different defects usually do not overlap entirely; hence, spectra from these defects can be measured separately. A correlation of optical properties with structural property is based on the same argument. Relaxed excited states of sufficient radiative lifetimes can be studied. The condition is that the microwave transition rate is of the same order or bigger than the reciprocal radiation lifetime. If this condition is not fulfilled, during the lifetime of the state no occupation changes of the Zeeman levels can be induced by the microwaves. Electron spin resonance spectra of excited states with lifetimes of about microseconds were detected several times. The actual technical limit, which is given mainly by the microwave amplitude to be fed into the cavity at low temperature, is probably at lifetimes of about 100 ns.

The optical detection of ESR is based on either the observation of a microwave-induced intensity change in luminescence or the observation of microwave-induced changes of the degree of polarization of absorption or emission. The Zeeman levels involved must be occupied differently, and the radiative transitions must differ in polarization properties. It should be noted that the Zeeman splitting of the levels is only about 10^{-4} eV, while the phonon broadening of the levels caused by electron phonon coupling is three orders of magnitude greater. Fortunately, the polarization of the optical absorption or emission transitions is usually only affected slightly by the phonons, so that microwave-induced population changes within the Zeeman levels can be observed.

The examples and methods following are chosen mainly with respect to defect identification. For example, interesting studies like radiative–nonradiative transitions or electron transfer in excited states for which ODMR is an excellent tool and the study of relaxed excited states are not mentioned if the defect model is known from ground-state ESR.

Optical detection of ESR using the polarization change of absorption bands was recently proven to be more sensitive than the simple population transfer picture predicts. In particular, the observed high signal intensity for ENDOR was quite surprising and is not yet understood. Its use, however, is very valuable in various areas of materials science.

There is a great interest in being able to measure ESR and ENDOR with spatial resolution, especially in semiconductor physics, where homogeneous wafers are needed for device applications. Methods modeled after the successful NMR tomography are less suitable, since with the large Bohr magneton the necessary magnetic field gradients are too high compared to the nuclear magneton. However, the optical detection of ESR and ENDOR easily achieves a rather good spatial resolution, which is limited by the diameter of the optical beam used and, of course, the S/N ratio, since with too little light intensity the signal becomes very weak. First experiments of this kind are discussed in Section 6.7.

6.2   Basic Features of Optical Transitions of Defects in Solids

In ionic and semiconductor crystals many defects have localized energy states within the band gap and possess optical absorption with energies below the band gap energy. Figure 6.38*a* shows this schematically for the transition from the defect ground state to the first excited state. The absorption is then observed within the *optical window* of the crystal, where no band-gap transitions ($h\nu < E_g$) and no transitions caused by lattice vibrations occur. This window ranges from several to about 0.1 eV.

The correlation of such a defect-induced optical absorption with a particular defect is usually a difficult task. This is demonstrated, for example, by the long history of the research on color centers in alkali halides [54]. Assignments are usually attempted by variation of the irradiation time, if the defect is intrinsic like a vacancy or electron or hole center formed as a consequence of radiation damage. However, this is not always unambiguous. Several defects can be produced simultaneously with overlapping absorption bands. Often there is one characteristic luminescence band. The measurement of its excitation spectrum can be used to identify the associated absorption band provided no other defect luminescence band overlaps it strongly.

In ionic and semiconductor crystals most defect states are sensitive to the positions of the nearby atoms or ions so that the form of the absorption and the form and the energy position of the emission depend on the vibrations of the

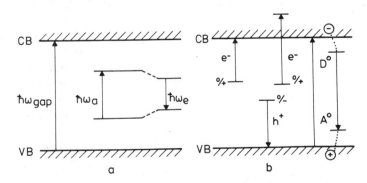

Figure 6.38 Schematic representation of several optical transitions in solids. (*a*) Intracenter absorption and emission with Stokes shift. Such transitions mostly occur in wide band-gap materials such as insulators. (*b*) Ionizing transitions to the conduction band and valence band, intracenter transitions into resonant states in the conduction band and D^0–A^0 donor–acceptor pair recombination luminescence after creating an electron-hole pair by band-gap illumination and capturing of electrons and holes by the D^+ donors and A^- acceptors.

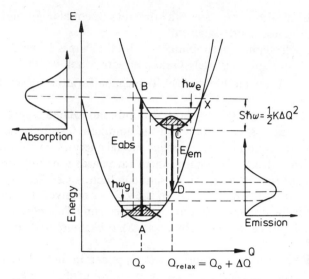

Figure 6.39   Configurational coordinate (CC) diagram of optical absorption and emission.

surrounding ions. This is conveniently discussed in the approximation of the configuration coordinate diagram (CC diagram), where the lattice vibrations are represented by a single localized mode with a configuration coordinate Q and linear coupling is assumed for the vibrational energy, which results in a parabolic energy curve for both the ground and excited states (see Figure 6.39). They have the form

$$E = \tfrac{1}{2}KQ^2 \tag{33}$$

where K is a force constant. Absorption and emission are vertical transitions (Born–Oppenheimer approximation). After the electrical dipole transition of the absorption has occurred, the electron distribution is changed and accordingly the lattice will relax and adjust to a new equilibrium position represented by $Q_0 + \Delta Q$ in Figure 6.39. From the vibrational ground state of the excited state parabola the emission takes the form of a transition to a ground-state configuration, which must relax back into the original configuration. The relaxation process involves phonon emission, which is the reason for the observed Stokes shift ΔE between the energies of absorption and emission ($\Delta E = E_{abs} - E_{em}$); ΔQ is directly related to the strength of the electron phonon coupling, which is characterized by the Huang–Rhys factor S

$$S \cdot \hbar\omega = \tfrac{1}{2}K \cdot \Delta Q^2 \tag{34}$$

where ω is the vibrational frequency (assumed to be equal in ground and excited states in the simplest approximation).

For large electron phonon coupling ($S > 10$) the shapes of the absorption and emission bands are approximately Gaussian with a half-width proportional to $\hbar\omega\sqrt{S}$. This is indicated in Figure 6.39. Typical half-widths are several tenths of an electron volt.

If the coupling is weak ($S < 1$), both ground- and excited-state parabolas are hardly displaced against each other and the dominant feature is the *zero phonon line* (ZPL). The transition involves no vibrational energies. Typical examples are transitions within the $4f^n$ configuration in rare earth ions in solids or for $3d^n$ configurations of transition metals in semiconductors (see Section 6.6).

In intermediate coupling ($1 < S < 6$) the ZPL is resolved, but the multiphonon structure is the dominant feature of the spectrum. The intensity of the ZPL relative to the whole band is given by $\exp(-S)$. It becomes undetectable for high values of S.

When a ZPL can be measured, the application of uniaxial stress and magnetic and electrical fields can cause a splitting of the ZPL, from which structural information such as the defect symmetry can be derived (e.g., [54] Chapters 5 and 6 and further references therein for ionic crystals).

Defects with a strong electron phonon coupling may not show a luminescence or may show only a very weak one. In the simple framework of the CC diagram a nonradiative deexcitation occurs, when the excited-state energy reached in a Franck–Condon absorption transition (point B in Figure 6.39) lies above the intersection of ground- and excited-state potential energy curves (point C in Figure 6.39). Here the system can be deexcited through the intersection directly into the ground state under phonon emission. In the approximation of linear coupling and equal vibration frequencies for ground and excited states this occurs when $E_{em} < \frac{1}{2}E_{abs}$ (Dexter–Klick–Russel rule) [55, 56].

In semiconductors the excited states of intracenter transitions can also be resonant states in the conduction band. Upon excitation with light of energy exceeding the band-gap energy electron hole pairs are created in the conduction and valence band and a donor–acceptor pair recombination luminescence can occur. An ionized donor D^+ captures an electron from the valence band to form D^0, and an ionized acceptor A^- captures a hole to form A^0. The paramagnetic pair D^0–A^0 recombines; that is, its electron and holes recombine and emit a fluorescence light. Its photon energy depends on the energy levels of the donors and acceptors and on their Coulomb energy [57] (Figure 6.38c). Defects in semiconductors can also be ionized by irradiation with light. Both electron emission into the conduction band and hole emission into the valence band occur. The photon energy at the onset of this process is determined by the energy levels. The ionization cross sections are usually smooth functions of the photon energy and look like broad absorption regions. They can be measured until the band edge is reached.

6.3   Optical Detection of Electron Spin Resonance by Optical Emission

6.3.1   Triplet-State Optically Detected Electron Spin Resonance

Optically detected electron spin resonance measured in the optical emission has been a standard tool in the study of organic crystals and molecules within the properties of excited triplet states. A recent review by Lynch and Pratt [58] lists specific reviews on ODMR of triplet states in many areas together with ODMR or inorganic and semiconductor crystals. To illustrate the method an example is described for a two-electron center in CaO.

Defects with two valence electrons, for example, in a ns^2 configuration, often have relaxed excited triplet states, in which the two electrons have parallel spins, $S = 1$. The optical absorption leads first to a singlet excited state with $S = 0$, which then relaxes into the triplet state by intersystem crossing. The two electrons then possess a fine structure interaction (e.g., caused by dipole–dipole interaction), so that in first-order perturbation theory the energy is given for an axially symmetric fine structure tensor and B_0 parallel to its principal z axis

$$E = g_z \mu_B B_0 m_s + D_z [m_s^2 - S(S + 1)] \tag{35}$$

Figure 6.40 shows the level scheme as a function of the magnetic field of the spin Hamiltonian containing fine structure and the electron Zeeman interaction ($D = D_z$). The three levels are all occupied as a result of relaxation from a higher singlet state. Radiative transitions from the triplet states into the singlet ground states are forbidden. However, the states $|+1\rangle$ and $|-1\rangle$ can mix with excited singlet states by the spin–orbit interaction; therefore, a finite radiative transition probability into the ground state is observed for these levels, while $|0\rangle$ cannot decay by radiative transitions. The radiative lifetimes of the $|+1\rangle$ and $|-1\rangle$ levels depend on the size of the spin–orbit interaction and the energy separation of excited singlet states from the triplet state.

Because of the radiative decay of the $|+1\rangle$ and $|-1\rangle$ levels, these levels are less populated in a stationary state compared to the level $|0\rangle$, where population is accumulated (see Figure 6.40). It is assumed that the spin lattice relaxation time T_1 is large compared to the radiative lifetime. Therefore, the microwave transition $|0\rangle \rightarrow |+1\rangle$ and $|0\rangle \rightarrow |-1\rangle$ shift population into the radiative levels, upon which the luminescence intensity is enhanced. This can easily be observed. On the other hand, one can also observe the magnetic circular polarization of the emission $(I_+ - I_-)$(MCPE), where for the low-field transition there is an enhancement, and for the high-field transition a decrease is observed (see Figure 6.40, where $D > 0$ is assumed).

The F centers in CaO, where an O^{2-} vacancy is occupied by two electrons, do have such relaxed excited triplet states, and their ODESR was observed in such a way [59, 60]. Figure 6.41 shows as an example the ODESR spectrum of

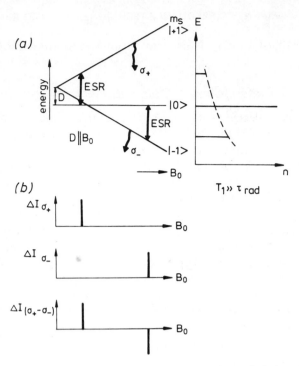

Figure 6.40 (a) Level scheme of triplet states to illustrate the optical detection of ESR by microwave-induced change in the luminescence intensity or by the magnetic circular polarization of the emission (MCPE). It is assumed that $B_0 \parallel D_z$, $T_1 \gg \tau_{rad}$. (b) Schematic representation of the microwave-induced intensity changes of circularly polarized light.

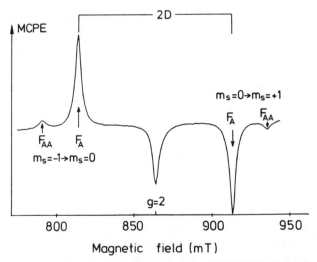

Figure 6.41 Magnetic circular polarization of the emission (MCPE) recorded at $\lambda > 700$ nm under microwave irradiation (24 GHz) as a function of the magnetic field ($B_0 \parallel \langle 100 \rangle$) in CaO containing $F_A(Mg)$ and $F_{AA}(Mg)$ centers. The $F_A(Mg)$ centers are two electrons trapped at an oxygen vacancy next to a Mg^{2+} impurity. The $F_{AA}(Mg)$ centers have two Mg^{2+} impurities opposite each other along a $\langle 100 \rangle$ direction next to the trapped electron; $T = 1.6$ K. Adapted with permission from *Solid State Commun.*, **43**, F. J. Ahlers, F. Lohse, and J.-M.Spaeth, "Application of Magnetic Resonance Techniques to the Study of Defects in Solids," Copyright 1982, Pergamon Journals, Ltd.

F_A and F_{AA} centers in CaO doped with Mg^{2+} that is observed at 700 nm. Both centers differ slightly in their fine structure constant D. Measurements with higher resolution show a superhyperfine interaction with the 10% abundant $^{25}Mg(I = \frac{5}{2})$. Six equidistant superhyperfine lines can be resolved [60, 61]. In the F_A center there is one Mg^{2+} next to the O^{2-} vacancy; in the F_{AA} center, two Mg^{2+} are opposite to each other along a $\langle 100 \rangle$ axis.

6.3.2 Optically Detected Electron Spin Resonance by Donor–Acceptor Pair Recombination Luminescence

Until recently, in semiconductor physics the usual way to observe ODESR has been by donor–acceptor pair recombination luminescence. There are many results in II–VI semiconductors, such as ZnS and ZnSe, and also a great deal in III–V semiconductors, such as GaP. The reader is referred to the review by Cavenett [62].

The fundamental process is illustrated in Figure 6.42. After a band–band

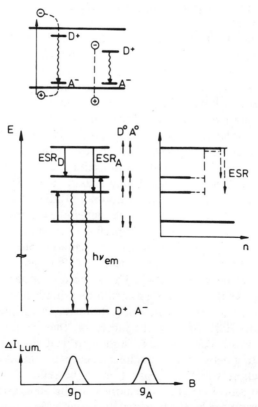

Figure 6.42 Schematic representation of donor–acceptor recombination in semiconductors and of the mechanism to observe the ESR as microwave-induced change in intensity of the donor–acceptor recombination luminescence.

excitation according to

$$D^+ + A^- + e^- + h^+ \rightarrow D^0 + A^0 \tag{36}$$

paramagnetic donors D^0 and acceptors A^0 are formed.

The unpaired electron at the donor and the unpaired hole at the acceptor can recombine with the emission of a luminescence radiation

$$D^0 + A^0 \rightarrow D^+ + A^- + h\nu \tag{37}$$

The level scheme of D^0 and A^0 in a magnetic field is shown in Figure 6.42. The two spins are assumed to be very weakly coupled [62, 63]. Therefore, the energy positions of the levels are determined only by their respective g-factors and the magnetic field. Since a radiative decay from the two triplet states into the singlet ground state is forbidden, a luminescence is observed only from the two singlet states, where the two spins are antiparallel. Therefore, in a stationary state, the population of the two triplet states is higher compared to the two singlet states, similar, as previously discussed, to the state $|0\rangle$ in the highly coupled triplet system.

Upon the microwave transitions indicated in Figure 6.42 the recombination luminescence is increased, and in principle one can observe the ESR lines of both the donor and acceptor provided their g-factors are different enough. Usually, one observes only the donor resonance since the p-type hole states of the acceptors experience a dynamical Jahn–Teller effect, which makes the resonance very difficult to observe. The resonance can be observed on application of uniaxial stress that is large enough to suppress the dynamical Jahn–Teller effect.

With this technique the ODESR of anion antisite defects in GaP: Zn (p type) were recently observed. After band–band excitation two strongly overlapping luminescence bands can be observed, one peaking at 0.95 eV and the other at 1.20 eV. In the 0.95-eV luminescence a doublet ODESR spectrum is observed, which is caused by a P atom on a Ga site (*antisite*) with one unpaired electron. The doublet splitting is caused by the hyperfine interaction with the ^{31}P nucleus ($I = \frac{1}{2}$) [64, 65]. Using a specific modulation technique one can also resolve the superhyperfine interaction with four nearest ^{31}P neighbors from which the structure model can be inferred [64]. The acceptor resonance was not seen.

Figure 6.43 shows the ODESR spectrum measured in the 1.20-eV luminescence, which is caused by an antisite defect (however, in a triplet state) and which also contains the ESR line of the acceptor that is participating in the luminescence process. The two low- and high-field lines are caused by the antisite; the splitting between the two lines of each doublet is caused by the ^{31}P hyperfine interaction, while the separation of the centers of the two doublets (between ca. 240 and 400 mT) corresponds to the separation of the lines in Figure 6.41 and is caused by the fine-structure interaction. It was concluded that

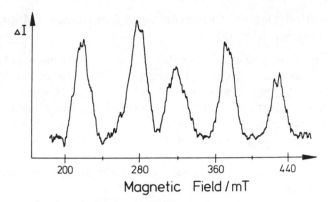

Figure 6.43 X-band ODESR spectrum observed as a microwave-induced change of the donor–acceptor recombination luminescence in GaP:Zn (excitation wavelength 514 nm). The spectrum shows the P-antisite donor $P_{Ga}P_3Y_P$ in a triplet state and the acceptor (central line), which is probably an interstitial Fe^{3+}. $B_0 \parallel \langle 111 \rangle$. Reprinted with permission from B. K. Meyer, Th. Hangleiter, J.-M. Spaeth, G. Strauch, Th. Zell, A. Winnacker, and R. H. Bartram, *J. Phys. C*, **18**, 1503 (1985).

the structure model is a PP_3Y_p defect with an unknown atom on one of the four nearest-neighbor P sites and that the acceptor may be a nearby interstitial Fe^{3+} impurity [66]. The level of the defect is at $E_v + 1.5$ eV, where E_v is the energy of the valence band. The PP_3Y_p defect can be excited into its triplet state by direct excitation from the valence band with subband gap light, a possibility also indicated schematically in Figure 6.42. This excitation enhances the luminescence, since the defects can be excited in the bulk of the crystal and not only in a thin surface layer, as is usually the case for band–band transitions caused by the high absorption constant.

When measuring the ODESR by recombination luminescence a common difficulty is that there is a spatial distribution between donors and acceptors and that the exchange interaction between them can be large and varying according to distances. It is especially large for shallow donors and acceptors. This results in a broadening of the ESR lines. Therefore, in most cases the donor resonances show no hyperfine or superhyperfine structures, and a defect identification can be based only on the g-factors.

There are many resonances known from which the defect cannot be identified. The broadening effect is less important for *deep level* defects. One way to overcome this difficulty is to measure the ODESR spectrum with time resolution. It was shown for shallow In-donors in ZnO that the resolution can be greatly enhanced by exciting the luminescence with pulsed light and by taking only the long lifetime tails of the luminescence for the ODESR measurement, which comes from the distant donor–acceptor pairs for which the exchange interaction is only small [67]. Little work has been done so far with time-resolved ODESR for solid-state defects.

6.4   Absorption-Detected Electron Spin Resonance (Magnetic Circular Dichroism Method)

The detection of ground-state ESR by measuring the microwave-induced change of the magnetic circular dichroism (MCD) is long known from color center physics [68], but it was not applied to materials science until very recently. It was originally devised to study excited states of F centers in alkali halides. Its recent use for ground-state ESR proved particularly useful in both semiconductor physics and inorganic systems like crystals with laser-active defects.

In Figure 6.44 the idea of the method is schematically indicated for the simplification of an *atomic* optical $s-p$ transition.

The allowed optical absorption transitions for right and left circular polarized light are indicated along with the relative matrix elements. The absorptions for both polarizations are split in energy by the spin−orbit splitting of the excited state. If this splitting is smaller compared to the phonon width of

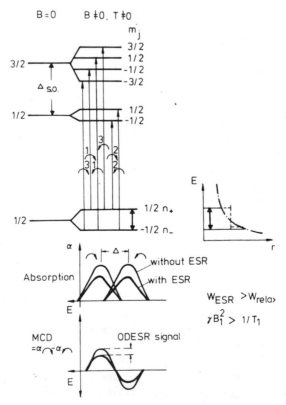

Figure 6.44 Simple *atomic* model to explain the magnetic circular dichroism (MCD) of the absorption and its microwave-induced decrease to detect ESR transitions.

the absorption bands the measurement of the MCD as a function of photon energy yields a derivative structure.

The MCD is defined as

$$\text{MCD} = \frac{d}{4}(\alpha^+ - \alpha^-) \tag{38}$$

where d is the crystal thickness and α^+, α^- are the absorption constants for right and left circular polarized light, respectively. If the absorption comes from a Kramers doublet with $S = \frac{1}{2}$, one obtains [68]

$$\text{MCD} = \frac{1}{2}\alpha_0 d \frac{\sigma^+ - \sigma^-}{\sigma^+ + \sigma^-} \frac{n_- - n_+}{n_- + n_+} \tag{39}$$

where α_0 is the absorption constant for unpolarized light, σ^+ and σ^- are the cross sections for right and left polarized light, respectively, and n_- and n_+ are the occupation numbers for the $m_s = \pm\frac{1}{2}$ states. The MCD is thus proportional to the occupation difference $n_+ - n_-$ and the cross-section difference $\sigma^+ - \sigma^-$.

The occupation difference $n_+ - n_-$ can be decreased by a microwave transition, provided the microwave transition rate is about the same as or larger than the spin–lattice relaxation rate $1/T_1$. Such an ESR transition thus results in a decrease of the MCD, which is monitored (see Figure 6.44). One measures the MCD as a function of the magnetic field under microwave irradiation. The decrease caused by the resonance is observed on the variation of the MCD as a function of the magnetic field according to

$$\text{MCD} \sim \tanh(g\mu_{\text{B}}B_0/2kT) \tag{40}$$

For low temperatures and the usual field variation this is practically a linear function of B_0.

Figure 6.45a shows the absorption spectrum of a Tl-doped KCl crystal after X irradiation at $-40°\text{C}$; Figure 6.45b is the MCD spectrum of this crystal after bleaching the F centers generated simultaneously. There are many centers created, partly diamagnetic, partly paramagnetic. Figure 6.46 shows the ODESR spectrum measured in the absorption band at 1040 nm of the laser-active $\text{Tl}^0(1)$ centers [69]. The absorption at 1040 nm has a negative paramagnetic MCD. Comparison with the ESR spectrum of $\text{Tl}^0(1)$ centers identified in Figure 6.9 [13] shows that the spectrum is identical if one allows for the two different microwave bands used (X band for conventional ESR, K band for ODESR). However, in ODESR, a direct correlation with an optical absorption band at 1040 nm was possible (see also Section 6.6.2). The centers consist of a Tl^0 atom next to an anion vacancy along $\langle 100 \rangle$ [13]. The ODESR spectrum contains allowed transitions of centers with their $\langle 100 \rangle$ axes parallel and perpendicular to the magnetic field orientation (indicated in Figure 6.46) as well as forbidden transitions, in which the Tl nuclear quantum numbers ($m_I = \frac{1}{2}$) also

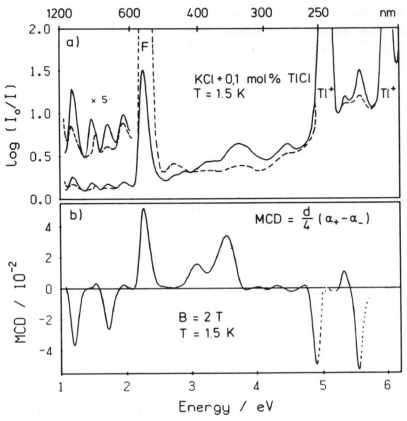

Figure 6.45 (*a*) Optical absorption spectrum of KCl–Tl after X irradiation at $-40°C$ (dashed line) and after *F*-center bleaching (solid line). (*b*) Magnetic circular dichroism (MCD) after *F*-center bleaching. Reprinted with permission from F. J. Ahlers, F. Lohse, J.-M. Spaeth, and L. F. Mollenauer, *Phys. Rev. B*, **28**, 1249 (1983).

change. Their intensity decreases with decreasing microwave power (Figure 6.46*a* and *b*) [69].

Figure 6.47 shows the optical absorption of the MCD in *as-grown*, undoped GaAs, which was grown by the liquid encapsulated Czochralski (LEC) method. In the absorption below the gap energy of 1.52 eV only a very weak band at 1.18 eV caused by an intracenter transition of the diamagnetic mid-gap defect EL2 is detectable. However, the MCD reveals the existence of further intracenter absorption bands caused by paramagnetic defects, which turned out to be the singly ionized state of the EL2 defects [70]. In direct measurement their optical absorption is not detectable. The sensitivity enhancement that is evident when measuring the MCD is a consequence of the applied form of modulation spectroscopy. The optical transitions are caused by transitions into two resonant states in the conduction band [71].

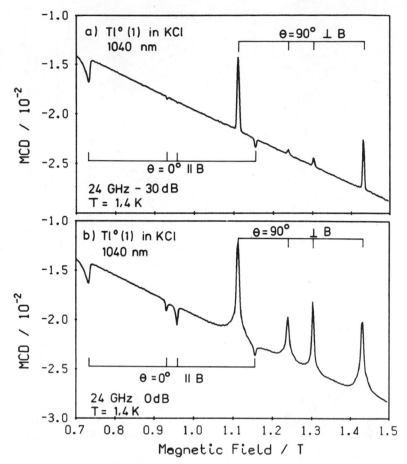

Figure 6.46 (a) An ODESR spectrum of $Tl^0(1)$ centers in KCl detected as a microwave-induced change of the MCD of the absorption band at 1040 nm, measured at low microwave power (24 GHz, 1.5 K). (b) Same for high microwave power.

When measuring the ODESR in this MCD, for example, at 1350 nm, the spectrum of Figure 6.48, which is due to paramagnetic EL2 defects, is observed. The defect structure was revealed by ODENDOR (see Section 6.5) to be an As antisite–As interstitial pair defect. The ESR spectrum shows the central hyperfine interaction between the unpaired electron with the ^{75}As nucleus ($I = \frac{3}{2}$). The conventional ESR of this defect in as-grown material is very weak; the S/N ratio is about 2:4 (the concentration of paramagnetic EL2 is ca. $10^{16} cm^{-3}$). In ODESR there is a gain in S/N of about two orders of magnitude.

Other examples of ODESR spectra measured recently are those of Pb^+ ions on cation sites next to an F^- vacancy in CaF_2, CrF_2 and BaF_2 [72, 73], which are analogous to the laser-active Tl defects previously discussed. Also Fe^{3+} in InP [74] and V^{2+} and V^{3+} on Ga sites in GaAs [74–76] can be measured. In

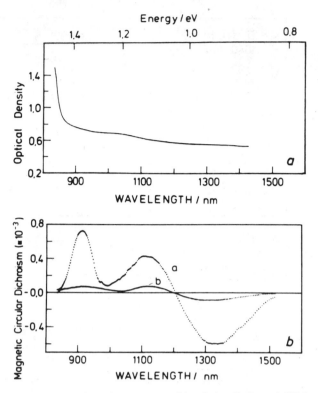

Figure 6.47 (*a*): Optical absorption of *as-grown* semi-insulating GaAs at 1.4 K (crystal thickness 0.3 mm). (*b*): curve a, MCD of the same sample. $T = 4.2$ K, $B = 2$ T; curve b, excitation spectrum of the ODESR lines of the paramagnetic EL2 defects (MCD tagged by the EL2 ESR lines). Reprinted with permission from B. K. Meyer, J.-M. Spaeth, and M. Scheffler, *Phys. Rev. Lett.*, **52**, 851 (1984).

the ODESR spectra of the transition metals in III–V semiconductors the ESR spectra were observed not only in their intracenter absorption bands, but also in the ionizing transitions to both the conduction and the valence bands [76]. This observation was unexpected. However, it renders the MCD technique particularly powerful, since ionizing transitions of defects into bands are almost always present with rather good transition probabilities.

6.5   Optically Detected Electron Nuclear Double Resonance

Except for triplet states in organic systems only very few experiments were performed, where ENDOR was observed in luminescence [77]. In ionic crystals it was Tm^{2+} in CaF_2 [78], a rare earth defect. A few observations were reported in semiconductors using the donor–acceptor recombination luminescence. There are observations in amorphous Si [79, 80] and in ZnSe [81]. The ENDOR lines were observed as an emission increase when NMR transitions

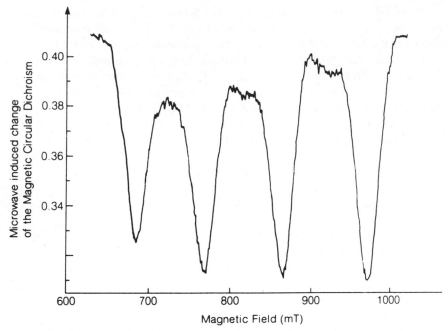

Figure 6.48 Optically detected ESR spectrum of paramagnetic EL2 defects in semi-insulating *as-grown* GaAs as microwave-induced decrease of the MCD of the absorption measured at 1350 nm $B_0 \parallel \langle 100 \rangle$, $T = 1.5$ K, $\nu_{ESR} = 24$ GHz. Reprinted with permission from B. K. Meyer, J.-M. Spaeth, and M. Scheffler, *Phys. Rev. Lett.*, **52**, 851 (1984).

were induced by a radio frequency applied in a small loop attached to the sample. In ZnSe, for example, two lines caused by ^{67}Zn and ^{77}Se were seen centered at the frequencies of the free nuclei. The emission enhancement effect was about 1%. The ENDOR lines were broad, about an order of magnitude broader than in conventional ENDOR. No angular dependence is reported [81]. Using time-resolved techniques ENDOR of ^{115}In in ZnO was observed [67].

The observation of ENDOR by emission has a principal disadvantage. If the lifetime of an excited state giving rise to the emission is 1 μs or shorter, then the homogeneous ENDOR line width is 1 MHz or more. Thus, one loses all the detailed information necessary for the structure determination. Furthermore, to be able to shift populations between nuclear Zeeman levels within the radiative lifetime one needs very high NMR transitions probabilities, requiring radio frequency field strengths of 10 mT or more, which at low temperatures causes quite a technical problem.

A better way to observe ODENDOR seems to be by the MCD technique. In principle, both ground-state and excited-state ENDOR can be observed, provided there is a closed optical pumping cycle. For the excited state the same restrictions apply as discussed previously for the emission method (Section 6.3).

Figure 6.49 shows the level scheme when $S = \frac{1}{2}$, and there is a central nucleus with $I_c = \frac{3}{2}$ and one ligand nucleus with $I = \frac{3}{2}$ (e.g., a simplified model for the paramagnetic EL2 defect in GaAs). In the experiment one sets the magnetic field onto a particular position of the ODESR line, for example, into the flank. Electron spin resonance transitions must obey the selection rule $\Delta m_I = 0$, $\Delta m_s = \pm 1$. Thus, when measuring the ESR in one of the four lines of the $I_c = \frac{3}{2}$ system, then at most only $\frac{1}{4}$ of all the spin packets can be involved; only $\frac{1}{4}$ of the MCD can be decreased when saturating the transition. However, if each line is inhomogeneously broadened by further superhyperfine interactions, only a fraction of a decrease should occur, since $\Delta m_{I,a} = 0$ must be obeyed for the ligand I_a. Thus, upon inducing NMR transitions between the nuclear Zeeman levels, one can include more $m_{I,a}$ substates into the ESR pumping cycle and thus increase the effect of decreasing the MCD. Therefore, the ENDOR transitions are detected as a further increase of the ODESR.

Figure 6.50 shows a section of the ODENDOR lines caused by the nearest [75]As neighbors of the paramagnetic EL2 defect in an as-grown, undoped GaAs crystal [70, 82, 83]. The ENDOR lines are as sharp and numerous as in conventional ENDOR. From their angular dependence the structure model of the EL2 defects can be derived.

A similar study was recently made for Pb^+ centers in CaF_2, SrF_2, and BaF_2, in which the structure was shown to be a Pb^+ ion on cation sites next to an F^- vacancy along $\langle 111 \rangle$. The ^{19}F ENDOR lines measured in the MCD and their angular dependence clearly showed the symmetry of the defect and its atomic structure [72]. Similarly, as for EL2, the S/N ratio of the ENDOR lines was very good.

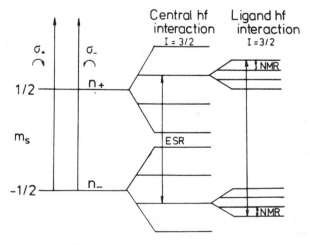

Figure 6.49 Level scheme to illustrate the detection of ENDOR by a radio frequency- and microwave-induced decrease of the MCD of the optical absorption.

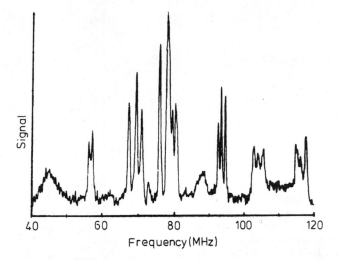

Figure 6.50 Section of the ODENDOR spectrum of the paramagnetic EL2 defects in semi-insulating *as-grown* GaAs. $T = 1.5\,K$, with B_0 in a (110) plane. The lines are caused by five ^{75}As neighbors of the As_{Ga}^+, which forms the core of the EL^2 defect. Reprinted with permission from B. K. Meyer, D. M. Hofmann, J. R. Niklas, and J.-M. Spaeth, *Phys. Rev. B*, **36**, 1332 (1987).

6.6   Correlation of Optically Detected Electron Spin Resonance and Optically Detected Electron Nuclear Double Resonance With Bulk Properties

6.6.1   Energy Levels

As with the photo-ESR and photo-ENDOR experiments discussed in Section 4.5, photo-ODESR and photo-ODENDOR experiments can be performed to correlate with energy levels and thus with electrical properties of the solid. An investigation of the EL2 defects in undoped GaAs also revealed another kind of correlation, which can be achieved with such an experiment; that is, two energy levels belong to the same defect in different charge states. In Figure 6.51a the onset of the MCD (or ODESR) of paramagnetic EL2 centers is shown to occur at a photon energy of 0.52 eV. Before occupying this D^+/D^{2+} level no MCD could be measured. At a photon energy of 0.74 eV the MCD (and again correspondingly the ESR signal) decreases, since the level of the mid-gap EL2, which is diamagnetic, is now reached, a second electron is raised into the defect. At higher energies more and more paramagnetic EL2 defects are transferred into paramagnetic EL2 defects with a corresponding decrease of the MCD. This experiment shows that the two levels belong to the two different charge states of the same defect. After occupying the mid-gap EL2 level its optical intracenter band at 1.18 eV appears, which can be used to calibrate the lost MCD as a consequence of the occupancy of the mid-gap EL2 level. The intracenter band was calibrated by deep level transient spectroscopy [70, 83].

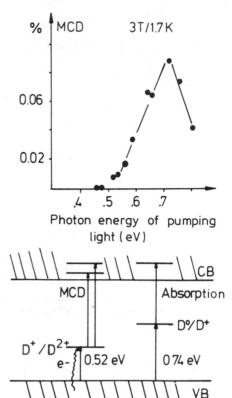

Figure 6.51 Excitation spectrum of the ODESR lines of the paramagnetic EL2 defects in semi-insulating *as-grown* GaAs:Zn (*p*-type). The paramagnetic EL2 level is occupied by raising the electrons from the valence band as from an exciting photon energy of $hv > 0.52\,\mathrm{eV}$. The diamagnetic EL2 level is occupied as from $hv > 0.74\,\mathrm{eV}$. The MCD of paramagnetic EL2 appears above the photon threshold of $0.52\,\mathrm{eV}$ and decreases again at $0.74\,\mathrm{eV}$, where paramagnetic EL2 is transformed into diamagnetic EL2 by capturing an electron; $T = 1.7\,\mathrm{K}$, $B = 3\,\mathrm{T}$. Upper half: experimental results; lower half: level scheme of paramagnetic EL2/diamagnetic EL2. Reprinted with permission from B. K. Meyer, D. M. Hofmann, J. R. Niklas, and J.-M. Spaeth, *Phys. Rev. B*, **36**, 1332 (1987).

6.6.2   Tagging of the Magnetic Circular Dichroism by Electron Spin Resonance and Electron Nuclear Double Resonance

By detecting the ground-state ESR or ENDOR with the MCD method it is possible to measure a kind of excitation spectrum of the ESR and ENDOR lines of a specific defect. The schemes of Figures 6.44 and 6.49 apply to all transitions to the excited states of the defect, that is, to every optical absorption band belonging to the defect of which the ground-state ESR and ENDOR can be measured. Therefore, one can set the ESR or ENDOR resonance conditions to a particular ESR or ENDOR line, vary the optical wavelength, and monitor the ODESR or ODENDOR signal intensity as a microwave- (and radio frequency) induced change of the MCD. Thus, from the total MCD of a sample one can measure in this way only the part that belongs to the selected ESR or ENDOR signal. This is especially useful when a superposition of optical absorptions that were caused by several defects is present, which is, for example, typical with radiation damage and impurity problems. Figure 6.45a shows the optical absorption after X-irradiation of KCl doped with Tl; Figure 6.52a shows the MCD measured as the excitation signal of one ESR line of the $\mathrm{Tl}^0(1)$ center for

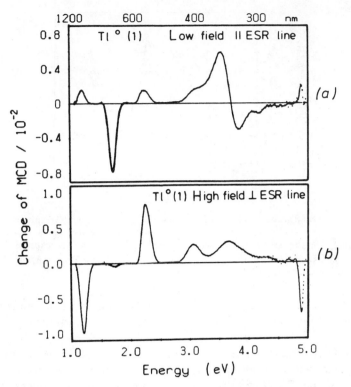

Figure 6.52 (a) Magnetic circular dichroism (MCD) at 1.4 K of the *parallel* Tl⁰(1) centers in KCl as tagged by the low-field ESR line. (b) The same for the *perpendicular* Tl⁰(1) centers tagged by their high-field ESR line. Reprinted with permission from F. J. Ahlers, F. Lohse, J.-M. Spaeth, and L. F. Mollenauer, *Phys. Rev. B*, **28**, 1249 (1983).

B_0 parallel to the center axis; B_0 perpendicular to the center axis is shown in Figure 6.52*b*. Thus, the MCD tagged by ESR [69] reveals only the optical transitions belonging to one center or center orientation. Altogether, eight optical Tl⁰(1) center absorptions were detected in this way. For the analogous Ga⁰(1) and In⁰(1) centers, the positions of the absorption bands can only be detected by this method, since they were buried in shoulders of other much stronger absorptions [84].

The MCD tagged by ESR of the V^{3+} defect on Ga sites in GaAs (Figure 6.53) reveals a rich optical absorption spectrum with several zero phonon lines and phonon replica, intracenter transitions into Jahn–Teller excited states, and an ionizing transition (at about 1.4 eV) into the conduction band. It is the first time that all the details of the optical spectrum, which were observed only partly by different authors, could unambiguously be assigned as belonging to one specific center. The sample also contained V^{2+} on Ga sites and another V-related complex defect, which make a clear assignment of optical bands very difficult otherwise [85, 86].

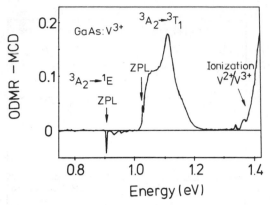

Figure 6.53 Excitation spectrum of the ODESR line of V_{Ga}^{3+} defects in high resistivity GaAs (MCD tagged by ESR). $T = 1.6$ K. Adapted from A. Görger, "Vanadiumzentren in GaAs und GaP: Magnetooptische und optisch nachgewiesene magnetische Resonanzuntersuchungen," Diplomarbeit, Universität-GH- Paderborn, 1987. J.-M. Spaeth, A. Görger, D. M. Hofmann, and B. K. Meyer, *Mater. Res. Sci. Symp. Proc.*, **104**, 363 (1988).

Similarly, for Fe^{3+} in InP such a rich optical spectrum can be identified. It was also found that the MCD and ESR can be observed in ionizing transitions to the bands.

As seen from Figure 6.52*a* this kind of experiment also gives information about the polarization of specific transitions; for example, the optical transition at 725 nm of the $Tl^0(1)$ centers in KCl is almost entirely polarized for the electrical vector perpendicular to the center axis since it is absent in Figure 6.52*b*.

6.7   Mapping With Optically Detected Electron Spin Resonance

In materials science the necessity to measure the presence of defects with spatial resolution is hard to meet by conventional methods. However, with emission and absorption, optical detection of ESR is possible within the limitations of the light beam diameter used to detect the resonances. Since, at least in principle, one can use (tunable) lasers, a resolution of approximately 10×10 μm should be obtainable.

Figure 6.54 shows the first experiment of this kind using the MCD of paramagnetic EL2 centers in GaAs (at 1350 nm) and the IR band of mid-gap EL2 (at 1 μm) to linearly map the concentration of the two charge states of the EL2 defect across $\frac{1}{2}$ a wafer (thickness 300 μm, resolution ca. 300 μm). As can be seen the two charge states occur almost in anticorrelation: the mid-gap EL2 follows a W shape; the paramagnetic EL2 an M shape across the wafer. However, the defect is distributed rather homogeneously and it is not W-shaped as was assumed previously. This implies that the shallow acceptors, regulating the occupation of the paramagnetic EL2 and the mid-gap EL2, must exist in an M-shaped distribution. Their nature is still unknown [87, 88].

Clearly, two-dimensional mapping can be developed, and it is currently being undertaken for the EL2 charge states [89].

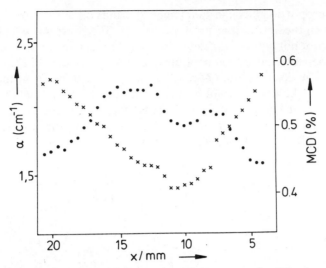

Figure 6.54 Spatially resolved measurement of the IR intracenter absorption band at 1.18 eV of mid-gap EL2 and of the MCD at 1350 nm of paramagnetic EL2 in semi-insulating *as-grown* GaAs across $\frac{1}{2}$ a wafer. Crosses represent paramagnetic EL2, dots represent mid-gap EL2. The spatial resolution was $0.3 \times 0.3 \, \text{mm}^2$. Adapted with permission from M. Heinemann, B. K. Meyer, J.-M. Spaeth, and K. Löhnert, in E. R. Weber, Ed., "The Occupation of the Two Charge States of EL2 in LEC-Grown GaAs Wafers—a Mapping Investigation," *Defect Recognition and Image Processing in III–V Compounds II*, Elsevier, New York, 1987.

6.8 Spin Determination by Measuring the Magnetic Circular Dichroism

For the determination of the charge state of a defect it is important to know its spin state. It is normally not easily determined from ESR spectra unless there is a fine structure splitting resolved in a multielectron system. Otherwise, one must resort to ENDOR. However, using the MCD technique, there is yet another way to determine the spin state, as was realized only recently [85, 86].

The following discussion is precise for an orbital singlet state. When there is orbital degeneracy, the same ideas can be followed; however, possible splittings and spin mixings must be taken into account [86, 90].

The MCD is composed of a diamagnetic (dia) and a paramagnetic (para) term: $\text{MCD} = \text{MCD}_{\text{dia}} + \text{MCD}_{\text{para}}$.

The diamagnetic term is proportional to the magnetic field B_0, and it is usually small compared to the paramagnetic part MCD_{para}, which is proportional to the Brillouin function B_S

$$\text{MCD}_{\text{para}} = C\langle S_z \rangle = CSB_S \tag{41}$$

with

$$B_S(\mu) = 1/S\{(S + \tfrac{1}{2}) \coth[(S + 1)/2\mu] - \tfrac{1}{2} \coth(\mu/2)\} \tag{42}$$

$$\mu = g_e \mu_B B_0 / kT \tag{43}$$

Here C is a proportionality constant (which depends on the nature of the excited states to which the optical transitions occur); g_e is the electronic g-factor. As will be seen, C does not enter in the determination of the spin state; B_S depends on the static magnetic field B_0, temperature T, and spin S of the defect as well as on its g-factor. The determination of S relies on the measurement of the Brillouin function, which, however, cannot be measured directly, because of the unknown diamagnetic contribution to the experimental MCD. However, since MCD_{dia} is linear in B, by measuring the total MCD for several magnetic fields (B_1, B_2) and temperatures (T_1, T_2) and following the ratio

$$R_{exp} = \frac{MCD(B_1, T_1) - MCD(B_1, T_2)}{MCD(B_2, T_1) - MCD(B_2, T_2)} \tag{44}$$

gives

$$R(S) = \frac{B_S(S, B_1/T_1) - B_S(S, B_1/T_2)}{B_S(S, B_2/T_1) - B_S(S, B_2/T_2)} \tag{45}$$

Equation (45) can be calculated theoretically by varying spin S. Thus, the spin of the defect is determined when the condition $R_{exp} = R(S)$ is fulfilled. Figure 6.55 shows this for V^{3+} defects on Ga sites in GaAs, which is a $3d^2$ configuration. As expected and known from ENDOR [91] $S = 1$. However, with the same method one can show that for V_{Ga}^{2+} $S = \frac{1}{2}$, a low-spin state exists [85, 86]. Although the ground state is ^2E, there is such a small orbital splitting, if any, that this does not influence the spin determination.

6.9   Experimental Aspects of Optically Detected Electron Spin Resonance and Optically Detected Electron Nuclear Double Resonance

The spectrometer to measure ODESR by a microwave-induced intensity change of the emission is very simple and was described earlier (e.g., [62]). Figure 6.56 shows a schematic description of the spectrometer that can measure microwave-induced MCD or MCPE. The sample is illuminated in a Faraday configuration; that is, the propagation direction of the circularly polarized light and B_0 are parallel. It is convenient for this configuration to produce the magnetic field with a split coil superconducting magnet.

The circularly polarized light is produced with the combination of a linear polarizer and a stress modulator [92]. This combination produces right and left circularly polarized light alternatively in one oscillating cycle of the modulator (frequency about 30 kHz). The absorption difference is detected by a lock-in detector and divided by the total light intensity. Thus, it is measured $(I^- - I^+)/(I^- + I^+)$.

The MCD can be measured to about 10^{-5} for an optical density of 1. For the measurement of the MCPE the stress modulator is placed behind the sample.

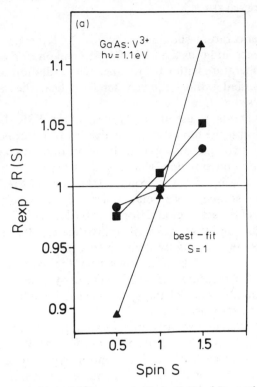

Figure 6.55 Spin determination of V_{Ga}^{3+} centers in GaAs. Ratio of R_{exp} and $R(S)$ as a function of S for V_{Ga} measured at 1.1 eV in high resistivity GaAs: V for the following pairs of magnetic field values: $B_1 = 1\,T$, $B_2 = 0.8\,T$ (circles), $B_1 = 0.8\,T$, $B_2 = 0.5\,T$ (squares), and $B_1 = 1.3\,T$, $B_2 = 0.5\,T$ (triangles) at temperatures $T_1 = 1.65\,K$ and $T_2 = 4.2\,K$ (the g-factor is 1.917). Adapted from A. Görger, B. K. Meyer, and J.-M. Spaeth, "Transition Metal Elements in III-V Semiconductors—A Study With Optically Detected Magnetic Resonance," in G. Grossman and L. Ledebo, Eds., *Semiinsulating III-V-Materials, Malmö 1988*, Hilger, Philadelphia, PA, 1988, p. 331; J.-M. Spaeth, A. Görger, D. M. Hofmann, and B. K. Meyer, *Mater. Res. Sci. Symp. Proc.*, **104**, 363 (1988).

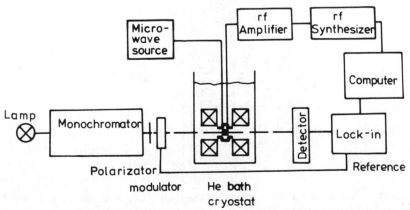

Figure 6.56 Block diagram of the apparatus for optical detection of magnetic resonance. For detection by MCPE the stress modulator and polarizer are placed behind the cryostat.

The sample temperature should be low (ca. 1.5 K), since the ground-state population difference and thus the ESR effect depend on B/T. It is advantageous to control the temperature within 0.1 K, since the irradiation of the sample with light, microwaves, and radio frequency tends to heat the sample and cause *background* signals.

Figure 6.57 presents a useful cavity design for ESR–ENDOR. It is a cylindrical cavity with the mode TE_{011}, which is open around the middle and can be described as two opposite top hats. It allows an optical access angle of 20° for both absorption and emission, which is necessary if one uses conventional light sources. The quality factor is around 5000. For the MCD one requires a highly stable light source, since small differences in absorption within the modulation phase of the stress modulator must be detected. Therefore, the use of lasers with a high amplitude noise of approximately 1% is not necessarily advantageous despite the higher light intensity. For ENDOR the radio frequency field is produced by four rods near the sample that act as two Helmholtz loops. To measure the MCD tagged by ESR and ENDOR it is convenient to computer control the spectrometer, including the positions of optical components, such as lenses.

: The ODESR effect found so far is much too large compared to the previous simple explanation. This is especially the case for the EL2 centers in GaAs (Figure 6.48), where in each of the four central hyperfine ESR transitions a decrease of the MCD by about 25% was found, which means that the lines are homogeneous. However, the observation of ENDOR contradicts this deduction. Moreover, the ENDOR effect observed is of the same order as the ESR effect, which is not true in conventional solid-state defect ENDOR, where a

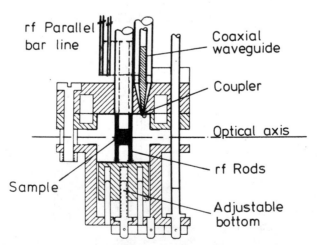

Figure 6.57   Cavity for optical detection of ESR. The cavity (TE_{011} mode) is open in the middle and allows optical access to about a 20° aperture. The coupling hook can be rotated during operation. The cavity can be floated with helium or be in a helium-exchange gas.

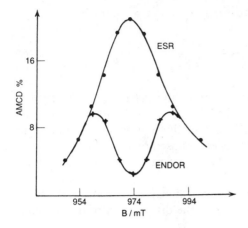

Figure 6.58 The ODESR and ODENDOR effect for paramagnetic EL2 defects in semi-insulating *as-grown* GaAs measured through one [75]As hyperfine ESR line. Adapted from D. M. Hofman, "Strukturaufklärung des EL2-Defektes in Galliumarsenid mit optisch nach-gewiesener Elektronen–Kern–Doppelreso-nanz," doctoral dissertation, Universität-GH-Paderborn, 1987.

stationary effect of about 1 to 2% of the ESR is usually observed. Figure 6.58 shows the experimental result on the EL2 defect in GaAs: the ODENDOR effect is high in the wings in each ODESR line [83]. Neither Figure 6.58 nor the size of the observed effects, which are also too large in other cases like P-antisite defects in GaP [83, 93] and $Pb^+(1)$ centers in the alkaline earth fluorides [73], are understood.

7 INTERPRETATION OF SUPERHYPERFINE AND QUADRUPOLE INTERACTION CONSTANTS

The atomistic structure of a defect is not all that is determined from ENDOR. The superhyperfine and quadrupole interaction constants and the orientation of the respective tensors give a detailed picture of the electronic structure of these defects. Therefore, they can be used to test a theoretical defect wave function to describe precisely the electronic structure.

However, the problem of the electronic defect structure is a complicated multielectron problem, and thus a theory is not in a position to calculate reliably the superhyperfine structure of defects, although in some simple cases there are quite good approximations available, such as in the field of color centers in alkali halides [94, 95].

It is beyond the scope of this chapter to discuss the various approximations made to interpret the superhyperfine and quadrupole constants. Only a few very simple remarks are made, which can be useful when analyzing defect structures.

If it is clear from the system, that the amplitude of the wave function for the defect decreases with distance from the core of the defect (wide-gap insulators and deep-lying defects), then the anisotropic interaction constants b of more distant nuclei are rather well described by the simple classical point dipole–dipole approximation. The unpaired electron is replaced by a point-dipole residing at the center of the defect, and the interaction constant becomes:

$$\frac{b}{h} = \frac{\mu_0}{4\pi} g_I \mu_n g_e \mu_B \frac{1}{h} \frac{2}{R_\alpha^3} \tag{46}$$

where R_α is the distance between the center of the defect in the nucleus in question. This simple observation is useful for an assignment of nuclei of a particular shell. The isotropic constants a, especially for more distant nuclei, are more difficult to interpret, since transfer effects caused by overlaps of ion cores or atomic cores play a decisive role. The near-neighbor interactions are determined mostly by covalency effects [95].

The wave function is often expanded into a linear combination of atomic orbitals (LCAO) to obtain a rough picture of the distribution of the unpaired electrons.

$$\psi = \eta_0 \psi_0 + \sum_i \eta_i \psi_i \tag{47}$$

where ψ_0 is the wave function of the unpaired electron at the central (impurity) atom, η_0 is the localization of the wave function at the central atom, and η_i represents the s and p wave functions of the neighbor atoms i. In semiconductors like Si or III–V compounds, they are usually described as the hybridized s and p functions

$$\psi_i = \alpha_i \psi_s + \beta_i \psi_p \tag{48}$$
$$\alpha_i^2 + \beta_i^2 = 1 \tag{49}$$

where s and p are the outer occupied s and p orbitals; α_i^2 describes the part of the unpaired electron in s orbitals, and β_i^2 is the part of the unpaired electron in p orbitals of the neighbor i.

To get a picture of how the unpaired electron is localized in the central atom and distributed about the neighbors, one uses the experimentally determined superhyperfine constants a_i, b_i to determine the coefficients η_0, η_i and α_i, β_i by comparison to values of a_f, b_f for a full electron residing in the respective orbitals; a_f and b_f are calculated for the *free atom* with a Hartree–Fock method.

$$\eta_i^2 \alpha_i^2 = a_i/a_f \tag{50}$$
$$\eta_i^2 \beta_i^2 = b_i/b_f \tag{51}$$

As an example for Si $a_f = 4150\,\mathrm{MHz}$, $b_f = 101\,\mathrm{MHz}$ [96].
Therefore,

$$\eta_i^2 \alpha_i^2 + \eta_i^2 \beta_i^2 = \eta_i^2 \tag{52}$$

and

$$\eta_i^2 = \left(\frac{|a_i|}{a_f} + \frac{|b_i|}{b_f}\right) N_i \tag{53}$$

where N_i is the number of equivalent nuclei in one shell; η_i^2 is called the *spin density*, and it is often plotted as a function of distance to show the localization or delocalization of the unpaired electrons.

This treatment is very rough and sometimes misleading. It is assumed that the electron (charge) distribution and the spin distribution are the same. This is generally not the case, as can be seen already from experiments. Often there are negative a values (see Table 6.4) observed indicating that there is a spin polarization [95], which can also reduce the observed b values, as was clearly shown for several O$^-$ defects in various oxides [97]. For substitutional chalcogens in Si there is a first calculation now available of the superhyperfine structure from first principles that shows the decisive influence of spin polarization, which is caused by the influence of the valence band on the deep level states [98].

Particular difficulties arise for shallow defects, that are defects, the energy levels of which are very close to the band energies. The mixing with band states leads to an oscillating wave function making it difficult, if not impossible, to assign an interaction tensor identified by ENDOR to specific nuclei about the defect. Such observations were made for shallow donors in Si in the very early ENDOR spectroscopy [99, 100].

To determine the charge state of a defect it is useful to apply a rough estimate of the quadrupole interaction constants. At a specific nucleus it consists mainly of two parts: one caused by the field gradient of point charges present in the defect structure and one caused by the unpaired spin density moving in p (or d) orbitals of the atom (ion) to which it belongs [101].

$$q = q(p) + q(b) \tag{54}$$

with

$$q(p) = \frac{\pm ne^2 Q(1 - \gamma_\infty)}{8\pi\varepsilon_0 I(2I - 1)} \frac{1}{R^3} \tag{55}$$

$$q(b) = \frac{8e^2 Q(1 - \gamma)}{2I(I - 1)\mu_n g_I \mu_B} b(p) \tag{56}$$

where Q is the quadrupole moment, γ and γ_∞ are the Steinheimer antishielding factors, n is the number of point charges, and R is the distance between the nucleus and the point charge. The second term is determined from the experimental values of the anisotropic superhyperfine constant b; $b(p)$ is that part of b caused by unpaired spin density in p orbitals. The charge-density distribution is at least roughly estimated from an LCAO analysis of the superhyperfine-interaction parameters, and it is approximated as a point-charge distribution to calculate $q(p)$. Such a simple estimate was used successfully to decide on the site of Ni^{3+} centers in GaP (whether a Ni^{3+} was substitutional or on the tetrahedral interstitial site) [22] and more recently on the charge state of the As-interstitial-forming part of the paramagnetic EL2 defect in GaAs [70, 83].

8   CONCLUSIONS AND OUTLOOK

Magnetic multiple resonance methods offer a very powerful tool for the investigation of defects in solids. The availability of modern experimental techniques makes it possible to tackle difficult problems of interest in materials research. The development of these methods is by no means complete. Efforts to improve spatial resolution and also to use the absorption method for thin layers are encouraged. Whether it is possible by microoptic techniques to lead the light beam parallel rather than perpendicular to a surface should be tested. Time-resolved experiments allow not only the study of dynamical phenomena, but also the use of donor–acceptor recombination luminescence to detect hyperfine and superhyperfine structures with better resolution. Furthermore, the correlation spectroscopy correlating defect properties with other properties should be further developed. For instance a combination of magnetic resonance investigations with the spectroscopy that detects level positions in semiconductors certainly would be very fruitful.

References

1.   G. Feher, *Phys. Rev.*, **114**, 1219, 1249 (1959).

2.   T. M. McKinney and I. B. Goldberg, "Electron Spin Resonance," in B. W. Rossiter and J. F. Hamilton, Eds., *Physical Methods of Chemistry*, Vol. IIIB, 1989, Chap. 4.

3.   A. Abragam and B. Bleaney, *Electron Paramagnetic Resonance of Transition Ions*, Clarendon, Oxford, 1970.

4.   C. P. Slichter, *Principles of Magnetic Resonance*, Harper & Row, New York, 1963.

5.   G. E. Pake and T. L. Estle, *The Physical Principles of Electron Paramagnetic Resonance*, 2nd ed., Benjamin, New York, 1973.

6.   A. Schweiger, "ENDOR Methodology," in P. B. Ayscough Ed., *Electron Spin Resonance*, Specialist Periodical Report, The Chemical Society, Vol. 10b, 1987.

7.   H. G. Grimmeiss, E. Janzen, H. Ennen, and O. Schirmer, *Phys. Rev. B*, **24**, 4571 (1981).

8.   S. H. Muller, M. Sprenger, E. G. Sievers, and C. A. J. Ammerlaan, *Solid State Commun.*, **25**, 987 (1978).

9.   J. Michel, J. R. Niklas, and J.-M. Spaeth, *Phys. Rev. Lett.*, **7**, 611 (1986).

10.   P. Studzinski, "Paramagnetische Ionen zur Untersuchung struktureller Phasenübergänge–Eine ENDOR Untersuchung," doctoral dissertation, Universität-GH-Paderborn, 1985.

11. U. Kaufmann, W. H. Koschel, J. Schneider, and J. Weber, *Phys. Rev. B*, **19**, 3343 (1979).

12. U. Kaufmann and J. Schneider, *Solid State Commun.*, **25**, 1113 (1978).

13. E. Goovaerts, J. A. Andriessen, S. V. Nistor, and D. Schoemaker, *Phys. Rev. B*, **24**, 29 (1981).

14. J.-M. Spaeth, "Atomic Hydrogen as a Model Defect in Alkali Halides," in V. M. Tuchkevich and K. K. Shvarts, Eds., *Defects in Insulating Crystals*, Springer, New York, 1981, p. 232; *Z. Phys.*, **192**, 107 (1966).

15. G. Heder, J. R. Niklas, and J.-M. Spaeth, *Phys. Status Solidi B*, **100**, 567 (1980).

16. H. Seidel, *Z. Phys.*, **165**, 218, 239 (1961).

17. G. Heder, "ENDOR-Untersuchungen an atomarem Wasserstoff auf Anionengitterplatz in KCl," doctoral dissertation, Universität-GH-Paderborn, 1979.

18. J. R. Niklas and J.-M. Spaeth, *Solid State Commun.*, **46**, 121 (1983).

19. F. Beeler, M. Scheffler, O. Jepsen, and O. Gunnarson, *Mater. Res. Soc. Symp. Proc.*, **46**, 117 (1985).

20. T. E. Feuchtwang, *Phys. Rev.*, **126**, 1628 (1962).

21. J. Hage, J. R. Niklas, and J.-M. Spaeth, *Mater. Sci. Forum*, **10–12**, 259 (1986).

22. Y. Ueda, J. R. Niklas, J.-M. Spaeth, U. Kaufmann, and J. Schneider, *Solid State Commun.*, **46**, 127 (1983).

23. K. Möbius and R. Biehl, "Electron Nuclear–Nuclear Triple Resonance of Radicals in Solutions," in M. N. Dorio and J. H. Freed, Eds., *Multiple Electron Resonance Spectroscopy*, Plenum, New York, 1979.

24. C. P. Poole, Jr., *Electron Spin Resonance*, Wiley, New York, 1973.

25. H. Seidel, *Z. Angew. Phys.*, **14**, 21 (1962).

26. J. R. Niklas, "Elektronen–Kern–Doppelresonanz-Spektroskopie zur Strukturuntersuchung von Festkörperstörstellen," Habilitationsschrift, Universität-GH-Paderborn, 1983.

27. Ch. Hoentzsch, J. R. Niklas, and J.-M. Spaeth, *Rev. Sci. Instrum.*, **49**, 1100 (1978).

28. S. Greulich-Weber, J. R. Niklas, E. Weber, and J.-M. Spaeth, *Phys. Rev. B*, **30**, 6292 (1984).

29. J. R. Niklas, private communication.

30. M. U. A. Bromba and H. Ziegler, *Anal. Chem.*, **51**, 1760 (1979).

31. H. Ziegler, *Appl. Spectrosc.*, **35**, 88 (1981).

32. M. U. A. Bromba and H. Ziegler, *Anal. Chem.*, **55**, 648 (1983).

33. M. U. A. Bromba and H. Ziegler, *Anal. Chem.*, **56**, 2052 (1984).

34. H. Söthe, P. Studzinski, and J.-M. Spaeth, *Phys. Status Solidi B*, **130**, 339 (1985).

35. P. Wagner, C. Holm, E. Sirtl, R. Oeder, and W. Zulehner, "Chalcogens as Point Defects in Silicon," in P. Grosse, Ed., *Festkörperprobleme: Advances in Solid State Physics*, Vieweg, Braunschweig, Vol. **24**, 1984, p. 191.

36. J. Michel, J. R. Niklas, and J.-M. Spaeth, *Mater. Res. Soc. Symp. Proc.*, **104**, 185 (1988).

37. J. Michel, J. R. Niklas, J.-M. Spaeth, and C. Weinert, *Phys. Rev. Lett.*, **57**, 611 (1986).

38. J.-M. Spaeth, *Cryst. Lattice Defects and Amorphous Mater.*, **12**, 381 (1985).

39. Ch. Hoentzsch and J.-M. Spaeth, *Phys. Status Solidi B*, **94**, 479 (1979).

40. P. Studzinski, J. R. Niklas, and J.-M. Spaeth, *Phys. Status Solidi B*, **101**, 673 (1980).

41. P. Studzinski and J.-M. Spaeth, *Radia. Eff.*, **73**, 207 (1983).

42. P. Studzinski and J.-M. Spaeth, *J. Phys. C.*, **18**, 6441 (1986).

43. J. Bourgoin and M. Lannoo, "Point Defects in Semiconductors II, Experimental Aspects," in M. Cardona, Ed., *Springer Series in Solid State Sciences*, Vol. 35, Berlin, 1983.

44. S. Greulich-Weber, "ENDOR-Untersuchungen an Chalkogenen in Silizium," doctoral dissertation, Universität-GH-Paderborn, 1987.

45. S. Greulich-Weber, J. R. Niklas, and J.-M. Spaeth, *J. Phys. Condensed Matter*, **1**, 35 (1989).

46. J.-M. Spaeth and J. R. Niklas, in J. T. Devreese, Ed., *Recent Developments in Condensed Matter Physics*, **1**, Plenum, New York, 1981, p. 393.

47. J. R. Niklas and J.-M. Spaeth, *Phys. Status Solidi B*, **101**, 221 (1980).

48. R. C. Barklie, J. R. Niklas, and J.-M. Spaeth, *J. Phys. C*, **13**, 1745, 1757 (1980).

49. R. C. DuVarney, J. R. Niklas, and J.-M. Spaeth, *Phys. Status Solidi B*, **97**, 135 (1980).

50. B. Biehl, M. Plato, and K. Möbius, *J. Chem. Phys.*, **63**, 3515 (1975).

51. N. S. Dalal and C. A. McDowell, *Chem. Phys. Lett.*, **6**, 617 (1970).

52. J. R. Niklas, R. U. Bauer, and J.-M. Spaeth, *Phys. Status Solidi B*, **119**, 171 (1983).

53. R. C. DuVarney, J. R. Niklas, and J.-M. Spaeth, *Phys. Status Solidi B*, **128**, 673 (1985).

54. W. B. Fowler, Ed., *Physics of Color Centers*, Academic, New York, 1968.

55. D. L. Dexter, C. C. Klick, and G. A. Russell, *Phys. Rev.*, **100**, 603 (1955).

56. R. H. Bartram and A. M. Stoneham, *Solid State Commun.*, **17**, 1593 (1975).

57. M. Lannoo and J. Bourgoin, "Point Defects in Semiconductors I, Theoretical Aspects," in M. Cardona, Ed., *Springer Series in Solid State Sciences*, Vol. 22, Berlin, 1981.

58. W. B. Lynch and D. W. Pratt, *Magn. Reson. Rev.*, **10**, 111 (1985).

59. P. Edel, C. Henniers, Y. Merle d'Aubigné, R. Romestain, and Y. Twarowsky, *Phys. Rev. Lett.*, **28**, 1268 (1972).

60. P. Dawson, C. M. McDonagh, B. Henderson, and L. S. Welch, *J. Phys. C*, **11**, L983 (1978).

61. F. J. Ahlers, F. Lohse, and J.-M. Spaeth, *Solid State Commun.*, **43**, 321 (1982).

62. B. C. Cavenett, *Adv. Phys.*, **30**, 475 (1981).

63. J. D. Dunstan and J. J. Davies, *J. Phys. C.*, **12**, 2927 (1979).

64. K. P. O'Donnell, M. K. Lee, and G. D. Watkins, *Solid State Commun.*, **44**, 1015 (1982).

65. N. Killoran, B. C. Cavenett, M. Godlewski, A. T. Kennedy, and N. D. Wilsey, *Phys. Status Solidi B*, **116**, 425 (1982).

66. B. K. Meyer, Th. Hangleiter, J.-M. Spaeth, G. Strauch, Th. Zell, A. Winnacker, and R. H. Bartram, *J. Phys. C*, **18**, 1503 (1985).

67. D. Block, A. Hervé, and R. T. Cox, *Phys. Rev. B*, **25**, 6049 (1982).

68. L. F. Mollenauer and S. Pan, *Phys. Rev. B*, **6**, 772 (1972).

69. F. J. Ahlers, F. Lohse, J.-M. Spaeth, and L. F. Mollenauer, *Phys. Rev. B*, **28**, 1249 (1983).

70. B. K. Meyer, D. M. Hofmann, J. R. Niklas, and J.-M. Spaeth, *Phys. Rev. B*, **36**, 1332 (1987).

71. B. K. Meyer, J.-M. Spaeth, and M. Scheffler, *Phys. Rev. Lett.*, **52**, 851 (1984).

72. M. Fockele, F. Lohse, J.-M. Spaeth and R. H. Bartram, *J. Phys. Condens. Matter*, **1**, 13 (1989).

73. M. Fockele, "Strukturaufklärung laseraktiver Bleizentren in Erdalkalifluoriden," doctoral dissertation, Universität-GH-Paderborn, 1987.

74. A. Görger, B. K. Meyer, and J.-M. Spaeth, "Transition Metal Elements in III–V Semiconductors—A Study With Optically Detected Magnetic Resonance," in G. Grossmann and L. Ledebo, Eds., *Semiinsulating III-V-Materials, Malmö 1988*, Hilger, Philadelphia, PA, 1988, p. 331.

75. A. Görger, "Vanadiumzentren in GaAs und GaP: Magnetooptische und optisch nachgewiesene magnetische Resonanzuntersuchungen," Diplomarbeit, Universität-GH-Paderborn, 1987.

76. J.-M. Spaeth, A. Görger, M. Heinemann, D. M. Hofmann, and B. K. Meyer, *Proceedings of the E-MRS Conference, Les Editions de Physique*, Paris, 1987, XVI, p. 421.

77. K. P. Dinse and C. J. Winscon, "Optically Detected ENDOR Spectroscopy," in R. H. Clarke, Ed., *Triplet State ODMR Spectroscopy*, Wiley, New York, 1984.

78. G. Strauch, Th, Vetter, and A. Winnacker, *Phys. Lett.*, **34**, 160 (1983).

79. Y. Sano, K. Morigaki, and I. Hirabayashi, *Phys. Status Solidi*, **117A** and **118B**, 923 (1983).

80. F. Boulitrop, *Phys. Rev. B*, **28**, 6192 (1983).

81. J. J. Davis, J. E. Nichols, and R. P. Barnard, *J. Phys. C*, **18**, L93 (1985).

82. D. M. Hofmann, B. K. Meyer, F. Lohse, and J.-M. Spaeth, *Phys. Rev. Lett.*, **53**, 1187 (1984).

83. D. M. Hofmann, "Strukturaufklärung des EL2-Defektes in Galliumarsenid mit optisch nachgewiesener Elektronen–Kern–Doppelresonanz," doctoral dissertation, Universität-GH-Paderborn, 1987.

84. F. J. Ahlers, F. Lohse, Th. Hangleiter, J.-M. Spaeth, and R. H. Bartram, *J. Phys. C*, **17**, 4877 (1984).

85. J.-M. Spaeth, A. Görger, D. M. Hofmann, and B. K. Meyer, *Mater. Res. Sci. Symp. Proc.*, **104**, 363 (1988).

86. A. Görger, B. K. Meyer, J.-M. Spaeth, and A. M. Hennel, *Semicond. Sci. Technol.*, **3**, 832 (1988).

87. M. Heinemann, B. K. Meyer, J.-M. Spaeth, and K. Löhnert, in E. R. Weber, Ed., "The Occupation of the Two Charge States of EL2 in LEC-Grown GaAs Wafers—a Mapping Investigation," *Defect Recognition and Image Processing in III–V Compounds II*, Elsevier, New York, 1987.

88. K. Krambrock, private communication.

89. A. Winnacker, private communication.

90. B. Clerjaud, private communication.

91. J. Hage, J. R. Niklas, and J.-M. Spaeth, *J. Electron. Mater.*, **14a**, 1051 (1985).

92. S. N. Jasperson and S. E. Snatterley, *Rev. Sci. Instrum.*, **40**, 761 (1969).

93. B. K. Meyer and J.-M. Spaeth, *Phys. Rev. B*, **32**, 1409 (1985).

94. G. Heder, J.-M. Spaeth, and A. H. Harker, *J. Phys. C*, **13**, 4965 (1980).

95. A. M. Stoneham, *Theory of Defects in Insulators and Semiconductors*, Clarendon, Oxford, 1975.

96. G. D. Watkins, "Point Defects in Solids," in J. H. Crawford, Jr. and L. M. Slifkin, Eds., *Semiconductors and Molecular Crystals*, Vol. 2, Plenum, New York, 1975, p. 349.

97. F. J. Adrian, A. N. Jette, and J.-M. Spaeth, *Phys. Rev. B*, **31**, 3923 (1985).

98. H. Overhof, C. M. Weinert, and M. Scheffler, *Proceedings of the 15th International Conference on Defects in Semiconductors, Budapest, 1988, Materials Science Forum*, Vol. 38–41, Part 1, Trans. Tech. Publications Aedermannsdort, Switzerland, 1989, p. 293.

99. E. B. Hale and R. L. Mieher, *Phys. Rev.*, **184**, 751 (1969).

100. J. L. Ivey and R. L. Mieher, *Phys. Rev. B*, **11**, 822 (1975).

101. P. von Engelen, *Phys. Rev. B*, **22**, 3144 (1980).

Chapter **7**

X-RAY DIFFUSE SCATTERING

Raymond P. Scaringe and Robert Comès

This chapter provides an introduction to those aspects of X-ray diffuse scattering that are useful in the determination of the atomic arrangement and chemical formulation of disordered crystalline compounds. Material has been selected with the solid state chemist in mind, but should also be useful to chemical crystallographers who are unfamiliar with the technique. Topics such as substitutional disorder and short-range order in alloys have been omitted in favor of a more detailed account of diffuse scattering from molecular crystals. Although quantitative discussions are limited to crystalline materials, the treatment of lattice distortions of the second kind should provide a useful introduction to ideas used in analyzing diffraction from polymers and liquid crystals.

The first section treats the geometric aspects of X-ray diffraction for various types of reciprocal lattices. In the second section the concepts used in describing disorder are discussed in an essentially nonmathematical fashion. The third section consists entirely of diffraction calculations. We have avoided the use of Fourier theory, and no mathematics beyond integral calculus is assumed. Many of the illustrative calculations have been devised especially for this chapter. The calculation of the diffuse scattering is usually reduced to a closed-form expression from which the effects of disorder can be clearly understood. In the fourth section, experimental results are given as an indication of the method's utility in resolving structural problems associated with disorder. Special attention has been given to the subject of conducting molecular compounds (one-dimensional metals) because recent interest in these materials has been remarkable.

The final section covers the experimental techniques used to survey and measure the scattered intensity throughout reciprocal space.

1   QUALITATIVE DESCRIPTION OF LATTICES AND X-RAY DIFFRACTION

In the following sections we will show that the X-ray diffraction pattern is greatly affected by the internal state of order of the substance being considered. It is because of its sensitivity to atomic arrangement that X-ray diffraction is of interest to chemists. However, the ideas that are useful in describing the atomic arrangement of an entire substance are somewhat different from those used to describe a single molecule within the substance. To understand the diffraction pattern, it is not only necessary to describe the molecules that comprise the substance, but also the mutual arrangement or ordering of the molecules. In this section we consider the concept of order from a particular viewpoint, that of an infinite space lattice. Although this is obviously appropriate for most crystals, it may seem too restrictive to be useful for amorphous substances or even for extremely perturbed crystalline materials. This is not, however, the case; the diffraction patterns of such substances often display definite similarities to those of their more ordered counterparts. This implies some similarity in the internal organization of the two types of materials. It should therefore not be surprising that some similarities in the description of this organization are also possible. We will find that the concept of a lattice is useful even for substances where the internal organization is essentially that of a liquid.

The treatment given below is aimed at providing a qualitative understanding of the effects of translational symmetry on the diffraction pattern. For several special cases we describe an idealized space lattice and give the form of the corresponding reciprocal lattice. The geometrical aspects of the diffraction pattern are deduced by using Ewald constructions. Proof of the stated results is deferred to later sections in which diffraction calculations are performed.

1.1   Lattices and Translational Symmetry

Consider the two-dimensional array of points in Figure 7.1, where we imagine the array to extend infinitely in two dimensions. This is an example of a plane lattice, and it is described completely by the two lattice vectors **a** and **b**. Any point on the lattice that is displaced an integral number of times by **a**, **b**, or any combination of such displacements will come to rest on some other lattice point. The vectors **a** and **b** are special, but they are not unique. For example, the vectors **a′** and **b′** would do just as well. For our purposes the choice is arbitrary, but it is often convenient or conventional to make a particular choice based on symmetry considerations.

The extension to three dimensions (Figure 7.2) requires a third lattice vector **c**, which is not in the plane defined by the first two. In general, any vector $\mathbf{t}_n$ of the form

$$\mathbf{t}_n = n_a \mathbf{a} + n_b \mathbf{b} + n_c \mathbf{c} \tag{1}$$

that emanates from a lattice point will terminate on a lattice point. If we arbitrarily choose one such point as the origin of the lattice, the position of any

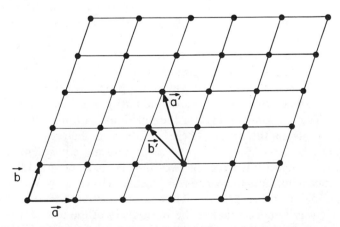

Figure 7.1   Two-dimensional space lattice with two alternate cell vectors.

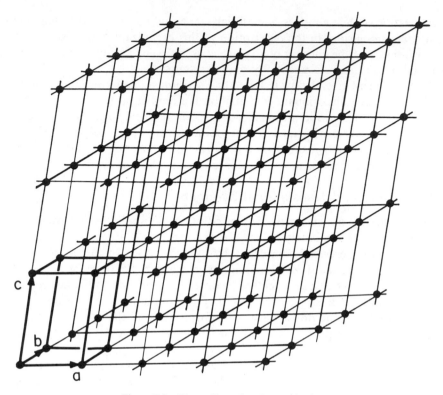

Figure 7.2   Three-dimensional space lattice.

lattice point n is given by the associated vector $\mathbf{t}_n$, where n refers to the set of three integers n_a, n_b, n_c. If the lattice points in Figure 7.2 are taken to represent an array of identical atoms, we have then described a simple crystal composed of a single element. In this case, the position of every atom in the crystal is given by a particular $\mathbf{t}_n$. Equation (1) expresses the translational symmetry of the crystal that is three-dimensionally periodic. When we speak of a system with perfect order, the periodicity expressed by (1) is implied.

Most crystalline solids are much more complex than the monoatomic one described above. As an example, consider the two-dimensional array of benzene molecules in Figure 7.3, which also shows an appropriate set of lattice points. In this case, the lattice points, or nodes of the lattice, take on a slightly different significance. They no longer represent the positions of atoms, but rather a set of equivalent points in the array. These points owe their equivalence to the translational symmetry of the benzene array and are equivalent in the same sense as those in Figure 7.1. Thus, the two-dimensional analogue of (1) applies to both the lattice points and the benzene molecules. However, while the molecules and the periodicity of their arrangement are physical realities, the lattice is only a convenient way to conceptualize the periodicity of the array. It is no longer possible to describe the position of all atoms in the array by using the vectors $\mathbf{t}_n$

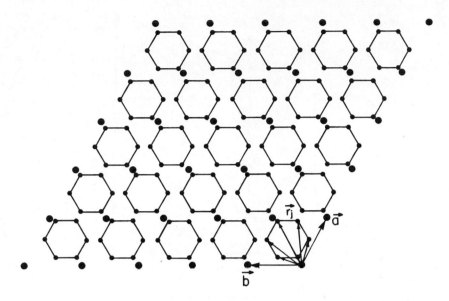

Figure 7.3   Periodic two-dimensional array of benzene molecules with lattice points.

alone. However, if we first define a set of vectors $\mathbf{r}_j$ (Figure 7.3) that connect each atom in a single molecule to a particular lattice point, then the $\mathbf{r}_j$ in addition to the $\mathbf{t}_n$ are sufficient to describe the position of every atom in the array.

The lattice vectors $\mathbf{a}$, $\mathbf{b}$, and $\mathbf{c}$ (Figure 7.2) describe a parallelpiped of volume,

$$V = (\mathbf{a} \times \mathbf{b}) \cdot \mathbf{c} \tag{2}$$

This, the unit cell volume, has a particular significance. It is possible to generate the entire crystal by replicating the contents of the unit cell at all positions lying at the termini of all possible $\mathbf{t}_n$. Usually, the unit cell volume is the smallest such volume for which this is possible, but sometimes a larger volume is chosen based on symmetry considerations. This theory is more fully discussed in Chapter 1. A lattice point lies at the origin of each unit cell, and therefore the vectors $\mathbf{t}_n$ connect the origins of all cells in the crystal.

As is further illustrated in Chapter 1, there is also associated with the direct space lattice vectors a set of reciprocal lattice vectors $\mathbf{a}^*$, $\mathbf{b}^*$, and $\mathbf{c}^*$, which are defined as follows:

$$
\begin{array}{ccc}
\mathbf{a}^* \cdot \mathbf{a} = 1 & \mathbf{b}^* \cdot \mathbf{a} = 0 & \mathbf{c}^* \cdot \mathbf{a} = 0 \\
\mathbf{a}^* \cdot \mathbf{b} = 0 & \mathbf{b}^* \cdot \mathbf{b} = 1 & \mathbf{c}^* \cdot \mathbf{b} = 0 \\
\mathbf{a}^* \cdot \mathbf{c} = 0 & \mathbf{b}^* \cdot \mathbf{c} = 0 & \mathbf{c}^* \cdot \mathbf{c} = 1
\end{array}
\tag{3}
$$

An often useful and equivalent definition is,

$$\mathbf{a}^* = \mathbf{b} \times \mathbf{c}/V \qquad \mathbf{b}^* = \mathbf{c} \times \mathbf{a}/V \qquad \mathbf{c}^* = \mathbf{a} \times \mathbf{b}/V \qquad V^* = 1/V \tag{4}$$

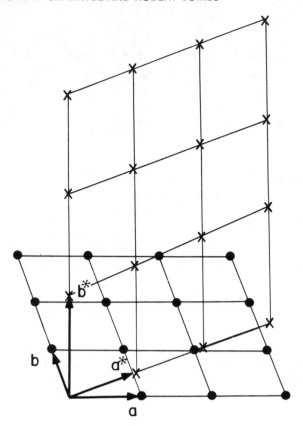

Figure 7.4   Geometric relationship between direct lattice and reciprocal lattice. All lattice points are in the plane of the drawing.

where V is defined in (2). The vectors $\mathbf{a}^*$, $\mathbf{b}^*$, and $\mathbf{c}^*$ define a lattice in reciprocal space that is completely analogous to the direct space lattices already discussed. The translational vector in reciprocal space is given by,

$$\mathbf{t}_H^* = h\mathbf{a}^* + k\mathbf{b}^* + l\mathbf{c}^* \tag{5}$$

where the Miller indices h, k, l are all integral. As with the direct space lattices, the vector connecting any two nodes is given by a particular $\mathbf{t}_H^*$. The geometric relationship between the direct and reciprocal lattices is depicted in Figure 7.4, where for clarity, we have only shown the a^*b^* net. The concept of the reciprocal lattice greatly simplifies X-ray diffraction calculations and the interpretation of diffraction patterns. There is a sharp distinction between the reciprocal lattice and reciprocal space. Reciprocal space is continuous. The reciprocal lattice defined above exists only at discrete points in this space and is therefore discontinuous.

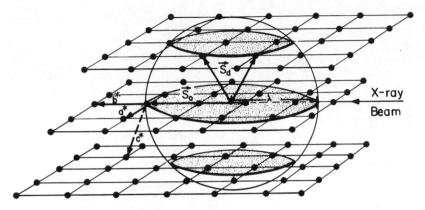

Figure 7.5 Interaction of a three-dimensional reciprocal lattice with the Ewald sphere.

1.2   Ewald Constructions

A common use of the reciprocal lattice is in the Ewald construction, the purpose of which is to predict diffraction patterns for a given set of experimental conditions. We first consider such a construction for the case (Figure 7.5) of a three-dimensional reciprocal lattice whose general translational vector is given by (5). One imagines that a crystal is located at the center of a sphere of radius $1/\lambda$. An X-ray beam of wavelength λ is directed along the unit vector S_0 and passes through the center of the sphere. The point at which the incident beam exits the sphere is considered to be the origin of the reciprocal lattice. Although for purposes of the diagram the origin of the reciprocal lattice appears to be separated from the crystal, the two are considered to be connected. The orientation of the crystal (and therefore of the direct lattice) determines the orientation of the reciprocal lattice; any rotation of the sample implies a similar rotation of the reciprocal lattice. Diffraction occurs whenever a point on the reciprocal lattice touches the surface of the Ewald sphere. The direction of the diffracted beam S_d is given by the vector from the center of the sphere to the point at which the sphere is contacted.

In Figure 7.6, **c** is normal to the plane of the diagram, and therefore the a^*b^* net is in this plane. The crystal is orientated such that the reciprocal lattice point $(2, 2, 0)$ is in contact with the sphere; the diffracted beam is in the plane of the drawing. If we suppose that c^* is normal to the plane of the drawing for this particular sample, it is evident that all reciprocal lattice points above and below the plane of the drawing (these are not drawn) will superimpose in projection onto one of the points already drawn. Furthermore, if the broken circle in the diagram represents the surface of the sphere at a height of $3c^*$ (i.e., $l = \pm 3$), it can be seen that the reciprocal lattice points $(1, -2, 3)$ and $(1, -2, -3)$ are also in contact with the sphere. For these points, the diffracted beams are above and below the plane of the diagram (broken lines).

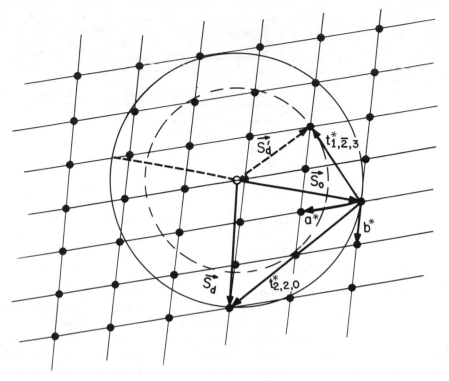

Figure 7.6 Ewald construction in projection on the a^*b^* plane. Dashed circle represents the surface of the sphere at a height of $3c^*$ above the plane of the drawing.

For any arbitrary orientation of the sample, there is usually no diffracted beam. Diffraction only takes place for each particular orientation of the crystal that puts a reciprocal lattice node in coincidence with the sphere. This is typical of Bragg diffraction and is a consequence of perfect three-dimensional translational symmetry in object space. Normally, some kind of moving crystal experiment is performed when sampling a large number of reciprocal lattice points.

The diffraction angle, 2θ, is defined as the angle between $\mathbf{S}_0$ and $\mathbf{S}_d$ and can be found in terms of the appropriate $\mathbf{t}_H^*$ directly. From Figure 7.6,

$$\mathbf{t}_H^* = (\mathbf{S}_d - \mathbf{S}_0)/\lambda$$

whenever $\mathbf{t}_H^*$ is in contact with the sphere. The modulus is given by,

$$|\mathbf{t}_H^*| = (1/\lambda)[(\mathbf{S}_d - \mathbf{S}_0)\cdot(\mathbf{S}_d - \mathbf{S}_0)]^{1/2}$$

Since $\mathbf{S}_0$ and $\mathbf{S}_d$ are both unit vectors,

$$|\mathbf{t}_H^*| = (1/\lambda)(2 - 2\mathbf{S}_d\cdot\mathbf{S}_0)^{1/2} = (1/\lambda)(2 - 2\cos 2\theta)^{1/2}$$

If the double-angle formula is used and rearranged,

$$|\mathbf{t}_H^*| = 2 \sin \theta / \lambda$$

Since for every vector $\mathbf{t}_H^*$ there is a corresponding set of lattice planes in direct space with an interplanar spacing of $1/|\mathbf{t}_H^*| = d$, the above takes on the more familiar form of Bragg's law,

$$2d \sin \theta = \lambda$$

It is possible to generalize the above diagram to take into account orientations of the sample for which no reciprocal lattice point is in contact with the sphere. We can define a general point in reciprocal space and bring this point into contact with the sphere. The general form of the vector between the origin and such a point can be expressed as

$$\mathbf{s} = s_1\mathbf{a}^* + s_2\mathbf{b}^* + s_3\mathbf{c}^* \tag{6}$$

where s_1, s_2, and s_3 are not necessarily integral. We further require that the vectors $\lambda\mathbf{s}$, $\mathbf{S}_0$, and $\mathbf{S}_d$ satisfy the triangular relationship

$$\mathbf{s} = (\mathbf{S}_d - \mathbf{S}_0)/\lambda \tag{7}$$

and, therefore, the direction of the diffracted beam is given by

$$\mathbf{S}_d = \mathbf{S}_0 + \lambda\mathbf{s}$$

Thus, once we have specified the orientation of the crystal with respect to the direct beam, we wish to know if the intensity diffracted in the direction $\mathbf{S}_d$ is nonzero. The answer will be, "Yes," only if $\mathbf{s}$ corresponds to a point on the reciprocal lattice. In the case of Bragg diffraction, $\mathbf{s}$ must satisfy the relationship,

$$\mathbf{s} = \mathbf{t}_H^* \tag{8}$$

as explained earlier. However, the reciprocal lattice defined through (3) is not appropriate for all periodic arrays. For other types of reciprocal lattices the solution to the construction will be different. Thus, once the reciprocal lattice is defined, all possible directions of diffracted beams for a particular orientation of the sample can be found through the use of the construction.

The scattering vector $\mathbf{s}$ occurs in all diffraction calculations, and its meaning should be well understood. It is always defined as a vector from the origin of reciprocal space to some point on the sphere. This places some restrictions on the orientation of the sample. Also, since we can calculate $|\mathbf{s}|$ in precisely the same way as was done for $|t_H^*|$,

$$|s| = 2 \sin \theta / \lambda \tag{9}$$

and the maximum value of $|s|$ is given by $|s|_{max} = 2/\lambda$. The volume of reciprocal space that can be investigated experimentally is dependent on the choice of wavelength.

One may well wonder how the definition of s given in (6) can be applied to anything other than Bragg diffraction. After all, the vectors $\mathbf{a^*}$, $\mathbf{b^*}$, and $\mathbf{c^*}$ represent proper translations of a reciprocal lattice that resulted from perfect periodicity in object space. As we will see in Sections 3 and 4, many substances not only give rise to Bragg diffraction, but also scatter in directions not given by (8). In such cases, it is convenient to express the scattering vectors s of the *extra* scattering in terms of the three-dimensional reciprocal lattice vectors. We will always refer to this *extra* scattering as diffuse scattering. However, even when there is no Bragg diffraction, we can still choose three reciprocal vectors of arbitrary length to indicate some meaningful directions relative to the problem at hand. In such cases, the three reciprocal vectors in (7) lose their original meaning.

1.3   The Reciprocal Lattices Corresponding to One- and Two-Dimensional Lattices in Direct Space

It is possible to define the reciprocal lattice of arrays that are not three-dimensional. From the foregoing, our interest in reciprocal lattices should be clear. If we can find the reciprocal lattice corresponding to some array in object space, the geometrical properties of the scattering pattern can be found through the Ewald construction. The reciprocal lattice only indicates from which positions in reciprocal space scattering can occur. It does not relate how much scattering to expect. The reciprocal lattice has a precise analogue in scattering calculations. This is called the interference function and we will encounter several of these in Section 3. In the following paragraphs we give two examples of reciprocal lattices of particular interest in the context of X-ray diffraction; these are the reciprocal lattices, which correspond to the one- and two-dimensional space lattices. Although substances that are strictly one- or two-dimensional cannot be studied experimentally, the diffraction patterns they would theoretically produce are useful in understanding the diffuse scattering from real materials. The reciprocals of objects that are not periodic arrays can also be defined. A discussion of this can be found elsewhere [1].

For two-dimensional lattices in direct space (Figure 7.7a), the reciprocal lattice corresponds to a series of parallel rods in reciprocal space (Figure 7.2b). Since we have defined only two proper translations in direct space (here, $\mathbf{a}$ and $\mathbf{b}$), it follows that only two translations can be defined in reciprocal space. These are $\mathbf{a^*}$ and $\mathbf{b^*}$ and represent the two shortest intercolumn vectors in reciprocal space. The extent of reciprocal lattice normal to the a^*b^* plane is considered to be infinite. The cross section of the rods corresponding to an infinite two-dimensional lattice is actually a point, and the rods should properly be referred to as lines in reciprocal space. However, the lattices describing real substances are never infinite, so the description in terms of rods allows us some latitude,

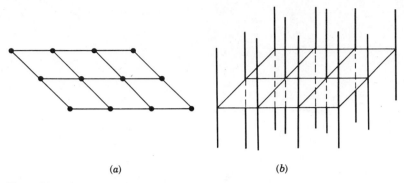

(a)   *(b)*

Figure 7.7   (*a*) Two-dimensional space lattice. (*b*) The corresponding reciprocal lattice.

should the need for a finite cross section arise. At this point we may also anticipate that the nodes of the three-dimensional reciprocal lattice, previously regarded as points, may actually occupy a small but finite volume.

The Ewald construction for the two-dimensional reciprocal lattice in Figure 7.7*b* is shown in projection in Figure 7.8*a*. Notice that while only one rod produces a diffracted beam (solid vector) in the plane of the diagram (other than the direct beam), all rods that are within the diameter of the sphere will pierce the sphere above and below the plane of the drawing; each of these will therefore give rise to two diffracted beams (dashed vectors). The figure also shows a cylindrically bent piece of photographic film. In Figure 7.8*b* the areas where diffracted beams have exposed the film are indicated. The infinite extent of the reciprocal lattice normal to the *a**b* plane somewhat relaxes the requirements for observing diffraction. Let us define **c*** in (6) as the unit vector normal to the *a**b* plane; all other quantities retain their original meaning. From the Ewald construction it is apparent that diffraction occurs when **s** is of the form,

$$\mathbf{s} = h\mathbf{a}^* + k\mathbf{b}^* + s_3\mathbf{c}^*$$

The only restriction on the value of s_3 arises from the fact that **s** must terminate on the sphere. This form of **s** is also the general vector of translational symmetry for the two-dimensional reciprocal lattice. Thus a vector from any part of any rod to any part of any other rod will be of this form.

For a linear lattice in direct space (Figure 7.9), the reciprocal lattice consists of a series of parallel sheets that are considered to extend infinitely in directions normal to the linear lattice in direct space. If the translational vector in direct space is defined as **c**, the spacing of the sheets in reciprocal space is given by $1/|c| = |c^*|$. The appropriate Ewald diagram is given in Figure 7.10. The form of **s** is now even less restricted. We require only that $s_3 = l\mathbf{c}^*$, where l is integral; s_1 and s_2 can take on any values consistent with **s** terminating on the surface of the sphere. The scattering vector is given by,

$$\mathbf{s} = s_1\mathbf{a}^* + s_2\mathbf{b}^* + l\mathbf{c}^*$$

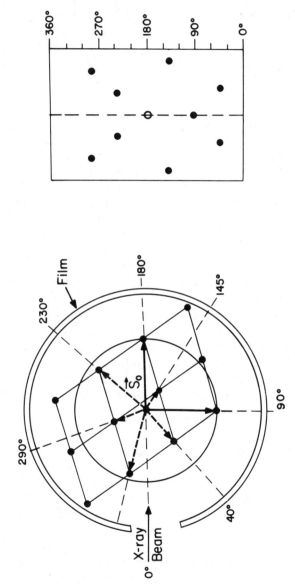

Figure 7.8 Ewald construction for the reciprocal lattice in Figure 7.7b, in projection. Dashed vectors represent diffracted beams above and below the plane of the drawing. Areas where photographic film has been exposed are indicated.

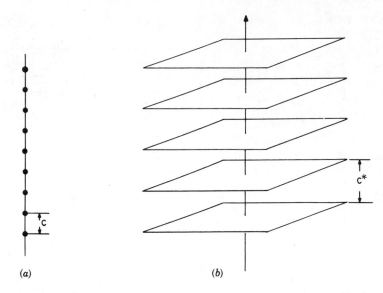

Figure 7.9 (a) One-dimensional space lattice. (b) Corresponding reciprocal lattice.

Each sheet gives rise to an infinite set of beams defining a cone. As shown in Figure 7.10, these beams trace a circle on the flat film (ellipses for the cylindrical film) if c^* is colinear with the incident beam. If c^* were shorter, we would see a series of concentric circles on the flat film. As is usually the case in diffraction, the pattern will depend on the orientation of the sample. If c^* is normal to the beam (Figure 7.11), we see curved lines (straight lines for the cylindrical film).

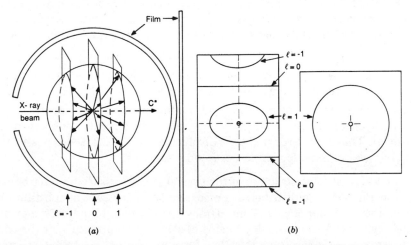

Figure 7.10 (a) Ewald construction for reciprocal lattice in Figure 7.9b. (b) Exposed areas on cylindrical film (left) and flat film (right).

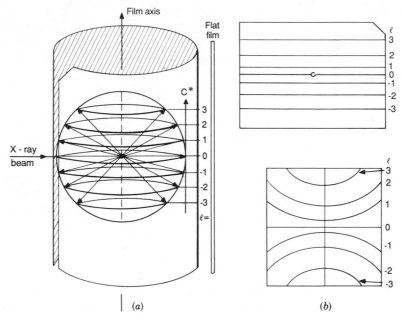

Figure 7.11 (a) Ewald construction for reciprocal lattice in Figure 7.9b. (b) Exposed areas on cylindrical film (top) and flat film (bottom).

2   TYPES OF LATTICE DISORDER

If the internal organization of a substance is such that the rigid periodicity implied by (1) is not present, the substance is said to be disordered. It is thus no longer possible to determine the precise position of every atom in the sample. The best that can be hoped for is a statistical description of the atomic arrangement, which preserves an average periodicity. For crystalline substances, the breakdown of translational symmetry can be described in one of two ways:

1. The contents of all unit cells in the crystal are not identical. This can occur, for example, when two isomorphous metal complexes or salts are cocrystallized and is referred to as substitutional disorder.

2. The atomic arrangement of all cells in the crystal is not identical. This usually occurs when two or more possible arrangements are energetically similar. This is often referred to as displacive disorder. It is also possible for both types of imperfection to occur simultaneously.

To fully describe the structure of the crystal it is necessary to describe how these imperfections are distributed. For example, in the case of substitutional disorder, the two components, A and B, may be distributed randomly; they may tend to aggregate forming small domains of separate components; or they may tend toward a new phase such that they form an alternating ABAB structure. Similarly, if there are two possible arrangements of the cell contents, q and r, then the presence of arrangement q in a particular cell may tend to make arrangement r in one of the nearest-neighbor cells less likely because of unfavorable intermolecular contacts. In either case the distribution of imperfections can usually be described statistically. The diffuse scattering from the sample provides important clues that guide the choice of the disorder model. The final test of any disorder model is how well it explains the experimentally observed diffuse scattering.

The type of imperfections alluded to above can be adequately described by relatively minor changes in our ideal view of perfect order. Such imperfections are said to preserve, on average, overall long-range order. This is not true for all conceivable types of disorder, and it is convenient to divide disorder into two classes based on this concept. We will find in Section 3.9 that the diffraction pattern is qualitatively different for the two classifications. In the first of these, it is possible to define an overall periodicity. The local atomic arrangement may display considerable deviations from rigorous translational symmetry, but whatever these deviations are they become no larger as we consider points at great distances from our arbitrary reference point. If the atomic arrangement adheres to such a description the substance is said to exhibit lattice disorder of the first kind. On the other hand, if no overall periodicity is preserved, the deviations from rigorous translational symmetry progressively worsen as we consider points at greater and greater distances from our reference point, and the substance is said to exhibit lattice disorder of the second kind. The distinction between the two types is most easily understood by comparing two linear lattices, one perturbed by imperfections of the first kind and one perturbed by those of the second kind. Since one must always consider an ensemble average in statistical descriptions, Figure 7.12 shows four linear atomic arrays for each kind of disorder. For both kinds of disorder, we designate the average nearest-neighbor distance as a. This implies that the average nth nearest-neighbor distance is given by na. Initially, the arrangements corresponding to the two types of disorder may seem similar. However, the linear lattice segments with disorder of the first type display a tendency to cluster around the average nth neighbor distance; this is true even for large n. The lattices with disorder of the second type show a smaller tendency to cluster, especially for large n. This is most easily seen by superimposing the individual arrays to obtain the overall distribution (Figure 7.12b and d). In essence this is the distinction between the two types of disorder.

Mathematically, this can be described as follows: Let $p_1(x)$ denote the distribution of nearest-neighbor distances; that is, the probability that two nearest-neighbors are separated by x is given by $p_1(x)$. If the average nearest-neighbor separation is a, then the maximum in the function $p_1(x)$ occurs at $x = a$. In a similar manner the distribution of nth neighbors is denoted $p_n(x)$. If $p_1(x)$ is symmetric about its maximum at $x = a$, it is frequently convenient to define all neighbor distributions in terms of their respective maxima $x = na$. These are denoted $p_n(na - x)$, where x ranges over all space. We can also define the probability $P(x)$ of finding an atom at some distance x from the *origin* atom,

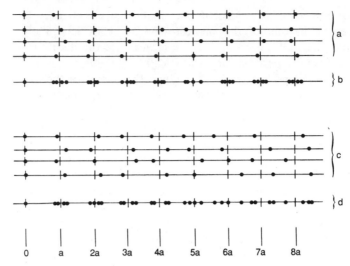

Figure 7.12 Segments of disordered linear arrays: (a) disorder of the first kind; (b) superposition of the four segments in (a); (c) disorder of the second kind; and (d) superposition of the four segments in (c).

no matter what order neighbor it may be. This is the overall distribution function.

For lattice imperfections of the first kind the essential requirement can be expressed as follows:

$$p_n(na - x) = p_1(a - x)$$

and this is valid for all values of n; that is, the probability that two nearest-neighbors are separated by some distance $a - x$ is no different from the probability that two nth neighbors are separated by $na - x$. The nth neighbor distribution function is therefore independent of how remote a neighbor is considered. The function $P(x)$ is a periodic function of n, which repeats the function $p_1(a - x)$ about each point $x = na$. This is what is meant by saying that overall long-range order is preserved. We can speak of an average lattice with a translational vector **a**, remembering that the actual translational symmetry is only approximate. The function $P(x)$ is plotted in Figure 7.13a. It can be considered to be the normalized superposition of many arrays like those in Figure 7.12a. To a first approximation, substances with disorder of the first kind exhibit diffraction patterns that are the same as a perfectly ordered crystalline substance. However, the lack of perfect translational symmetry implies that there will also be diffuse scattering.

The concepts of the average lattice and the average structure are central in describing the atomic arrangement of a substance. The overall distribution function in Figure 7.13a is a representation of the average structure of a

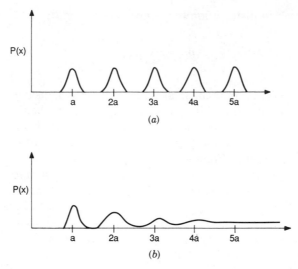

Figure 7.13   Overall atomic distribution functions for: (*a*) disorder of the first kind and (*b*) disorder of the second kind.

hypothetical, one-dimensional material. An equally valid approach would be to represent the functions in Figure 7.13*a* by a series of atoms in *partially occupied* positions along the line. The average structure given by $P(x)$ may or may not represent a reasonably accurate description of the actual structure, which in this case is represented by the lattice segments in Figure 7.12*a*. For real crystalline solids with disorder of the first kind, there is an average three-dimensional lattice and also the related average structure. The distinction between the average structure and the actual structure is important because the Bragg intensities only contain information about the average structure. This information is necessary, but not sufficient to describe the actual structure. The additional necessary information is present in the diffuse scattering. When differences between the average structure and the actual structure are great, investigation of the diffuse scattering is essential. Failure to take into account the diffuse scattering properly can result in an inadequate, or even a completely incorrect determination of the atomic arrangement. We will discuss how this can occur in the following sections.

In the case of disorder of the second kind, overall periodicity is not preserved, and the function $p_n(x)$ is not a constant for all n. It is then necessary to formulate some other rule for the mutual arrangement of atoms. The most common assumption is that nearest-neighbor distribution is the same for all atoms and that there are no further constraints governing the overall ordering. If $p_1(x)$ displays a maximum at $x = a$, then $p_2(x)$ will generally show a maximum at $x = 2a$, as was true for disorder of the first kind. However, in this case the maximum at $x = 2a$ will be smaller and the distribution about this point will be broader. With increasing n, the maxima of $p_n(x)$ become increasingly weaker and broader. The overall distribution function $P(x)$ is given by

$$P(x) = \sum_n p_n(x)$$

which generally reaches some constant value for large x. Thus if we now describe the atomic distribution in terms of an average lattice, then for some large number of *lattice* spacings it will be just as likely to find an atom halfway between lattice sites as it would be to find an atom on the lattice position. The function $P(x)$ is plotted in Figure 7.13b.

It is generally not trivial to obtain a closed form expression for $p_n(na - x)$ in the case of disorder of the second kind. To obtain these functions the nth self-convolution of the function $p_1(a - x)$ must be found; a brief discussion is given in [2] and a more extensive one in [3]. If the nearest-neighbor distribution is Gaussian, a closed form expression $p_n(na - x)$ can be found. If we define the nearest-neighbor distribution as follows:

$$p_1(a - x) = K \exp[-(a - x)^2/w]$$

then the nth neighbor distribution is found to be,

$$p_n(na - x) = (n)^{-1/2} K \exp[-(na - x)^2/nw] \tag{10}$$

where K is the normalization constant for the nearest-neighbor distribution. For this distribution, it is easy to see how the general features of $P(x)$ (Figure 7.13b) arise. The nearest-neighbor distribution has a maximum value of K, which occurs at $x = a$, while the maximum value of the nth neighbor distribution is only $(n)^{-1/2}K$, which occurs for $x = na$. The width of the nearest-neighbor distribution is w, while that for the nth neighbor distribution is nw. For very large values of n, the maxima of the nth, and $(n + 1)$th distribution will not be resolved, and $P(x)$ will tend to become constant.

For disorder of the second kind, we have only short-range order. It is this type of order that is found in liquids and amorphous solids. The concept of an average lattice is only useful over a limited range of neighbors. However, even this limited type of order produces distinctive features in the X-ray scattering.

2.1   Anisotropy in the State of Order

In the early stages of crystal structure determination, the Bragg intensities can be used in a variety of ways to obtain a trial atomic arrangement. This can be considered distinct from the process of determining how well the model agrees with experiment and is often the most difficult stage of a structure deter-mination. For materials exhibiting diffuse scattering, one must also consider trial models, but there are no generally applicable methods for obtaining such models. In this regard, the general distribution of diffuse scattering in reciprocal space is most helpful in formulating the general features of the model. If the distribution of diffuse scattering is anisotropic, this implies that there is a corresponding anisotropy in the local atomic arrangement. This immediately limits the number of models that need be considered and is therefore of great practical value. Two types of anisotropy occur so frequently in diffuse scattering patterns that it is worthwhile to examine the disorder that gives rise to them. In this chapter we refer to these disorders as linear and planar. In the literature one will also encounter the terms one- and two-dimensional disorder. This leads to some difficulties since the latter terms are not as descriptive. For simplicity, the subject will be introduced from the standpoint of displacive imperfections. An equivalent discussion can also be given for substitutional imperfections.

2.2   Planar Disorder

Consider a crystal to be constructed by stacking a series of identical layers, one upon the next, keeping all layers parallel. The atomic arrangement within each layer is perfectly ordered, with lattice constants of a and b. If the stacking of the layers is in any way irregular, we have the case of planar disorder.

For example, suppose that there are irregularities in the interplanar spacings, or that the layers are slipped, in an irregular fashion, with respect to each other. The diffuse scattering will be confined to rods in reciprocal space. The rods run normal to the layers, and their projection on the ab plane is the same as that for a perfect lattice of the same dimensions. Experimentally, the diffuse scattering shows up as streaking along the festoons of a Weissenberg photograph or along the lattice rows of precession photographs. It is therefore similar to the scattering produced by an ideal two-dimensional lattice (Figures 7.7 and 7.8). If the layers have no thickness (sheets of atoms), the intensity on any segment of any rod is the same as that on any other rod. If on the contrary, one considers layers of three-dimensional molecules (a two-dimensional array of three-dimensional objects) the intensity will vary along a given rod in a way that reflects the *intramolecular* spacings.

This is true only when there is absolutely no correlation between the irregularity introduced by the displacement of a given layer and the displacements of the neighboring layers. If the displacements of neighboring layers are somewhat interdependent and result in weak local ordering between layers, the intensity along a given rod will vary in a way that depends on the details of the interlayer correlations.

The distribution of diffuse intensity, in the general case, can become very complex, but can ordinarily be calculated for any case of interest (Section 3).

2.3   Linear Disorder

Consider a crystal to be made of a series of identical parallel columns. A column may represent a linear array of atoms or a linear array of molecules of arbitrary size and shape. Further assume that each column is perfectly ordered in one dimension, with a lattice constant c. Initially, we consider the case of chains with one atom per repeat distance. Eventually, we will consider the changes that occur when linear arrays of molecules are introduced. We now suppose that the chains are packed together such that in projection on the ab plane, they form a perfect two-dimensional lattice. If relative to the origin of the crystal, all chains contain atoms at $z = 0, c, 2c, \ldots$, the crystal is perfectly ordered. Consider the case where there are many chains in the crystal that contain atoms at $z = 0, c, 2c, \ldots$, but there are also the same number of chains with atoms at positions $z = 0 + d, c + d, 2c + d, \ldots$; note that the two are still identical except for a relative shift by an amount d. If the two *kinds* of chains occur randomly throughout the crystal, we have a particular case of linear disorder. If instead of only one kind of relative shift, there are many such shifts $d_1, d_2, \ldots$, such that any two given chains can be shifted by any amount, this is a similar but distinct case of linear disorder. In both cases the diffuse scattering is confined to sheets in reciprocal space that are spaced by an amount $1/|c|$; the scattering will be identical to that of the ideal linear lattice (Figures 7.10 and 7.11), with one exception. There will be Bragg diffraction in the $(h, k, 0)$ plane as for a perfect crystal. This follows from the fact that the projection of the structure on the ab plane is identical to that of a perfect crystal. In the case where the chains are composed of linear molecules, the intensity will vary strongly from sheet to sheet. For all of these, the intensity within any given sheet will still be constant (except for the $l = 0$ sheet).

If we now relax the condition of linear molecules and allow linear arrays of molecules of any arbitrary shape, there is only one essential change in the diffuse scattering. The scattering within a given sheet can vary. This modulation is caused by interference of the waves scattered by the atoms of a single molecule.

Many other irregular arrangements of parallel linear arrays are possible. Although each may present some unique features, they are all examples of linear disorder. As such, the associated diffuse scattering is always confined to sheets in reciprocal space, which are normal to the direction of translational symmetry in direct space.

3 DIFFRACTION CALCULATIONS

Diffuse scattering is a subject that cannot be discussed on a useful level without considering theoretical expressions for the scattered intensity. For many practical problems, the theory is more tedious than complicated and this can be largely overcome by working through simple examples to become familiar with the formalism. In this section we will consider several such examples. These have been chosen either because of their general importance or because they are simple enough to be of some illustrative value. In each case, we derive an algebraic formula that describes the intensity of the diffuse scattering. The derivation of such expressions is necessary if one is to arrive at a quantitative structural model from the experimental diffuse scattering. Since a primary goal of structural chemistry is the description of atomic arrangement, the calculations are presented in some detail. Preliminary topics such as the X-ray scattering from a single electron, the atomic scattering factors, and the assumptions underlying the use of the kinematic theory of X-ray diffraction, presented in Chapter 1 will not be repeated here.

For our purposes it is necessary to use a somewhat different form of the structure factor than that used in ordinary structure analysis. The structure factor of the cell that lies at the terminus of $\mathbf{t}_n$ is given by

$$F_n(\mathbf{s}) = \sum_j f_{nj} \exp(-2\pi i \mathbf{s} \cdot \mathbf{r}_{nj}) \tag{11}$$

The sum extends over all the atoms in the unit cell. The vectors $\mathbf{r}_{nj}$ connect each atom in the cell to the origin of the cell and are usually expressed in the form:

$$\mathbf{r}_{nj} = x_{nj}\mathbf{a} + y_{nj}\mathbf{b} + z_{nj}\mathbf{c} \tag{12}$$

where x_{nj}, y_{nj}, and z_{nj} are the fractional coordinates; s has been defined in (6). The index n allows for each cell in the crystal to vary somewhat in content or arrangement. If for all n, $f_{nj} \neq f_j$, there is substitutional disorder. If $\mathbf{r}_{nj} \neq \mathbf{r}_j$, there is displacive disorder. If all cells are identical, (11) can be put into the form usually used for structure determinations by dropping the subscript n, setting $\mathbf{s} = \mathbf{t}_H^*$, and taking the dot product explicitly.

The amplitude of the scattered wave is found by summing the contributions from all cells in the crystal:

$$E = \sum_n F_n \exp(-2\pi i \mathbf{s} \cdot \mathbf{t}_n)$$

where the sum is over the N cells of the crystal and is shorthand for the triple sum over n_a, n_b, and n_c. With this, the intensity is given by,

$$I_N(\mathbf{s}) = EE^* = \sum_n \sum_{n'} F_n F_{n'}^* \exp[-2\pi i \mathbf{s} \cdot (t_n - t_{n'})] \tag{13}$$

We can now group together all terms containing the vector $t_m = t_{n'} - t_n$, or more explicitly, $t_m = m_a\mathbf{a} + m_b\mathbf{b} + m_c\mathbf{c}$, where $m_a = (n_a' - n_a)$, and so on. With this (13) can be rewritten as follows:

$$I_N = \sum_m \left(\sum_n F_n F_{n+m}^* \right) \exp(2\pi i \mathbf{s} \cdot \mathbf{t}_m) \tag{14}$$

Equation (14) is quite complicated since the sums take into account the size and shape of the crystal as well as any disorder if all $F_n \neq F$. The sum in parentheses can be made independent of the index n by using the average value of $F_n F_{n+m}^*$.

We may then symbolically write,

$$\sum_n F_n F^*_{n+m} = K_m \langle F_n F^*_{n+m} \rangle$$

where K_m is numerically equal to the number of terms (unit cells) in the sum for a given t_m. This approximation only applies if the defects inherent in the F_n are distributed homogeneously throughout the crystal and are not, for example, all located at the surface. Also, the approximation will be best when K_m is large so that a large number of cells are considered. However, errors arising from using this expression for terms with small K_m will be small since these will make a

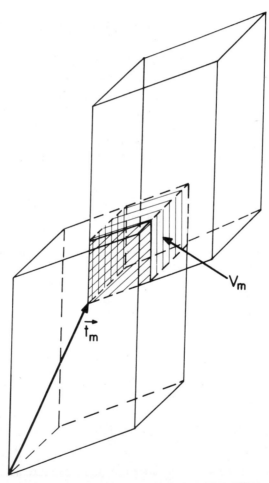

Figure 7.14   Volume common to two crystals with a relative displacement of t_m.

proportionally small contribution to (14). Then, K_m can be found by considering two identical crystals displaced by $\mathbf{t}_m$, as shown in Figure 7.14. The volume common to the two, the shaded volume, is denoted V_m. Since the volume of a single unit cell is V, the number of unit cells in the shaded volume K_m is given by:

$$K_m = V_m/V$$

It is convenient to make some explicit assumptions regarding the shape of the crystal. We will assume that the crystal is a parallelpiped of dimensions $N_a \times N_b \times N_c$ along the primary crystallographic directions. This is not the most general formulation possible, but it is usually adequate for reasons to be discussed later in this section. With this, the total number of cells in the crystal is

$$N = N_a N_b N_c$$

The determination of an expression for K_m is straightforward. Let us calculate the number of cells in the shaded volume when the two crystals have only been displaced along the **c** direction. Then, $\mathbf{t}_m$ is of the form

$$\mathbf{t}_m = m_c \mathbf{c}$$

where $m_c = 0, 1, 2, \ldots, N_c - 1$. For $m_c = +1$, $K_{0,0,1} = N_a N_b (N_c - 1)$, for $m_c = +2$, $K_{0,0,2} = N_a N_b (N_c - 2)$, and so, $K_{0,0,mc} = N_a N_b (N_c - |m_c|)$. A similar expression holds for each of the other two directions so that,

$$K_m = (N_a - |m_a|)(N_b - |m_b|)(N_c - |m_c|) = (N - |m|)$$

where the last term in parentheses is an implicit triple product suppressing the a, b, and c indices. If this expression is used, (14) becomes

$$I_N = \sum_m (N - |m|)\langle F_n F_{n+m}^* \rangle \exp(2\pi i \mathbf{s} \cdot \mathbf{t}_m) \tag{15}$$

Equation (15) is a useful starting place for many diffraction calculations. The general formula is still somewhat complicated but simplifies considerably for many special cases of practical significance. In the balance of this section we use (15) to predict the scattering pattern for a variety of such cases.

3.1   Diffraction From a Crystal With Perfect Order

This calculation will serve to introduce the interference function and the size effect that are important for crystals with or without disorder. For perfect order we have $F_n = F$ for all cells so that, $\langle F_n F_{n+m}^* \rangle = F F^*$. When this is used in (11), we have

$$I_N(s) = FF^* \sum_m (N - |m|) \exp(2\pi i s \cdot t_m) \tag{16}$$

$$= FF^* \left[\sum_{ma} (N_a - |m_a|) \exp(2\pi i s \cdot a m_a) \right]$$

$$\times \left[\sum_{mb} (N_b - |m_b|) \exp(2\pi i s \cdot b m_b) \right]$$

$$\times \left[\sum_{mc} (N_c - |m_c|) \exp(2\pi i s \cdot c m_c) \right]$$

where we have written out the triple sum, and expressed t_m in terms of its components. The evaluation of the sums in brackets is straightforward but lengthy; it can be found in the appendixes. Let us denote any of the sums in (16) as $S(N_i, s \cdot a_i)$, where $a_1 = a$, $a_2 = b$, $a_3 = c$, $N_1 = N_a$, and so on; then we have from the appendixes,

$$S(N_i, s \cdot a_i) = \sum_{m_i = -(N_i - 1)}^{(N_i - 1)} (N_i - |m_i|) \exp(2\pi i s \cdot a_i m_i)$$

$$= \sin^2 (N_i \pi s \cdot a_i)/\sin^2 (\pi s \cdot a_i) \tag{17}$$

With (17) we can rewrite (16) as follows:

$$I_N(s) = FF^* \frac{\sin^2(N_a \pi s \cdot a)}{\sin^2(\pi s \cdot a)} \frac{\sin^2(N_b \pi s \cdot b)}{\sin^2(\pi s \cdot b)} \frac{\sin^2(N_c \pi s \cdot c)}{\sin^2(\pi s \cdot c)} \tag{18}$$

The notation $S(N_i, s \cdot a_i)$ will often be used as shorthand for the functions on the right side of (17). The product of the three $S(N_i, s \cdot a_i)$ is the interference function of a perfect three-dimensional finite lattice. For large N_i, each function $S(N_i, s \cdot a_i)$ is small except when $s \cdot a_i$ is nearly integral. When the dot product is exactly integral, the function reaches its maximum value of N^2. Note that the final expression in (17) becomes indeterminate for integral values of the dot product; one can confirm that the maximum value is N^2 by applying L'Hospital's rule twice in succession. For sufficiently large N_i, which is normally the case, we can neglect all scattering except when the three $S(N_i, s \cdot a_i)$ are simultaneously at their respective maxima. We then seek a general form for s for which all dot products are integral. From the definitions of a^*, b^*, and c^* [(3) and (5)] it is clear that all dot products will be integral if s is of the form $s = t_H^*$. This conclusion was already discussed in the context of Ewald constructions. In this case, however, the definition of the reciprocal lattice (3) could be deduced directly from the form of (18). The idea of a reciprocal lattice comes about naturally from the emergence of the interference function in diffraction calculations. Since in this case X-ray diffraction is essentially limited to the points defined by $\mathbf{t}_H^*$, the intensity for the Bragg reflections is often written as follows:

$$I_{hkl} = F_{hkl} F_{hkl}^* \tag{19}$$

where the whole expression has been divided by N^2 to yield the intensity per unit cell. In final form then, the usual expression does not contain any information on the size of the crystal, and it is not capable of predicting the intensity away from the points (h, k, l).

If instead of the triple sum in (16) there is only a single sum, the calculation would be appropriate for a linear lattice. Only a single $S(N_i, \mathbf{s} \cdot \mathbf{a}_i)$ would appear in (18), and this is therefore the interference function of a perfect linear finite lattice. This requires only a single dot product to be integral and therefore defines a set of equally spaced sheets in reciprocal space. If there were only two sums in (16), we would have the two-dimensional case. The intersection of two sets of sheets in reciprocal space corresponds to a series of rods, which is the geometric form of the interference function for a two-dimensional lattice.

The significance of the quantities N_i in (18) is not transparent. Normally, they do not correspond to the dimensions of the experimental *single crystal*. The macroscopic crystal actually comprises many smaller regions, each of which is considered to be perfect. These smaller regions are usually described as mosaic blocks; a crystal made up of mosaic blocks is depicted in Figure 7.15. The size of an individual mosaic block can vary depending on the substance being considered, the crystal growth conditions, and the thermal history of the crystal. For many substances, the blocks are about $10^3 - 10^4$ Å per side.

The mosaic structure of crystals can lead to several experimental effects. The blocks, or crystallites, are never in perfect mutual alignment. The apparent orientation of the crystal actually represents the average orientation of the crystallites within. As the crystal is rotated through a particular setting angle during a moving crystal photograph or on a diffractometer, each crystallite

Figure 7.15 Mosaic structure of a crystal.

comes into the diffracting position at a somewhat different value of the angle, which leads to a broadening of the diffraction profile. This broadening is easily observed by examining either the spot length along the cylinder direction of a Weissenberg photograph or the widths of the profiles obtained by the Ω-scan technique on a single crystal diffractometer. Good quality single crystals usually exhibit full widths at half-height in the range $0.10°-0.20°$ in Ω. If the width becomes 5 to 10 times this value, the diffraction maxima become very weak, and intensity measurements are more time-consuming. If the widths are very narrow, the crystallites are not very misaligned. This can invalidate one of the assumptions of kinematic theory. Such crystals are said to exhibit both extinction and an apparent systematic error in the stronger intensities at low diffraction angles. As a practical matter, these effects are usually small and can be corrected for. In extreme cases, the crystal can usually be made more mosaic by dipping it into liquid nitrogen.

The mosaic structure of real crystals also plays a role in diffraction calculations. The translational symmetry of the lattice is considered to extend only over the volume of a single block. No interference is considered to occur between waves scattered by different crystallites. The quantities N_i, in (15)–(18), then refer to the extent of a single crystallite along the principal crystallographic directions.

3.2   The Size Effect

Aside from the misalignment of crystallites, the finite size of a single crystallite can also give rise to a broadening of the diffraction profile. This can be understood by examining Figure 7.16, where $S(N_i, s_i)$ is plotted for two small values of N_i. Relative to the maximum height, the function becomes broader with decreasing N_i. This is referred to as size effect broadening. There are also numerous secondary maxima, which are displaced approximately by integral multiples of $1/N_i$, and which become stronger, relative to the main maxima, with decreasing N_i. These are not normally observed in X-ray diffraction for several reasons, the large size of the usual crystallite being an important one. Even if this were not the case, there is no reason to believe that all blocks are the same size; hence, the secondary maxima whose positions depend on the size would tend to form an unresolved envelope if an ordinary macroscopic crystal were used. The secondary maxima have been observed [4] in the electron diffraction patterns of very small crystals. Since there is normally no way to determine the exact size and shape of the crystallites, little generality is lost by assuming a specific shape as was done to obtain (15).

The half-width of the main maximum can be found by iterative solution of the following:

$$\sin^2[N_a\pi(h + \Delta s)]/\sin^2[\pi(h + \Delta s)] = N_a^2/2$$

where the profile in the $(h, 0, 0)$ direction has been chosen as a concrete example, and Δs is in units of fractions of a repeat distance. Approximate solution of the

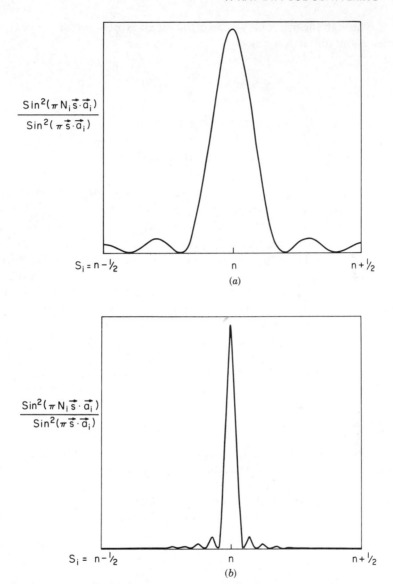

Figure 7.16 Peak function for a perfect crystal with N_i cells along the $\mathbf{a}_i$ direction ($n = h, k,$ or l): (a) $N = 5$ and (b) $N = 20$.

above yields $\Delta s_{1/2} = \pm 0.444/N_a$, so the full width is given roughly by,

$$\Delta s \cong 1/N_a \tag{20}$$

Since the expression does not contain h, the size effect broadens all diffraction maxima equally. This is a distinguishing feature that is not true of all broadening sources. Through the use of (9) we can also find an expression for the peak width in terms of the diffraction angle:

$$\Delta s/a = \frac{\cos \theta}{\lambda} \Delta(2\theta)$$

With (20) we have the following rough expression:

$$\Delta(2\theta) = \lambda/N_a a \cos \theta$$

For $\cos \theta \simeq 1$, $\lambda = 1.5$ Å, and $a = 15$ Å, a crystallite dimension of 150 Å ($N = 10$) gives an angular width of 0.6° while for 1500 Å ($N = 100$), the value is roughly 0.06°. Thus, a relatively small number of cells gives rise to quite narrow profiles.

Crystals with moderate mosaic spreads are often referred to as *ideally imperfect*. This should not be confused with the concept of lattice imperfections that is used throughout this chapter. Mosaic structure does not constitute a lattice imperfection since it in no way implies an irregularity of the internal atomic arrangement of a single crystallite. A somewhat more obvious but similar misconception sometimes occurs with regard to powdered samples. A powder is simply a collection of randomly oriented particles. If the particles are crystalline, each may contain from one to many crystallites, depending on the particle size and sample preparation. Again, this implies nothing about the internal structure of the crystallites and does not constitute lattice disorder.

3.3   Diffraction From Crystals With Disorder

In crystal diffraction both sharp Bragg peaks and diffuse scattering are often evident. For this reason it is desirable to develop an expression from (15) that separates the two types of scattering.

The average structure factor is defined as follows:

$$\langle F \rangle = \frac{1}{N} \sum_n F_n \tag{21}$$

where the sum extends over all cells of the crystal and reflects the average effect of all the disorder that may be present. For any given cell, we can define the structure factor as follows:

$$F_n = \langle F \rangle + \varphi_n \tag{22}$$

where with (21) and (22) we have, $(1/N) \sum_n \varphi_n = \langle \varphi_n \rangle = 0$. With this, the quantity $\langle F_n F_{n+m}^* \rangle$ in (15) becomes,

$$\langle F_n F_{n+m}^* \rangle = \langle F \rangle \langle F^* \rangle + \langle F \rangle \langle \varphi_{n+m}^* \rangle + \langle F^* \rangle \langle \varphi_n \rangle + \langle \varphi_n \varphi_{n+m}^* \rangle$$

Since $\langle \varphi_n \rangle = \langle \varphi_{n+m} \rangle = 0$,

$$\langle F_n F_{n+m}^* \rangle = \langle F \rangle \langle F^* \rangle + \langle \varphi_n \varphi_{n+m}^* \rangle \tag{23}$$

Equation (15) can now be written in the form

$$I(\mathbf{s}) = I_1(\mathbf{s}) + I_2(\mathbf{s}) \tag{24}$$

where

$$I_1(\mathbf{s}) = \langle F \rangle \langle F^* \rangle \sum_m (N - |m|) \exp(2\pi i \mathbf{s} \cdot \mathbf{t}_m) \tag{25}$$

$$I_2(\mathbf{s}) = \sum_m (N - |m|) \langle \varphi_n \varphi_{n+m}^* \rangle \exp(2\pi i \mathbf{s} \cdot \mathbf{t}_m) \tag{26}$$

Equation (25) is identical to that for a perfect crystal (16), except that F has been replaced by $\langle F \rangle$. This makes intuitive sense if one is familiar with the ideas used in structure determinations. For example, if an atom appears to occupy two positions in the unit cell with equal probability, one places a *half-atom* in each site to calculate the structure factor for the cell. A half-atom is obtained by dividing the scattering factor by two. It is easy to show that this is completely equivalent to the assumption that there are only two types of cells in the crystal, each containing a full atom in one of the two positions and averaging the structure factors of the two to obtain $\langle F \rangle$ for use in (25). Since the sum in (25) is identical to that in (16), disorder does not broaden the peak profiles of the Bragg diffraction. Equation (26) contains all of the information on diffuse scattering. From (23), the coefficients containing the structural information are given by

$$\langle \varphi_n \varphi_{n+m}^* \rangle = \langle F_n F_{n+m}^* \rangle - \langle F \rangle \langle F^* \rangle \tag{27}$$

Once one has defined a particular disorder model, these coefficients can be calculated and summed in (26) to obtain the diffuse scattering. For displacive disorder, F_n can often be expressed as follows:

$$F_n = F \exp(2\pi i \mathbf{s} \cdot \Delta \mathbf{r}_n) \tag{28}$$

where $\Delta \mathbf{r}_n$ is small compared to internuclear distances, and represents the displacement of the unit cell, away from the average position. For one atom per unit cell, (28) always holds and is often appropriate for one molecule per cell as well. For more than one molecule per cell one must usually develop an expression for each molecule independently; this will be illustrated for the case of thermal disorder later in this section. From (28) and (22) we have

$$\varphi_n = F \exp(2\pi i \mathbf{s} \cdot \Delta \mathbf{r}_n) - \langle F \rangle \tag{29}$$

This can be thought of as the structure factor for diffuse scattering and is often used in the case of linear disorder (Section 3.8). The coefficients given by (19) not only depend on the displacements $\Delta \mathbf{r}_n$, but also on the correlations between the displacements. By this we mean that the occurrence of a particular displacement $\Delta \mathbf{r}_n$ in cell n may affect the possible values of the displacement $\Delta \mathbf{r}_{n+1}$ in a neighboring cell $n + 1$. This is equivalent to saying that the position of a given molecule may affect the positions of the molecules around it, which is, of course, completely reasonable. For example, suppose all displacements in the crystal are only along the c axis and therefore are of the form $\mathbf{r}_n = \Delta z_n \mathbf{c}$. For the cell (n_a, n_b, n_c), let us call the magnitude of this displacement d. If the molecules are tightly packed, this may lead to an identical displacement d in the cell $(n_a, n_b, n_c + 1)$, and it implies that the same will happen in the cell $n_c + 2$, and so on. Hence, the entire line of cells would be shifted in unison by an amount d in the $\mathbf{c}$ direction. This is a case of very strong correlations. If on the other hand the shift of the first cell only necessitates some smaller shift d' in the cell $n_c + 1$, and therefore a yet smaller shift d'' in the cell $n_c + 2$, the effect of the first shift diminishes along the chain and the correlations of the $\Delta \mathbf{r}_n$ are weaker. In the extreme limit where the initial displacement d causes no disruption of the next cell, the displacements are said to be uncorrelated, which implies that the molecules in the crystal are capable of a small amount of independent motion about their mean position defined by the average cell.

It should now be clear from (25) why the Bragg intensities only yield the average structure. They depend only on the average value of the structure factor and are completely insensitive to any correlations among the displacements of various cells. A particularly strong cause of correlations is chemical bonding. It is in fact true that crystals can be disordered in such a way as to obscure the effects of chemical bonding in the average structure. In such cases, it is not even possible to determine the structure of the molecules in the crystal from the information provided by the Bragg intensities.

In the extreme limit of no correlations the sum in (26) contains only one term, the $m = (0, 0, 0)$ term, and the diffuse scattering is given by

$$I_2(\mathbf{s}) = \langle \varphi_n \varphi_n^* \rangle = \langle F_n F_n^* \rangle - \langle F \rangle \langle F^* \rangle \tag{30}$$

This is a slowly varying function of $\mathbf{s}$, and it displays no sharp maxima like those for Bragg diffraction. If there is only one atom per cell and the $\Delta \mathbf{r}_n$ are isotropic, the diffuse scattering is related to the atomic scattering factor and therefore has spherical symmetry in reciprocal space. For a single molecule per cell it is related to the structure factor of the molecule and can become somewhat modulated. We will encounter this phenomenon in our discussion of thermal diffuse scattering. The case of no correlations occurs to some extent in all crystals, but is not very interesting from a structural standpoint. It is when correlations are present that the diffuse scattering yields structural information that is not obtainable by studying the Bragg reflections.

3.4 Thermal Diffuse Scattering

The experimental X-ray scattering from even the most simple crystal disagrees with that predicted by (18), which was derived on the premise of perfect order. In real crystals the atoms are always in motion because of thermal agitation. All crystals therefore display a type of displacive disorder, which will affect the intensities of the Bragg reflections as well as give rise to diffuse scattering. Thermal motion in crystalline materials has been extensively studied because of its importance in understanding the thermal and mechanical behavior of solids. The subject is treated here only because thermal motion is an ever-present source of diffuse scattering. As such, a basic understanding of the phenomenon is necessary when investigating experimental patterns for diffuse scattering from other sources of disorder. The following treatment of thermal disorder given below is sufficient for this purpose; however, more extensive treatments can be found elsewhere [5]. Unlike most cases of disorder considered in this chapter, thermal disorder is dynamic; the positions of the atoms change as a function of time. At constant temperature, this aspect of the problem has no measurable effect on the X-ray scattering, but must be taken into account for neutron scattering. For X-ray scattering we need only consider the instantaneous positions of all the atoms in the solid. Moreover, since the time scale of the experiment is long compared to vibrational frequencies, each atom, during the course of the experiment, will undergo all possible displacements many times. It is therefore not even necessary to perform a time average, since in this case an average over all cells (a space average) gives equivalent results.

3.5   Thermal Displacement Waves

If all the atoms in a simple monoatomic crystal vibrate about their equilibrium positions independently of the vibrations of their neighbors, we can use (30) to calculate the diffuse scattering. If we are concerned with molecular crystals, (30) can still be used for the case of one molecule per cell, providing that we assume that the molecule is a rigid body. This implies that the molecule has no conformational degrees of freedom so that all atoms in the molecule move as a unit. The idea of a rigid molecule is a generally useful one since the valence forces that bind the atoms of a single molecule are much stronger than the forces that bind different molecules in the crystal. The scattering produced from the independent thermal motion of molecules is readily observable and will be treated later. There is also a type of thermal motion for which the displacements of the rigid molecules (or atoms for a monoatomic crystal) are strongly correlated. This can give rise to diffuse scattering that is stronger and more localized in reciprocal space. The correlations can be described in terms of thermal displacement waves. At any given time there are many such waves propagating through the crystal. It is possible to calculate the diffuse scattering that would result from one such wave and then obtain the entire scattering pattern by superimposing the scattering produced by many such waves. This is equivalent to assuming that the waves propagate independently through the crystal, and we will make this assumption in our treatment of the diffuse scattering. It should be understood, however, that this is not rigorous. If one takes into account that two waves can be coupled, this can give rise to the so-called second-order thermal diffuse scattering; higher orders are also possible. An explicit calculation of the second-order case is given in Warren [6]. In what follows, we assume that there is only one atom or one rigid molecule per unit cell and therefore (28) is applicable. We will refer to the displacement of the lattice nodes to avoid unnecessary repetition of this restriction.

A single displacement wave moving through the crystal can be visualized by first decomposing the crystal into a series of parallel layers of equal thickness. The vector normal to these layers is denoted **k** and need not be parallel to any of the primary crystallographic directions; it is referred to as the propagation vector, and has the dimensions of inverse length. Each layer is then displaced according to a plane wave equation with an amplitude **A** and wavelength L, where $L = 1/|k|$. At any instant, the nodes of the lattice appear as shown in Figure 7.17. The amplitude **A** is a vector quantity since the waves can be either transverse or longitudinal. From Figure 7.17 it should be clear that this is an example of planar disorder. The instantaneous displacement of the nth cell is given by,

$$\Delta \mathbf{r}_n = \mathbf{A} \, \cos(2\pi \mathbf{k} \cdot \mathbf{t}_n) \tag{31}$$

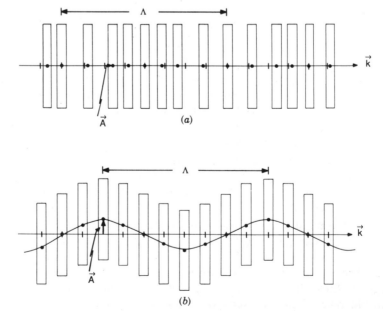

Figure 7.17 Thermal displacement waves in a crystal: (a) longitudinal (compression) wave and (b) transverse wave.

From (28) we can calculate the average value of the structure factor:

$$\langle F \rangle = F \langle \exp(-2\pi i s \cdot \Delta \mathbf{r}_n) \rangle \tag{32}$$

Since from Figure 7.17 it is clear that for every displacement of $\Delta \mathbf{r}_n$ there is also one of $-\Delta \mathbf{r}_n$,

$$\langle F \rangle = F \langle \cos 2\pi s \cdot \mathbf{r}_n \rangle$$

If we assume that the displacements are small, we can expand the cosine, while retaining only the first of the terms in $\Delta \mathbf{r}_n$.

$$\langle F \rangle = F[1 - 2\pi^2 \langle (s \cdot \Delta \mathbf{r}_n)^2 \rangle] \tag{33}$$

Substitution of the explicit form of the displacement wave from (31) for $\Delta \mathbf{r}_n$ yields,

$$\langle F \rangle = F[1 - 2\pi^2 (s \cdot \mathbf{A})^2 \langle (\cos 2\pi \mathbf{k} \cdot \mathbf{t}_n)^2 \rangle] \tag{34}$$

From (34) it is clear that the structure factor for the average cell is diminished by an amount dependent on the amplitude of the wave projected on the scattering vector. The cosine term is most easily evaluated by using the half-angle formula, from which

$$\langle (\cos 2\pi \mathbf{k} \cdot \mathbf{t}_n)^2 \rangle = \tfrac{1}{2} - \tfrac{1}{2} \langle \cos 4\pi \mathbf{k} \cdot \mathbf{t}_n \rangle$$

Since for a sufficiently large number of cells the argument $4\pi \mathbf{k} \cdot \mathbf{t}_n$ can take all values modulo 2π with equal probability, $\langle (\cos 4\pi \mathbf{k} \cdot \mathbf{t}_n) \rangle = 0$, and the original term is $\langle (\cos 2\pi \mathbf{k} \cdot \mathbf{t}_n)^2 \rangle = \tfrac{1}{2}$. With this, (34) is now independent of n and becomes becomes

$$\langle F \rangle = F[1 - \pi^2 (s \cdot \mathbf{A})^2] \tag{35}$$

Using this expression in (25) we have an expression analogous to that for a perfect crystal. For $s \cdot \mathbf{A} \neq 0$, or more explicitly, if $s \neq 0$ and the amplitude has some nonzero projection on s, the intensity for the crystalline reflections will always be smaller than for the crystal with perfect order. It is also apparent that the intensity difference becomes more pronounced at high values of the scattering angle (large $|s|$).

We now wish to calculate the coefficients of (26) to obtain the diffuse scattering. Again using (28) we have

$$F_n F^*_{n+m} = FF^* \exp[-2\pi i (s \cdot \Delta \mathbf{r}_n - s \cdot \Delta \mathbf{r}_{n+m})]$$

Assuming the $\Delta \mathbf{r}_n$ values are small, we can expand the exponential, neglect higher order terms, and average term by term,

$$\langle F_n F_{n+m}^* \rangle = FF^*[1 - 2\pi i \langle \mathbf{s} \cdot \Delta \mathbf{r}_n \rangle - 2\pi^2 \langle (\mathbf{s} \cdot \Delta \mathbf{r}_n - \mathbf{s} \cdot \Delta \mathbf{r}_{n+m})^2 \rangle$$

Since $\langle \mathbf{s} \cdot \Delta \mathbf{r}_{n+m} \rangle = \langle \mathbf{s} \cdot \Delta \mathbf{r}_n \rangle$, and by definition $\langle \mathbf{s} \cdot \Delta \mathbf{r}_n \rangle = \mathbf{s} \cdot \langle \Delta \mathbf{r}_n \rangle = 0$, the imaginary terms vanish. When the real term is squared,

$$\langle F_n F_{n+m}^* \rangle = FF^*[1 - 4\pi^2 \langle (\mathbf{s} \cdot \Delta \mathbf{r}_n)^2 \rangle + 4\pi^2 \langle (\mathbf{s} \cdot \Delta \mathbf{r}_n)(\mathbf{s} \cdot \Delta \mathbf{r}_{n+m}) \rangle]$$

Only the third term in the above expression requires information concerning correlations between displacements. From (33),

$$\langle F \rangle \langle F^* \rangle = FF^*[1 - 4\pi^2 \langle (\mathbf{s} \cdot \Delta \mathbf{r})^2 \rangle + 4\pi^4 \langle (\mathbf{s} \cdot \Delta \mathbf{r}_n)^2 \rangle^2]$$

Neglect of the higher order term and use of the above expression for $\langle F_n F_{n+m}^* \rangle$ yields,

$$\langle \varphi_n \varphi_{n+m}^* \rangle = \langle F_n F_{n+m}^* \rangle - \langle F \rangle \langle F^* \rangle$$
$$= FF^* 4\pi^2 \langle (\mathbf{s} \cdot \Delta \mathbf{r}_n)(\mathbf{s} \cdot \Delta \mathbf{r}_{n+m}) \rangle \qquad (36)$$

Thus far we have made no explicit assumptions concerning the correlation of the displacements, and so the above is general for any type of displacive disorder providing that the displacements are small. Using the explicit form (31) for $\Delta \mathbf{r}_n$, we obtain

$$\langle \varphi_n \varphi_{n+m}^* \rangle = FF^* 4\pi^2 \langle (\mathbf{s} \cdot \mathbf{A})^2 \cos(2\pi \mathbf{k} \cdot \mathbf{t}_n) \cos(2\pi \mathbf{k} \cdot \mathbf{t}_{n+m}) \rangle$$

If $\mathbf{t}_{n+m} = \mathbf{t}_n + \mathbf{t}_m$ and the sum angle formula is used, the following results:

$$\langle \varphi_n \varphi_{n+m}^* \rangle = FF^* 4\pi^2 (\mathbf{s} \cdot \mathbf{A})^2 [\langle \cos^2(2\pi \mathbf{k} \cdot \mathbf{t}_n) \rangle \cos(2\pi \mathbf{k} \cdot \mathbf{t}_m)$$
$$- \sin(2\pi \mathbf{k} \cdot \mathbf{t}_m) \langle \cos(2\pi \mathbf{k} \cdot \mathbf{t}_n) \sin(2\pi \mathbf{k} \cdot \mathbf{t}_n) \rangle]$$

where quantities not containing $\mathbf{t}_n$ have been removed from the angular brackets since the average is only over all possible values of n. The second term in the angular brackets is zero. This can be seen by noting that $\sin x \cos x = \frac{1}{2} \sin 2x$. Since for every vector $\mathbf{t}_n$ there is a vector $\mathbf{t}_{-n} = -\mathbf{t}_n$, the sine terms average to zero in pairs. The first term in angular brackets was encountered in (34) and is equal to $\frac{1}{2}$, with which the expression for the coefficients of the diffuse term is,

$$\langle \varphi_n \varphi_{n+m}^* \rangle = FF^* 2\pi^2 (\mathbf{s} \cdot \mathbf{A})^2 \cos 2\pi \mathbf{k} \cdot \mathbf{t}_m \qquad (37)$$

Substitution of this into (26) yields

$$I_2(s) = FF^*\pi^2(s \cdot A)^2 \sum_m (N - |m|) \exp[2\pi i(s + k) \cdot t_m]$$

$$+ FF^*\pi^2(s \cdot A)^2 \sum_m (N - |m|) \exp[2\pi i(s - k) \cdot t_m]$$

where we have used the relationship,

$$\cos(2\pi k \cdot t_m) = \tfrac{1}{2}[\exp(2\pi i k \cdot t_m) + \exp(- 2\pi i k \cdot t_m)]$$

in obtaining the above form. Both sums are identical in form to (16), the expression for a crystal with perfect order, except that s is replaced by $s \pm k$, and $FF^*\pi^2(s \cdot A)^2$ replaces FF^*. We can therefore write the expression for the diffuse scattering immediately:

$$I_2(s) = FF^*\pi^2(s \cdot A)^2 S[N_a, (s + k) \cdot a]S[N_b, (s + k) \cdot b]S[N_c, (s + k) \cdot c]$$
$$+ FF^*\pi^2(s \cdot A)^2 S[N_a, (s - k) \cdot a]S[N_b, (s - k) \cdot b]S[N_c, (s - k) \cdot c] \quad (38)$$

The above gives rise to two sets of reflections analogous to those for the Bragg case. For every Bragg reflection with $s = t_H^*$, there is a pair of diffuse reflections that are displaced by the vector k in reciprocal space. The situation is depicted in Figure 7.18 for a representative t_H^*. It is as if there are three separate

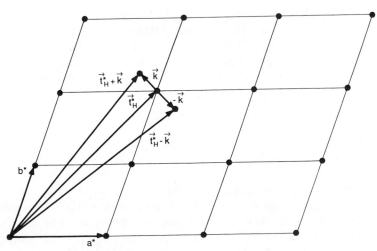

Figure 7.18 A pair of satellite reflections caused by a single thermal displacement wave. A similar pair occurs for each reciprocal lattice point.

crystals, characterized by the general reciprocal vectors of translational symme-try, t_H^*, $t_H^* + k$, and $t_H^* - k$. From the form of the interference function in (38) it is clear that the diffuse scattering is as sharp as the Bragg scattering; the term diffuse is not particularly descriptive in this case. From (35), the intensity for the Bragg reflection is, $I_1(s) = N^2FF^*[1 - 2\pi^2(s \cdot A)^2]$, where we have dropped the term quartic in $s \cdot A$. The total intensity in the two satellite reflections is $I_2(s + k) + I_2(s - k) = N^2FF^*2\pi^2(s \cdot A)^2$. Therefore, the intensity missing in any particular Bragg reflection is present in diffuse peaks; the intensity is simply scattered at a different angle.

We now must consider the net effect of the many thermal displacement waves traveling through the crystal, each with a different value of k and A. Each wave will produce two satellite reflections around each node of the average lattice, displaced by k and with an intensity proportional to $(s \cdot A)^2$. Experimentally it is found that the intensity of the satellites generally decreases with increasing $|k|$, which implies that the amplitude decreases with decreasing L. This can be qualitatively understood by considering the effect of two such waves that are identical in all respects except wavelength. The unit vector normal to both is $\hat{k}$; the propagation vectors are $k_1 = \hat{k}/L_1$ and $k_2 = \hat{k}/L_2$. The minimum energy arrangement of the crystal is considered to be that of the average structure. A displacement wave represents an increase in the energy that depends on the total amount of distortion caused by the wave. A measure of distortion caused by a single wavelength is $|A|$, the amplitude of the wave. The total amount of distortion must take into account how many wavelengths can be present simultaneously in the finite crystal. For shorter wavelengths, the number is obviously higher than for longer ones. Therefore, for a given amplitude, the short wavelength displacement wave (large $|k|$) represents a higher energy con-figuration of the crystal. Since the thermal energy available for any wave is on the order of kT, those with large $|k|$ will tend to have smaller amplitudes and thus, from (38), produce satellites of lower intensity. Since there is a very large number of possible values for k, it is clear that individual satellite reflections will superimpose in the vicinity of each Bragg peak and give rise to a continuous scattering with a broad maximum. The profile of a diffraction maxima containing both the Bragg and diffuse contributions is shown in Figure 7.19.

The temperature diffuse scattering is found experimentally to be anisotropic around each node in reciprocal space. This is because all crystals, even those of cubic symmetry, are more easily deformed in some directions than in others. Generally, the diffuse envelope is elongated along those directions that are most easily distorted.

At high values of the scattering angle, the diffuse scattering can become large enough to cause appreciable errors in the measurement of the Bragg intensities. In very precise work, for example, when these intensities are to be used to study the valence electron density [7, 8], corrections must be made for these errors.

Since the decrease in the Bragg intensities is the net result of many thermal displacement waves, it is convenient to develop an expression that is not

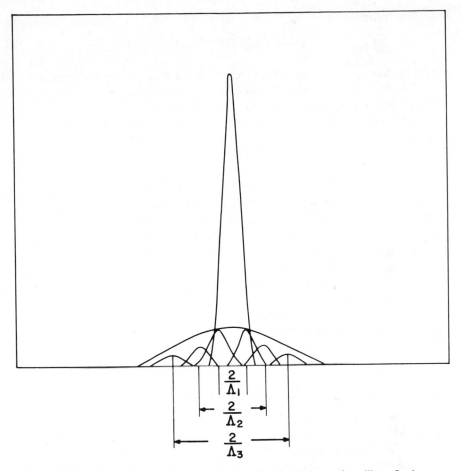

Figure 7.19   Peak profile for crystal reflection showing envelope of satellite reflections.

dependent on the amplitude of a particular wave. From (33),

$$\langle F \rangle = F[1 - 2\pi^2 \langle (\mathbf{s} \cdot \Delta \mathbf{r}_n)^2 \rangle] = F[1 - 2\pi^2 |s|^2 \langle u^2 \rangle]$$

where u is the component of $\Delta \mathbf{r}_n$ on $\mathbf{s}$. Substitution of the value for $|s|$ given in (9) yields the following:

$$\langle F \rangle = F[1 - 8\pi^2 \langle u^2 \rangle (\sin \theta / \lambda)^2]$$

where this explicitly assumes that the mean square amplitude of vibration caused by all waves is isotropic. The above is often used in one of the two following forms:

$$\langle F \rangle = F \exp[-B(\sin \theta/\lambda)^2] = FD \tag{39}$$

where the temperature factor B is given by $B = 8\pi^2 \langle u^2 \rangle$. The Debye factor D is a useful shorthand for the entire exponential. An often-used extension of this formula allows a separate value of B for each atom in crystals with more than one atom per cell. This is sometimes included in the scattering factor,

$$f'_j = f_j \exp[-B_j(\sin \theta/\lambda)^2] \tag{40}$$

Further extensions of this idea allow for anisotropic motion of each atom, where (40) would be replaced by a more complicated formula. Small wave vector thermal displacement waves, or acoustic phonons, as they are termed if there is only one atom or rigid molecule per cell, are the most important form of thermal *disorder* in all crystals. In molecular crystals, or crystals with several atoms per cell, there exist not only the small wave vector acoustic phonons, but also a large number of so-called optical phonons. For molecular crystals, one finds, for example, the molecular vibrations and librations among the optical phonons. The temperature factor B is actually a composite term that takes into account the average effect of all of these motions on the Bragg intensities. Often, the diffuse scattering from optical phonons is very weak and can be neglected. A notable exception to this will be considered next.

3.6   The Continuous Diffuse Scattering of Molecular Crystals

We have already pointed out that correlations in the displacements of the atoms should be expected because of intermolecular attractions and repulsions. However, at some temperature that is an appreciable fraction of the melting point, one should also expect that the molecules of the crystal would have some degree of independent motion. For molecular crystals, room temperature is often within 100 K of the melting point, and a particular type of diffuse scattering is found experimentally. This scattering can still be accounted for by the general expression (38) if one superimposes waves of equal amplitude for each wave vector, demonstrating that such motion adheres to the general description of thermal waves. However, for the case of independent motion, it is simpler to perform a direct calculation. This also provides more direct physical insight into a most relevant aspect of many molecular crystals: An important part of the thermal diffuse scattering takes the form of large clouds of weak intensity in reciprocal space. This is referred to as the continuous diffuse scattering of molecular crystals.

We will calculate the diffuse scattering for the case of a single molecule that is undergoing small isotropic thermal displacements independently of its neighbors. The calculation is then extended to two or more molecules per cell. The continuous diffuse scattering from real molecular crystals is often explained quite well from such simple calculations.

For simplicity we assume that no other form of thermal displacements exist, and therefore the expression for $\langle F \rangle$ is given by (39). For the coefficients of the diffuse term we have,

$$\langle \varphi_n \varphi_{n+m}^* \rangle = \langle FF^* \exp(-2\pi i s \cdot \Delta r_n) \exp(-2\pi i s \cdot \Delta r_{n+m}) \rangle - FF^* D^2$$

We are considering the case of uncorrelated displacements, which can be thought of as the molecules rattling about their mean positions independently. For $m \neq 0$, the two exponentials may be averaged independently; and since these averages are by definition D the above vanishes. Thus there is only one term, the $m = 0$ term on (26), and this is,

$$\langle \varphi_n \varphi_n^* \rangle = FF^* (1 - D^2)$$

Since $D^2 \cong [1 - 8\langle u^2 \rangle \pi^2 (\sin \theta / \lambda)^2]^2$, for small mean-square displacements, the diffuse scattering is very weak. Its distribution in reciprocal space is determined by the intensity function of the single molecule of the unit cell. It is as if the molecule were in free space; there are no interference effects from neighboring molecules. The intensity scattered per unit cell is given by,

$$I_2(s) = FF^* (1 - D^2) \tag{41}$$

This can be considered to be the difference between the intensity scattered by the molecules at rest and the intensity for the molecule undergoing isotropic displacements.

As an example, consider the intensity scattered by a benzene molecule in a cell of arbitrary dimensions. We will neglect the term $(1 - D^2)$, since its effect is easily predicted. It will make the pattern weak at low values of 2θ, and progressively stronger at higher angles; it is spherically symmetric. Also, since the molecule has a center of symmetry, F is real, and we can plot it as a function of s. Although F can be positive or negative, the intensity is simply given by the square of F anywhere in reciprocal space. The structure factor of the benzene molecule is shown in Figure 7.20, there the variation of the atomic scattering factor f with scattering angle is also neglected. Assuming that the molecule sits in the ab plane of a rectangular cell of dimensions 10×10 Å, the grid marks correspond to $h, k = 2n$. Contours are plotted every 0.5 e, and the largest peak at $s = 0$ has been scaled to 1.0 e. We therefore can obtain the intensity scattered *per atom* (with a fictitious scattering factor of $f = 1$) by squaring the plotted value at any point s, where s is considered to be somewhere in the $a^* b^*$ plane. It can be seen that on the scale of reciprocal lattice spacings, the structure factor is a slowly varying function of s. It is sixfold symmetric, which reflects the molecular symmetry. The large positive peaks at $|s| = 0.82$ Å^{-1}, are caused by interference between atoms normal to the mirror planes that pass through two atoms. Since at $|s| = 0$, $1 - D^2 = 0$, the central peak would be almost absent in the actual diffuse scattering pattern. Also, since the atomic scattering factor falls off with

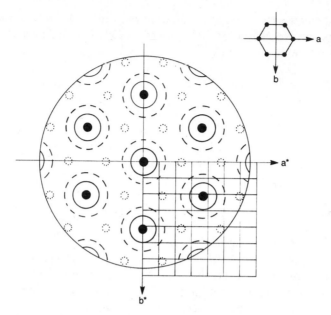

Figure 7.20 Structure factor for a cell containing a single benzene molecule. Scattering factor taken to be $f = 1$; ●, 1.0e; ———, 0.5e; - - -, 0.0e; · · ·, 0.5e; take $a = b = 10\,\text{Å}$, grid marks are for $h, k = 2n$.

increasing angle, the diffuse scattering will eventually decrease with increasing angle as well. The intensity in reciprocal space also extends indefinitely above and below the plane of the drawing. Thus, any section of reciprocal space parallel to the one drawn would be very similar in appearance. In this respect, the scattering is somewhat similar to that for a plane lattice. For the most part, the intensity is localized in very broad columns.

A correspondence between the intensities of the Bragg peaks and the structure factor of the molecule for the case of one molecule per cell can also be drawn. From our earlier discussion [see (18)], it is clear that the interference function is essentially zero, except at the nodes, where it is N^2 for *all* nodes in reciprocal space. It therefore does not create intensity differences among the Bragg reflections. In the present case, those nodes that fall within the columns corresponding to high values of $|F|$ in Figure 7.20 will be strong, while reflections corresponding to low values will be weak.

An interesting side effect of this is sometimes encountered in the process of phase determinations by direct methods (see Chapter 1). Essentially, this method assumes that the signs S_H ($H = hkl$) of the structure factors F_H adhere to the relationship $S_H S_{H'} S_{H-H'} = 1$, provided that the intensities I_H, $I_{H'}$, and $I_{H-H'}$, are all strong. From Figure 7.20 it is clear, however, that some strong reflections will correspond to areas of negative F. Thus, if the triple product in the above expression contains two sets of indices corresponding to positive regions, and one set corresponding to a negative region, the value of the triple product will actually be -1. In such cases the process of phase determination goes wrong. This has recently been observed in several structures containing many benzene rings [9, 10]. Since the continuous diffuse scattering is closely related to the structure factor of the molecule, it could conceivably be used to advantage [11] in the process of structure solution in difficult cases.

We shall now calculate the continuous diffuse scattering for two molecules per cell. We first set

$$F_n = M_{1;n} + M_{2;n}$$

where $M_{1;n}$ and $M_{2;n}$ each contain a sum similar to that of (11) except that $M_{1;n}$ contains only the $\mathbf{r}_{nj}$ for the first molecule of the nth cell, and $M_{2;n}$ contains only those for the second molecule. The average value of the structure factor is given by

$$\langle F \rangle = \langle M_{1;n} + M_{2;n} \rangle$$

Since the two molecules are being displaced independently,

$$\langle F_n \rangle = \langle M_{1;n} \rangle + \langle M_{2;n} \rangle$$
$$= M_1 \langle \exp(-2\pi i \Delta \mathbf{r}_{1;n}) \rangle + M_2 \langle \exp(-2\pi i \Delta \mathbf{r}_{2;n}) \rangle$$

Assuming that the two molecules undergo identical displacements on average (which is actually required if the two are related by a symmetry element),

$$\langle F \rangle = (M_1 + M_2)\mathbf{D} = F\mathbf{D} \tag{42}$$

where we have assumed the displacements are isotropic. This is identical to our expression for a single molecule, although F now contains contributions from two molecules. We also have,

$$\langle F_n F^*_{n+m} \rangle = \langle (M_{1;n} + M_{2;n})(M^*_{1;n+m} + M^*_{2;n+m}) \rangle$$
$$= \langle M_{1;n} M^*_{1;n+m} \rangle + \langle M_{1;n} M^*_{2;n+m} \rangle$$
$$+ \langle M_{2;n} M^*_{1;n+m} \rangle + \langle M_{2;n} M^*_{2;n+m} \rangle$$

The averages containing the structure factors of two different molecules are always of the form $M_i M^*_j \mathbf{D}^2$, since the two vibrate independently. Also, for $m \neq 0$, the other terms will be of the form $M_i M^*_i \mathbf{D}^2$, as for the case of a single molecule. Thus, for $m \neq 0$, we have

$$\langle F_n F^*_{n+m} \rangle = (M_1 M^*_1 + M_1 M^*_2 + M_2 M^*_1 + M_2 M^*_2)\mathbf{D}^2$$
$$= FF^*\mathbf{D}^2$$

With (42) the coefficients of the diffuse scattering are the following:

$$\langle \varphi_n \varphi_{n+m}^* \rangle (m \neq 0) = FF^*D^2 - FF^*D^2 = 0$$

As was the case for a single molecule, the coefficients for the diffuse scattering are all zero except the $m = 0$ term. For this,

$$\langle F_n F_n^* \rangle = (M_1 M_1^* + M_2 M_2^*) - D^2(M_1 M_2^* + M_2 M_1^*)$$

and with (42),

$$\langle \varphi_n \varphi_n^* \rangle = (M_1 M_1^* + M_2 M_2^*)(1 - D^2) \tag{43}$$

The diffuse scattering will now be a superposition of the scattering from the two molecules. It is not generally equal to $2M_1 M_1^*$, since the two molecules may be orientated differently in the cell. Equation (43) represents the intensity from two molecules that scatter completely incoherently. The situation is quite different for the Bragg reflections since $\langle F \rangle \langle F^* \rangle$ contains cross terms. The above calculation is easily extended to more than two molecules per cell, and the diffuse scattering per cell is given by,

$$I_2(s) = \langle \varphi_n \varphi_n^* \rangle = (1 - D^2) \sum_p M_p M_p^* \tag{44}$$

where the sum is over all molecules in the cell. In Figure 2.21, we show $I_2(s)$ for the two-molecule case. The pattern is for the actual benzene structure and was obtained by optical diffraction of a suitable model. In this case there are four molecules, but in this projection there are only two different orientations. Even though the molecules are not quite parallel to the plane of projection, the diffuse scattering can be recognized as the superposition of two slightly distorted patterns, similar to that in Figure 7.20. In Figure 7.21 the Bragg reflections are also shown.

In the following calculations of planar and linear disorder it will be convenient to discuss the diffuse scattering without considering the effects of thermal motion. This omission should cause no difficulty, since these effects have been discussed above in some detail. If one wishes to compare experimental patterns to the calculated results obtained by the methods described below, the effects of thermal motion are easily included. One simply replaces the usual atomic scattering factor with the form given in (40).

What has been said above about the continuous diffuse scattering of molecular crystals illustrates the effect of the molecular structure factor for the independent motions of molecules close to the melting point. It should be noted, however, that the molecular structure factor is a general feature of molecular crystals that also affects the diffuse scattering from ordinary thermal waves. Large amplitude waves usually correspond to the acoustic part of the dispersion

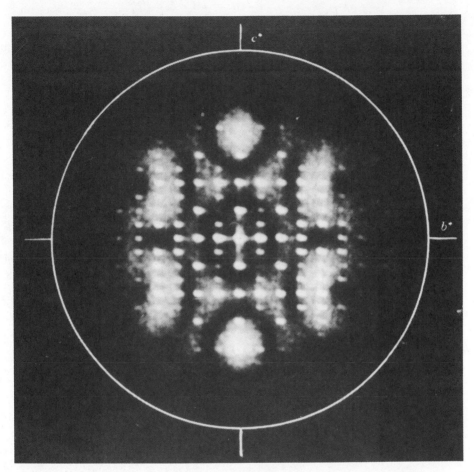

Figure 7.21 Optical simulation of X-ray scattering for actual benzene structure, with thermal agitation. Large areas of continuous diffuse scattering are evident. Reproduced, with permission, from J. L. Amorós and M. Amorós, *Molecular Crystals*, Wiley, New York, 1968, pp. 300–397 and references cited therein. Copyright © 1968 by John Wiley & Sons, Inc.

spectrum and to displacements of rigid molecules (or rigid parts of molecules). This also gives rise to large clouds of diffuse intensity corresponding to the intensity modulations of the molecular structure factor; a simple example will be described below (Figure 7.33, Section 4.2). This type of scattering can be observed far from the melting point.

3.7   Planar Disorder

Many crystalline structures can be described in terms of the stacking of layers. These layers may be only a single atom in thickness or may correspond to one or more molecular thicknesses. Often, the distinguishing feature of a layer structure is that the forces binding the components within a given layer are in some respect different from those binding one layer to another. When the forces between layers are weaker than those within, some irregularity in the stacking may occur. If these irregularities leave all layers in the structure parallel, the crystal is said to display planar disorder.

When all layers are identical, we define the two lattice vectors within any layer as $\mathbf{a}$ and $\mathbf{b}$. The average interlayer spacing is $1/|c^*|$, which is equal to $|c|$ only if $\mathbf{c}$ is normal to both $\mathbf{a}$ and $\mathbf{b}$. Perfect order within a given layer implies that $\langle F_n F^*_{n+m} \rangle$ is not a function of m_a or m_b, but still depends on m_c. From (27), the same is also true for $\langle \varphi_n \varphi^*_{n+m} \rangle$. In the following, when m_c occurs as a subscript, we will instead use mc to avoid the subscripting of subscripts. Eventually, we will drop the c altogether, but the meaning of the index remains unaltered. For planar disorder (26) can then be written as follows:

$$I_2(\mathbf{s}) = \left[\sum_{mc} (N_c - |m_c|)\langle \varphi_{nc}\varphi^*_{nc+mc} \rangle \, \exp(2\pi i \mathbf{s} \cdot \mathbf{c}m_c) \right]$$

$$\times \left[\sum_{ma} (N_a - |m_a|) \exp(2\pi i \mathbf{s} \cdot \mathbf{a}m_a) \right]$$

$$\times \left[\sum_{mb} (N_b - |m_b|) \exp(2\pi i \mathbf{s} \cdot \mathbf{b}m_b) \right]$$

where the coefficients of the diffuse scattering have been removed from under the summations over m_a and m_b. The sums over m_a and m_b are identical to (17), and therefore identical to those of a perfect crystal. Assuming that N_a and N_b are large allows us to neglect the size effect and the above reduces to,

$$I_2(s) = N_a^2 N_b^2 \left[\sum_{mc} (N_c - |m_c|)\langle \varphi_{nc}\varphi^*_{nc+mc} \rangle \, \exp(2\pi i \mathbf{s} \cdot \mathbf{c}m_c) \right] \qquad (45)$$

The diffuse scattering is therefore limited to the same reciprocal lattice rows as the Bragg diffraction. Diffuse scattering can only occur when the scattering vector is of the form $\mathbf{s} = h\mathbf{a}^* + k\mathbf{b}^* + s_3\mathbf{c}^*$. The modulation of the diffuse scattering along a given row (h, k) will depend on the remaining summation in (45).

Consider the case where each layer may be shifted by an amount $\mathbf{v}_{nc} = \Delta x_{nc}\mathbf{a} + \Delta y_{nc}\mathbf{b}$, so that $\mathbf{v}_{nc}$ represents a slip of the n_cth layer of the stack. The interlayer spacing, $1/|c^*|$, is a constant for all layers. The coefficients of the diffuse scattering are given by,

$$\langle \varphi_{nc}\varphi^*_{nc+mc} \rangle = FF^*[\langle \exp(-2\pi i \mathbf{s} \cdot \mathbf{v}_{nc}) \exp(-2\pi i \mathbf{s} \cdot \mathbf{v}_{nc+mc}) \rangle$$

$$- \langle \exp(2\pi i \mathbf{s} \cdot \mathbf{v}_{nc}) \rangle \langle \exp(-2\pi i \mathbf{s} \cdot \mathbf{v}_{nc}) \rangle] \qquad (46)$$

where we have used (28). Depending on the possible values of the $\mathbf{v}_{nc}$ and the correlations between them, a large variety of disorder schemes is possible. The three cases detailed below represent a brief introduction.

CASE 1

The slip vectors $\mathbf{v}_{nc}$ can assume any magnitude or direction parallel to the layers. The quantities $2\pi\mathbf{s} \cdot \mathbf{v}_{nc}$ can therefore take on any value modulo 2π provided that not both $\mathbf{s} \cdot \mathbf{a}$ and $\mathbf{s} \cdot \mathbf{b}$ are zero. With this restriction, the second term in (46) is zero. The only row for which this is not true is the row $h = 0, k = 0$. For this row, $2\pi\mathbf{s} \cdot \mathbf{v}_{nc} = s_3$, irrespective of the value of $\mathbf{v}_{nc}$. Then (46) is equal to zero, and there is no diffuse scattering. Also, $\langle F \rangle = F \exp(2\pi i s_3) = F$ when s_3 is integral, and therefore the Bragg scattering for the $(0, 0, l)$ reflections is identical to that for a perfect crystal. On the other hand, $\langle \exp(2\pi i \mathbf{s} \cdot \mathbf{v}_{nc}) \rangle = 0$ for any other row, and there is no Bragg diffraction for these. If there is no correlation between the layers, the exponentials in the first term in (46) may be averaged independently and (46) is zero for $m_c \neq 0$. The sum in (45) then contains only one term, the $m_c = 0$ term, which is equal to FF^*. The expression for the diffuse scattering then becomes,

$$I_2(\mathbf{s})\{\mathbf{s} \cdot \mathbf{v} \neq 0\} = N_a^2 N_b^2 N_c FF^*$$

which is simply N_c times the intensity scattered by a single layer. For any given orientation of the crystal, the diffuse scattering can be understood in terms of the reciprocal lattice in Figure 7.7 and the Ewald construction in Figure 7.8, with the exception of the $(0, 0, l)$ row.

Scattering similar to that described here is observed at low temperatures for some forms of β-alumina [12]. In these compounds, the cations reside in parallel sheets that are isolated from each other by an ordered framework of Al_2O_3. Communication between the sheets is weak, and there is little or no correlation between the displacement of successive sheets, which produces diffuse scattering localized in columns normal to the sheet direction. The smectic phases of certain liquid crystals [12–17] also gives rise to this type of diffuse scattering. In both types of materials, the structure within the layers is also somewhat imperfect.

CASE 2

There is some correlation between the layers. The vectors $\mathbf{v}_n$ are still as described above. Starting from the first layer in the crystal, the structure can be described by a perfect layer structure over some number of layers until the first slip occurs. The slip can take on any magnitude or direction parallel to the layers with equal probability. Then, from the perspective of the first slipped layer, the structure is again perfect for succeeding layers until the next slip occurs, and so on. Let us denote the probability that the layer $n_c + m_c$ is the same as the n_cth layer (i.e., $\mathbf{v}_{nc} = \mathbf{v}_{nc+mc}$) as p_m. As for the case of no correlations, $\langle F \rangle$ is zero [except for $(0, 0, l)$], since $-2\pi i\mathbf{s} \cdot \mathbf{v}_c$ can still assume any value modulo 2π. The quantity $\langle F_n F_{n+m}^* \rangle$ will be nonzero only if the $n + m$th layer is correlated with the nth. If we suppose that only nearest neighbors are correlated, $\langle F_n F_{n+m}^* \rangle$ is nonzero only because of the propagation of neighbor interactions. The quantity p_1 is the probability that no slip will occur between nearest neighbors. The probability that three layers in succession are the same is given by $p_1 \cdot p_1 + \Delta p$, which is simply the product of the probabilities that the two independent events will occur simultaneously, plus some small correction term that takes into account the fact that the third layer could be the same as the first by totally random chance. In the present case, we have an unlimited number of v_{nc}, and therefore the accidental coincidence of two layers is essentially zero; thus, $\Delta p = 0$. If we now consider the probability that the mth layer in the crystal is the same as the first, we have $p_m = p^m$, the product of the probabilities of m equally probable independent events occurring simultaneously. Since any layer can be the first, p^m also represents the probability that the layer $n + m$ is the same as the nth. We now have $\langle \varphi_{nc} \varphi_{nc+mc}^* \rangle = \langle F_n F_{n+m}^* \rangle = p^{|m|} FF^*$, and (45) becomes,

$$I_2(\mathbf{s}) = FF^* \sum_m (N - |m|) p^{|m|} \exp(2\pi i \mathbf{s} \cdot \mathbf{cm})$$

where we have divided by $N_a^2 N_b^2$ to obtain the scattering power per row and dropped the subscripts c. As long as the probability of a fault is not too small, we can make the approximation that $(N - |m|)p^m = Np^m$. For example, let $p = 0.90$, which represents a relatively large tendency toward ordering, and suppose that for the number of layers in the crystallite $N = 300$, which is a reasonable approximation. The 30th term in the sum should be $0.9^{30}(300 - 30)$, although we approximate it as $0.9^{30}(300)$. The difference is only 10%, and in either case this term is only 5% of the first, which makes the relative error less than 1%. For smaller p, the approximation becomes even better. In order to obtain a workable expression, it is often assumed that the crystallites are very thick and the summation is extended to an infinite number of layers. The expression for the diffuse scattering is

$$I_2(\mathbf{s}) = NFF^* \left[1 + 2 \sum_{m=1}^{\infty} p^m \cos(2\pi \mathbf{s} \cdot \mathbf{cm}) \right] \tag{47}$$

where use has been made of the fact that for each layer m there is also a layer $-m$. Sums of this form are quite common in diffraction calculations. From its evaluation, which can be found in the appendixes,

$$I_2(\mathbf{s}) = NFF^* \frac{1 - p^2}{1 + p^2 - 2p \cos(2\pi \mathbf{s} \cdot \mathbf{c})} \tag{48}$$

For any particular problem, the numerator is a constant, so that the maxima and minima of (48) correspond to the minima and maxima of the denominator, respectively. The denominator is a periodic function of l, the Miller index, since the denominator minimizes when $\mathbf{s} \cdot \mathbf{c}$ is integral, or for $s_3 = l$. The maxima occur at the same position as for a perfect crystal. The maximum value of $I_2(\mathbf{s})$ is given by

$$I_2(l) = FF^*(1 + p)/(1 - p)$$

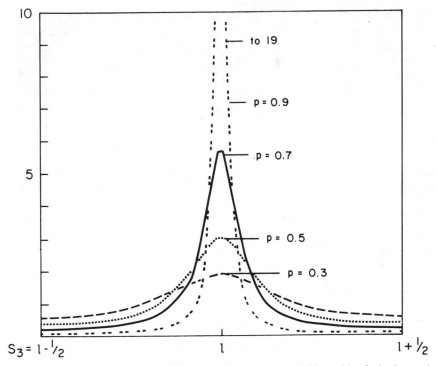

Figure 7.22 Variation of peak shape of diffuse scattering for crystal with stacking faults. Increasing values of p indicate an increasing tendency toward perfect order.

and is independent of the index l, except for the slow variation of F. The peak profile along the row direction for $s_3 = l - \frac{1}{2}$ to $l + \frac{1}{2}$ is shown in Figure 7.22 for a number of values of p. The profile is similar to the case of size broadening (Figure 7.16). However, there the $(0, 0, l)$ reflections have the same profile as those in any other row, while for the present case, the width of the $(0, 0, l)$ reflection depends only on the total number of layers. If p is too close to unity, or if n_c is small, (48) is no longer valid. Under these conditions, size effect broadening will modify the profiles predicted by (48). When the widths predicted by (48) and (18) are comparable, one can approximate that they are quadratically additive. Otherwise, the sum above (47) must be evaluated without approximations.

CASE 3

The slip vectors $\mathbf{v}_{nc}$ can take on only two values. We suppose that the nodes of the reference layer are as shown by the dots in Figure 7.23, but that there are an equal number of layers slipped such that their nodes are represented by x's. If we let F be the structure factor of the reference layer,

$$\langle F \rangle = (F/2) + (F/2) \exp[-2\pi i \mathbf{s} \cdot (a/2 + b/2)]$$
$$= (F/2)\{1 + \exp[\pi i(h + k)]\}$$

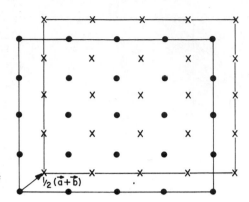

Figure 7.23 Planar disorder: reference lattice O; displaced lattice X.

With this we have

$$\langle F \rangle \langle F^* \rangle = \tfrac{1}{2} FF^* \{1 + \cos[\pi(h + k)]\}$$

Since this is zero for $h + k = 2n + 1$, the average structure is C centered. The coefficients of the diffuse scattering are given by

$$\langle \varphi_{nc} \varphi^*_{nc+mc} \rangle = FF^* \{ \langle \exp(-2\pi i s \cdot \mathbf{v}_{nc}) \exp(-2\pi i s \cdot \mathbf{v}_{nc+mc}) \rangle$$
$$- [1 + \cos \pi(h + k)]/2 \}$$

where $\mathbf{v}_{nc}$ can only take on the two values, 0 and $(\mathbf{a} + \mathbf{b})/2$. If we examine any two adjacent layers, the possible cell configurations are shown in Figure 7.24. The probability that the nth layer is the same as the $(n + m)$th layer is denoted p_m. The coefficients of the diffuse scattering are given by,

$$\langle \varphi_n \varphi^*_{n+m} \rangle = p_m FF^* + (1 - p_m) FF^* \cos[\pi(h + k)]$$
$$-(FF^*/2)\{1 + \cos[\pi(h + k)]\}$$

Since $p_m = \tfrac{1}{2}$ represents the random probability that the layer n is the same as the layer $n + m$, it is convenient to set $p_m = \tfrac{1}{2} + \tfrac{1}{2} q_m$. With this the expression for the coefficients of the diffuse scattering reduces to,

$$\langle \varphi_n \varphi^*_{n+m} \rangle = \tfrac{1}{2} \{1 - \cos[\pi(h + k)]\} q_m$$

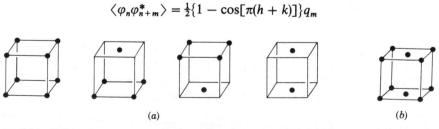

(a)   (b)

Figure 7.24 Cell configurations resulting from planar disorder as in Figure 7.23: (a) actual cells and (b) average cell.

and therefore, the diffuse scattering is limited to the rows $h + k = 2n + 1$, for which there is no diffraction from the average structure. The quantity q_1 is a measure of the tendency of the layers to form an ordered arrangement. If $q_m (m \neq 0)$ is zero, $p_m = \frac{1}{2}$, and the arrangement of layers is completely random. Thus, only the $m = 0$ term in (45) is nonzero, and the diffuse scattering along the $h + k = 2n + 1$ rows is not modulated, except for the variation of F. If there is a tendency to order only because of nearest-neighbor interactions, p_1 may be quite different from the random value of $\frac{1}{2}$. However, we expect that p_m will approach $\frac{1}{2}$ as m increases. It will be shown later that for nearest-neighbor interactions only, $p_m = \frac{1}{2} + \frac{1}{2} q^{|m|}$ and, therefore, $q_m = q^{|m|}$. Using this for q_m in the expression for the coefficients (45) becomes,

$$I_2(\mathbf{s}) = (FF^*/2)\{1 - \cos[\pi(h + k)]\} \sum_n (N - |m|)q^{|m|} \exp(-2\pi i \mathbf{s} \cdot \mathbf{c}m)$$

The sum has precisely the same form as that above (47). If we again consider thick crystallites, and approximate $(N - |m|)$ by N, the following expression for the diffuse scattering along the rows is obtained:

$$I_2(\mathbf{s}) = NFF^*\{1 - \cos[\pi(h + k)]\}/2 \left[\frac{1 - q^2}{1 + q^2 - 2q \cos(2\pi \mathbf{s} \cdot \mathbf{c})} \right] \qquad (49)$$

The quantity q is known as the order parameter. If $q = 0$, there is complete disorder, and the diffuse scattering along the rows $h + k = 2n + 1$ is a constant. If there is a strong preference for the primitive cell arrangements shown in Figure 7.24a, q tends toward one. Consequently, the diffuse scattering becomes modulated along the row, and the maxima occur for integral values of $\mathbf{s} \cdot \mathbf{c}$ and therefore at the nodes of the perfect primitive lattice. The behavior of (49) for $h + k = 2n + 1$ is in all respects similar to (48), which was derived for stacking faults with correlations; thus, the widths of the diffuse maxima are a strong function of the order parameter q (Figure 7.22). The main result of this calculation is that there are reflections everywhere one would expect by considering only the primitive cell arrangements in Figure 7.24a. However, these fall into two classes: sharp Bragg reflections for $h + k = 2n$ and broad diffuse reflections for $h + k = 2n + 1$. This is the first case considered in which confusion would arise in a structural study if the diffuse scattering is not taken into account properly. To appreciate this, assume that there is only one atom per cell, and therefore Figure 7.24b represents the average atomic arrangement. Then, the structure contains *atoms* separated by a, b, $[(a/2)^2 + (b/2)^2]^{1/2}$, and so on. In the actual structure, there are only atoms separated by a, b, $[(a/2)^2 + (b/2)^2 + c^2]^{1/2}$. The average structure, which is determined by the sharp Bragg reflections, does not correspond to reality! Thus, any analysis of the properties of the solid based on this structure may also be erroneous. However, we have considered this problem theoretically. Experimentally, one may fail to notice or choose to ignore that the $h + k = 2n + 1$ reflections are broader than those with $h + k = 2n$, an oversight more likely to happen now than in the past. With the availability of computer-driven diffractometers with autoindexing routines it is now possible to determine lattice constants and to measure intensity data without preparing single crystal photographs. Under these circumstances the weaker diffuse scattering is easily overlooked. If, however, it is detected, the intensities will often be measured without checking the widths. If one considered all *reflections* as Bragg reflections, it would invariably be true that as a class, the $h + k = 2n + 1$ reflections are weaker than the $h + k = 2n$ reflections. This is because the intensity predicted by (49) is not proportional to N_c^2 as are the $h + k = 2n$ reflections. Even if q is relatively close to unity so that the approximations leading to (49) are no longer valid, the $h + k = 2n + 1$ will still be somewhat weaker. One can of course attempt to compensate for this in the structure refinement by assuming that the atomic position $x = \frac{1}{2}$, $y = \frac{1}{2}$ is only partially occupied. This would lead to the conclusion that the interatomic distance $[(a/2)^2 + (b/2)^2]^{1/2}$ occurs less frequently than the primitive separations. Since this separation never occurs in the actual structure, treating the diffuse peaks as though they were Bragg peaks is just as incorrect as neglecting them altogether.

In determining the average structure from the Bragg intensities, one can circumvent such problems, to some extent, by comparing the derived interatomic distances with those published for similar compounds. In the present case, for example, one may find that an atomic separation of $[(a/2)^2 + (b/2)^2]^{1/2}$ corresponds to a much smaller separation than normally observed. However, when there are one or more molecules in the cell, stacking disorder can lead to an average structure for which the interatomic separations are just as reasonable as the actual ones. In any case, the actual structure can only be experimentally determined if the diffuse scattering is properly taken into account.

From the foregoing, one may have the impression that the order parameter q is a probability; this is not the case. Suppose, for example, that the probability that the nearest-neighbor layers are the same is $p_1 = 0.10$. This is considerably less than the random value of $\frac{1}{2}$. It indicates a strong preference for nearest neighbors to be different, and therefore the body-centered arrangements in Figure 7.24a are favored. Since $p_m = \frac{1}{2} + \frac{1}{2}q^{|m|}$, we find that for $m = 1$, $q = -0.90$; and so, unlike a probability, the order parameter can be negative. For $q = -1$ we have perfect order and the cell is body centered, which is the opposite of the case $q = 1$ for which we have perfect order and the cell is primitive. When q is not too close to -1, the diffuse scattering is again given by (49), but since q is negative the denominator minimizes at half-integral values of $\mathbf{s} \cdot \mathbf{c}$ and therefore the diffuse scattering displays maxima along the $h + k = 2n + 1$ rows that are midway between reciprocal layers l. In the limit that $q = -1$, the ordered body-centered arrangement will have $c^*_{\text{body centered}} = \frac{1}{2}c^*_{\text{primitive}}$, so that this behavior is easily understood. Hence, all reflections occur as predicted by an ordered body-centered arrangement, but the reflections again fall into two classes, broad and sharp, based on the parity of $h + k$. The difficulties arising from improper treatment of the diffuse scattering for the case of $q \geq 0$ also apply to the case $q \leq 0$. The total scattering pattern is

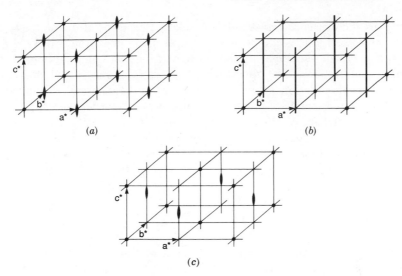

(a)

(b)

(c)

Figure 7.25   Schematic representation of diffuse scattering and Bragg diffraction for planar disorder as in Figure 7.23 and the three cases of the order parameter: (a) $1 \leqslant q < 0$; (b) $q = 0$; and (c) $0 > q \geqslant -1$.

shown schematically in Figure 7.25, for the three cases of the order parameter.

We now wish to indicate how one can derive an expression for cases where the layers are correlated such as in the relationship, $p_m = \frac{1}{2}(1 + q^{|m|})$, which was used to obtain (49). A generally applicable [18, 19] method involves developing recursion relationships among the p_m by standard probability arguments. It is also possible to generate the necessary recursion relationships for any general case by applying a result from the matrix formulation of the scattering problem [20, 21]. From this method [21] it has been shown that

$$\langle F_n F^*_{n+m} \rangle = \langle F_n F^*_{n-m} \rangle^* = \mathrm{Tr}(\mathbf{FEP}^m)$$

where the quantities in parentheses are matrices and $\mathbf{P}^m$ implies matrix multiplication, m times. For a crystal that is made up of R types of layers, the matrices are of order $R \times R$. The elements of each are defined as follows:

$$\mathbf{F} = \begin{bmatrix} F_1 F_1^* & F_1 F_2^* & \cdots & F_1 F_R^* \\ F_2 F_1^* & & & \vdots \\ \vdots & & & \\ F_R F_1^* & F_R F_2^* & \cdots & F_R F_R^* \end{bmatrix} \qquad \mathbf{E} = \begin{bmatrix} e_1 & 0 & 0 & \cdots & 0 \\ 0 & e_2 & 0 & \cdots & 0 \\ \vdots & & & & \vdots \\ 0 & 0 & 0 & \cdots & e_R \end{bmatrix}$$

$$\mathbf{P} = \begin{bmatrix} p_{11} & p_{12} & \cdots & p_{1R} \\ p_{21} & & & \vdots \\ \vdots & & & \\ p_{R1} & p_{R2} & \cdots & p_{RR} \end{bmatrix}$$

where e_i is the existence probability, or relative proportion of the ith type of layer. The quantity p_{ij} is the probability that if the nth layer is of type i, the $(n + 1)$th layer will be of type j. In this particular form we can treat only nearest-neighbor correlations. The quantities F_i are the same as we have used throughout. The notation Tr indicates the sum of the diagonal elements.

As a concrete example of how this method works, consider case 3 of planar disorder discussed above. We then have, $F_1 = F$, $F_2 = F \exp[-\pi i(h + k)]$, and $e_1 = e_2 = \frac{1}{2}$ for two types of equally probable layers. The quantity $p_{11} = p_{22}$ is the probability that the $n + 1$ layer will be the same as the nth, and is therefore in our previous description, p_1. The matrix $\mathbf{P}$ is given by,

$$\mathbf{P} = \begin{bmatrix} p & (1 - p) \\ (1 - p) & p \end{bmatrix}$$

The probability that the mth layer is the same as the $(n + m)$th was p_m in our previous notation; this will be the diagonal element of the matrix $\mathbf{P}^m$. The probability that the nth and the $(n + m)$th will be different is given by $(1 - p_m)$, and is therefore the off-diagonal element of $\mathbf{P}^m$. If we set $p = (1/2)(1 + q)$, where $1 \leqslant q \leqslant 1$, as was done previously, the matrix becomes,

$$\mathbf{P} = \frac{1}{2} \begin{bmatrix} (1 + q) & (1 - q) \\ (1 - q) & (1 + q) \end{bmatrix}$$

and

$$\mathbf{P}^2 = \mathbf{P} \cdot \mathbf{P} = \frac{1}{4} \begin{bmatrix} [(1 + q)^2 + (1 - q)^2] & [2(1 + q)(1 - q)] \\ [2(1 + q)(1 - q)] & [(1 + q)^2 + (1 - q)^2] \end{bmatrix}$$

$$\mathbf{P}^2 = \frac{1}{2} \begin{bmatrix} 1 + q^2 & 1 - q^2 \\ 1 - q^2 & 1 + q^2 \end{bmatrix}$$

After several such multiplications, it becomes clear that

$$\mathbf{P}^m = \frac{1}{2} \begin{bmatrix} 1 + q^m & 1 - q^m \\ 1 - q^m & 1 + q^m \end{bmatrix}$$

Then,

$$\mathrm{Tr}(\mathbf{FEP}^m) = \langle F_n F_{n+m}^* \rangle$$

$$= (FF^*/2)(1 + q^m) + (FF^*/2) \cos[\pi(h + k)](1 - q^m)$$

Since in this case, $\mathrm{Tr}(FEP^m)^* = \mathrm{Tr}(FEP^m)$, we can replace q^m by $q^{|m|}$ to obtain $\langle F_n F_{n-m}^* \rangle$.

Our final expression for the coefficients of the diffuse scattering is

$$\langle \varphi_n \varphi_{n+m}^* \rangle = \langle F_n F_{n+m}^* \rangle - \langle F \rangle \langle F^* \rangle$$

$$= (FF^*/2)\{1 - \cos[\pi(h + k)]\}q^{|m|}$$

as we found in the discussion preceding (49). The matrix method therefore allows the generation of recursion relationships without relying on problem-dependent probability arguments. However, it is possible in difficult cases that the recursion relations do not lead to a simple closed form expression for a general element of $\mathbf{P}^m$. In such cases it is still possible to evaluate and sum the $\mathbf{P}^m$ numerically on a digital computer.

Stacking faults, such as discussed in cases 2 and 3, are quite common in a wide range of materials. The phenomena was first studied for close-packed structures of metals and alloys, and this led to a fairly extensive development [20–29] of the theory. In these structures one must not only consider nearest-neighbor interactions, but also next-nearest-neighbor interactions. An introduction to calculations for these cases can be found elsewhere [30, 31]. Next-nearest-neighbor and higher order correlations can also be treated by the matrix method [27, 28].

Reports of planar disorder in molecular crystals are numerous [32–35]. For these compounds, the disorder is similar but somewhat more complicated than in case 3 above. Often there are four *kinds* of layers, which we will refer to as $A_1, A_2, B_1,$ and B_2. These are all identical except for a translation, as in case 3, or are related by some nonspace group symmetry element. A discussion of the possible symmetry relationships [36] is of interest in this context. The average structure consists of a double layer AB arrangement where there is some definite proportion A_1/A_2 and B_1/B_2 for a particular crystal. The actual structure can display a preference for an $A_1B_1A_1B_1 \ldots$ arrangement until the first fault occurs, and then may go on $A_2B_2A_2B_2 \ldots$ until the next fault occurs. Another possibility is a preference for $A_1B_1A_2B_2A_1B_1 \ldots$ and there are others as well. The diffuse scattering for two disordered polytypes of o-chlorobenzamide has been studied [37]. A detailed treatment of setting up the equations for such a case can be found in [38].

3.8   Linear Disorder

In linearly disordered materials there are parallel linear arrays that are somehow arranged irregularly. In crystalline materials this irregularity can often be traced to the fact that the forces responsible for the ordering of a single linear array are stronger than the forces between linear arrays.

Let the vector of translational symmetry for the linear arrays be in the $\mathbf{c}$ direction of the average structure and assume that there are no correlations between the displacements of different arrays. This implies that the coefficients $\langle \varphi_n \varphi_{n+m}^* \rangle$ are only nonzero if $m_a = m_b = 0$. Equation (26) then takes the form,

$$I_2(\mathbf{s}) = N_a N_b \sum_{mc} (N_c - |m_c|) \langle \varphi_{nc} \varphi_{nc+mc}^* \rangle \exp(2\pi i \mathbf{s} \cdot \mathbf{c} m_c)$$

The scattering power per linear array is given by,

$$I_2(\mathbf{s}) = \sum_m (N - |m|) \langle \varphi_n \varphi_{n+m}^* \rangle \exp(2\pi i \mathbf{s} \cdot \mathbf{c} m) \qquad (50)$$

where we have dropped the subscript c. Initially, we treat the case for which the vector of translation symmetry for the linear arrays is the same length as that for the average structure. If each array, as a unit, can undergo a small displacement, ($\langle \Delta \mathbf{r}_n \rangle = 0$), we have from (29), $\varphi_n = F \exp(2\pi i \mathbf{s} \cdot \Delta \mathbf{r}_n) - \langle F \rangle$. Since the linear array has the same period as the average structure, the same expression holds for φ_{n+m}. The coefficients of the diffuse scattering are given by,

$$\langle \varphi_n \varphi_{n+m}^* \rangle = FF^* - \langle F \rangle \langle F^* \rangle$$

where $\langle F \rangle = F \langle \exp(2\pi i \mathbf{s} \cdot \Delta \mathbf{r}_n) \rangle$. The quantity $\langle F \rangle$ takes into account the average displacement of all linear arrays. With this (50) becomes,

$$I_2(\mathbf{s}) = [FF^* - \langle F \rangle \langle F^* \rangle] \sum_m (N - |m|) \exp(2\pi i \mathbf{s} \cdot \mathbf{c}m)$$

where the sum should be recognized as the interference function for a finite linear lattice. With (17) the final expression becomes

$$I_2(\mathbf{s}) = [FF^* - \langle F \rangle \langle F^* \rangle] \frac{\sin^2(N\pi \mathbf{s} \cdot \mathbf{c})}{\sin^2(\pi \mathbf{s} \cdot \mathbf{c})}$$

The diffuse scattering is therefore restricted to sheets in reciprocal space that are normal to the repeat direction of the array (Figures 7.10 and 7.11). The sheets are spaced by an amount $1/|c^*|$ and have a thickness on the order of $1/N$. The intensity within a given sheet is constant if there is a single atom per repeat distance. If the arrays are comprised of molecules, the intensity will vary from sheet to sheet and be modulated within a given sheet according to the variation of FF^*. If, for example, the arrays are formed by the stacking of benzene molecules with their planes normal to $\mathbf{c}$, the modulation of each sheet would be as in Figure 7.20. This type of linear disorder is observed in dicarboxylic acids of the general formula $(CO_2H)-(CH_2)_n-(CO_2H)$ and has been studied systematically [39]. In these structures the acid molecules are arranged to form infinite chains. The molecules within a given chain are joined to adjacent members by hydrogen bonds, while only van der Waals interactions operate between chains. Small independent displacements, thermal in origin, of entire chains produce the diffuse scattering.

We now suppose the state of order within the chain is best described in terms of dimers. The true vector of translation symmetry for the linear array is $2\mathbf{c}$. A portion of the ac plane of a crystal containing a linear chain of diatomic molecules is shown in Figure 7.26a. We assume that any other lattice plane containing $\mathbf{c}$ looks much the same; there are no correlations normal to the chain direction.

The average lattice (Figure 7.26b) displays an average repeat of c, and the electron density map resulting from the analysis of the Bragg reflections would reveal equally spaced atoms that are slightly elongated along the $\mathbf{c}$ direction.

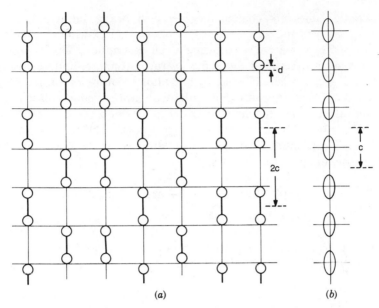

Figure 7.26 Crystal with linear disorder: (a) section of actual structure and (b) average structure.

This is true as long as the displacements d, which result from dimer formation, are smaller than the experimental resolution. It is possible, but inconvenient, to use (50) directly to calculate the diffuse scattering. This is because the indices m in (50) refer to a repeat of c, while the true repeat of the linear chains is $2c$. It would therefore be necessary to find an expression for the coefficients such that, $\langle \varphi_n \varphi_{n+m}^* \rangle = \langle \varphi_n \varphi_{n+m+2}^* \rangle$. An easier method can be used by considering the derivation of (50), which started with (13). We can define the average cell as containing two lattice points along the **c** direction. Normally there is no advantage in doing this since the single cell contains all the unique information available from the average structure. Taking $x = y = 0$ for the average cell, the functional dependence of the dot products in the structure factor expressions on s reduce to a dependence on s_3. Anticipating the final result, we will use only the Miller index l. From Figure 7.26a, the two possible values for the displacements are $\Delta \mathbf{r}_n = \pm d\mathbf{c}$. The average structure factor for the single cell is

$$\langle F \rangle = f \cos(2\pi l d)$$

where d is in fractions of an average repeat. In the double cell we have $\mathbf{c}' = 2\mathbf{c}$, and $|c'^*| = \frac{1}{2}|c^*|$. The fractionally occupied positions are $z = \frac{1}{4} \pm d'c'$, and $z = -\frac{1}{4} \pm d'c'$, where $d' = d/2$. This description gives average positions separated by $c'/2 = c$, as required. The average structure factor is given by

$$\langle F \rangle = (f/2)\{\cos[2\pi l'(\tfrac{1}{4} + d')] + \cos[2\pi l'(\tfrac{1}{4} - d')]\}$$

$$= f \cos(2\pi l' d') \cos(\pi l'/2)$$

This description does not change the predicted Bragg diffraction, since for $l' = 2n + 1$, $\langle F \rangle = 0$, and for $l' = 2n$, $\langle F \rangle = 2f \cos(2\pi l'd')$, which represents the same contribution per atom as the single cell description. This is simply an alternative way of describing the same average structure, and can easily be extended to any case where $c' = nc$, by including n equivalent lattice points in the average cell. We can now use (50), with all quantities being replaced by their primed equivalents. Since the index m' refers to the double cell, we have for any chain, $F_{n'}F^*_{n'+m'} = F_{n'}F^*_{n'}$; therefore, $\varphi_{n'}\varphi^*_{n'+m'} = \varphi_{n'}\varphi^*_{n'}$. This is the advantage of the double cell description. We can choose any cell as the reference cell and define the structure factor as follows:

$$F = f\{\exp[-2\pi i(-\tfrac{1}{4} + d')] + \exp[-2\pi i(\tfrac{1}{4} - d')]\}$$
$$= 2f \cos[2\pi l'(\tfrac{1}{4} - d')]$$

From Figure 7.26a, there are only two types of chains, those where the structure factor is as above, or those for which there is a displacement of $c'/2$. Therefore, we have

$$\langle F_{n'}F^*_{n'} \rangle = FF^*$$
$$\langle F \rangle = (\tfrac{1}{2}F[1 + \exp(\pi i l')]$$
$$\langle F \rangle \langle F^{**} \rangle = FF^* \cos^2(\pi l'/2)$$

Thus, the final form of the coefficients is

$$\langle \varphi_{n'}\varphi^*_{n'} \rangle = FF^*[1 - \cos^2(\pi l'/2)]$$

Using this in (50) results in

$$I_2(\mathbf{s}) = FF^*[1 - \cos^2(\pi l'/2)] \sum_{n'} (N' - |m'|) \exp(2\pi i \mathbf{s} \cdot c'm')$$

$$= FF^*[1 - \cos^2(\pi l'/2)] \frac{\sin^2(N_c \pi \mathbf{s} \cdot \mathbf{c}')}{\sin^2(\pi \mathbf{s} \cdot \mathbf{c}')} \tag{51}$$

As expected, we again obtain the interference function for a finite linear lattice, but with a reciprocal repeat $\tfrac{1}{2}$ that of the average structure. If the chains are long, we can neglect the size effect, and the intensity of a sheet l' is given by

$$I_2(l') = N^2 FF^*[1 - \cos^2(\pi l'/2)] \tag{52}$$

where N is the number of cells in the chain. The intensity of a given sheet is constant, but the intensity changes from sheet to sheet, according to the variation of FF^*. If we considered the scattering from a linear array of arbitrarily shaped dimerized molecules, the expression for the structure factor of the double cell would contain $\mathbf{s} = s_1\mathbf{a} + s_2\mathbf{b} + l'\mathbf{c}'$. The intensity within a sheet l' would vary as a function of s_1, s_2, but the thickness and relative intensity of diffuse sheets would still be given by (51). Equation (52) is also interesting in that for $l' = 2n$, the intensity becomes zero because of the cosine term. Hence, the diffuse scattering only shows up between the reciprocal lattice planes l of the average cell. This is a frequent occurrence for cases like the present one where the average displacement of the atoms of each chain individually is zero. For arrays with $c' = 3c$, the sheets with $l' = 3n$ would be absent; for $c' = 4c$, $l' = 4n$ would be absent, and so on. This is not, however, a feature of linear disorder in general. For example, consider a crystal as in Figure 7.26, but with each chain, as a unit, shifted by some amount d_n, which is different for various chains. The average displacement from the nodes of the average lattice is still zero if for every chain shifted by d_n, there is another chain somewhere in the crystal shifted by $-d_n$. However, it is no longer true that the average shift for atoms within a given chain is zero. If the additional shifts can assume only two values $\pm d_2\mathbf{c}'$, the intensity of the diffuse scattering is given by

$$I_2(l') = N^2 FF^*[1 - \cos^2(\pi l'/2)\cos^2(2\pi l'd_2)]$$

This expression no longer vanishes for $l' = 2n$ and now the distribution of diffuse intensity is qualitatively similar to that expected from a single chain. An experimental example with $c' = 3c$, and with an additional shift $\pm d_2$ is reported for $Ni(bqd)_2I_1$ [40] (bqd = 1,2-benzoquinonedioximato). In the limit where the displacement of an entire chain can assume any value $\langle F \rangle = 0$ (for $l' \neq 0$), and the second term in the above vanishes. The intensity would then be precisely that of an isolated chain except for the $l' = 0$ sheet.

There are apparently no experimental examples of this type of disorder for which the linear arrays are the sole structural unit of the crystal. However, there are many examples where the arrays reside in channels provided by an ordered arrangement of a host lattice. In such cases, the chains do not contribute to the Bragg scattering (again except in the $l = 0$ zone). The repeat spacing of the chain need not be an integral multiple of host lattice spacing. Two lattices for which the repeat spacings cannot be related by $cn' = nc'$, where n and n' are small integers, are said to be incommensurate. Experimental examples include hydrocarbon clathrates of urea [41] and iodine chains in a variety of matrices [42–46]. The experimental pattern for the suberic acid clathrate of urea is shown in Figure 7.27.

Thus far we have only treated perfectly ordered linear arrays. However, it is possible to describe systems that, on the average, are no more disordered than the diatomic chains described above, but the linear arrays have no true translational symmetry. For example, consider a chain comprising a random mixture of atoms and diatomic molecules (Figure 7.28). If we build up a crystal of such chains as in Figure 7.26a, the average structure is similar to that in Figure 7.26b, and it consists of a chain of equally spaced atoms, somewhat elongated in the $\mathbf{c}$ direction. It is possible to treat this and all similar cases by

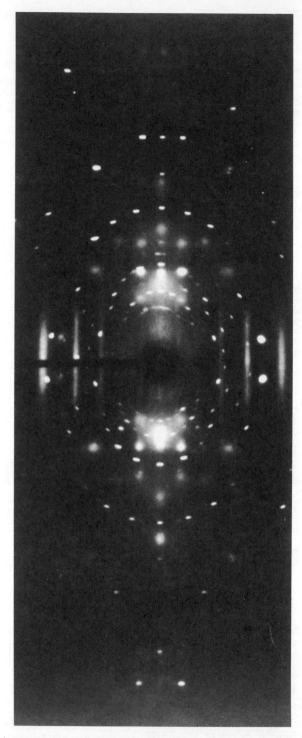

Figure 7.27 Stationary crystal, cylindrical film photograph of linearly disordered urea–suberic acid clathrate. Streaks (as in Figure 7.11b) resulting from linear disorder are clearly evident. Sharp spots are Bragg diffraction from noncharacteristic components of the radiation. Diffuse spots are from an envelope of thermal wave satellites. Reproduced, with permission, from J. L. Amorós and M. Amorós, *Molecular Crystals*, Wiley, New York, 1968, pp. 300–397 and references cited therein. Copyright © 1968 by John Wiley & Sons, Inc.

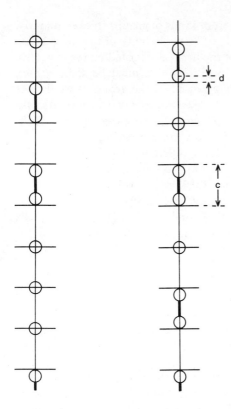

Figure 7.28 Linear chains containing a random sequence of monoatomic and diatomic units.

using (50). However, since there is no perfect periodicity, developing expressions and recursion relationships among the coefficients of the diffuse scattering can become tedious. As mentioned earlier in connection with planar disorder, an alternative formulation of the scattering problem that makes use of matrix algebra is possible. This formalism is especially useful for cases where the length of the groups in a linear lattice (or the thickness of the layers in planar disorder) are not equal. A special result of the theory for linear disorder [47] is

$$I_2(\mathbf{s}) = \langle F \rangle \langle F^* \rangle + \frac{\langle F^* \rangle \langle FQ \rangle}{1 - \langle Q \rangle} + \frac{\langle F \rangle \langle F^* Q^* \rangle}{1 - \langle Q^* \rangle} \tag{53}$$

where,

$$\langle FF^* \rangle = \sum_r e_r F_r F_r^*$$

$$\langle F \rangle = \sum_r e_r F_r$$

$$\langle Q \rangle = \sum_r e_r \exp(-iq_r)$$

$$\langle FQ \rangle = \sum_r e_r F_r \exp(-iq_r)$$

The array is considered to comprise R different kinds of groups stacked linearly. The structure factor of a given group is F_r. The probability of finding a given kind of group at some random position of the array is denoted e_r, the existence probability of the group. Since at every position there must be some group, $\sum_r e_r = 1$. The quantity Q_r is the phase shift caused by the group. If we denote the *cell vector* of the rth group c_r, the phase angle associated with the phase shift Q_r is $q_r = 2\pi s \cdot c_r$. Equation (53) holds only for large N and if the sequence of the R types of cells in the array is random. If there are correlations between the groups, the sequence will not be random, and a different formula applies [47]. Although the diffuse scattering predicted by (53) is often localized in sheets normal to the c direction, they can be much thicker than the value $1/N$ that is characteristic of ordered chains and unequally spaced. If for some particular problem and value of s, $s \cdot c$ is integral for all R kinds of cells, $Q_r = 1$ for all cells and (53) tends to infinity. However, the limit of (53) is finite and equal to

$$I_2(s) = (1/\langle c_r \rangle^2)[\langle FF^* \rangle \langle c_r \rangle^2 + \langle F \rangle \langle F^* \rangle \langle c_r^2 \rangle$$

$$- \langle c_r \rangle (\langle F \rangle \langle F^* c_r \rangle + \langle F^* \rangle \langle F c_r \rangle)] \tag{54}$$

Both (53) and (54) give the scattering power per cell; for a linear array of N cells, the intensity per array is N times this value.

We now wish to apply this formalism to a random admixture of atoms and diatomic molecules. Let the existence probability of the monoatomic cells be e_1 and that of the diatomic cells be $e_2 = 1 - e_1$. From Figure 7.28, $c_1 = c$, $c_2 = 2c$, $F_1 = F$, and $F_2 = f\{\exp(-2\pi i s \cdot cd) + \exp[-2\pi i s \cdot c(1 - d)]\}$. Using these quantities in (53), we can plot the diffuse scattering for two values of e_1 in Figure 7.29. In these calculations we have taken d to be 5% of the average atomic separation, $|c|$. We have also used a fictitious scattering factor, $f = 1$, and neglected thermal motion. In the actual scattering pattern the decrease of these factors with increasing angle can eventually cause a decline in the scattering intensity. It can be seen that the cross section of the diffuse sheets is strongly dependent on the value of e_1. The peaks are broad for $e_1 = \frac{1}{2}$, for which $\frac{1}{2}$ of the cells in the array contain diatomic molecules; the other $\frac{1}{2}$ are single atoms. For larger values of e_1, the peaks will broaden a bit more, but the diffuse scattering becomes very weak. The peaks in Figure 7.29a are approximately six times the height of those in Figure 7.29b. As e_1 approaches unity, long segments of the chains consist of an ordered atomic array. For small values of e_1, large segments of the chain consist of an ordered array of diatomic molecules and the scattering approaches that for infinite chains of diatomic molecules. The peaks are located approximately where one would expect for an ordered diatomic array, but not exactly. The extent to which they are shifted also depends on the existence probabilities.

An experimental example of this type of linear disorder is found in the one-dimensional ionic conductor, Hollandite [48], in which there are linear chains of disordered potassium atoms contained in an ordered MgO framework. The

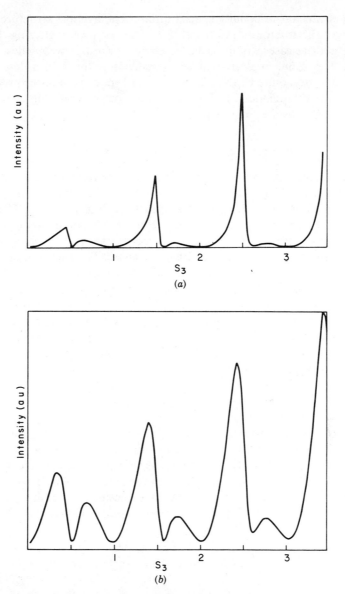

Figure 7.29 Diffuse scattering from chains in Figure 7.28, with the scattering factor taken to be $f = 1$: (a) 10% monoatomic cells and (b) 50% monoatomic cells.

order within the chains is deduced from the diffuse scattering and can be described in terms of a random admixture of four different linear *cells* of lengths $3c$, $4c$, $5c$, and $6c$. Application of (53) for this material has been discussed in detail [47].

In several of the preceding cases of linear disorder, the average structure, which represented the only information available from the Bragg intensities, consisted of a chain of equally spaced atoms. The average structure gives no indication of the true nature of the chains. The result of neglecting the diffuse scattering in these cases would be an incomplete or totally wrong description of the solid. The problem is particularly difficult for linear disorder, since the diffuse scattering is quite weak and spread over broad areas of reciprocal space. There is very little chance of observing this type of scattering unless single-crystal photographs are prepared.

3.9   Lattice Distortions of the Second Kind

In Section 2.3, we described a simple monoatomic lattice with lattice distortions of the second kind. The essential feature of the atomic arrangement is that the width of the nth neighbor distribution increases with increasing n. The problem of linear arrays with disorder of the second kind can be reformulated by considering each atom to reside in a *cell* of repeat vector c_r. Initially, we take all c_r to be parallel, but of somewhat different length. We can then speak of the distribution of cell lengths in the array, and this is equivalent to the distribution of interatomic distances, which also allows the treatment of linear arrays of identical molecules. The distribution of intermolecular separations is described by the distribution of cell lengths. We can always define an average cell length $\langle c \rangle$ by dividing the total length of the array by the number of cells comprising it. In terms of the average cell vector $\langle \mathbf{c} \rangle$ we can define the repeat vector for any cell in the array as follows:

$$\mathbf{c}_r = \langle \mathbf{c} \rangle + \Delta c_r \hat{\mathbf{c}} \tag{55}$$

where $\hat{\mathbf{c}}$ is a dimensionless unit vector in the $\langle \mathbf{c} \rangle$ direction. If we assume that the distribution of cell lengths is symmetric about the mean, we require that $\langle \Delta c \rangle = 0$. With (55), the distribution $p(\Delta c)$ is sufficient to define the distribution of cell lengths $p(c_r)$. If the sequence of cells in the array is random, we can use (53) to calculate the diffuse scattering. Since we are treating a linear array of identical atoms or molecules, $F_r = F$, and upon rearrangement (53) reduces to

$$I_2(\mathbf{s}) = FF^* \frac{1 - \langle Q \rangle \langle Q^* \rangle}{1 + \langle Q \rangle \langle Q^* \rangle - (\langle Q \rangle + \langle Q^* \rangle)} \tag{56}$$

As with (53), the above is valid only for very long arrays. The sequence of cells is assumed to be random but in the proportions dictated by the existence probabilities. If the linear arrays are part of a three-dimensional solid, it is also assumed that there are no correlations between arrays. In the original derivation of (55), it was assumed that the N cells of the array were of R distinct types, with existence probabilities of $e_1, e_2, \ldots, e_R$. However, if R is very large, we can approximate the discrete e_r by the normalized continuous distribution $p(\Delta c)$.

For example, suppose the quantities Δc are distributed exponentially about zero, as in Figure 7.30. Then, the probability of finding a cell in the array whose

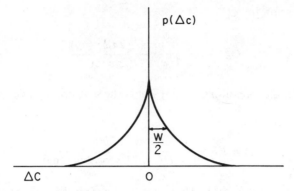

Figure 7.30 Schematic representation of exponential distribution of cell length deviations.

length deviates from the mean length by Δc decreases as Δc becomes larger than or smaller than zero. How quickly this probability diminishes is determined by the width of the distribution w. The functional form of the distribution is given by

$$p(\Delta c) = \tfrac{1}{2}w \, \exp(-|\Delta c|/w)$$

and the function is normalized so that

$$\int_{-\infty}^{\infty} p(\Delta c)\, d(\Delta c) = 1$$

With (55) this defines the nearest-neighbor distribution of interatomic or intermolecular distances. The overall distribution of intermolecular distances would be similar in appearance to Figure 7.13b, but with peak shapes characteristic of the exponential distribution.

The probability that a cell with phase shift $Q = \exp[-2\pi i s \cdot (\langle c \rangle + \Delta c \hat{c})]$ is given by $p(\Delta c)$. The mean phase shift for all cells is given by

$$\langle Q \rangle = \tfrac{1}{2}w \, \exp(-2\pi i s \cdot \langle c \rangle) \int_{-\infty}^{\infty} \exp\left(-\frac{|\Delta c|}{w}\right) \exp(-2\pi i s \cdot \hat{c}\, \Delta c) d(\Delta c)$$

Using the Euler relationship and noting that the integral over the resulting sine term vanishes, we find

$$\langle Q \rangle = \frac{1}{w} \, \exp(-2\pi i s \cdot \langle c \rangle) \int_{0}^{\infty} \exp\left(-\frac{\Delta c}{w}\right) \cos(2\pi s \cdot \hat{c} \Delta c) d(\Delta c)$$

Integrals of the above form can be found in standard compilations, $\int \exp(ax)\cos(bx)dx = a/(a^2 + b^2)$, $a > 0$. Evaluation of the integral yields the following expression for $\langle Q \rangle$

$$\langle Q \rangle = \frac{(1/w)^2 \exp(-2\pi i s \cdot \langle c \rangle)}{(1/w)^2 + (2\pi s \cdot \hat{c})^2}$$

Substitution of this expression into (56) yields the scattering power per cell of the linear array.

$$I_2(s) = FF^* \frac{1 - U^2}{1 + U^2 - 2U \cos(2\pi s \cdot \langle c \rangle)} \tag{57}$$

where $U = 1/[1 + (2\pi w s \cdot \hat{c})^2]$. This expression for the diffuse scattering is similar to that encountered in our discussion of planar disorder (48). The quantity in brackets is the interference function of a linear lattice with disorder of the second kind. The quantities U are not independent of s, and the scattering is not a truly periodic function of the inverse average cell length. For any constant values of s_1, s_2, the diffuse scattering is plotted in Figure 7.31, for $w = 0.1\langle c \rangle$, as a function of s_3. If we take $|c^*| = 1/\langle c \rangle$, there are maxima in the scattering at integral values of s_3, which correspond to the points at which the denominator of (57) reaches a minimum. As s_3 increases, intensity begins to build up between the maxima, and the sheets of diffuse intensity become less distinct. If there is only one atom per cell, the intensity becomes constant at very large values of s_3, but nonzero. For linear arrays of molecules the intensity at large intervals would still be modulated by the molecular structure factor. The

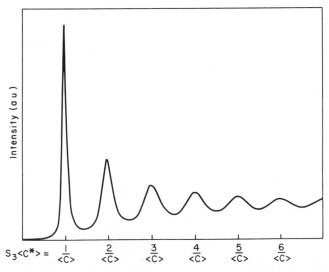

Figure 7.31   Diffuse scattering for linear disorder of the second kind (structure factor taken to be $F = 1$), $w = 0.1\langle c \rangle$.

intensity at the maxima decreases as the square of the order so that the second maximum is approximately $\frac{1}{4}$ the height of the first. The width of the first maximum is proportional to w and therefore strongly dependent on the width of the distribution of cell lengths in the linear array. The widths of the maxima increase approximately as the square of the order for those maxima that are well resolved. It should be noted that there is a certain similarity between the scattering produced by linear arrays with disorder of the second kind and the overall distribution of interatomic distances (Figure 7.13b).

If we had initially assumed that the distribution of cell lengths is Gaussian, U in (57) would be of the form $U = \exp[-(w\mathbf{s} \cdot \hat{\mathbf{c}})^2]$, where w is proportional to the width of the Gaussian distribution $p(\Delta c)$. For small widths, $\exp[-(w\mathbf{s} \cdot \hat{\mathbf{c}})^2] \sim 1/[1 + (w\mathbf{s} \cdot \hat{\mathbf{c}})^2]$, which is the same form of U found earlier for an exponential distribution of Δc. Thus, for small distribution widths, the scattering is not very sensitive to the details of the distribution.

A detailed analysis of the diffuse scattering from linear chains with disorder of the second kind has been reported [49] for N,N'-diethyl-N,N'-dihydrophenazinium iodide and N,N'-dibenzyl-N,N'-dihydrophenazinium iodide. In these compounds, the width of the distribution is only a few percent of the average cell length, and therefore the sheets broaden less quickly than those in Figure 7.30. Similar effects are observed in $Hg_{1-x}AsF_6$ [50].

The ideas discussed above can be extended to study disorder of the second kind in three dimensions. The molecules or atoms making up the solid are considered to reside in a cell with lattice vectors, $\mathbf{a}_1$, $\mathbf{a}_2$, and $\mathbf{a}_3$, where each lattice vector is considered to have a distribution of lengths and directions. This results in an overall distribution of neighbor distances in three dimensions analogous to the linear case. A three-dimensional array with disorder of the second kind is often referred to as a paracrystal; the theory of X-ray scattering from paracrystalline arrays has been extensively developed [51]. A general introduction to applications of the theory [52] includes a number of examples for macromolecules. A treatment of diffraction from polymers [53] also contains some discussion of paracrystalline theory.

4   EXPERIMENTAL RESULTS

In this section we discuss experimental results for a number of crystalline compounds of interest to chemists. For each one, the chemical nature or physical properties of the material can only be understood if the diffuse scattering is analyzed.

4.1   Nickel Phthalocyanine Iodide

The compound $Ni(Pc)I_{1.0}$ (Pc = phthalocyanine) crystallizes in the tetragonal space group $P4/mcc$ with two formula units per cell [54]. The metal complexes are stacked linearly along the c direction (Figure 7.32). These arrays are packed

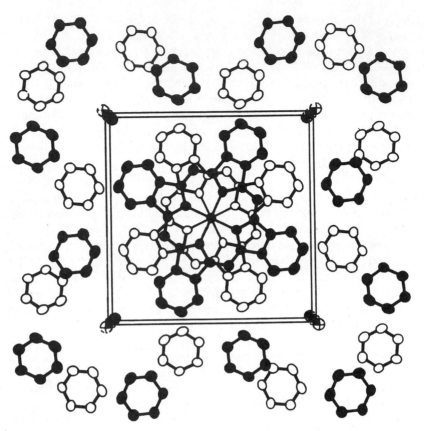

Figure 7.32 Perspective view of the Ni(Pc)I$_{1.0}$ structure showing the average position of iodine atoms in the channels.

together such that the benzo groups of the phthalocyanine form channels that are occupied by the iodine. The average iodine sublattice consists of linear chains of equally spaced iodine atoms with an average separation of approximately 3.2 Å. From the results of the average structure, the compound could be reasonably formulated as [Ni(III)(Pc)]I. However, the apparent thermal motion of the iodine ion along the chain direction is twice as large as that normal to the chain. Moreover, the average I—I separation of 3.2 Å, is shorter than the value of approximately 4 Å that is normally observed in compounds containing the iodide ion.

Oscillation photographs of single crystals clearly exhibit diffuse scattering characteristic of linear disorder. The diffuse sheets indicate a true linear repeat distance of three times the average I—I separation ($c' = 9.6$ Å). Also, based on the true repeat, the diffuse sheets with $l' = 3n$ are absent. These observations are therefore analogous to the linearly disordered diatomics discussed in Section 3.8. The crystal was considered to comprise uncorrelated chains, each chain being perfectly ordered and containing three average iodine sites per cell. Each chain can be shifted by 0, $\pm c'/3$, relative to the arbitrary reference chain, with equal probability. Experimental intensity data were measured using CuK_α radiation and a doubly bent graphite monochromator. The data were fitted to the following equation:

$$I_2(l') = f_I \exp[-B(\sin\theta/\lambda)^2]\tfrac{1}{9}\{1 + 2\cos[2\pi l'(\tfrac{1}{3} - d)]\}^2$$
$$\times \{9 - [1 + 2\cos[2\pi l'/3)]^2\}$$

which was obtained by the methods illustrated in Section 3; d is in fractions of the actual (triple) cell. The quantity f_I is the atomic scattering factor for iodine, and B was taken to be the equivalent of that found in the average structure for vibrations *normal* to the chains. Note that the value of the scattered intensity is zero if $d = 0$ (i.e., no disorder) or $l' = 3n$ (as observed). The equation was considered to contain only two adjustable parameters, a scale factor and the displacement from the average site d. The best fit was determined by the method of least squares. The experimental values and those predicted by the above equation (the square root was taken in both cases) are given in the following table. The agreement index from the one-dimensional structure analysis is $R = 0.08$, which is satisfactory for a study of this nature. The derived value of the displacement is $d = 0.0245(2)$. This corresponds to a bonded, I—I, triiodide distance of 3.00 Å and a nonbonded I $\cdots$ I separation of 3.72 Å. Both values are within the range found in ordered triiodide structures. On the basis of the information provided by the diffuse scattering analysis, the compound should be formulated as $[Ni^{+2.33}(pc)](I_3^-)_{1/3}$, which corresponds to a formally fractional or mixed-valence state for the nickel ion and an oligomerized form of iodine. The former is consistent with the physical properties of the substance and the latter is in accord with resonance Raman data [54]. Quantitative diffuse scattering results have been obtained for several other iodine chain compounds [40, 46, 49, 55].

Observed and Calculated Intensities of
Diffuse Sheets in Ni(pc)I.

l'	$[I_0(l')]^{1/2}$	$[I_c(l')]^{1/2}$
1	19.4	14.4
2	25.3	23.0
4	43.7	46.0
5	33.3	31.7
7	55.7	53.9
8	20.4	21.8
10	42.3	44.0
11	4.5	8.1

4.2   Tantalum Pentaiodide

The compound tantalum pentaiodide crystallizes in a cell of average symmetry *Ccmm* [56]. The average structure consists of close-packed layers of iodide ions separated by layers of tantalum ions. Only $\frac{2}{5}$ of the octahedral holes provided by the iodine sublattice are occupied by tantalum; in each case the occupancy factor of the tantalum ion is found to be 0.5. On the basis of the average structure, the only possible structural conclusion is that all tantalum ions are octrahedrally coordinated. Whether the tantalum moieties form chains, sheets, or an oligomer of some kind cannot be determined from the average structure. The diffraction photographs display diffuse spots characteristic of planar disorder. The diffuse scattering has been interpreted in terms of a slipped layer structure somewhat similar to that discussed in Section 3.7 (case 3). In the present case the slip vector $\mathbf{v}_n$ can take on four different values: $\mathbf{v}_n = 0, \mathbf{a}/2, \pm\mathbf{a}/4$. If we refer to the resulting layer types as A_1, A_2, B_1, and B_2, respectively, the sequence of permitted layers is constrained to be of the general type *ABAB*, where *A* may be A_1 or A_2 and similarly for *B*, depending on correlations between layers. Each layer was considered to consist of a series of discrete

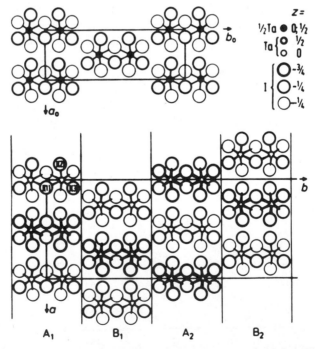

Figure 7.33 Structure of dimeric $(TaI_5)_2$ as deduced from analysis of the diffuse scattering. Reproduced, with permission, from U. Muller, *Acta Crystallogr. Sect. B*, **35**, 2502 (1979).

dimeric $(TaI_5)_2$ molecules (Figure 7.33). Each dimer is formed by edge sharing and therefore the structural formula is

$$
\begin{array}{c}
I \\
\diagup \quad \diagdown \\
Ta \qquad TaI_4 \\
\diagdown \quad \diagup \\
I
\end{array}
$$

A detailed analysis of the diffuse scattering based on discrete dimers is in good agreement with experiment, and thus TaI_5 is actually a molecular solid.

4.3   Molybdenum Tetrachloride

A similar and rather surprising example is that of β-$MoCl_4$ [57]. This compound also exhibits diffuse scattering characteristic of planar disorder. The average structure gives little information other than octahedral coordination for

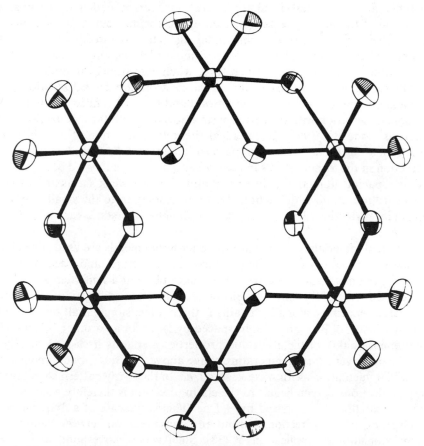

Figure 7.34   Molecular structure of β-$MoCl_4$ proposed on the basis of diffuse scattering. Reproduced, with permission, from U. Muller, *Angew. Chem. Int. Ed. Engl.*, **20**, 692 (1981).

the molybdenum ions. An analysis of the diffuse scattering suggests that the solid consists of discrete hexameric molecules and should be formulated $(MoCl_4)_6$. The hexamer is cyclic, and all molybdenum atoms are coplanar; there are no metal–metal bonds (Figure 7.34). Each molybdenum octahedra is doubly edge shared. Apparently, the diffuse scattering analysis is the only structural evidence supporting the existence of this unusual molecule.

4.4   One-Dimensional Metals

A very rich series of examples of linear *disorder* was provided during the past 10 years by the extensive study of *one-dimensional metals* [58]. Since these materials provide a pedagogical progression from the simplest case of monoatomic chains to the more complex situation of stacks of organic molecules, we describe them in some detail.

One-dimensional metals were studied theoretically in the late 1950s and early 1960s. For our purposes, the important point is that a one-dimensional metal is unstable; at 0 K a metal–insulator transition is associated with a sinusoidal distortion of the initially equally spaced atoms (or molecules) of each metallic chain. The wavelength of this sinusoidal distortion is directly related to the number of electrons (per atom or per molecule) present in the electronic conduction band. At finite temperature, when the atoms (or molecules) are equally spaced with a lattice constant c, the electrons are delocalized along each chain; at 0 K for an electronic conduction band that is $1/n$ filled, a longitudinal sinusoidal distortion with the period nc is associated with the localization of the electrons along the chains. In the ideal theoretical situation of a strictly one-dimensional metal a diffraction pattern taken at 0 K would show strong diffraction in the form of sheets in reciprocal space, spaced by $1/c$ because of the average spacing along the chain, and in addition very weak sheets at a distance of $1/nc$ from each main diffraction sheet corresponding to the small amplitude sinusoidal distortion. This is just the one-dimensional static case of (38) with $\mathbf{k} = 1/nc$.

With real materials, only quasi-one-dimensional metals are used. The chains (or stacks) are coupled regarding the average structure, and because of this coupling, the metal–insulator transition takes place at a finite temperature T_c. As a consequence of this three-dimensional coupling, the diffraction arising from the average spacing c along the chains is three-dimensional at all temperatures and gives layers of Bragg reflections spaced by $1/c$ as for an ordinary crystal. The low dimensionality of the electronic properties manifests itself structurally by the existence over a large temperature range above T_c of quasi-one-dimensional-sinusoidal fluctuations with wave vector $1/nc$. In the X-ray pattern each layer of Bragg reflections perpendicular to the chain direction is therefore accompanied by weak satellite sheets spaced by $\pm 1/nc$. This is the case of a dynamic wave propagating in the c direction, without any phase relation between neighboring chains and another application of (38). Such waves correspond in terms of phonon dispersion curves to the giant Kohn anomaly of one-dimensional metals.

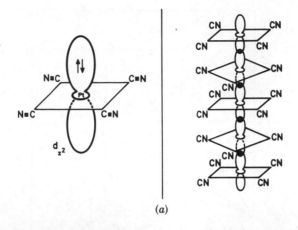

(a)

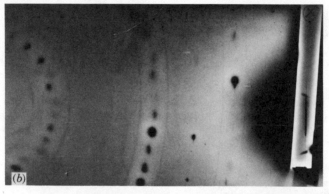

Figure 7.35 (a) Schematic representation of KPC structure showing the platinum chains and the overlap of electronic orbitals along them (after [58] and references therein). (b) X-ray pattern at 300 K from $K_2Pt(CN)_4Br_{0.30} - 2.3H_2O$ showing scattering in the form of satellite diffuse sheets caused by precursor one-dimensional fluctuations that at lower temperatures leads to the Peierls transition of one-dimensional metals. The pattern was taken with the fixed film, fixed crystal technique [61].

The first example of such structural effects was provided by $K_2Pt(CN)_4Br_{0.30} - 2.3H_2O$, where for the purposes of X-ray scattering, the metallic chains can be essentially viewed as single rows of platinum atoms (Figure 7.35a). An X-ray pattern taken at room temperature is shown in Figure 7.35b; in particular, note the uniform intensity in the satellite diffuse sheets. This uniform intensity distribution results from single rows of Pt atoms. Figure 7.36 reveals the corresponding dynamical $2k_F$ anomaly of the acoustic phonons.

Another well-known example is provided by tetrathiafulvalene–tetra-cyanoquinodimethane (TTF–TCNQ) or its selenium analogue tetraselena-fulvalene–tetracyanaquinodimethane (TSeF–TCNQ) [58]. In this case the metallic chains are built of stacks of molecules (Figure 7.37); interference between the X-rays scattered from the different atoms (in particular the electron-rich selenium) of each molecule *modulate* the intensity of the satellite diffuse

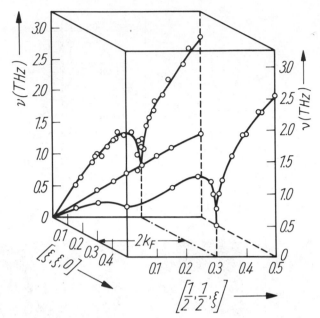

Figure 7.36 The giant Kohn anomaly of the longitudinal acoustic dispersion branch of KCP measured by inelastic neutron scattering. The minimum frequency in the reciprocal sheet of reduced wave vector $(h, k, 2k_F)$ corresponds to a maximum amplitude of the thermal motion that is at the origin of the enhanced intensity along the diffuse lines shown in the pattern of Figure 7.35b (after [58] and references cited therein).

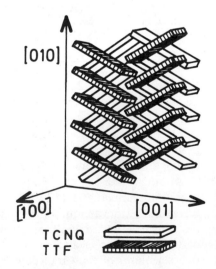

Figure 7.37 Schematic representation of the TTF–TCNQ structure and its selenium analogue showing the molecular stacks in the [010] crystallographic direction. Most other related compounds have a similar structure (after [58] and references cited therein).

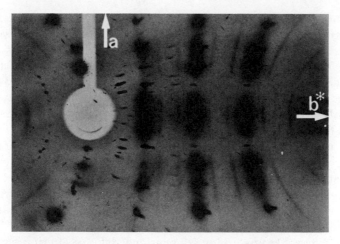

Figure 7.38 X-ray pattern at 30 K from TSeF–TCNQ (after Megtert and co-workers [60]). Compare with the pattern of Figure 7.35*b* and note the intensity modulation along each diffuse sheet, which is caused by the molecular form factor of TSeF. Despite the existence of the electron-rich selenium in this organic compound, the exposure time is 15 h with the fixed crystal–fixed film technique.

sheets shown in Figure 7.38. A full analysis of this case has allowed determination of the nature of the molecular movements associated with the sinusoidal fluctuations [59, 60].

The examples above are in the metallic regime of such systems; the one-dimensional nature of the diffuse X-ray scattering complements data on the anisotropy of the plasma edge as determined from polarized reflectance

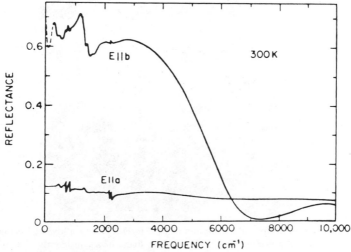

Figure 7.39 Polarized reflectance spectrum of TTF–TCNQ single crystals: (*a*) transverse (100) direction and (*b*) parallel to stacking direction (010).

measurements shown in Figure 7.39 for TTF–TCNQ [58]. In fact, it gave the most significant demonstration of the existence of an extremely well-defined, one-dimensional Fermi surface in such systems.

Below T_c in the insulating state, the distortion waves on the different chains are coupled and the intensity of the diffuse scattering is condensed in satellite reflections of an ordered, modulated structure. Figure 7.40 shows two patterns from another one-dimensional conductor tetramethyl tetraselenafulvalene–dimethyl tetracyanoquinodimethane, (TMTSF–DMTCNQ) above and below its 42 K metal–insulator transition. Above T_c it shows in particular the beginning of the evolution from the one-dimensional scattering toward well-defined satellite reflections as the distortions between parallel molecular stacks progressively couple [61]. In Figure 7.41 one can find the corresponding conductivity data from the same compound together with that of other related systems [62].

A quite different example of one-dimensional diffuse scattering is provided by partially polymerized bis(p-toluene sulfonate) 2,4-hexadiyne-1,6-diol (common-

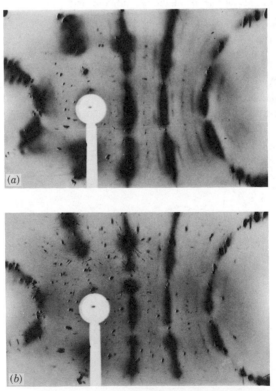

Figure 7.40   X-ray patterns from the compound TMTSF–DMTCNQ, above and below transition temperature: (*a*) Pattern at 100 K (T_c + 58 K). Diffuse lines show considerable intensity concentration in the form of broad spots indicating the onset of coupling between chains. (*b*) Pattern below T_c. Diffuse spots have condensed into well-defined satellite reflections (after [61]).

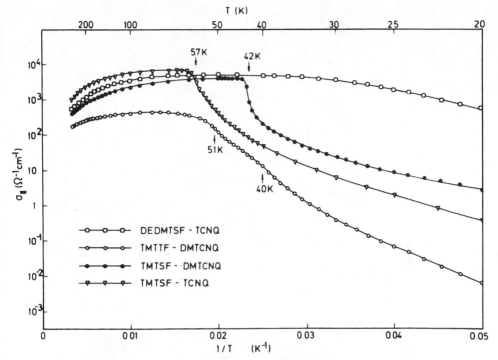

Figure 7.41 Conductivity data from a variety of related one-dimensional conductors showing in particular the sharp drop of conductivity from TMTSF–DMTCNQ at 42 K, the temperature at which this compound undergoes its Peierls transition (after [62]).

ly called PTS). This compound belongs to a series of related diacetylenes that undergo a topochemical polymerization in the crystalline state. The samples remain single crystals throughout the whole transformation, which proceeds as shown schematically in Figure 7.42. One of the questions that remained open for some time regarding these compounds was whether the polymerization proceeds by the formation of small polymer clusters or the growth of chains within the monomer matrix.

Figure 7.43 shows a pattern corresponding to a conversion of 50%. One clearly observes diffuse lines corresponding to intensity distributed in sheets perpendicular to the stacking direction [63]. The one-dimensional nature of this diffuse scattering clearly shows that the polymerization proceeds by way of the growth of isolated polymer chains within the monomer matrix. This is an example of very intense diffuse scattering that allows us to use a precession camera. Photographs prepared using such cameras have the advantage of providing an undistorted picture of reciprocal space.

Recently, a class of one-dimensional conductors that display superconductivity (usually under high pressure) have been studied [64–66]. These compounds are formulated (TMTSF)$_2$X and differ structurally from the TTF–TCNQ type of materials in that they contain only one type of conducting chain, namely, the

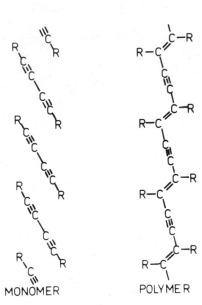

Figure 7.42 Schematic representation of the stacking of monomer units (left) and of a polymer chain (right) along the crystallographic b direction (after [63]).

MONOMER   POLYMER

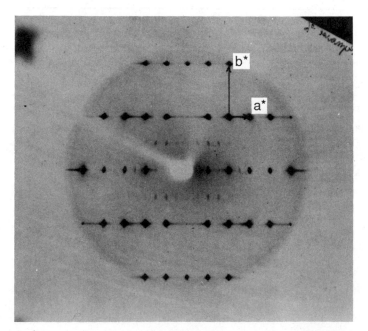

Figure 7.43 Precession X-ray pattern from a polydiacetylene single crystal (PTS) taken at a conversion state of about 50% monomer and 50% polymer. The diffuse lines perpendicular to b^* correspond to polymer chains within the monomer matrix (from [64]). Note the undistorted representation of reciprocal space and the weak spots on the layers separated by $\pm\frac{1}{2}b^*$ from the equator; these spots arise from the $\lambda/2$ wavelength from the continuous spectrum also diffracted by the monochromator and which provide an estimate of the intensity of the diffuse scattering.

592

organic TMTSF stacks (Figure 7.44); the counterions (e.g., ReO_4^-) act only as electron acceptors. For noncentrosymmetric counterions, several of these systems undergo metal–insulator transitions at atomspheric pressure, but at temperatures that are too high to be readily accounted for by the instability of the one-dimensional electron gas. As a consequence, the precursor scattering to such a phase transition should not have a one-dimensional nature in contrast to the other examples already shown above. Indeed, as illustrated in Figure 7.45a and b one observes in such cases broad and isotropic diffuse spots above T_c that progressively concentrate into well-defined superstructure reflections below T_c. This provides an example of three-dimensional diffuse scattering. An analysis of 160 such superstructure reflections has shown that the phase transition mainly corresponds to an ordering of the ReO_4^- ions but also reveals a small distortion of the organic stack [67]. Since the superstructure doubles the lattice constant in the chain direction, it also coincides with the distortion required to open a gap at the Fermi level [$a*/2 = 2k_F$ for $(TMTSF)_2X$] and therefore explains the occurrence of the metal–insulator transition. This example illustrates in a vivid

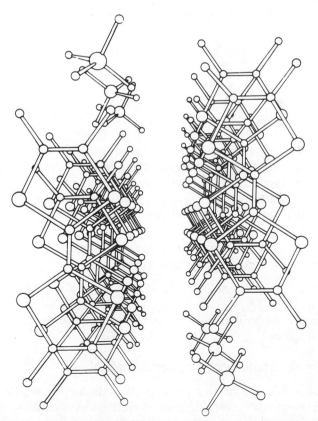

Figure 7.44   Schematic view of $(TMTSF)_2ReO_4$-type compounds (after [65]).

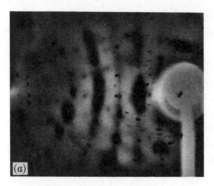 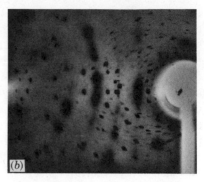

Figure 7.45   X-ray patterns from $(TMTSF)_2ReO_4$ above and below the transition temperature: (a) pattern at 177.5 K (just above T_c) showing precursor three-dimensional diffuse scattering where the superstructure reflections appear in low temperature phase and (b) pattern at 150 K (below T_c) showing well-defined superstructure reflections (after [66]).

way how combined studies of the qualitative and quantitative (three-dimensional nature of the precursor effect) aspects of the diffuse scattering can be essential in understanding the physical properties of materials.

5   EXPERIMENTAL TECHNIQUES

Experimental techniques for the investigation of X-ray diffuse scattering follow the same general principles as ordinary single-crystal diffraction experiments. Commercially available equipment is not, however, generally adequate unless it is suitably modified. For this reason, it is often more appropriate to build a dedicated instrument. Even under these circumstances a given instrument cannot be used for all types of diffuse scattering problems, and usually must be matched to each specific class of experiments, sometimes even to a particular material. Consequently, a detailed description of experimental techniques is extremely difficult. Some general ideas can be found in [68] and the following sections.

Two basic features that differentiate diffuse scattering from the better-known X-ray diffraction also determine the instrumentation:

1.   Diffuse scattering is weak or even extremely weak compared to Bragg reflections.
2.   Diffuse scattering is diffuse; that is, it usually extends over a large portion of reciprocal space.

5.1   The Small Intensity of Diffuse Scattering

As stated above, diffuse scattering gives rise to considerably weaker intensities than ordinary X-ray diffraction. To give some orders of magnitude, a scattering of 10^{-3} compared to the average Bragg reflection intensity can be considered as fairly strong, while examples such as shown in Figures 7.35*b* and 7.38 correspond to intensities of 10^{-6} or 10^{-7}. Compared to Bragg reflections and on an absolute scale this can become particularly weak for organic materials. Recall that the continuous spectrum emitted by ordinary X-ray tubes is about 10^{-3} compared to the intensity of the characteristic K_α radiation; clearly to investigate diffuse scattering one must first eliminate the Bragg reflections arising from this continuous spectrum by using a monochromator [69].

Flat monochromators bring about an additional loss in intensity of the incident K_α radiation beam. Classical single-crystal monochromators using the (101) reflection from quartz, can be cut at the appropriate angle with respect to the reflection plane, and singly bent; this, respectively, allows one to maximize the solid angle of the incident beam on the monochromator, and by focusing, to increase the intensity of the reflected beam used for the experiment. The (200) reflection of single crystals of lithium fluoride and the (002) reflection of pyrolytic graphite provide better reflectivities. These materials (especially pyrolytic graphite) are used in plates parallel to the reflection plane. If curved, this imposes symmetric source-to-monochromator, and monochromator-to-detector distances, but with the advantage that double curvatures are possible, leading to approximately point-focused beams and accompanied by a considerable gain in intensity. This type of monochromator is curved in the form of a lune [70].

In addition to the use of monochromators, several other sources of background should be eliminated. For example, scattering by air can be eliminated by using vacuum cameras or cameras filled with helium gas. Fluorescence by the constituent atoms of the sample can often, but not always, be eliminated by the choice of the incident K_α radiation or filters of appropriate elements applied on the photographic film.

It is also advantageous to use larger samples than normally used for ordinary diffraction experiments. To fully utilize the intensity gain obtained through focusing, the sample should ideally intercept the entire beam. For transmission experiments the thickness should be optimized to obtain the maximum scattered intensity. If,

$$I = x \exp(-\mu\rho x)$$

where x is the thickness, μ is the mass-absorption coefficient for the wavelength used, and ρ is the density of the sample; the optimum thickness is

$$x_0 = 1/\mu\rho$$

Since μ depends on the wavelength, this can also provide a motivation for the use of a particular K_α radiation.

Aside from the optimization of crystal thickness, at least one dimension of the studied material should also be large enough to avoid the incident beam coming in contact with glue, wax, or any other system used to hold the sample. Even in extremely small amounts diffraction or scattering by wax or glue can completely shadow the diffuse scattering from the investigated sample.

5.2   Diffuse Scattering Is Diffuse

By definition in any diffuse scattering problem, the intensity is spread over a large volume of reciprocal space (sometimes over all reciprocal space). This is another important difference from Bragg diffraction, where except for instrumental resolution effects, the intensity is concentrated on each reciprocal lattice node. The advantage of such a feature is that it allows one to relax the resolution by using looser collimation or by focusing techniques (such as produced by bent monochromators) that increase the scattered intensity.

From another point of view, it almost completely eliminates the possibility of using classical diffraction techniques such as the powder method or the single-crystal–rotating crystal method, where at a given scattering angle, intensity originating from different points of reciprocal space is collected.

For measurements of diffuse scattering with classical detectors, four-circle automatic diffractometers, which allow one to measure the intensity originating from any point of reciprocal space, can be effectively adapted with limited software work, the use of monochromators, and a loose collimation. The main advantage of diffractometers is the ease with which quantitative measurements of the scattered intensity can be made. The disadvantage is the time required for the step-scan technique; not only is the intensity originating from each point extremely weak (one sometimes needs several hours of counting for a given point of reciprocal space), but in addition one must collect the intensity on many points because of the slow variation of the scattering in reciprocal space.

Generally, for a first exploratory experiment, it is advisable to use photographic techniques. In this case, one can simultaneously collect the intensity originating from very large portions of reciprocal space. The simplest and most efficient method by far is the fixed crystal, fixed film method, sometimes improperly called the monochromatic Laue method [68], which has already been described earlier (e.g., see Figure 7.10). Patterns obtained by this method (Figures 7.35b and 7.38) give a somewhat complex and deformed cut through reciprocal space, but each point of the pattern can be assigned to a unique point in reciprocal space. In addition, since the crystal is in a fixed orientation throughout the experiment, each point of reciprocal space investigated contributes to the scattered intensity during the entire exposure (typical exposure times are from several hours to a few days). The method is therefore particularly suited for very weak scattering, as shown in Figure 7.38. The method is also well suited for variable temperature measurements since there is no motion involved. Preliminary investigation by this method allows one to determine the location of the diffuse scattering in reciprocal space rapidly. Specific regions of interest can then be scanned efficiently using a step-by-step detector technique.

Other photographic techniques such as Weissenberg cameras or precession cameras are also effective provided that monochromators, a loose collimation, and possibly a helium atmosphere are used. As is well known, they yield patterns corresponding to planes of reciprocal space, which give a more directly understandable view of the distribution of the diffuse scattering in reciprocal space (in particular the precession method). However, this simplication of the patterns is paid for by longer exposure times, each point of reciprocal space being in the reflecting position during only a small fraction of the total time of the experiment. Weissenberg cameras or precession cameras are therefore only suited for relatively intense diffuse scattering (Figure 7.43); weak effects may necessitate exposures of several weeks or more.

5.3 Future Developments

We may conclude from these general considerations on experimental techniques that the quantitative measurement of X-ray diffuse scattering can be a rather lengthy process, and is quite susceptible to interference from background effects. For this reason, X-ray diffuse scattering investigations will greatly benefit from the development of synchrotron radiation and position-sensitive detectors.

Synchrotron radiation provides high intensities and its tunable wavelength capabilities afford additional flexibility in optimizing the experiment to each specific investigation. The possibility of using the variations of the complex X-ray scattering factor in the vicinity of the absorption edges (anomalous scattering) should also open new possibilities in this area by allowing investigations to focus on a particular element of a given compound.

Position-sensitive detectors, which are already in use in one-dimensional form and which are in development in two dimensions, should soon combine the advantages of detectors (quantitative and fast) with those of photographic techniques. Diffuse scattering, because it varies relatively slowly in reciprocal space, should in fact be a first application of these detectors since it suits their limited spatial resolution.

Appendix A

Evaluation of $S_N = \sum_{m=-(N-1)}^{(N-1)} (N - |m|) \exp(iym)$.

The sum is most easily evaluated by noting that it is equivalent to the double sum (13)–(15),

$$S_N = \left[\sum_{n=0}^{N-1} \exp(-iyn) \right] \left[\sum_{n'=0}^{N-1} \exp(iyn') \right] = R_N R_N^*$$

Let $\exp(-iy) = x$; the sum R_N is then

$$R_N = 1 + x + x^2 + \cdots x^{N-1} \qquad \text{and}$$

$$xR_N = \qquad x + x^2 + \cdots x^{N-1} + x^N$$

Subtract the two results in the expression:

$$R_N = (1 - x^N)/(1 - x)$$

and multiply by the complex conjugate to get,

$$S_N = R_N R_N^* = \frac{1 - [x^N + (x^N)^*] + (x^* x)^N}{1 - (x + x^*) + x^* x}$$

Substitution for x and subsequent use of the sum angle formula gives the final result:

$$S_N = \frac{2 - 2 \cos Ny}{2 - 2 \cos y} = \frac{\sin^2(Ny/2)}{\sin^2(y/2)}$$

The sum can also be evaluated directly, in its original form (Appendix B, [2]), but the calculation is more involved.

Appendix B

Evaluation of $Z = \displaystyle\sum_{m=-\infty}^{\infty} p^{|m|} \exp(iym)$.

The sum over finite limits can be written as follows:

$$Z_N = Q_N + Q_N^* - 1$$

where

$$Q_N = \sum_{m=0}^{N} x^m$$

and $x = p \exp(iy)$. Since Q_N is of the same form as R_N in Appendix A,

$$Q_N = \frac{1 - x^N}{1 - x} = \frac{1 - p^N \exp(iyN)}{1 - p \exp(iy)}$$

Multiplication of the numerator and denominator by $[1 - p \exp(-iy)]$ yields,

$$Q_N = \frac{1 - p \exp(-iy) - p^N \exp(+iyN) + p^{N+1} \exp[+iy(N-1)]}{1 + p^2 - 2p \cos y}$$

Addition of $Q_N^* - 1$ results in an expression for the finite summation:

$$Z_N = \frac{1 - p^2 - p^N \cos Ny + p^{N+1} \cos(N - 1)y}{1 + p^2 - 2p \cos y}$$

For $|p| < 1$, the limit as $N \to \infty$ is,

$$Z = \frac{1 - p^2}{1 + p^2 - 2p \cos y}$$

References

1. M. J. Buerger, *Crystal Structure Analysis*, Wiley, New York, 1960, pp. 407–430.

2. A. Guinier, *X-ray Diffraction*, Freeman, San Francisco, CA, 1963, pp. 297–298.

3. R. Hosemann and S. N. Bagchi, *Direct Analysis of Diffraction by Matter*, North-Holland, Amsterdam, 1962, pp. 131–147.

4. A. L. G. Rees and J. A. Spink, *Acta Crystallogr.*, **3**, 316 (1950).

5. J. L. Amorós and M. Amorós, *Molecular Crystals*, Wiley, New York, 1968, pp. 300–397 and references cited therein.

6. B. E. Warren, *X-ray Diffraction*, Addison-Wesley, Reading, MA, 1969, pp. 165–169.

7. D. Feil, *Isr. J. Chem.*, **16**, 149 (1977).

8. E. Barnighausen, *J. Appl. Crystallogr.*, **8**, 477 (1975).

9. D. Britton and J. D. Dunitz, *Acta Crystallogr. Sect. A*, **37**, 131 (1981).

10. D. Britton and J. D. Dunitz, *Acta Crystallogr. Sect. A*, **38**, 171 (1982).

11. J. L. Amorós and M. Amorós, *Molecular Crystals*, Wiley, New York, 1968, pp. 289–299 and references cited therein.

12. J. P. Boilot, G. Collin, R. Comès, J. Thèry, R. Collongues, and A. Guinier, "Superionic Conductors," in G. D. Mahan and W. L. Roth, Eds., Plenum, New York, 1976.

13. A. M. Levelut, J. Doucet, and M. Lambert, *J. Phys. (Paris)*, **35**, 773 (1974).

14. A. DeVries, "Liquid Crystals," in F. D. Saeva, Ed., Dekker, New York, 1979. pp. 1–67.

15. A. DeVries, *Pramana*, Suppl. **1**, 93 (1975).

16. I. Chistyakov, *Adv. Liq. Cryst.*, **1**, 143 (1975).

17. B. K. Vainshtein and I. G. Chisyakov, *Pramana*, Suppl. **1**, 79 (1975).

18. A. Guinier, *X-ray Diffraction*, Freeman, San Francisco, CA, 1963, pp. 262–270.

19. R. W. James, *The Optical Principles of the Diffraction of X-rays*, Bell, London, 1954, pp. 557–559.

20. S. Hendricks and E. Teller, *J. Chem. Phys.*, **10**, 147 (1942).

21. J. Kakinoki and Y. Komura, *J. Phys. Soc. Jpn.*, **7**, 30 (1952).

22. A. J. C. Wilson, *Proc. R. Soc. London*, **A180**, 227 (1942).

23. H. Jagodzinski, *Acta. Crystallogr.*, **2**, 201 (1949).

24. H. Jagodzinski, *Acta. Crystallogr.*, **2**, 208 (1949).

25. H. Jagodzinski, *Acta. Crystallogr.*, **2**, 298 (1949).

26. H. Jagodzinski, *Acta. Crystallogr.*, **7**, 17 (1954).

27. J. Kakinoki and Y. Komura, *J. Phys. Soc. Jpn.*, **9**, 169 (1954).

28. J. Kakinoki and Y. Komura, *J. Phys. Soc. Jpn.*, **9**, 177 (1954); J. Kakinoki, *Acta Crystallogr.*, **23**, 827 (1967).

29. J. Kakinoki and Y. Komura, *Acta Crystallogr.*, **19**, 137 (1965).

30. A. Guinier, *X-ray Diffraction*, Freeman, San Francisco, CA, 1963, pp. 226–237.

31. B. E. Warren, *X-ray Diffraction*, Addison-Wesley, Reading, MA, 1969, pp. 298–312.

32. A. M. Mathieson, *Acta Crystallogr.*, **6**, 399 (1953).

33. M. G. Northolt and L. E. Alexander, *Acta Crystallogr. Sect. B*, **27**, 523 (1971).

34. L. M. Pant, *Acta Crystallogr.*, **17**, 219 (1964).

35. Y. Kato, Y. Takaki, and K. Sakurari, *Acta Crystallogr. Sect. B*, **30**, 2683 (1974).

36. K. Dornberger-Schiff and H. Grell-Niemann, *Acta Crystallogr.*, **14**, 167 (1961).

37. Y. Takaki, Y. Kato, and K. Sakuri, *Acta. Crystallogr. Sect. B*, **31**, 2753 (1975).

38. Y. Takaki and K. Sakuri, *Acta. Crystallogr. Sect. B*, **32**, 657 (1976).

39. J. L. Amorós and M. Amorós, *Molecular Crystals*, Wiley, New York 1968, pp. 250–267 and references cited therein.

40. H. Endres, H. J. Keller, H. J. Megnamisi-Belcombe, W. Marini, H. Pritzkow, J. Weiss, and R. Comès, *Acta Crystallogr. Sect. A*, **32**, 954 (1976).

41. A. E. Smith, *Acta Crystallogr.*, **5**, 224 (1952).

42. D. Chasseau, J. Gaultier, C. Hauw, S. Lefrant, J. Rivory, E. Rzepka, and H. Strzelecka, *Solid State Commun.*, **34**, 873 (1980).

43. H. R. Luss and D. L. Smith, *Acta Crystallogr. Sect. B*, **36**, 1580 (1980).

44. T. E. Phillips, R. P. Scaringe, B. M. Hoffman, and J. A. Ibers, *J. Am. Chem. Soc.*, **102**, 3435 (1980).

45. D. L. Smith and H. R. Luss, *Acta Crystallogr. Sect. B*, **33**, 1744 (1977).

46. F. H. Herbstein and M. Kapon, *Acta Crystallogr. Sect. A*, **28**, 574 (1972).

47. R. P. Scaringe and J. A. Ibers, *Acta Crystallogr. Sect. A*, **35**, 803 (1979).

48. H. U. Beyler, *Phys. Rev. Lett.*, **37**, 1557 (1976).

49. H. Endres, J. P. Pouget, and R. Comès, *J. Phys. Chem. Solids, Supp.*, **43**, 739 (1982).

50. R. Spal, C. E. Chen, T. Egami, P. J. Nigrey, and A. J. Heeger, *Phys. Rev.*, **B21**, 3110 (1980).

51. R. Hosemann and S. N. Bagchi, *Direct Analysis of Diffraction by Matter*, North-Holland, Amsterdam, 1962, pp. 302–353.

52. R. Hosemann, *Endeavour*, **32**, 99 (1973).

53. M. Kakudo and N. Kasai, *X-ray Diffraction by Polymers*, Elsevier, New York, 1972, pp. 114–133.

54. C. J. Schramm, R. P. Scaringe, D. R. Stojakovic, B. M. Hoffman, J. A. Ibers and T. J. Marks, *J. Am. Chem. Soc.*, **102**, 6702 (1980).

55. K. Huml, *Acta Crystallogr.*, **22**, 29 (1967).

56. U. Muller, *Acta Crystallogr. Sect. B*, **35**, 2502 (1979).

57. U. Muller, *Angew. Chem. Int. Ed. Engl.*, **20**, 692 (1981).

58. J. T. Devreese and R. P. Evrard, Eds., *Highly Conducting One Dimensional Solids*, Plenum, New York, 1979.

59. S. Megtert, J. P. Pouget, R. Comès, A. F. Garito, K. Bechgaard, J. M. Fabre, and L. Giral, *J. Phys. Lett. (Paris)*, **39**, L-118 (1978).

60. K. Yamaji, S. Megtert, and R. Comès, *J. Phys. (Paris)*, **42**, 1327 (1981).

61. J. P. Pouget, R. Comès, and K. Bechgaard, *The Physics and Chemistry of Low Dimensional Solids*, in L. Alcacer, Ed., Reidel, Hingham, MA, 1980.

62. K. Bechgaard, *The Physics and Chemistry of Low Dimensional Solids*, in L. Alcacer Ed., Reidel, Hingham, MA, 1980.

63. P. Robin, J. P. Pouget, R. Comès, and A. Moradpour, *J. Phys. (Paris)*, **41**, 415 (1980).

64. D. Jerome, A. Mazaud, H. Ribault, and K. Bechgaard, *J. Phys. Lett. (Paris)*, **41**, L-95 (1980).

65. K. Bechgaard, K. Carneiro, F. B. Rasmussen, M. Olsen, G. Rindorf, C. S. Jacobsen, M. J. Pedersen, and J. C. Scott, *J. Am. Chem. Soc.*, **103**, 2440 (1981).

66. J. P. Pouget, R. Moret, R. Comès, and K. Bechgaard, *J. Phys. (Paris)*, **42**, 543 (1981).

67. R. Moret, J. P. Pouget, R. Comès, and K. Bechgaard, *Phys. Rev. Lett.*, **49**, 1008 (1982).

68. B. Dorner and R. Comès *Dynamics of Solids and Liquids by Neutron Scattering*, in T. Springer, Ed., Springer, New York, 1977.

69. *International Tables of Crystallography*, Vol. III, Kynoch, Birmingham, Great Britain, 1972.

70. L. H. Schwartz, L. A. Morrison, and J. B. Cohen, *Advances in X-ray Analysis*, Vol. 7, Plenum, New York, 1977.

INDEX